AF767243

Applied MANOVA and Discriminant Analysis

WILEY SERIES IN PROBABILITY AND STATISTICS

Established by WALTER A. SHEWHART and SAMUEL S. WILKS

Editors: *David J. Balding, Noel A. C. Cressie, Nicholas I. Fisher, Iain M. Johnstone, J. B. Kadane, Geert Molenberghs, Louise M. Ryan, David W. Scott, Adrian F. M. Smith, Jozef L. Teugels*
Editors Emeriti: *Vic Barnett, J. Stuart Hunter, David G. Kendall*

A complete list of the titles in this series appears at the end of this volume.

Applied MANOVA and Discriminant Analysis

Second Edition

CARL J. HUBERTY

The University of Georgia
Department of Educational Psychology
Athens, Georgia

STEPHEN OLEJNIK

The University of Georgia
Department of Educational Psychology
Athens, Georgia

WILEY-INTERSCIENCE

A JOHN WILEY & SONS, INC. PUBLICATION

This book is printed on acid-free paper. ♾

Copyright © 2006 by John Wiley & Sons, Inc., Hoboken, New Jersey. All rights reserved.
Published simultaneously in Canada.

No part of this publication may be reproduced, stored in a retrieval system or transmitted in any form or by any means, electronic, mechanical, photocopying, recording, scanning or otherwise, except as permitted under Sections 107 or 108 of the 1976 United States Copyright Act, without either the prior written permission of the Publisher, or authorization through payment of the appropriate percopy fee to the Copyright Clearance Center, 222 Rosewood Drive, Danvers, MA 01923, (978) 750-8400, fax (978) 750-4744. Requests to the Publisher for permission should be addressed to the Permissions Department, John Wiley & Sons, Inc., 605 Third Avenue, New York, NY 10158-0012, (212) 850-6011, fax (212) 850-6008, E-Mail: PERMREQ@WILEY.COM.

For ordering and customer service, call 1-800-CALL-WILEY.

Library of Congress Cataloging-in-Publication Data:

Huberty, Carl J., 1934-
Applied MANOVA and discriminant analysis.—2nd ed./Carl J. Huberty, Stephen Olejnik.
p. cm
Rev. ed. of: Applied discriminant analysis. c1994.
Includes bibliographical references and index.
ISBN-13: 978-0-471-46815-8 (acid-free paper)
ISBN-10: 0-471-46815-0 (acid-free paper)
1. Discriminant analysis. 2. Multivariate analysis. I. Olejnik, Stephen. II. Huberty, Carl J., 1934- Applied discriminant analysis. III. Title.

QA278.65.H83 2006
519.5'35—dc22

2005054231

To our wives, Sandy and Sherrie

In memory of Clem (father), Rosalie (mother), and
Ken (brother) Huberty
Harry (father), Ann (mother), and
Tony (brother) Olejnik

Brief Contents

Contents

List of Figures

List of Tables

Preface to Second Edition

Over the past 10 years or so, some ideas related to *discrimmant analysis* (DA) have been refined and extended by statisticians and quantitative methodologists to some extent. In this second edition, we have made attempts to report some of these refinements and extensions so as to help guide researchers and methodologists in the conduct and reporting of studies involving multivariate analysis of (co)variance, descriptive discriminant analysis, and predictive discriminant analysis. The goals for this second edition are the same as those for the fist edition.

We heeded the many comments and suggestions made by reviews of the first edition. Seven major changes were made for this second edition:

- Deletion of appendix with computer output
- Basic SPSS and SAS computer syntax and output are embedded in the text
- Two applications chapters were deleted; easy access to websites should suffice for the interested researcher
- Inclusion of detailed discussions of multivariate analyses of variance and covariance
- Recent references (from a variety of disciplines) given in the text proper and for further reading
- Addition of a chapter on analyses related to predictive discriminant analysis
- The website (obtained through Wiley) from whiib computer programs and data sets may be obtained

In Appendix A we include the descriptions of five data sets that are used in the text proper and in chapter exercises; three of these sets contain "real" data. Many statisticians/statistical methodologists have contributed to DA-related concepts and methods over the past eight decades. These researchers have contributed many "original" ideas related to DA. A list of what we term DA-related "originators" is given in Appendix B.

More readings related to many topics in the current edition may be found in the first edition under the heading Additional Readings (Chapters I–IV, VI–X, XIV–XVI, and XX).

Acknowledgments

We would like to thank four academics for making their computer programs available for our use: **J. Algina** (University of Florida), **G. P. McCabe** (Purdue University), **C. E. McHenry** (Tennessee Tech University), and **J. D. Morris** (Florida Atlantic University).

Three other academics gratefully provided data sets for our use: **J. F. Baumann** (University of Georgia), **C. A. Ethington** (University of Memphis), and **M. S. Poudevigne** (Clayton State University).

There are several students who attended The University of Georgia and gave us considerable assistance in preparing this second edition. Jason Lemley was very patient and persistent in helping to process some of the text. Veena Dhankher and In-Suk Kim were very helpful in providing computer assistance for several data sets used in this book. And special thanks to the students in the 2005 Fall section of the multivariate class, particularly Hye-Jeong Choi and Chen-Yao Koa, who persistently requested further clarification of the text and identified omissions and mistypes in an earlier draft of this book.

We would also like to thank our University of Georgia Colleague, Seock-Ho Kim, for his assistance in the use of the LaTex text formating software.

The patience and efforts of Wiley editors, including Christine Punzo, Susanne Steitz, and Stephen Quigley, are greatly appreciated.

Preface to First Edition

Three types of users were kept in mind while preparing the manuscript for this book: (1) graduate students who want to expand their background in multivariate data analysis methods (for purposes of subsequently producing and consuming applied empirical research), (2) experienced applied researchers who want to enhance or update their quantitative background, and (3) (experienced or budding) methodologists who want to learn about some of the details and, perhaps, some unresolved problems in applied discriminant analysis. For the first two types, frequent references are made to computer printout information, and many suggestions are given for conducting various analyses. For the third type, references are made to research *on* discriminant analysis methods, some technical but not too theoretical notes are given near the ends of some chapters, and some issues and problems in discriminant analysis are discussed in the final two chapters.

Two general goals are suggested for the user of this book (1) to learn how to talk, read, and write about discriminant analysis; and (2) to further develop a personal approach to, or philosophy of, empirical research and data analysis. Somewhat more specific objectives for the user are:

- To become aware of the types of research questions that may be addressed using discriminant analysis
- To learn the meaning of concepts and terms associated with discriminant analysis
- To be able to read, understand, and interpret various computer package printouts that pertain to discriminant analysis
- To be able to critically read and evaluate reports of applied discriminant analysis
- To be able to design a study that uses discriminant analysis, carry out the analysis, and write up the report

Four real data sets are utilized to illustrate various analysis results, in both text examples and exercises. The illustrative results were obtained via the three statistical computer packages: BMDP, SAS, and SPSS. A number of printouts are included in an appendix. Although the latest available releases of these packages were used, it is recognized that new releases will become available. It is conjectured that the computational results of the packages will, for the most part, remain fairly constant. New formats and possibly additional results may, however, become obtainable. It is also conjectured that subsequent to the study of discriminant analysis results in this book, interpretation problems with new format and additional results will be minimal.

Because references are made to multiple regression and multiple correlation, some familiarity with associated concepts and parameter estimators would be desirable. Knowledge of multiple regression and multiple correlation particulars is not, however, essential to developing an understanding of aspects of discriminant analysis.

Although the expression *discriminant analysis* is used in different ways by different people, two different aspects of discriminant analysis are emphasized in this book predictive discriminant analysis (PDA) and descriptive discriminant analysis (DDA). In the behavioral sciences what is generally referred to by the expression "discriminant analysis" is DDA, while researchers in most other fields of study, as well as statisticians, generally think of PDA. In this book, PDA is presented first (in Part Two). Part Three (i.e., DDA) may, however, be studied first if desired.

There are three data analysis "themes" emphasized explicitly or implicitly in this book: (1) Look at your data prior to your final analysis; (2) use judgment and common sense in conducting analyses and in interpretations of results; and (3) do not hesitate to conduct multiple analyses (within or across computer packages) of your data. Two additional emphases are suggested with respect to discriminant analysis in particular: (1) Expend considerable effort in the initial selection and definition of variables to be studied; and (2) remember that although PDA and DDA are related, they are used for different purposes. Finally, it should be recognized that issues and problems associated with the conduct and interpretation of results of discriminant analysis still exist.

Notation

Symbol	Meaning	Section in First Use
p	Number of response variables	2.3
J	Number of groups	2.4
n_j	Number of units in Group j	2.4
	Subscripts	
$i = 1, 2, \ldots, p$	Variable	2.4
$j = 1, 2, \ldots, J$	Group	2.4
$u = 1, 2, \ldots, n_j$	Analysis unit	2.4
$Y_{i_{uj}}$	Score on variable i for unit u in Group j	2.4
N	Total number of units ($=\Sigma n_j$)	2.4
$\overline{Y}_{i.j}$	Mean of variable i for Group j	2.4
SS_j	Sum-of-squares for Group j	2.4
s_j^2	Variance for Group j	2.4
$\text{cov}_j(Y_i Y_{i'})$	Covariance for variables i and i'	2.4
$r_{j_{ii'}}$	Correlation of variables i and i' for Group j	2.4
CP	Sum of cross products	2.4
$\mathbf{y}_{.j}$	$(p \times 1)$ vector of means for Group j	2.4
$\mathbf{Y}_{uj}$	$(N \times p)$ matrix of p response variable scores for Group j	2.4
$\mathbf{Y}'_{uj}$	Transpose of matrix of response variable scores	2.4
$\mathbf{S}_j$	$(p \times p)$ covariance matrix for Group j	2.4
$\mathbf{C}_j$	$(n_j \times p)$ mean-centered matrix for Group j	2.5
$\mathbf{SSCP}_j$	$(p \times p)$ sum-of-squares and cross-products matrix for Group j	2.5.1
$\lvert\mathbf{S}_j\rvert$	Determinant of covariance matrix for Group j	2.5.2

$\mathbf{S}_j^{-1}$	Inverse of covariance matrix for Group j	2.5.3
λ	Eigenvalue	2.5.4
$\mathbf{b}$	Column vector of response variable weights	2.5.4
c	Euclidean distance	2.6
$\mathbf{E}$	$(p \times p)$ error **SSCP** matrix	2.6
$\mathbf{S}_e$	$(p \times p)$ error covariance matrix	2.6
D	Mahalanobis distance	2.6
d	Cohen standardized distance	2.6
b_i	Linear combination weight for variable i	2.7
b_0	Linear combination constant	2.7
$P(H)$	Probability of event H	2.8
μ_j	Population mean for Group j	3.2.1
$\mu_.$	Population mean across all groups	3.2.1
α_j	$\mu_j - \mu_.$	3.2.1
s_e^2	Error variance	3.2.1
df_e	Error degrees of freedom	3.2.1
r_{pb}^2	Squared point-biserial correlation	3.2.1
$\boldsymbol{\mu}_j$	$(p \times 1)$ vector of means for Population j	3.2.2
T^2	Hotelling statistic	3.2.2
$\boldsymbol{\Sigma}_j$	$(p \times p)$ covariance matrix for Population j	3.3
T^{*2}	Yao test statistic	3.4
η_{pop}	Population correlation of outcome variable and grouping variable	3.5.1
df_h	Hypothesis degrees of freedom	3.5.1
$\mathbf{H}$	$(p \times p)$ hypothesis **SSCP** matrix	3.5.2
Λ	Wilks lambda statistic	3.5.2
r	$\min(p, \text{df}_h)$	3.5.2
ν	$= 1, 2, \ldots, r$; subscript for eigenvalues	3.5.2
λ_ν	νth eigenvalue for $\mathbf{E}^{-1}\mathbf{H}$	3.5.2
U	Bartlett–Pillai statistic	3.5.2
b	$\max(p, \text{df}_h)$	3.5.2
Θ	Roy statistic	3.5.2
V	Hotelling–Lawley trace	3.5.2
η^2	Univariate effect size	4.2.1
ω^2	Univariate effect size	4.2.1
η_{mult}^2	Multivariate effect size	4.2.2
τ^2	Multivariate effect size	4.2.2
ξ^2	Multivariate effect size	4.2.2
ζ^2	Multivariate effect size	4.2.2
ω_{mult}^2	Multivariate effect size	4.2.2
ψ	Population group contrast	4.4.1
$\boldsymbol{\psi}$	Population vector for a group contrast	4.4.2
ξ_{adj}^2	Adjusted multivariate effect size	4.4.2

$\mathbf{b}$	$(p \times 1)$ vector of weights	5.2.1
LDF	Linear discriminant function	5.2.2
b_j^*	Standardized weight	5.4
$\Lambda_{(i)}$	Wilks Λ with variable i deleted	6.3.1
$F_{(i)}$	F-to-remove for variable i	6.3.1
$\boldsymbol{\mathcal{A}}$	Matrix of orthonormal contrast coefficients	8.6
$\mathcal{A}$	Matrix of contrast coefficients	10.4
W	Mauchly statistic	10.7
$\varepsilon, \varepsilon^*, \varepsilon', \tilde{\varepsilon}$	Adjustments for degrees of freedom in repeated measures	10.7
$\tilde{d}^2$	Squared sample Euclidian distance	12.2
Δ^2	Squared population Mahalanobis distance	12.3
$f(\mathbf{x}_u \mid j)$	Value of density function at $\mathbf{x}_u$, given membership of unit u in Population j	12.4.1
$P(\mathbf{x}_u \mid j)$	Probability of unit u having $\mathbf{x}_u$ vector, given membership in Population j – typicality probability	12.4.2
$P(j \mid \mathbf{x}_u)$	Probability that unit u belongs to Population j – posterior probability	12.4.3
π_j	Prior probability for Population j	12.4.4
q_j	Estimated prior probability for Population j	12.5
$\hat{f}(\mathbf{x})$	Estimated value of density function f at $\mathbf{x}$	13.1
$\bar{\mathbf{x}}_j$	$(p \times 1)$ vector of means for Group j	13.2
D_{uj}	Sample (Mahalanobis) distance of unit u and Group j centroid (using $\mathbf{S}_j$)	13.2
$\hat{P}(j \mid \mathbf{x}_u)$	Estimated posterior probability that u belongs to Group j	13.3
D_{uj}^*	Sample distance between unit u form Group j centroid (using $\mathbf{S}_e$)	13.3
QCF	Quadratic classification function	13.4.1
Q_{uj}	QCF score for unit u in Group j	13.4.1
LCF	Linear classification function	13.4.2
L_{uj}	LCF score for unit u in Group j	13.4.2
c_j	LCF constant for Group j	13.4.2
$C(j \mid j')$	Cost of assigning a Group j' unit to Group j	13.7
$n_{jj'}$	Number of units in cell (j, j') of $J \times J$ classification table	14.5
ρ	True correlation coefficient	15.1
ρ_v	True validity coefficient	15.1
$P^{(o)}$	Optimal hit rate	15.2
$P^{(a)}$	Actual hit rate	15.2
$P^{(e)}$	Expected actual hit rate	15.2
$\tilde{D}^2$	Estimator for Δ^2	15.3.1

Q_j	McLachlan's estimator $= 1 - \hat{P}_j^{(a)}$	15.3.1
L-O-O	Leave-one-out method of classification	15.3.3
M-P-P	Maximum-posterior-probability method of hit rate estimation	15.3.4
e_j	Expected frequency in cell (j, j) of classification table	16.2.1
e	Σe_j	16.2.1
H_o	Observed hit rate	16.2.1
H_e	Expected hit rate	16.2.1
o	Σn_{jj}	16.2.2
I	Improvement-over-chance hit rate	16.4
ϕ	Standard normal distribution function	18.5
$R(X)$	Rank of score on X	19.2.1

PART I

Introduction

How discriminant analysis fits in the arena of multivariate statistical methods is reviewed. An introduction of the *two* aspects of discriminant analysis (DA), descriptive discriminant analysis (DDA) and predictive discriminant analysis (PDA), are prefaced with a little history of DA. Aspects of the design of a DA study involving a DDA or PDA are discussed.

Statistical preliminaries to a discussion of DA are presented. This presentation includes some of the basic mathematical and statistical concepts involved in a DA: matrix operations, distance, linear combination, probability, and statistical testing. The role of *researcher judgment* in data analyses and interpretations thereof is emphasized.

Applied MANOVA and Discriminant Analysis, Second Edition, by Carl J. Huberty and Stephen Olejnik
Copyright © 2006 John Wiley & Sons, Inc.

CHAPTER 1

Discriminant Analysis in Research

1.1 A LITTLE HISTORY

Some of the ideas associated with *discriminant analysis* go back to around 1920. The English statistician Karl Pearson (1857–1936) proposed what was called the coefficient of racial likeness (CRL), a type of intergroup distance index. The CRL was studied extensively by G. M. Morant (1899–1964) in the 1920s. In the 1920s, too, study of another distance index started in India, to be formalized by P. C. Mahalanobis (1893–1972) in the 1930s. The idea of multivariable intergroup distance was translated to that of a linear composite of variables derived for the purpose of two-group classification by R. A. Fisher (1890–1962) in the 1930s. The distance and variable composite ideas appeared in print prior to Fisher's seminal discriminant analysis article in 1936 ("The use of multiple measurements in taxonomic problems," which appeared in *Annals of Eugenics*). At the suggestion of Fisher, M. M. Barnard applied two-group (predictive) discriminant analysis in a 1935 study involving seven Egyptian skull characters. The extension of two-group classification to multiple groups was given by C. R. Rao in 1948. Many other extensions and refinements of Fisher's ideas have appeared since the 1940s.

A detailed presentation of all historical developments pertaining to discriminant analysis will not be attempted here. There are at least 14 excellent sources of references related to discriminant analysis developments (listed chronologically):

Hodges (1950)
Tatsuoka and Tiedeman (1954)
Tatsuoka (1969)
Anderson et al. (1972)
Cacoullos (1973)
Das Gupta (1973)

Applied MANOVA and Discriminant Analysis, Second Edition, by Carl J. Huberty and Stephen Olejnik
Copyright © 2006 John Wiley & Sons, Inc.

Subrahmaniam and Subrahmaniam (1973)
Toussaint (1974)
Huberty (1975)
Lachenbruch (1975)
Hand (1981)
Panel on Discriminant Analysis, Classification, and Clustering (1989)
McLachlan (1992)
Giri (2004)

Although the initial study of discriminant analysis involved applications in the biological and medical sciences, considerable interest was aroused by statisticians/methodologists in areas of study such as business, education, engineering, and psychology. This interest led to the writing of textbooks that covered discriminant analysis in various forms and from various perspectives. Some pre-1970 books with an applied flavor are:

Rao (1952)
Kendall (1957)
Cooley and Lohnes (1962)
Rulon et al. (1967)

The potential for the application of discriminant analysis in education and psychology (and in other areas of study?) may be attributed to methodologists associated, in one way or another, with Harvard University during the 1950s and 1960s. This is evidenced by publications cited above (Cooley and Lohnes, 1962; Rulon et al., 1967; Tatsuoka, 1969; Tatsuoka and Tiedeman, 1954), a symposium on discriminant analysis at Harvard (Tiedeman et al., 1951), an application (Tiedeman and Sternberg, 1952), a report for the Educational Research Corporation (Tiedeman et al., 1953), and two reports for the Office of Education (Gribbons and Lohnes, 1969; Hockersmith, 1969).

The writings about discriminant analysis for the first three or four decades focused on the prediction of group membership, labeled *predictive discriminant analysis* (PDA) in the current book. In the nonbehavioral sciences, this focus has continued to this day. Giri (2004, pp. 477–482) lists 88 references related to PDA; the publication years range from 1921 to 1991. Although Fisher considered linear variable composites (i.e., linear discriminant functions, LDFs) from a mathematical standpoint in the 1930s, it was not until the 1960s that LDFs were considered seriously for purposes of interpreting effects revealed via a multivariate analysis of variance (MANOVA) (e.g., Cooley and Lohnes, 1968; Jones and Bock, 1960); this aspect of discriminant analysis is labeled *descriptive discriminant analysis* (DDA) in the current book. In the view of some methodologists, the study of structure (through LDFs) in the context of MANOVA has considerable potential for substantive theory exploration and development. As important as such study may be considered, its use has been very limited in applied research settings over the past four decades.

1.2 OVERVIEW

Empirical research in nearly every discipline is rarely confined to the study of a single response variable, a characteristic or attribute or trait on which the researcher obtains scores or responses for a collection of *analysis units*.[1] Data sets typically involve measures on a number of variables, and it may be desirable to consider (i.e., to analyze and interpret the analysis for) all the variables *simultaneously*. Data analysis methods used to conduct such analyses are in the general domain of multivariate statistical methods.

The expression *multivariate methods* covers quite an expanse of data analysis methods. If we think of multivariable methods, simple analysis of variance (ANOVA), which involves two variables (one grouping and one outcome variable), would be included, as would the three-variable two-factor ANOVA situation. Also, multiple correlation (as opposed to multiple regression) involving $X_1, X_2, \ldots, X_p$ on the one hand and Y on the other hand would clearly be a multivariable or multivariate situation. There are, of course, many other multivariate types of analyses, two of which are included under the expression *discriminant analysis* (DA).

To get a rough idea of how DDA and PDA fit into a general scheme of analyses of multivariate data, consider Figure 1.1. When a research question pertains directly to the study of a comparison among, say, three groups of analysis units for each of which we have, say, 10 *outcome* variable scores, then the analyses of interest would be MANOVA along with DDA. On the other hand, if the single set of response variables play the role of predictors, and there is a single grouping variable, then the associated research question would pertain to how well group membership of analysis units may be predicted—a PDA.

If a study involves a single set of response (*not* outcome or predictor) variables, the analysis to be applied would be, for example, a cluster analysis or a principal component analysis. For two sets of response variables, analyses of interest may be multiple regression, multiple correlation, or canonical correlation.

1.3 DESCRIPTIVE DISCRIMINANT ANALYSIS

As mentioned above, when a multivariate study design involves a single set of response variables that are outcome variables, along with one or more grouping variables, then a MANOVA would be conducted. This analysis would, depending upon "real" MANOVA effects, be followed up with a DDA. In DDA, the basic question of interest pertains to grouping variable effects on the multiple outcome variables or, more

[1]The term, analysis unit, or, simply, unit, is used in the current book to indicate an object or individual or subject being studied.

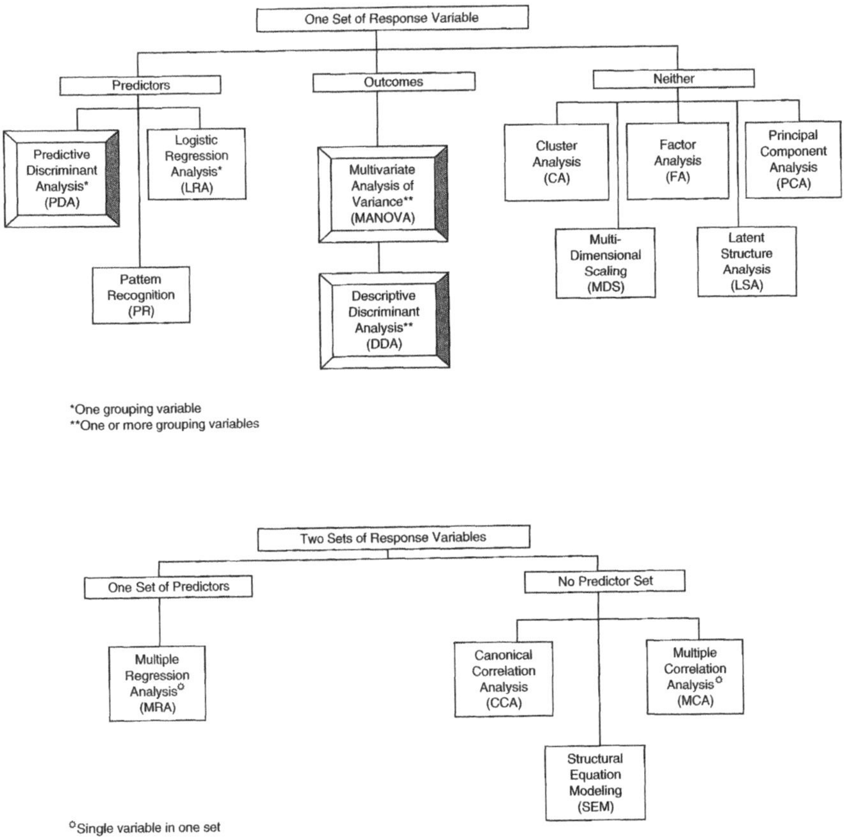

Figure 1.1 Classification of multivariate methods.

specifically, to group separation or group differences with respect to the set of outcome variables. Techniques of DDA are, therefore, closely aligned to the study of effects determined by a MANOVA. It is assumed that the reader is familiar with the use of univariate analysis of variance (ANOVA) to study the effects of one or more grouping variables on a *single* outcome variable. Having familiarity with univariate ANOVA, the reader undoubtedly can think of research situations involving one factor (or grouping variable) and involving multiple factors with a single outcome variable. To come up with MANOVA situations, then, one need only think of multiple outcome variables for one-factor or multiple-factor designs.

It may be noted that an ANOVA null hypothesis may be stated as a correlational hypothesis, the correlation between the grouping variable and the outcome variable. In MANOVA, too, we may consider the relationship between the grouping variable, on the one hand, and the *set* of outcome variables on the other.

As in most other multivariate contexts, linear combinations/composites of response (i.e., outcome) variables in a MANOVA/DDA context are determined. A two-group MANOVA may be viewed as a (univariate) two-group ANOVA where the single outcome variable consists of a linear combination of the original multiple outcome variables. It is these variable combinations that are the center of attention in DDA. That is, the main reason for conducting a MANOVA/DDA is to interpret, in some cases, the resulting variable combination(s) that is(are) associated with group differences. The interpretation pertains to the effects judged generalizable via a MANOVA. One attempts to associate substantively interpretable *constructs* with the variable combinations underlying the effects of interest.

An example of an application of MANOVA and DDA is given by Schuab and Tokar (1999). A cluster analysis was conducted using 240 undergraduate students and 17 variables pertaining to counseling expectations. It was decided to go with five clusters. To interpret cluster differences, scores on 4 linear composites—based on the original 17 variables—were used as MANOVA outcome variable scores. Two meaningful constructs (via DDA)—Optimism, and Neuroticism and Closedness—were defined, and considered in interpreting differences among the 5 clusters of 240 students.

As implied from this example, the primary questions addressed in a DDA are:

- How many constructs (dimensions) characterize group separation?
- What (latent) constructs characterize group separation?

1.4 PREDICTIVE DISCRIMINANT ANALYSIS

The processes of prediction and identification are very common in our society. Some examples involve the prediction or identification of the following:

- Life expectancy
- Length of time to check out at a grocery store
- Economic growth
- Number of attendees at a social function
- Academic achievement
- Length of time for mail service
- Success in a gifted education program
- Voting support of a candidate
- State revenue income
- Sales revenues

In a more academic setting, one might be interested in predicting family planning devices, fish site rainfall, physical/mental disorder type, marital outcome, or deficient taxpayers. Each of these examples involves one or more predictor or explanatory variables along with one criterion or outcome variable. In some instances the criterion variable is quantitative and is measured using at least an ordinal scale of measurement. For such a situation, a multiple regression analysis would be conducted. In other instances the criterion is categorical (i.e., a grouping variable), measured with a nominal scale; sometimes the criterion is dichotomous, sometimes polytomous. It is in the latter type of situation with a categorical criterion (and, usually, with unordered categories) that a predictive discriminant analysis (PDA) is applicable.

In multiple regression analysis, a *prediction rule* is developed that involves a linear combination/composite of the predictors. A linear combination of predictors is also used in PDA; however, the rule consists of as many linear combinations as there are categories (or groups). Such a rule enables the researcher to predict membership of an analysis unit in one of the criterion groups; or, viewing it another way, to determine the group with which a unit is identified. For example, suppose that it is of interest to identify a high school student who would potentially drop out of school. With two groups of students, a set of graduates and a set of dropouts, a prediction rule would be formulated using such predictors as overall grade average, absenteeism, family structure, family social status, and gender. Using the five predictor measures for each student in each group, predictor weights for two linear combinations, one associated with each group, are determined. The two linear combinations would then be used for a subsequent student to predict group membership (graduate or dropout); a student would be assigned to the group with which is associated the larger linear combination score. (As we shall see in Section 13.3, two probabilities of group membership for each student may be determined.)

As might be implied from the example above, the primary question addressed in a PDA is:

- How accurately can group membership be predicted?

Two related PDA questions are:

- Is the resulting "hit rate" better than that obtainable by chance?
- If so, how much better?

An example of an application of a PDA is given by Kumar and Sahai (1993). The prediction of four family planning devices was of interest, using 12 predictors based on family socioeconomic and demographic characteristics. (Nine of the predictors were continuous, two were unordered categorical, and one dichotomous.) The four separate-group hit rates ranged from .77 to .83—no prior probabilities were specified.

[There are some studies in which both a DDA and PDA are applied using the same set of response variables (e.g., Andreev, 2003; Spatz et al., 2003); why this is done is not really clear because of the distinct purposes for the two analyses.]

In discussing DDA and PDA, and from Figure 1.1, reference is made to multiple correlation analysis (MCA) and to multiple regression analysis (MRA). The following analogy may be made:

$$\text{DDA} : \text{MCA} :: \text{PDA} : \text{MRA}.$$

That is, DDA and MCA are conducted for relationship purposes, whereas PDA and MRA are for prediction purposes.

1.5 DESIGN IN DISCRIMINANT ANALYSIS

1.5.1 Grouping Variables

Basic to the designs of many studies involving only a single response variable—so-called *univariate designs*—is the *grouping variable*. On the one hand, the grouping variable may be one that can be manipulated in that the researcher has control in assigning levels of the grouping variable to the analysis units. For example, if the grouping variable is "Method of Instruction," before the treatments (i.e., the methods) are implemented, the researcher can (randomly, perhaps) assign a unit to a method. On the other hand, the grouping variable may be defined by levels that the researcher cannot assign to the analysis units. Such grouping variables are sometimes labeled "organismic" or "subject" variables. Examples of nonmanipulable grouping variables are gender, mental age, ability, personality type, and educational attainment. The same notions apply to grouping variables in multivariate designs.

Of course, just as in univariate designs, researchers must strive for high degrees of both internal and external validity in studies with which discriminant analysis is associated. [Briefly, *internal validity* pertains to causal relationships or the absence of confounding variables while *external validity* pertains to generalizability. Cook and Campbell (1979, Chapter 2) provide a detailed discussion of validity issues in the behavioral sciences.] External validity concerns are not very different for the two types of grouping variables. Such is not the case with regard to internal validity. There is greater potential for low internal validity with nonmanipulable grouping variables than with manipulable ones, where random assignment of units to a level of the grouping variable is possible. In group comparison problems involving nonmanipulable grouping variables, internal validity is of particular concern. If group differences result, we would like to attribute these differences to the grouping variable and not to some intervening variable(s). A careful descriptive analysis of the units in the groups is expected in assessing internal validity.

1.5.2 Response Variables

As a prelude to a discussion about response variables, let us briefly discuss what it means to say that a "multivariate analysis"—an expression, according to David (1998), that was originated by M. S. Bartlett (1910–2002) in 1939—was conducted. There are at least two meanings of what a multivariate analysis consists (Bernstein 1988, pp. 2–8): (1) an inquiry into the structure of relationships among multiple variables and (2) the study of linear representations or composites of relationships among variables. The first meaning pertains to linear combinations or composites of variables that may be employed in the study of *structure* in multiple correlation, canonical correlation, DDA, and (in some senses) factor analysis. Linear composites are also employed when structure is not of focal interest—in multiple regression and PDA, for example.

The second meaning applies to both DDA and PDA in that for both analyses a linear composite of the response variables has a central (but different) purpose. The first meaning, on the other hand, applies to DDA but not usually to PDA. [See Huberty (1994a) for more on purposes of multivariate analyses.]

No matter which meaning one considers or which analysis one applies, for any multivariate analysis the choice of response variable domain and the variables themselves is of utmost importance. (The choice involves not only what unit attributes to study, but also the measurement of those attributes, as discussed in the next paragraph.) Serious thought ought to be given to variable choice when a discriminant analysis is to be used. In DDA, the researcher employs a collection, or multiple collections, of outcome variables, linear composites of which have potential to lead to a definable and interpretable *variable structure* underlying the resultant group differences. When a prediction rule in PDA is sought, the researcher, of course, seeks a rule that he or she hopes will yield relatively high predictive accuracy. Among other things, the goodness of a rule is highly dependent on linear composites of relevant predictors, predictors "related" in some way to the categorical criterion.

In identifying analysis unit attributes of interest, we need to keep in mind not only which variables to measure, but how to measure them. A determiner of the extent to which we can identify meaningful *structure* (i.e., *latent variables*) is our ability to measure the variables in our chosen *system*. That is, are the proxies for (or indicators of) the chosen variables sufficient for sound variable measurement? Are we doing an adequate job of measuring the chosen variables? Assuming that the chosen system of variables has an appropriate structure base, do we need additional proxies or indicators? Do we need different proxies or indicators? For example, suppose that one is considering "learning" as a potential latent variable. Supposedly, one would include, as a minimum, a battery of achievement tests that yield measures of learning. But how about also considering measures of self-confidence, sense of responsibility, creativeness, social competence, motivation, teacher capability, and so on? Basic, also, to the measurement problem is the instrumentation used. Are we using instruments that yield, in any relevant sense, valid and reliable measures?

In a given research situation (that would involve DDA), it *may* make sense to hypothesize an underlying structure before data are collected. Such an hypothesis would be pertinent to the initial response variable choice and to the proxies and indicators utilized. Then, via data-based methods, one determines the emergent latent variable(s). But, as pointed out by O. Kempthorne (1919–2000) over 30 years ago, seldom does one have a "model" in mind before the collection of data (Kempthorne 1971, p. 759). (We have more to say on this issue in Section 22.5.)

After deciding on one or more variable domains (e.g., classroom climate, interpersonal behavior, personality), appropriate response variables need to be chosen.

Inevitably, the number of initially chosen variables will need to be reduced. The number of variables included in most discriminant analysis studies might be limited to something on the order of 10 or 12 unless there exist compelling reasons for including more. To start, the initial variable list should be *logically screened*, based on substantive theory, prior research, and reliability of measures, as well as on practical grounds. Next, the list can be *statistically* screened, although with some caution. If two variables are very highly correlated for the data on hand, one of them might be dropped. Certainly, if two variables are essentially the same trait or characteristic, there is no need to enter scores on both into the analysis. Statistical screening can sometimes be accomplished via multiple univariate analyses (e.g., univariate ANOVA F tests). A caution: Variables that do not yield statistical significance should *not necessarily* be dropped. But, if a variable contributes nothing but "noise" in a univariate sense (a univariate F value of less than 1.0, say), it is recommended that consideration be given to dropping the variable prior to a discriminant analysis. Variable screening is appropriate in both DDA and PDA.

Variable reduction can also be accomplished by employing some type of dimension analysis (e.g., component analysis or factor analysis). Such an analysis is appropriate when all or a substantial portion of the initial variables are from a single domain. This variable reduction approach would be appropriate when dealing with a number of test or inventory items. The dimension scores derived would then be used as input for a discriminant analysis. The use of a principal component analysis (PCA) is suggested as a prelude to a DDA or a PDA when some (or all) of the response measures are single item scores. Such a data reduction process would make sense if the item response options were the same for a collection of items. Of course, for a "large" set of items, multiple PCAs may be conducted. Detailed discussions of PCA are given by Flury (1995, pp. 20–27), McLachlan (1992, pp. 197–201), Rencher (2002, Chapter 12), and Webb (2002, pp. 319–329). There is another research situation where "variable" reduction is suggested. Suppose responses to a 50-item questionnaire are obtained. Surely, a researcher would *not* claim to have measures on 50 "variables." What might be considered, then, is to sum responses on 5 (or 4 or 6?) items that have the same response options. (An example of this approach is apparent in the 9-variable data set described in Appendix A.)

Another concern regarding response variables pertains to *when* measures on these variables are obtained. If a predictive discriminant analysis is the analysis of interest, it is important that the response measures be obtained *before* the groups are defined. Invalid predictions and interpretations of results are likely if group membership causes systematic differences in the response measures. This may be a problem, for example, when distinguishing college graduates from college dropouts on the basis of locus of control measures obtained *after* the students graduate/drop out.

A comment regarding the roles of the grouping variable(s) and the response variables is in order. In a group comparison problem, a grouping variable plays the role of an "independent" variable, whereas response variables play the role of "dependent" variables. (This common usage is unfortunate and potentially misleading and is *not* followed in the current book.)[2] In a prediction problem, the roles of the variables are reversed, with the response variables now playing the role of *predictor* variables, and a grouping variable playing the role of a *criterion* variable.

There is a research situation where the use of the term "outcome variable" or the term "predictor variable" *may* not be appropriate. Suppose analysis unit measures on a set of p response variables were obtained *prior to* the determination of unit group membership. This may apply, for example, to the situation of the study of College Major. Suppose further that the research interest is on the comparison of groups (e.g., Major). Such a comparison may be accomplished through the study of the *relationship* between Major on the one hand and a linear composite of a set of p response variables on the other hand. [MANOVA and DDA techniques (see Part II) could be used to determine the number of dimensions one could associate with the resulting relationship, and to identify potentially substantively meaningful constructs underlying the dimensions.]

Finally, we have a few comments about sampling of analysis units and about response variable measures. The basic concern in sampling is that of *representativeness* of some meaningful population (defined in terms of analysis units). The degree of representativeness is assessed in great part by researcher judgment; a description and documentation of the sampled units should help. Random sampling is championed by some methodologists, but it is difficult to implement and sometimes not entirely appropriate in practice. *Simple random selection does not necessarily ensure representativeness.* That is, by chance a "random" sample could be atypical of the target population. Measures of response variables should be obtained in the most reliable way possible, so that independence of measures across units is assured. Suggestions about desirable sample sizes are provided in Sections 15.7 and 18.5.

With respect to the research design of a study that would involve a PDA, a final comment follows. When designing a PDA study, it is assumed that the measures on the predictor variable are obtained *before* the grouping variable is determined.

Suggestion It is suggested that the reader of this book return to this chapter for a reread when designing a study that would involve a DDA (Sections 1.3 and 1.5) or a PDA (Sections 1.4 and 1.5).

Further Reading

Hand (1997) provides a very thorough discussion of a variety of concepts related to classification (i.e., PDA) rules (for the reader who has already studied the PDA basics). See Huberty (2000a) for a review.

Hawkins (1993) and Rayens (1993) provide comprehensive reviews of McLachlan (1992).

[2] F. Mosteller (1909–1990) and J. W. Tukey (1915–2000) state: "These [terms] have been extremely effective in producing confusion when dealing with data" (1977, p. 31). See, also, Dodge (2003, p. 197) and Moore (2004, p. 80).

Huberty (1994a) presents a review and conceptual discussion of two primary purposes (prediction and structure identification) and a secondary purpose (response variable ordering) of a multivariate analysis.

Huberty and Lowman (1998) illustrate rather complete applications of DDA and PDA in contexts of higher education research; considerations, issues, and problems related to the two analyses are discussed.

Huberty and Morris (1989) contend that conducting multiple ANOVAs in a multivariate group comparison context is of limited utility; rather, a multivariate analysis should focus on the study of outcome variable constructs.

Huberty (2002) discusses the basics of DDA and PDA (with a very limited use of formulas).

McLachlan (1992) presents an outstanding review of discriminant analysis in general, from a fairly technical point of view. Over 1200 references on applications as well as methods are given. Specific reference will be made to this monograph—as the author calls it—repeatedly throughout the present volume.

Ragsdale and Stam (1992) give references for a number of published data sets in business applications.

Rencher (1998, Chapters 5 and 6) and Rencher (2002, Chapters 8 and 9) provide some detailed discussions of DDA and PDA in the respective chapters; adequate statistical formulation is given in both books.

Therrien (1989) discusses PDA in a (fairly theoretical) context of pattern recognition; in this context the term *classifier* is used rather than *classification/allocation rule*.

Definition **Standard deviation**: A married man whistling at a girl.

EXERCISES

1. This is an exercise calling for the design of a group comparison study. The research purpose pertains to group separation (for which MANOVA would be conducted) and the identification of some underlying structure (via the use of DDA). Define a research setting involving three or more groups. So, reasonable group and corresponding population identification is to be done first. Next, identify 6 to 12 outcome variables and how each would be measured. Finally, fill in a data matrix; it would be an N (total number of analysis units) $\times$ p (number of outcome variables) matrix. Obtain the data matrix from a colleague, a book, a simulation, or other source. [Note: You may want to relabel some variables and rescale some variable measures in an available data set so as to make the data set more relevant to you. Also, you may want to consider at least one categorical outcome variable (with more than two unordered categories?).]

2. This exercise is similar to Exercise 1, but now the research purpose pertains to group membership prediction (i.e., a PDA would be conducted). For this data set, the response variables are considered predictors.
(*Note:* Reader-generated data will be referred to in some exercises in subsequent chapters.)

CHAPTER 2

Preliminaries

2.1 INTRODUCTION

In this chapter we will present basic matrix operations useful when analyzing multivariate data sets. While we recognize that in actual practice much of the analyses of multivariate data will be completed with the assistance of computer software programs, we do not believe that a firm understanding of multivariate results can be obtained without some basic understanding of matrix notations and operations. To introduce the matrix operations we will begin with a research context and a small data set. Using these data we will demonstrate how the data may be manipulated, analyzed, and summarized. The matrix operations we present in this chapter will be used throughout the book to examine group separation and classification.

2.2 RESEARCH CONTEXT

We begin with the context for a research study by Baumann, Seifert-Kessell, and Jones (1992). (Hereafter, this will be referred to as the "Baumann study.") The researchers were interested in comparing three strategies for teaching reading comprehension to fourth-grade students (our "analysis units"). One strategy was to teach students a number of reading comprehension monitoring strategies. This approach was called "Think Aloud" (TA). A second strategy was labeled "Directed Reading and Thinking Activity" (DRTA), which required students to make predictions and evaluate their predictions as they read stories. The third strategy, labeled "Directed Reading Activity" (DRA), was an instructed control condition using a common approach to teaching reading comprehension. Following the intervention period, measures on three outcome variables were obtained. The first variable was "Error Detection Task" (EDT, Y_1) where students were asked to identify inconsistencies (errors) in a story passage. The second variable was measured via the "Degrees of Reading Power" (DRP, Y_2),

Applied MANOVA and Discriminant Analysis, Second Edition, by Carl J. Huberty and Stephen Olejnik
Copyright © 2006 John Wiley & Sons, Inc.

TABLE 2.1 Scores on the Error Detection Task (Y_1) and Degrees of Reading Power (Y_2) for the Think Aloud (TA) and Directed Reading Activity (DRA) Groups

TA		DRA	
Y_1	Y_2	Y_1	Y_2
4	43	5	34
4	34	9	36
4	45	5	42
3	39	7	37
8	40	4	44
1	27	9	49
7	46	3	38
7	39	4	38
9	31	2	38
6	39	5	50
4	40	7	31
12	52	8	49
14	53	10	54
12	53	9	52
7	41	12	50
5	41	5	35
9	46	8	36
13	52	12	46
11	55	4	42
5	36	8	47
11	50	6	39
15	54	5	38

a standardized test of reading comprehension. The third variable was based on a comprehension monitoring questionnaire that asked students questions on the strategies they used while reading to increase their comprehension. In this chapter we introduce several useful procedures to use when analyzing multivariate data sets. To demonstrate these procedures we will use Y_1 and Y_2 scores from the TA and DRA groups. These data are presented in Table 2.1. (An SPSS data file containing these data, labeled BAUMANN2g2v, is available at the Wiley website.)

2.3 DATA, ANALYSIS UNITS, VARIABLES, AND CONSTRUCTS

Data constitute the raw material for methods of statistical analysis. In this book the word *data* is used to represent the numerical values that are in some way manipulated and analyzed. A *datum* may be considered as a *score* value for some analysis unit on some variable. All data collected for a particular study may be referred to as a *data set*. An example of a data set is given in Table 2.1.

Various terms are used to indicate the objects being studied: element, individual, subject, case, or unit. In this book, the neutral term *analysis unit*, or simply, *unit*, is used. In Table 2.1 the scores 4 and 43 belong to a single unit and the scores 5 and 34 belong to a different unit. A *variable* is a characteristic or attribute or trait of interest about a unit that can take on different score values. In Table 2.1, Y_1 (Error Detection Task) and Y_2 (Degrees of Reading Power) are considered as two different variables. In the study and use of discriminant analysis, there are fundamentally two types of variables. *One* type is a *grouping variable* (or group indicator variable); "scores" on such a variable simply indicate group membership. In Table 2.1, for example, the variables Y_1 and Y_2 are grouped together under either TA (Think Aloud strategy) or DRA (Directed Reading Activity strategy). These are two *levels* of a grouping variable that might be called "Instructional Method." The researchers were interested in comparing the relative effectiveness of the two instructional strategies. Other examples of grouping variables—called "independent" variables by some—are type of school (elementary, secondary), geographic area, sports participation, post-secondary experience, and residence (rural, urban, suburban). A grouping variable is a special type of categorical variable. "Scores" on a grouping variable are used to identify groups (of units), differences between or among which may be of interest to study. The group differences are studied in terms of some characteristic(s) of the units under consideration. Such a characteristic is called an *outcome variable* (or a criterion variable).

Another role a grouping variable may play is that of a criterion. That is, it may be of interest to predict membership in a group in terms of, again, some characteristic(s) of the units under consideration. In this context such a characteristic is called a *predictor variable*. It should be noted that both outcome variables and predictor variables may be categorical in nature. A generic expression used in this book for an outcome or predictor variable is *response variable*. This is the *second* type of variable to which we refer in the previous paragraph. In a multiple regression context with p X variables and one Y variable there would be $p + 1$ response variables, p predictor variables, and one criterion variable. In this book the letter X is used to represent a predictor variable, and the letter Y is used to represent an outcome variable.

A *construct* is a *latent* or unobservable variable that is represented or defined by a linear combination of observable variables. Because no single observable variable is likely to capture the true meaning of the construct, we will typically be interested in multiple outcome variables or *indicators* of the construct. For example, in the Baumann study, the outcome of interest, reading comprehension, is unobservable, but each observable variable, Y_1 and Y_2, may be considered as an indicator of the construct. Because we have more than one indicator of the construct (typically, several observable variables will be considered), one of our goals in descriptive discriminant analysis (DDA) will be to determine the linear combination of two (or more) variables that maximizes the difference between (or among) the groups being compared; here, the TA and DRA groups. An examination of how the variables are combined (or weighted) will provide a definition of the construct(s) under investigation. This issue is discussed in Chapter 5. Alternatively, we may seek the best linear combination of Y_1 and Y_2 to predict group membership—this pertains to predictive discriminant analysis (PDA), which is discussed in Part IV of this book. Because we will seek linear combinations of observable variables to represent one or more constructs, it is *essential* that researchers carefully choose the variables to be included in a study. The variables that are included in a study should be chosen on a theoretical or intuitive basis because they are believed to represent one or more dimensions on which the groups may differ. Consequently, the variables chosen should be expected to be intercorrelated in one or more variable clusters. Choosing variables simply because variable scores are available or easy to obtain will likely lead to uninterpretable or meaningless results.

2.4 SUMMARIZING DATA

The data set in Table 2.1 presents two scores for each individual within each of two groups. To represent these scores, we use the following notation: $Y_{i_{uj}}$. The subscript i indicates the variable, $i = 1, 2, \ldots, p$. The subscript j indicates group membership, $j = 1, 2, \ldots, J$. And the subscript u indicates a specific unit within Group j, $u = 1, 2, \ldots, n_j$, where n_j indicates the number of units in Group j. The total sample size in a study is represented by $N (N = \sum_{j=1}^{J} n_j)$. To summarize and describe a data set, researchers typically report the mean and variance (or standard deviation) for each variable within each group. A group mean is an aggregation of scores over all of the units for a variable in a specific group. To represent a group mean on variable i in Group j we use $\overline{Y}_{i.j}$. The dot (.) replaces the u subscript to indicate that the statistic is aggregated over the n_j units. For example, using the data in Table 2.1, the mean Error Detection Task, $i = 1$, for the TA Group, $j = 1$, is computed as:

$$\begin{aligned}\overline{Y}_{1.1} &= \sum_{u=1}^{n_1} \frac{Y_{1_{u1}}}{n_1}\\ &= \frac{171}{22}\\ &\doteq 7.77.\end{aligned}$$

The variance of the $Y_{i_{uj}}$ scores on variable i in Group j is represented by $s^2_{jY_i}$. The variance of scores $Y_{i_{uj}}$ is computed as the ratio of the sum-of-squared deviations (SS_j) of each score within a group from the group mean score to the number of scores within the group minus 1:

$$s^2_{jY_i} = \sum_{u=1}^{n_j} \frac{(Y_{i_{uj}} - \overline{Y}_{i.j})^2}{n_j - 1}.$$

The variance of the scores on variable 1 (Error Detection Task) in Group 1 (TA) found in Table 2.1 is $s^2_{1Y_1} \doteq 323.86/21 \doteq 15.422$. Table 2.2 summarizes the group mean ($\overline{Y}_{i.j}$), sum-of-squares (SS_j), and variance (s^2_j) for each variable in each group. From these results we find that the TA group had mean scores a little higher on both variables than the DRA group, and were also more dispersed.

TABLE 2.2 Mean, Sum-of-Squares, and Variance for Test Scores on the Error Detection Task (Y_1) and Degrees of Reading Power (Y_2) for the Think Aloud (TA) and Directed Reading Activity (DRA) Groups ($n_1 = n_2 = 22$)

	TA		DRA	
	Y_1	Y_2	Y_1	Y_2
Mean ($\overline{Y}_{.j}$)	7.77	43.45	6.68	42.05
Sum-of-Squares (SS_j)	323.864	1297.455	160.773	918.955
Variance (s_j^2)	15.422	61.784	7.656	43.760

Up to this point we have presented the data for each variable separately. That is, we have examined the data from a *univariate* perspective. With multiple outcome variables it is often useful to take a *multivariate* approach to summarize the data; that is, to consider two or more outcome variables together. One such descriptive statistic is the correlation between two variables. The Pearson correlation is the most popular statistic that summarizes the linear relationship between two variables. If each variable reflects a common construct, we would expect some correlation between the two variables. In the present context, both Y_1 and Y_2 are believed to reflect a student's ability to comprehend a reading passage. The Pearson correlation is computed as the ratio of the covariance, $\text{Cov}_j(Y_1Y_2)$, to the product of the two standard deviations. That is,

$$r_{jY_1Y_2} = \frac{\text{Cov}_j(Y_1Y_2)}{(s_{jY_1})(s_{jY_2})}, \tag{2.1}$$

where

$$\text{Cov}_j(Y_1Y_2) = \frac{\sum_{u=1}^{n_j}(Y_{1_{uj}} - \overline{Y}_{1.j})(Y_{2_{uj}} - \overline{Y}_{2.j})}{n_j - 1},$$

$$s_{jY_1} = \sqrt{s_{jY_1}^2},$$

and

$$s_{jY_2} = \sqrt{s_{jY_2}^2}.$$

Using the data in Tables 2.1 and 2.2 for the TA group, $j = 1, \text{Cov}_1(Y_1Y_2) \doteq 24.679$, $s_{1Y_1} \doteq \sqrt{15.422}$, and $s_{1Y_2} \doteq \sqrt{61.784}$, the correlation between Y_1 and Y_2 variables for students in the TA program is

$$r_{1Y_1Y_2} = \frac{\text{Cov}_1(Y_1Y_2)}{(s_{1Y_1})(s_{1Y_2})} \doteq \frac{518.273/21}{(3.927)(7.860)} \doteq .799.$$

Similarly, for the DRA group, the correlation between Y_1 and Y_2 is .498. For both groups there is a positive relationship between Y_1 and Y_2, students who scored high on Y_1 tend to score high on Y_2, and students scoring low on Y_1 tend to score low on Y_2. But, the relationship appears to be stronger among students in the TA group. In the formula for computing the $\text{Cov}_1(Y_1Y_2)$, the numerator,

$$\sum_{u=1}^{n_1}(Y_{1_{u1}} - \overline{Y}_{1.1})(Y_{2_{u1}} - \overline{Y}_{2.1}),$$

is called the sum-of-cross products (CP). This statistic turns out to be very useful in the calculation of several statistics used in multivariate analyses.

Using a multivariate approach, group means are reported as a set and referred to as the group *centroid*. The TA group centroid is

$$\begin{bmatrix} 7.77 \\ 43.45 \end{bmatrix}.$$

This column of means is called a *vector* and will be represented by a **bold** lowercase letter, for example, $\mathbf{y}_{.1}$. We will use the previously defined subscript notation to identify the elements of the vector. A dot (.) represents the aggregation of the units (e.g., group means), and j identifies group membership (e.g., $j = 1, 2$). This column vector can be *transposed* to a row vector, $\mathbf{y}'_{.1} \doteq [7.77 \quad 43.45]$. Writing each column as a row, or each row as a column, transposes a vector.

Two or more vectors presented together is referred to as a *matrix*. For example, the first two columns of Table 2.1 can be thought of as a data matrix for the TA group presenting individual scores on each of the two outcome variables. We represent a matrix using a bold capital letter, for example, $\mathbf{Y}_{uj}$. (We use the subscripts, u and j, to indicate that the elements of the matrix include unit u scores for Group j. The i notation is dropped because the matrix will represent several variables.) The number of rows and the number of columns determines the *order* of a matrix. The matrix of individual test scores for the TA group ($j = 1$) is of 22×2 or $n_j \times p$ order, where n_j is the sample size in Group j and p is the number of outcome variables. The transpose of $\mathbf{Y}_{u1}$, ($\mathbf{Y}'_{u1}$) is written as:

$$\mathbf{Y}'_{u1} = \begin{bmatrix} 4 & 4 & \dots & 15 \\ 43 & 34 & \dots & 54 \end{bmatrix}.$$

A square matrix is one with the number of rows equal to the number of columns. An example of a symmetric square matrix is the *covariance matrix*, $\mathbf{S}_j$, where the elements on the main diagonal represent the variances of the variables under investigation, and the off-diagonal elements represent the covariances. For the TA group ($j = 1$) the covariance matrix is written as:

$$\mathbf{S}_1 = \begin{bmatrix} s^2_{1_{Y_1}} & \text{Cov}_1(Y_1Y_2) \\ \text{Cov}_1(Y_1Y_2) & s^2_{1_{Y_2}} \end{bmatrix} \doteq \begin{bmatrix} 15.422 & 24.679 \\ 24.679 & 61.784 \end{bmatrix}.$$

The covariance matrix, $\mathbf{S}_j$, is described as a *symmetric matrix* because the entries are reflected about the main diagonal. That is, the first row is identical to the first column, the second row is identical to the second column, and so forth.

2.5 MATRIX OPERATIONS

The covariance matrix for each group may be computed from the respective data matrix using matrix operators. To begin, a mean-centered data matrix, $\mathbf{C}_j$, is computed by subtracting from each score the corresponding variable group mean. To subtract group means from the $n_j \times p$ data matrix, $\mathbf{Y}_{uj}$, an $n_j \times p$ matrix of means for Group j, $\mathbf{Y}_{.j}$, is needed. For the TA group, the (22×2) $\mathbf{Y}_{.1}$ matrix may be written as:

$$\mathbf{Y}_{.1} \doteq \begin{bmatrix} 7.77 & 43.45 \\ 7.77 & 43.45 \\ \vdots & \vdots \\ 7.77 & 43.45 \end{bmatrix}.$$

The (22×2) mean-centered data matrix, $\mathbf{C}_1$, for the TA group is

$$\mathbf{C}_1 = \mathbf{Y}_{u1} - \mathbf{Y}_{.1} \doteq \begin{bmatrix} 4 & 43 \\ 4 & 34 \\ \vdots & \vdots \\ 15 & 54 \end{bmatrix} - \begin{bmatrix} 7.77 & 43.45 \\ 7.77 & 43.45 \\ \vdots & \vdots \\ 7.77 & 43.45 \end{bmatrix} \doteq \begin{bmatrix} -3.77 & -.45 \\ -3.77 & -9.45 \\ \vdots & \vdots \\ 7.23 & 10.55 \end{bmatrix}.$$

Matrix addition or subtraction is obtained by adding or subtracting each element of one matrix to or from the corresponding element of a second matrix. Because addition and subtraction is done element by element, the two matrices must be of the same order.

From the mean-centered data matrix, $\mathbf{C}_j$, the sum-of-squares, SS_j, and cross-products, CP_j, can be obtained through multiplication. Two matrices (e.g., **A** and **B**) can be multiplied if they are *compatible*. That is, we can multiply matrix **A** by matrix **B**, **AB**, if the number of columns of matrix **A** equals the number of rows in matrix **B**. Consider the following two matrices:

$$\mathbf{A} = \begin{bmatrix} 3 & 2 & 1 \\ 1 & 2 & 3 \\ 2 & 3 & 1 \end{bmatrix} \mathbf{B} = \begin{bmatrix} 1 & 5 \\ 2 & 3 \\ 2 & 1 \end{bmatrix}$$

Matrix **A** is of order 3×3 and matrix **B** is of order 3×2. Matrix **A** and matrix **B** are compatible for multiplication; **AB** but not for **BA**. To multiply **A** by **B** each entry in a row of matrix **A** is multiplied by the entry in a corresponding column in matrix **B** and summed:

$$\begin{bmatrix} 3 & 2 & 1 \\ 1 & 2 & 3 \\ 2 & 3 & 1 \end{bmatrix} \times \begin{bmatrix} 1 & 5 \\ 2 & 3 \\ 2 & 1 \end{bmatrix}$$

$$= \begin{bmatrix} (3 \times 1) + (2 \times 2) + (1 \times 2) & (3 \times 5) + (2 \times 3) + (1 \times 1) \\ (1 \times 1) + (2 \times 2) + (3 \times 2) & (1 \times 5) + (2 \times 3) + (3 \times 1) \\ (2 \times 1) + (3 \times 2) + (1 \times 2) & (2 \times 5) + (3 \times 3) + (1 \times 1) \end{bmatrix} = \begin{bmatrix} 9 & 22 \\ 11 & 14 \\ 10 & 20 \end{bmatrix}.$$

2.5.1 SSCP Matrix

The product of the transpose of the mean-centered matrix for Group j, $\mathbf{C}'_j$, and the mean-centered matrix, $\mathbf{C}_j$, yields the matrix of SS_j on the main diagonal and CP_j on the off-diagonal entries. This matrix is called the sum-of-squares and cross-products (**SSCP**) matrix . That is, the entries of the $\mathbf{C}'_j\mathbf{C}_j$ matrix contain SS_j for each variable on the main diagonal, and CP_j's are the off-diagonal entries. Using our $\mathbf{C}_j$ matrix for the TA group, the $\mathbf{SSCP}_1$ matrix can be computed as:

$$\mathbf{C}'_1\mathbf{C}_1 = (\mathbf{Y}_{u1} - \mathbf{Y}_{.1})'(\mathbf{Y}_{u1} - \mathbf{Y}_{.1})$$

$$\doteq \begin{bmatrix} -3.77 & -3.77 & \cdots & 7.23 \\ -.45 & -9.45 & \cdots & 10.55 \end{bmatrix} \times \begin{bmatrix} -3.77 & -.45 \\ -3.77 & -9.45 \\ \vdots & \vdots \\ 7.23 & 10.55 \end{bmatrix}.$$

Because each entry of $\mathbf{C}'_1$ and $\mathbf{C}_1$ are deviation scores, multiplying the entries of the first row of $\mathbf{C}'_1$ by the entries of the first column of $\mathbf{C}_1$ and summing the products results in the sum-of-squares, $\mathrm{SS}_{1_{Y_1}} = \sum_{u=1}^{n_1}(Y_{1_{u1}} - \overline{Y}_{1_{.1}})^2$ for Y_1. Similarly, summing the products resulting from multiplying the entries of the second row of $\mathbf{C}'_1$ with the entries of the second column of $\mathbf{C}_1$ result in $\mathrm{SS}_{1_{Y_2}}$. The sum-of-products of the entries of the first row of $\mathbf{C}'_1$ and the second column of $\mathbf{C}_1$ or the second row of C'_1 and first column of C_1 result in $\mathrm{CP}_{1_{Y_1Y_2}} = \sum_{u=1}^{n_1}(Y_{1_{u1}} - \overline{Y}_{1_{.1}})(Y_{2_{u1}} - \overline{Y}_{2_{.1}})$. The results are

$$\mathbf{C}'_j\mathbf{C}_j = \mathbf{SSCP}_j$$

$$= \begin{bmatrix} \sum_{u=1}^{n_j}(Y_{1_{uj}} - \overline{Y}_{1_{.j}})^2 & \sum_{u=1}^{n_j}(Y_{1_{uj}} - \overline{Y}_{1_{.j}})(Y_{2_{uj}} - \overline{Y}_{2_{.j}}) \\ \sum_{u=1}^{n_j}(Y_{1_{uj}} - \overline{Y}_{1_{.j}})(Y_{2_{uj}} - \overline{Y}_{2_{.j}}) & \sum_{u=1}^{n_j}(Y_{2_{uj}} - \overline{Y}_{2_{.j}})^2 \end{bmatrix}.$$

For the TA group, ($j = 1$),

$$\mathbf{SSCP}_1 \doteq \begin{bmatrix} 323.864 & 518.273 \\ 518.273 & 1279.455 \end{bmatrix}.$$

Multiplying $\mathbf{SSCP}_j$ by the ratio $1/(n_j - 1)$, or the reciprocal of the degrees of freedom, results in a matrix of variances on the main diagonal and covariances elsewhere. That is, $[1/(n_j - 1)]\mathbf{SSCP}_j = \mathbf{S}_j$. The ratio, $1/(n_j - 1)$, is a *scalar*, or single number. When any matrix is multiplied by a scalar, each entry in the matrix is multiplied by the scalar value. In our case we are multiplying each sum-of-squares and sum-of-cross-products by the reciprocal of the degrees of freedom, or equivalently dividing each entry in the $\mathbf{SSCP}_1$ matrix by $n_1 - 1$. The results are the variances and covariance of the measures. For our data with $n_1 = 22$, the covariance matrix for Group 1 is

$$\mathbf{S}_1 = \frac{1}{n_1 - 1}\mathbf{SSCP}_1 = \begin{bmatrix} s^2_{1_{Y_1}} & \mathrm{Cov}_1(Y_1Y_2) \\ \mathrm{Cov}_1(Y_1Y_2) & s^2_{1_{Y_2}} \end{bmatrix};$$

that is,

$$\mathbf{S}_1 \doteq \frac{1}{22-1}\begin{bmatrix} 323.864 & 515.273 \\ 518.273 & 1297.455 \end{bmatrix} \doteq \begin{bmatrix} 15.422 & 24.679 \\ 24.679 & 61.784 \end{bmatrix}.$$

2.5.2 Determinant

Earlier we stated that with multivariate analyses the means of several variables are presented as a vector of means called a *centroid.* The purpose of the centroid is to represent the convergence of several dimensions in a single point. While several variables may reflect a single dimension, for now we will consider each variable as a single dimension. The mean of each outcome variable provides the "typical" observation in a single dimension. In addition to knowing the "typical" observation in the system of variables, we will be interested in some indicator of the variance of the observations for the system of p variables. That is, we will be interested in reducing the square covariance matrix, **S**, into a single index representing a *generalized variance.* A statistic that accomplishes this objective is called a *determinant* and is represented by the symbol $|\cdot|$ where $\cdot$ represents the matrix whose determinant is of interest. In the case of the covariance matrix, the determinant is represented as $|\mathbf{S}|$. The calculation of the determinant of a matrix can be computationally intensive. But in the case of a 2×2 matrix it is a simple calculation. Consider the (2×2) matrix

$$\mathbf{B} = \begin{bmatrix} a & b \\ c & d \end{bmatrix}.$$

The determinant of **B**, $|\mathbf{B}|$, equals $ad - bc$. For example, if

$$\mathbf{B} = \begin{bmatrix} 7 & 8 \\ 3 & 3 \end{bmatrix},$$

then $|\mathbf{B}| = 7(3) - 8(3) = -3$. Now consider our covariance matrix, $\mathbf{S}_1$, for the TA group:

$$\begin{bmatrix} 15.422 & 24.679 \\ 24.679 & 61.784 \end{bmatrix}.$$

Then, $|\mathbf{S}_1| \doteq 15.422(61.784) - 24.679(24.679) \doteq 343.780$. The determinant represents a measure of variability in our system of two variables. The determinant of a covariance matrix $|\mathbf{S}|$ is called the *generalized variance.* The generalized variance is a very useful statistic to summarize the overall variability in our data set.

Let us examine the calculation of the determinant of the covariance matrix a little closer to get a better feel for its meaning. Recall the elements of the two-variable covariance matrix are

$$\begin{bmatrix} s_{Y_1}^2 & \mathrm{Cov}(Y_1 Y_2) \\ \mathrm{Cov}(Y_1 Y_2) & s_{Y_2}^2 \end{bmatrix}.$$

So, $|\mathbf{S}| = (s^2_{Y_1})(s^2_{Y_2}) - [\mathrm{Cov}(Y_1 Y_2)]^2$ and, from Eq. (2.1), $[\mathrm{Cov}(Y_1 Y_2)]^2 = (s^2_{Y_1})(s^2_{Y_2})(r^2_{Y_1 Y_2})$. With substitution and factoring, the determinant of the covariance matrix can be written as: $|\mathbf{S}| = s^2_{Y_1}[s^2_{Y_2}(1 - r^2_{Y_1 Y_2})]$. Because $(1 - r^2_{Y_1 Y_2})$ can be interpreted as the proportion of variation in the second outcome variable (Y_2) that is *not* shared with the first outcome variable (Y_1), the value in the brackets is the unique variance associated with Y_2. That is, the bracketed term is the amount of variation contributed to the system that is independent of the first variable. Which measure is considered first or second is totally arbitrary, but the total variation in the system of variables is constant. If the variables are completely confounded, having unity as the correlation ($r_{Y_1,Y_2} = 1$), then Y_2 would contribute no additional variation to the system beyond the variation attributed to Y_1. If, on the other hand, Y_1 and Y_2 were independent of each other ($r_{Y_1,Y_2} = 0$), then the determinant in our system of variables would be the product of the variable variances. Generally, the correlation between the two variables used in our study is somewhere between these two extremes. Note that if the two variables are perfectly correlated, the determinant would equal 0.

2.5.3 Inverse

Up to this point we have discussed the addition, subtraction, and multiplication of matrices. We have not mentioned division. The matrix counterpart of division in scalar arithmetic is a much more involved process. The matrix process is called *inversion*. We will concern ourselves with inversion of square matrices only. The inverse of a matrix $\mathbf{B}$ is denoted by $\mathbf{B}^{-1}$. An inverse in matrix arithmetic is analogous to a reciprocal in scalar arithmetic. The reciprocal of a real number b is the number b^{-1} such that $b(b^{-1}) = 1$, the multiplicative *identity* in real numbers. Analogously, the inverse of a matrix $\mathbf{B}$ (of order $m \times m$) is the matrix $\mathbf{B}^{-1}$ (of order $m \times m$) such that $\mathbf{BB}^{-1} = \mathbf{I}$, where the entries on the main diagonal of $\mathbf{I}$ equal 1, all other entries are 0, and is order $m \times m$. The $\mathbf{I}$ matrix is called an *identity matrix*.

Finding the inverse of a 2×2 matrix is fairly straightforward. Consider matrix $\mathbf{B}$ presented earlier:

$$\mathbf{B} = \begin{bmatrix} a & b \\ c & d \end{bmatrix}.$$

Then, if $|\mathbf{B}| = a(d) - b(c) \neq 0$, the inverse of $\mathbf{B}$ equals

$$\mathbf{B}^{-1} = \frac{1}{|\mathbf{B}|}\begin{bmatrix} d & -b \\ -c & a \end{bmatrix}.$$

For example,

$$\mathbf{B} = \begin{bmatrix} 7 & 8 \\ 3 & 3 \end{bmatrix},$$

$$\mathbf{B}^{-1} = \frac{1}{21-24}\begin{bmatrix} 3 & -8 \\ -3 & 7 \end{bmatrix}$$
$$= \begin{bmatrix} -1 & 8/3 \\ 1 & -7/3 \end{bmatrix}.$$

By matrix multiplication it can be verified that

$$\mathbf{B}\mathbf{B}^{-1} = \mathbf{B}^{-1}\mathbf{B} = \begin{bmatrix} 1 & 0 \\ 0 & 1 \end{bmatrix} = \mathbf{I}.$$

If $|\mathbf{B}| = ad - bc = 0$, the inverse of $\mathbf{B}$ is not possible, and we say that $\mathbf{B}$ is *singular*. Finding inverses of higher-order matrices involves much calculation and is not illustrated here (see Johnson and Wichen, 2002, pp. 96–97; Rencher, 2002, pp. 23–25). Further knowledge of the inversion process is not necessary for our study. For an additional example, consider the covariance matrix $\mathbf{S}_1$ presented earlier:

$$\mathbf{S}_1 \doteq \begin{bmatrix} 15.422 & 24.679 \\ 24.679 & 61.784 \end{bmatrix}.$$

Recalling $|\mathbf{S}_1| \doteq 343.780$, the inverse of $\mathbf{S}_1$ is

$$\mathbf{S}_1^{-1} \doteq \begin{bmatrix} .179 & -.072 \\ -.072 & .045 \end{bmatrix}.$$

2.5.4 Eigenanalysis

There is another process of assigning scalars to matrices that is central to a number of multivariate methods. Let us start with a preliminary definition. If $\mathbf{B}$ is a square matrix of order $m \times m$, then $|\mathbf{B} - \lambda\mathbf{I}| = 0$ is called the *eigenequation* (or *characteristic* equation) for the matrix $\mathbf{B}$. The lowercase Greek letter lambda, λ, denotes an *eigenvalue*. This equation involves a determinant of a matrix difference that yields a polynomial of the mth degree in λ. For example, with $m = 2$ if

$$\mathbf{B} = \begin{bmatrix} 7 & 8 \\ 3 & 3 \end{bmatrix},$$

then

$$|\mathbf{B} - \lambda\mathbf{I}| = \left|\begin{bmatrix} 7 & 8 \\ 3 & 3 \end{bmatrix} - \lambda\begin{bmatrix} 1 & 0 \\ 0 & 1 \end{bmatrix}\right|$$
$$= \left|\begin{bmatrix} 7-\lambda & 8 \\ 3 & 3-\lambda \end{bmatrix}\right|$$
$$= (7-\lambda)(3-\lambda) - 24$$
$$= \lambda^2 - 10\lambda - 3.$$

The m roots of the characteristic equation are the *eigenvalues* (or *characteristic roots*) of **B**. The m eigenvalues (not necessarily distinct, and some may be zero) are denoted $\lambda_1, \lambda_2, \ldots, \lambda_m$ and are the scalars referred to in the preceding paragraph. To obtain the eigenvalues, the following equation can be used for a quadratic equation of the form $a\lambda^2 + b\lambda + c = 0$. Here

$$\lambda = \frac{-b \pm \sqrt{b^2 - 4ac}}{2a}.$$

Using our eigenequation,

$$\begin{aligned}\lambda &= \frac{10 \pm \sqrt{10^2 - 4(1)(-3)}}{2(1)}\\ &\doteq \frac{10 \pm 10.583}{2}\\ &\doteq -.292 \quad \text{and} \quad 10.292.\end{aligned}$$

In our study of discriminant analysis we will be seeking a weight for each variable of a set of variables. In matrix terms we will be seeking a *vector* of weights. Specifically, we will be seeking the vector solution of the messy matrix equation

$$\underset{(m \times m)}{(\mathbf{B} - \lambda\mathbf{I})} \cdot \underset{(m \times 1)}{\mathbf{b}} = \underset{(m \times 1)}{\mathbf{0}},$$

where **b** is a (column) vector of m weights and **0** is a vector of m zeros. It turns out that there is one vector solution $\mathbf{b}_m$ corresponding to each of the m eigenvalues of **B**—some vector solutions may be the trivial zero vector. The vector $\mathbf{b}_m$ is called the *eigenvector* (or *characteristic vector*) of the matrix **B** associated with the eigenvalue λ_v $(v = 1, 2, \ldots, m)$.

The computation of eigenvalues and eigenvectors in nearly all discriminant analyses is quite complicated, indeed. As we will see, eigenequations involve inverses as well as determinants, all of which are manipulatively complicated. We will, as nearly everyone does, leave the manipulations for the computers.

2.6 DISTANCE

It is often useful to describe differences between groups in terms of a distance measure that was suggested by Euclid of Alexandria (c. 325 B.C. to c. 265 B.C.). Recalling the well-known geometric theorem for the distance between two points as $a^2 + b^2 = c^2$, the distance between two centroids can be described. Letting a equal the difference between two group means on the first variable Y_1, $(\overline{Y}_{1_{.1}} - \overline{Y}_{1_{.2}})$ and b is the difference between two group means on the second variable Y_2, $(\overline{Y}_{2_{.1}} - \overline{Y}_{2_{.2}})$. The Euclidean distance between the two groups can be computed as:

$$c = \sqrt{(\overline{Y}_{1_{.1}} - \overline{Y}_{1_{.2}})^2 + (\overline{Y}_{2_{.1}} - \overline{Y}_{2_{.2}})^2}.$$

In the Baumann study the two group centroids are

$$\mathbf{y}_{.1} \doteq \begin{bmatrix} 7.77 \\ 43.45 \end{bmatrix} \quad \text{and} \quad \mathbf{y}_{.2} \doteq \begin{bmatrix} 6.68 \\ 42.05 \end{bmatrix}.$$

So, the distance between the two groups can be computed as:

$$\begin{aligned} c &\doteq \sqrt{(6.68 - 7.77)^2 + (42.05 - 43.45)^2} \\ &\doteq 1.774. \end{aligned}$$

Figure 2.1 can help in presenting the concept.

A potential problem with this measure of distance is that it assumes that the variables are measured with the same metric scale. With the current data example, this assumption is not tenable. That is, the score distributions of the two variables have different variances. To standardize the variables we will divide the squared mean differences by an error covariance matrix. The error covariance matrix, $\mathbf{S}_e$, is obtained by summing the separate group $\mathbf{SSCP}_j$ matrices across the J groups, $(\mathbf{E} = \sum_{j=1}^{J} \mathbf{SSCP}_j)$ and dividing each element by $N - J$, $(\mathbf{S}_e = 1/(N - J)\mathbf{E})$. The result of this division is the squared distance measure, known as the *Mahalanobis squared distance*:

$$D^2 = (\mathbf{y}_{.1} - \mathbf{y}_{.2})'\mathbf{S}_e^{-1}(\mathbf{y}_{.1} - \mathbf{y}_{.2}),$$

where $\mathbf{y}_{.1}$, and $\mathbf{y}_{.2}$ are vectors of p means (centroids) for Groups 1 and 2, respectively, and $\mathbf{S}_e^{-1}$ is the inverse of the error covariance matrix. The positive square root of Mahalanobis D^2 is a measure of distance in a standardized scale. The positive square root of D^2 is analogous to the standardized mean difference, d, [suggested by J. Cohen (1923–1998) in 1988, p. 20], used in the univariate context to estimate an effect-size measure for the difference between two groups. We will consider the Mahalanobis D^2 statistic further in Chapter 3. It should be noted that this measure of distance is not limited to just two variables, and other distance measures may also be of interest. For example, the distance between two data points, or the distance between a mean vector for a group and the grand mean vector across all groups, may also be of interest.

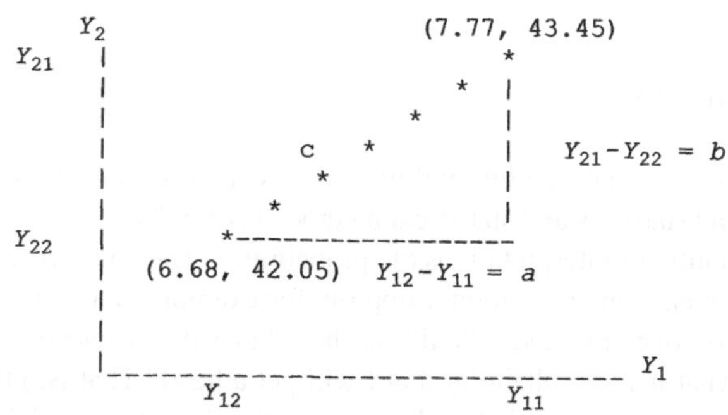

Figure 2.1 Distance in a plane.

2.7 LINEAR COMPOSITE

Consider a set of variables X_1, X_2, X_3, and X_4. A *linear composite* of these variables may be expressed as:

$$b_1X_1 + b_2X_2 + b_3X_3 + b_4X_4, \tag{2.2}$$

where the b values are real numbers. For example, suppose that the variable scores for a unit are $X_1 = 19$, $X_2 = 23$, $X_3 = 117$, and $X_4 = 46$, and the b values are $b_1 = 1.5$, $b_2 = -0.9$, $b_3 = 12.7$, and $b_4 = 3.6$. Then the unit's linear combination score is

$$(1.5)19 + (-0.9)23 + (12.7)117 + (3.6)46 \doteq 1659.3.$$

As such, a new "variable" composed of the four original variables is constructed, and the unit's score on this new variable is 1659.3.

You may ask: From where do those b values come? This is a reasonable question, but the answer to be advanced may appear to be a bit obtuse. In a number of multivariate analyses, linear composites play a central role. The b values or *weights* for the original variables are determined so that some criterion is obtained. The criterion varies across predictive discriminant analysis, multivariate analysis of variance, canonical correlation, multiple regression, and principal component analysis. A few criteria will be discussed explicitly in subsequent chapters. We need not be concerned about the calculation of the b values because reliance will be placed on the use of computer programs.

Two additional comments about linear combinations may be made. *First*, in addition to variable weights—the b's—in some data analyses call for an additive *constant*. Such a constant is often denoted as b_0. The *second* comment pertains to a linear composite of variable means rather than of the variables themselves: for example,

$$b_0 + b_1\overline{X}_1 + b_2\overline{X}_2 + b_3\overline{X}_3 + b_4\overline{X}_4. \tag{2.3}$$

[The numerical values of these b's would typically be different from those of the b's in (2.2).] Note that the composite in (2.3) is that of elements in a centroid. An example of such a mean composite is discussed in Chapter 5.

2.8 PROBABILITY

A formal discussion of probability will not be presented. Formal presentations may be found in many statistics and statistical methods textbooks.

There is a handful of interpretations of "probability." The one favored in this book is a relative frequency interpretation. Suppose, for example, that one has a fair coin with a "head" on one side and a "tail" on the other side. A question may be: If I flip this coin, what is the probability that I will get a head? That is, given that H is the event of flipping a head, what is the numerical value of $P(H)$? To answer this question theoretically, one would need to flip the coin an indefinite number of times (using identical flipping motions). One may obtain an *estimate* of the probability,

$\hat{P}(H)$, by repeatedly conducting the "experiment" of flipping the coin, say, 1000 times. Again the repeated flipping action is assumed to be identical (and reasonably done, of course). So an estimate of the probability would be

$$\hat{P}(H) = \frac{\text{number of } H\text{'s}}{1000}.$$

That is, the estimated probability is found by dividing the number of favorable outcomes (an H in the example above) of experiment repetitions divided by the total number of experiments conducted.

An estimated probability to which many empirical researchers have been exposed is a P value. This is a probability value associated with the value of a test statistic, say an ANOVA $F(3, 60)$ value of 3.30. Theoretically, we are seeking the probability of getting an $F(3, 60)$ value of 3.30 or greater under the condition that the null hypothesis is true. This is a *conditional probability* that may be denoted as $P = P[F(3, 60) \geq 3.30|H_0]$. Such a probability is obtained by making two assumptions. *First*, it is assumed that data conditions requisite for the use of the test statistic are met: namely, independence of observations, normality of scores in populations, and score variance homogeneity in the populations. *Second*, the probability is obtained assuming that the ANOVA "study" is replicated an indefinite number of times in the very same manner as was done in the study under consideration. A probability distribution is conceptually simulated that would theoretically be an F distribution with 3 and 60 degrees of freedom. Then the P value would be the proportion of studies (i.e., the area under the curve for the simulated F distribution) for which the F(3, 60) value obtained would be greater than or equal to 3.30, given that H_0 is true. (Incidentally, this particular P value is about .026.)

An estimated conditional probability that will be of interest later pertains to the probability of a unit belonging to a particular group, given the unit's vector of predictor scores, $\hat{P}(j|\mathbf{x}_u)$. This value is based on assumptions of theoretical data distributions and on the idea that if a unit with a score vector $\mathbf{x}_u$ is selected repeatedly from a meaningful population an indefinite number of times, $\hat{P}(j|\mathbf{x}_u)$ is the proportion of times that the unit emanated from Group j.

2.9 STATISTICAL TESTING

It is assumed that the reader of this book is familiar with some basic concepts related to statistical testing: null hypothesis, alternative hypothesis, test statistic, Type I error, Type II error, referent probability distribution, critical/rejection region, and P value (see the preceding subsection). Two popular approaches to the testing of a statistical null hypothesis are significance testing atributed to R. A. Fisher and hypothesis testing attributed to J. N. Neyman (1895–1981) and E. S. Pearson (1895–1980)—see Huberty (1987). The former is sometimes called the *P-value approach*, and the latter may be termed the *fixed-α approach*. The P-value approach, which is favored in this book, does not call for setting an α level or establishing a region of rejection. Therefore, there is no need for tables of probability distribution (e.g., F, t, χ^2) critical values; such tables are *not* provided in this book.

In many empirical studies, multiple null hypotheses are of interest and therefore multiple statistical tests need to be conducted. (This is of particular concern when multiple tests are conducted using data on a given collection of units.) Multiple testing should be taken into consideration when assessing statistical significance for an individual test in terms of Type I error rate. This can be accomplished in a number of ways. When using the P-value approach to testing, one can adjust the *tail probability* for each test using the method suggested by C. E. Bonferroni (1892–1960)—see Maxwell and Delaney (2000, pp. 177–180). What is done in this method is simply to multiply each test tail probability by the number of tests conducted. This product yields a P value for each test, a judgment about which needs to be made to determine statistical significance. (With the fixed-α approach, the application of the Bonferroni method involves divvying the overall α—not necessarily equally—among all of the tests conducted.) If one is uncomfortable with this straight Bonferroni method, one can use a modified Bonferroni method (Wright, 1992).

2.10 JUDGMENT IN DATA ANALYSIS

When talking with, and reading writings of, some beginning empirical researchers, one might get the impression that the use of statistical methods in analyzing data is cut and dried. The use of quantitative methods implies to some a high level of objectivity. But the experienced empirical researcher is, or should be, well aware of the potential high level of subjectivity that enters into quantitative research. Just as an example of misperception, it may be inferred that some researchers believe statistical test results are "definitive" in that one hypothesis or the other may be *proved* to be correct (or incorrect). True, statistical analysis can quantify uncertainty, but it cannot eliminate uncertainty (Warren, 1986).

Data analysis and interpretation abound with subjectivity. Subjectivity is necessarily so prevalent because of the many judgments needed in the research process. What are some of these many judgments to be made? Some judgments have to do with:

- Small versus large P value
- Small versus large effect-size value
- Small versus large structure r
- Initial choice of response variables
- Labeling of latent variables
- Response variable subset selection
- Difference of variable ordering index values
- Extent to which requisite data conditions are satisfied

Undoubtedly, the reader can think of more. Additional instances in the context of discriminant analysis appear in this book.

On what basis may judgments be made? Schaafsma and van Vark (1979) suggest that an important part of applying data analysis methods is played by "a priori professional knowledge." They describe this as "a mixture of (1) interpretations of previous data, e.g., data recorded in the literature, (2) common sense, (3) intuition, (4) ideas with respect to the availability of certain measurements, or their costs, (5) etc." (p. 108): That is, judgment is based on experience, previous research, and common sense. The latter term will not be discussed; it will be assumed that common sense is self-evident. When selecting a research method, Shulman (1988, p. 13) says that "the choice requires an act of judgment, grounded in both knowledge of methodology and the substantive area of the investigation." In the context of statistical testing, M. M. Tatsuoka (1920–1996) suggests to researchers: "Let common sense and your set of values be the judge!" He goes on to suggest that "it is about time that we outgrew the notion that science is value-free." And if "statistics is quantified common sense," as R. A. Fisher once said, "then its quantification, in turn, must be guided by common sense," (1982, p. 1782). Paulos (1991, p. 58) adds: "Statistics, more than most other areas of mathematics, is just formalized common sense, quantified straight thinking."

Finally, it seems reasonable that some advice of Kempthorne (1977) regarding terms be heeded. In analyzing data and in interpreting results of data analysis, there is no way of avoiding words such as "natural," "reasonable," "plausible," and the like. To quote Kempthorne: "Whatever we do, we are forced in the last resort to depend on an unanalyzable notion of 'reasonable' or whatever" (p. 759). So, when using discriminant analysis, for example, you should feel free to express your judgments as being "reasonable" and rely on "common sense."

2.11 SUMMARY

In this chapter we introduced a considerable number of new terms and have demonstrated the application of several matrix operations. These terms and operations will be used throughout the text to demonstrate and explain the analysis of multivariate data. While we hope the reader will become comfortable with this new vocabulary, and feel reasonably confident in the use of operations with small data sets, it will not be necessary to have mastered all of these operations to achieve a good understanding of multivariate data analysis. Throughout the text we will rely heavily on the use of SPSS and SAS to carry out the complex and tedious matrix operations. Our intention is to provide sufficient background into the operations used for multivariate analyses, but we will emphasize the interpretation of the results of the analyses.

Further Reading

Barnett (1990) presents a detailed discussion of matrix manipulations, matrix rank, eigenanalysis, and quadratic forms. Some mathematics background is needed for use of this book.

Healy (1986) has a monograph that succinctly covers the basics of matrices applied to the study of statistical methods.

Huberty (2000b) discusses a number of analysis contexts in which researcher judgment is relevant.

Kempthorne (1977) discusses various points of view regarding probability, statistical inference, and the use of data analysis.

Lad (1996) provides detailed discussions of the use subjectivity and philosophy in statistical methods.

Rothman (1990) and Saville (1990) question a common practice of adjusting statistical test probabilities in a multiple test situation.

Wang (1993) reviews many analysis and reporting contexts in which subjectivity and common sense should be used.

Definition ***Raw scores***: Data before being cooked by statisticians.

EXERCISES

1. If matrix **A** is of order $n \times p$ what must be the order of matrix **B** for the following matrix operations to be valid?
   (a) $\mathbf{A} + \mathbf{B}$
   (b) **AB**
   (c) **BA**

2. Table 2.2 provides the mean, sum-of-squares, and variance for variables Y_1 and Y_2. If, for the DRA group, the correlation between Y_1 and Y_2 equals .498, what would the **SSCP** ($\mathbf{SSCP}_2$) and covariance ($\mathbf{S}_2$) matrices equal?

3. Determine the generalized variance ($|\mathbf{S}_2|$) for the DRA group.

4. What does the inverse of the covariance matrix ($\mathbf{S}_2^{-1}$) for the DRA group equal?

5. In Section 2.5.1 the **SSCP** ($\mathbf{SSCP}_1$) was provided. Sum the **SSCP** matrices for the TA and DRA groups to obtain the total **SSCP**. Use this result to obtain $\mathbf{S}_e$.

6. Determine the generalized variance using $\mathbf{S}_e$, ($|\mathbf{S}_e|$).

7. Determine the inverse of $\mathbf{S}_e$, ($\mathbf{S}_e^{-1}$).

8. Verify your results for Exercise 7 by calculating the product $\mathbf{S}_e\mathbf{S}_e^{-1}$. Do the results equal the identity matrix (**I**)?

9. Let $\mathbf{A} = \begin{bmatrix} 10 & 15 \\ 3 & 6 \end{bmatrix}$. Provide the eigenvalues for **A** by solving the eigenequation, $|\mathbf{A} - \lambda\mathbf{I}| = 0$.

10. Using the results from Exercise 7 and the group centroids reported in Section 2.6, compute the Mahalanobis D^2 statistic.

PART II

One-Factor MANOVA/DDA

The techniques to be discussed in Part II pertain to the analysis and description of effects of a grouping variable on a collection of outcome (Y) variables. The design here involves a single factor (or grouping variable). Response variables are termed outcome variables in this part. These variables are the characteristics or attributes or traits of the analysis units under study, in addition to group membership. The "analysis" of the effects pertains to a statistical test of the comparison of levels of a factor. Such a test is typically associated with a multivariate analysis of variance (MANOVA) or with a contrast analysis involving particular levels of the factor. The "description" of effects pertains to a set of techniques, many of which are collectively called descriptive discriminant analysis (DDA). Methods of outcome variable deletion and outcome variable ordering, which may or may not be considered part of DDA, are also covered. Part II is concluded with suggestions for reporting MANOVA/DDA results.

The goals of the reader for this part are to be able to (1) critically evaluate an application of MANOVA/DDA and (2) write up a report of a study in which a MANOVA/DDA is applied. References may be made to an overview given in the form of a flowchart in Figure 7.2 while studying Part II.

Applied MANOVA and Discriminant Analysis, Second Edition, by Carl J. Huberty and Stephen Olejnik
Copyright © 2006 John Wiley & Sons, Inc.

CHAPTER 3

Group Separation

3.1 INTRODUCTION

The intent of this chapter is to acquaint the reader with statistical criteria used in conducting a one-way multivariate analysis of variance (MANOVA). MANOVA criteria may be viewed as extensions of univariate criteria. Thus, the concepts and formulations used in univariate analyses are reviewed first. In particular, two-group univariate analyses and a one-way analysis of variance (ANOVA) are reviewed; the former is reviewed specifically because of later emphasis on multivariate group contrasts. Emphasis is also later placed on calculating and reporting an effect-size index value when grouping variable effects—omnibus or contrast—are assessed. The main topic in this chapter pertains to four popular statistical test criteria used in MANOVA.

3.2 TWO-GROUP ANALYSES

3.2.1 Univariate Analysis

Hypothesis Test In a univariate context, when a comparison of two group means is carried out, the researcher may have one or more of the following questions in mind:

Question 1 Are the two "treatments" differentially effective? That is, is the effect of belonging to Population 1 different from the effect of belonging to Population 2?

In terms of an hypothesis test, the null hypothesis may be written as:

$$H_0\colon \alpha_1 = \alpha_2 (= 0)$$

where $\alpha_1 = \mu_1 - \mu_{.}$ $(= 0)$. That is, the deviation of the mean for Population 1 from the overall grand mean (across both populations) equals 0.

Applied MANOVA and Discriminant Analysis, Second Edition, by Carl J. Huberty and Stephen Olejnik
Copyright © 2006 John Wiley & Sons, Inc.

Question 2 How strong is the relationship between the grouping variable and the outcome variable? Or, knowing the population to which an analysis unit belongs, how accurately can the outcome variable be predicted?

The null hypothesis for this research question may be written as:

$$H_0\colon \rho = 0.$$

That is, the correlation between the outcome variable and the grouping variable equals 0.

Question 3 Is the mean outcome variable score for Population 1 different from the mean for Population 2?

For this question the null hypothesis may be written as:

$$H_0\colon \mu_1 = \mu_2.$$

These hypotheses are statistically and substantively equivalent and may be tested using one statistic, the Student t for two independent samples,

$$t = \frac{\overline{Y}_{.1} - \overline{Y}_{.2}}{\sqrt{s_e^2(1/n_1 + 1/n_2)}}, \tag{3.1}$$

where s_e^2 is the *error variance* computed as:

$$s_e^2 = \frac{\sum_{j=1}^{J}(n_j - 1)s_j^2}{N - J} = \frac{(n_1 - 1)s_1^2 + (n_2 - 1)s_2^2}{n_1 + n_2 - 2}, \tag{3.2}$$

where s_j^2 is the variance of the scores in Group j ($j = 1, 2$), n_j is the number of analysis units in Group j, and N is the total sample size, $N = \sum_{j=1}^{J} n_j$. Under the assumption of homogeneity of variance, that is, the variance of Population 1 equaling the variance of Population 2, $\sigma_1^2 = \sigma_2^2$, s_e^2 provides an estimate of the common population variance, σ^2.

In Chapter 2, a study by Baumann was described; means and variances for two reading programs on two outcome variables were provided in Table 2.2. A question in this context is: Is the mean Y_1 score for the TA population different from the mean Y_1 score for the DRA population? The t statistic would be computed with

$$s_e^2 \doteq \frac{(22 - 1)15.422 + (22 - 1)7.656}{44 - 2} \doteq 11.539,$$

and

$$t \doteq \frac{7.77 - 6.68}{\sqrt{11.539(1/22 + 1/22)}} \doteq 1.064.$$

The t statistic has a central Student t distribution (assuming the Y_1 scores are independent of each other, the two population distributions of Y_1 scores are normal, the two population variances of the Y_1 scores are equal, and the null hypothesis is true) with degrees of freedom, $\text{df}_e = n_1 + n_2 - 2 = N - 2$. (In this book, df_e denotes *error degrees of freedom*. In general, df_e is equal to the total number of units minus the number of groups or cells.) For the Baumann study, $\text{df}_e = 44 - 2 = 42$. For the current study, the tail probability of the observed computed t statistic [$t(42) \doteq 1.064$] under the null hypothesis is $P \doteq .293$. We would conclude that there is insufficient evidence to reject the null hypothesis. That is, there is insufficient evidence to indicate that the TA population mean differs from the DRA population mean on Y_1. Note, we are not saying that the population means are identical, only that there is insufficient evidence to conclude that the population means are different.

The square of the t statistic provides a useful alternative statistic that has a central F distribution:

$$t^2 = \frac{\left(\overline{Y}_{.1} - \overline{Y}_{.2}\right)^2}{s_e^2[(n_1 + n_2)/n_1 n_2]},$$

or as

$$t^2 = \frac{n_1 n_2}{n_1 + n_2}\left(\overline{Y}_{.1} - \overline{Y}_{.2}\right)(s_e^2)^{-1}\left(\overline{Y}_{.1} - \overline{Y}_{.2}\right). \tag{3.3}$$

Under the assumptions of independent units, normal population distributions, variance homogeneity, and a true null hypothesis, this statistic has a central F distribution with 1 and $N - 2$ degrees of freedom. Equation (3.3) is useful because an analogous statistic can be used to test a multivariate hypothesis.

It might be noted that the sample variance of the TA group is approximately twice the sample variance of the DRA group ($15.422/7.656 \doteq 2.014$). This *may* be an indication that the homogeneity of variance assumption is violated. Although the Student t test is generally robust to this assumption when sample sizes are equal, an alternative approach may be taken to test the null hypothesis H_0: $\mu_1 = \mu_2$. The test statistic is computed as:

$$t' = \frac{\overline{Y}_{.1} - \overline{Y}_{.2}}{\sqrt{s_1^2/n_1 + s_2^2/n_2}}, \tag{3.4}$$

and the degrees of freedom are computed as:

$$\text{df}' = \frac{(s_1^2/n_1 + s_2^2/n_2)^2}{\dfrac{(s_1^2/n_1)^2}{(n_1 - 1)} + \dfrac{(s_2^2/n_2)^2}{(n_2 - 1)}}. \tag{3.5}$$

Using the Baumann data, the results are

$$t' \doteq \frac{7.77 - 6.68}{\sqrt{15.422/22 + 7.656/22}} \doteq 1.064,$$

and

$$\mathrm{df}' \doteq \frac{(15.422/22 + 7.656/22)^2}{\dfrac{(15.422/22)^2}{(22-1)} + \dfrac{(7.656/22)^2}{(22-1)}} \doteq 37.7.$$

In this example, the computed test statistic, t', is identical to t computed from (3.1) because the sample sizes are equal. The degrees of freedom, however, are less (37.7 vs. 42) to "compensate" for the differences in sample variances. The P value is .293. The conclusion remains the same because the difference in degrees of freedom is minimal. The SPSS t test program computes both t tests, when variance homogeneity is assumed and when it is not assumed to be met.

Effect Size One of two indices of effect size is often reported along with the test of statistical significance. *One* index of effect size is a measure of association between the grouping variable and the outcome variable. When the grouping variable is dichotomous (e.g., TA vs. DRA) the point-biserial correlation coefficient, r_{pb}, which is an estimate of ρ, provides the desired measure of association. The square of r_{pb} reflects the proportion of outcome variable variance that may be attributed to the grouping variable, and is given by:

$$r_{\text{pb}}^2 = \frac{t^2}{t^2 + \mathrm{df}_e}. \tag{3.6}$$

Using the Baumann data, the squared point-biserial correlation is computed as:

$$r_{\text{pb}}^2 \doteq \frac{(1.064)^2}{(1.064)^2 + 42} \doteq .026.$$

These results indicate very little of the variation in the Error Detection Task (EDT) data may be attributed to the two reading programs.

A *second* index of effect size is the standardized mean difference, d (Cohen, 1988, p. 20). The absolute value of the difference of sample means is divided by the error standard deviation, $\sqrt{s_e^2}$.

$$d = \frac{|\overline{Y}_{.j} - \overline{Y}_{.j'}|}{\sqrt{s_e^2}}, \tag{3.7}$$

where j and j' represent two different groups. Using the Baumann data to compare the TA group with the DRA group, d is computed as:

$$d \doteq \frac{|7.77 - 6.68|}{\sqrt{11.539}} \doteq .321.$$

The t statistic and the standardized mean difference are related as follows:

$$t = \frac{\overline{Y}_{.1} - \overline{Y}_{.2}}{\sqrt{s_e^2(1/n_1 + 1/n_2)}} = \frac{\overline{Y}_{.1} - \overline{Y}_{.2}}{\sqrt{s_e^2}} \frac{1}{\sqrt{1/n_1 + 1/n_2}}.$$

With

$$\frac{1}{n_1} + \frac{1}{n_2} = \frac{n_1 + n_2}{n_1 n_2},$$

$$t = d\sqrt{\frac{n_1 n_2}{n_1 + n_2}}, \tag{3.8}$$

and

$$d = t\sqrt{\frac{n_1 + n_2}{n_1 n_2}}. \tag{3.9}$$

It should be noted that the two effect-size indices, r_{pb}^2 and d, are only applicable when $\sigma_1^2 = \sigma_2^2$.

3.2.2 Multivariate Analysis

Hypothesis Test The Baumann study had two reading outcome variables, Error Detection Task (Y_1) and Degrees of Reading Power (Y_2). While another univariate analysis of the second outcome could be conducted, the two outcome variables were shown to be correlated—see Section 2.4—indicating they may reflect a common construct. A more appropriate analysis would be to compare the two populations on that common construct through a multivariate analysis.

A generalization of the squared Student t statistic was advanced by H. Hotelling (1895–1979). The null hypothesis may be expressed as H_0: $\boldsymbol{\mu}_1 = \boldsymbol{\mu}_2$. Here $\boldsymbol{\mu}$ represents a vector of p population outcome variable means, that is, a population centroid. The number of variables must be less than the total sample size minus 1, $p < N - 1$. The null hypothesis states that the centroids for the two populations are identical. To test this hypothesis, the Hotelling statistic is used:

$$T_{12}^2 = \frac{n_1 n_2}{n_1 + n_2}\left[(\mathbf{y}_{.1} - \mathbf{y}_{.2})' \mathbf{S}_e^{-1} (\mathbf{y}_{.1} - \mathbf{y}_{.2})\right], \tag{3.10}$$

where $\mathbf{y}_{.j}$ is the $p \times 1$ vector of outcome variable means for Group j, and $\mathbf{S}_e$ is the $p \times p$ *error covariance matrix*. [This is a generalization of Eq. (3.3).] Table 2.2 contains the sample group means for the Y_1 and Y_2 outcomes in the Baumann study. The two group centroids can be presented as:

$$\mathbf{y}_{.1} \doteq \begin{bmatrix} 7.77 \\ 43.45 \end{bmatrix}, \qquad \mathbf{y}_{.2} \doteq \begin{bmatrix} 6.68 \\ 42.05 \end{bmatrix}.$$

The $\mathbf{SSCP}_1$ for the TA group was computed in Chapter 2 and found to be

$$\mathbf{SSCP}_1 \doteq \begin{bmatrix} 323.864 & 518.273 \\ 518.273 & 1297.455 \end{bmatrix}$$

For the DRA group $\mathbf{SSCP}_2$ was found in Exercise 2 of Chapter 2 to be

$$\mathbf{SSCP}_2 \doteq \begin{bmatrix} 160.773 & 191.318 \\ 191.318 & 918.955 \end{bmatrix}.$$

The sum of the separate $\mathbf{SSCP}_j$ matrices, $\mathbf{SSCP}_1 + \mathbf{SSCP}_2$, is

$$\mathbf{E} \doteq \begin{bmatrix} 484.637 & 709.591 \\ 709.591 & 2216.41 \end{bmatrix}.$$

And the error covariance matrix, $\mathbf{S}_e$ is obtained by multiplying $\mathbf{E}$ by $1/\text{df}_e$.

$$\mathbf{S}_e \doteq \frac{1}{44-2}\begin{bmatrix} 484.637 & 709.591 \\ 709.591 & 2216.41 \end{bmatrix} \doteq \begin{bmatrix} 11.539 & 16.895 \\ 16.895 & 52.772 \end{bmatrix}.$$

The inverse of the error covariance matrix is

$$\mathbf{S}_e^{-1} \doteq \begin{bmatrix} .158 & -.051 \\ -.051 & .035 \end{bmatrix}.$$

Using these results, T^2 is computed as:

$$T^2 \doteq \frac{(22)(22)}{22+22}\begin{bmatrix} 7.77-6.68 & 43.45-42.05 \end{bmatrix}\begin{bmatrix} .158 & -.051 \\ -.051 & .035 \end{bmatrix}\begin{bmatrix} 7.77-6.68 \\ 43.45-42.05 \end{bmatrix}$$
$$\doteq 1.10.$$

Critical values of the distribution of T^2 have been tabled; however, a simple transformation of T^2,

$$F = \frac{\text{df}_e - p + 1}{p(\text{df}_e)}T^2, \tag{3.11}$$

has a central F distribution with degrees of freedom $\nu_1 = p$ and $\nu_2 = \text{df}_e - p + 1 = N - p - 1$. This is true assuming independent units, bivariate normality, equal population covariance matrices, and null hypothesis, H_0: $\boldsymbol{\mu}_1 = \boldsymbol{\mu}_2$, being true. For the present data

$$F \doteq \frac{42-2+1}{2(42)}(1.10) \doteq .537.$$

The P value for the observed F statistic with 2 and 41 degrees of freedom equals .589, indicating that if the population centroids are identical, the probability of obtaining the observed $F(2, 41)$ statistic of .537 or something larger is quite high (i.e., .589).

You might note that the expression in brackets in Eq. (3.10) is the Mahalanobis D^2 statistic that was discussed in the previous chapter as a standardized squared distance measure between two sample centroids. That is,

$$D_{12}^2 = (\mathbf{y}_{.1} - \mathbf{y}_{.2})' \mathbf{S}_e^{-1} (\mathbf{y}_{.1} - \mathbf{y}_{.2}).$$

The positive square root of this distance measure is analogous to the standardized mean difference, d, presented in the univariate context. For the Baumann data, $D^2_{12} \doteq .098$ and $D_{12} \doteq .313$. Analogous to Eq. (3.8), Eq. (3.10) can be written as:

$$T^2_{12} = D^2_{12}\left[\frac{n_1 n_2}{n_1 + n_2}\right].$$

This expression demonstrates that as the distance between two sample centroids increases, T^2_{12} increases.

3.3 TEST FOR COVARIANCE MATRIX EQUALITY

The statistic in Eq. (3.11) has a central F distribution with ν_1 and ν_2 degrees of freedom, provided that the two population covariance matrices are the same; that is, $\boldsymbol{\Sigma}_1 = \boldsymbol{\Sigma}_2$. In 1949, G. E. P. Box (1919–2002) proposed a statistical procedure to test the null hypothesis that two or more covariance matrices are identical, H_0: $\boldsymbol{\Sigma}_1 = \boldsymbol{\Sigma}_2 = \cdots = \boldsymbol{\Sigma}_J$. The Box M statistic is given by:

$$M = \mathrm{df}_e \ln|\mathbf{S}_e| - \sum_{j=1}^{J} \mathrm{df}_j \ln|\mathbf{S}_j|,$$

where ln denotes a natural logarithm. A transformation of M results in a statistic having a central chi-squared distribution with degrees of freedom, $\nu = (J-1)(p+1)p/2$, under the assumptions that the unit scores are independent and multivariate normal in each population. The transformation is CM where

$$C = 1 - \frac{2p^2 + 3p - 1}{6(p+1)(J-1)}\left(\sum_{j=1}^{J} \mathrm{df}_j^{-1} - \mathrm{df}_e^{-1}\right),$$

if sample sizes are unequal. If sample sizes are equal,

$$C = 1 - \frac{(2p^2 + 3p - 1)(J+1)}{6(p+1)(N-J)},$$

where p = number of outcome variables
J = number of groups
N = total sample size, $\sum n_j$
df_j = degrees of freedom for Group j, $n_j - 1$
df_e = error degrees of freedom, $\sum \mathrm{df}_j$
$\mathbf{S}_j$ = covariance matrix for Group j
$\mathbf{S}_e$ = error covariance matrix, $\mathbf{E}/\mathrm{df}_e$

Statistic M can also be transformed to have a central F distribution, but we do not provide the details here; Rencher (2002, pp. 255–259) provides this alternative transformation. The SPSS MANOVA program reports both the chi-squared and F transformations. Using the Baumann data and the results from Chapter 2, the separate and error covariance matrices are

$$\mathbf{S}_1 \doteq \begin{bmatrix} 15.422 & 24.679 \\ 24.679 & 61.784 \end{bmatrix}, \qquad \mathbf{S}_2 \doteq \begin{bmatrix} 7.656 & 9.110 \\ 9.110 & 43.760 \end{bmatrix},$$

and

$$\mathbf{S}_e \doteq \begin{bmatrix} 11.539 & 16.895 \\ 16.895 & 52.772 \end{bmatrix}.$$

Using these data with degrees of freedom, $\nu = (2-1)(2+1)2/2 = 3$,

$$|\mathbf{S}_1| \doteq 343.780, \qquad |\mathbf{S}_2| \doteq 252.034, \qquad |\mathbf{S}_e| \doteq 323.495,$$

$$\ln|\mathbf{S}_1| \doteq 5.840, \qquad \ln|\mathbf{S}_2| \doteq 5.530, \qquad \ln|\mathbf{S}_e| \doteq 5.779,$$

$$M \doteq [(44-2)(5.779)] - [(22-1)(5.840) + (22-1)(5.530)] \doteq 3.948,$$

$$C \doteq 1 - \frac{(2(2)^2 + 3(2) - 1)(2+1)}{6(2+1)(44-2)} \doteq .948,$$

and

$$\chi^2(3) \doteq (3.948)(.948) \doteq 3.743.$$

The P value is .288. Based on these results, there is little evidence that the two population covariance matrices differ. There are two limitations associated with the Box test. *First*, it can be sensitive to multivariate nonnormality. *Second*, the degrees of freedom for this test can often be very large, resulting in an extremely sensitive/powerful test of the null hypothesis. Both of these limitations could lead a researcher to question the validity of the multivariate test on the vectors of means when the multivariate test is valid. As a result, researchers often do not rely heavily on the results of the test but rather rely on the robustness of the multivariate test on the equality of mean vectors when sample sizes are equal. When sample sizes differ substantially (e.g., one sample is at least twice the size of another), Hakstain et al. (1979, p. 1262) give the following advice:

> It is when $n_1 \neq n_2$ that problems arise. If the test of covariance matrix homogeneity is nonsignificant no problem exists; but if this test is significant, the user is faced with the multivariate extension of the Behrens–Fisher problem. The following strategy—in the order of steps listed—seems to be a reasonable approach.
>
> 1. Ascertain whether one is in the positive or negative condition. A direct assessment of this comes from comparison of the determinants (understood as generalized variances) of $\mathbf{S}_1$ and $\mathbf{S}_2$. If either $n_1 > n_2$ and $|\mathbf{S}_1| > |\mathbf{S}_2|$ or $n_1 < n_2$ and $|\mathbf{S}_1| < |\mathbf{S}_2|$, we have the positive condition, whereas if the opposite is obtained, we are in the negative condition. It is, of course, possible for $|\mathbf{S}_1|$ to be equal to $|\mathbf{S}_2|$, although $\mathbf{S}_1 \neq \mathbf{S}_2$. In such a case, however, the above taxonomy does not apply, and the effects of such heterogeneity may not be serious.

2. Given the positive condition, the T^2 test will be conservative. Thus the user should run the test and, if it is significant, reject the null hypothesis a fortiori.

3. Given the negative condition, the T^2 test will be liberal. Thus the user should run the test and, if it is nonsignificant, retain the null hypothesis.

4. If in the positive condition, T^2 is nonsignificant or if in the negative condition, T^2 is significant, two possibilities exist: (a) If the n's are not extremely different, they can be equalized by random deletion of subjects from the larger group. If, in fact, the null hypothesis is false, the loss of power may not be too great, and the previously significant T^2 (negative condition) may still be significant. The previously nonsignificant T^2 (positive condition) may not be significant. (b) If the n's are substantially different so that equalization would result in a massive loss of power or if the equalization performed when n's are not extremely different results in nonsignificant results, the user can employ one of several solutions to the multivariate Behrens–Fisher problem—reasonably precise approximations that do not require deletion of subjects.

One of the solutions to the multivariate Behrens–Fisher problem (heterogeneous covariance matrices with unequal n) to which Hakstain et al. (1979) refer was suggested by Yao (1965) and is discussed in the next section.

3.4 YAO TEST

If the assumption of equal covariance matrices is not met and the researcher doubts the validity of the Hotelling T^2 test, a procedure analogous to Eq. (3.4) may be used:

$$T^{*2} = (\mathbf{y}_{.1} - \mathbf{y}_{.2})' \left(\frac{\mathbf{S}_1}{n_1} + \frac{\mathbf{S}_2}{n_2} \right)^{-1} (\mathbf{y}_{.1} - \mathbf{y}_{.2}). \tag{3.12}$$

A transformation of T^{*2} to a central F has been suggested by Yao (1965):

$$\frac{f - p + 1}{pf} T^{*2} \stackrel{\cdot}{\longrightarrow} F(p, f - p + 1), \tag{3.13}$$

where

$$W_j = \frac{\mathbf{S}_j}{n_j}, \qquad W = \sum_{j=1}^{2} W_j,$$

$$V_j = (\mathbf{y}_{.1} - \mathbf{y}_{.2})' W^{-1} W_j W^{-1} (\mathbf{y}_{.1} - \mathbf{y}_{.2}),$$

and

$$\frac{1}{f} = \sum_{j=1}^{2} \frac{1}{n_j - 1} \left(\frac{V_j}{T^{*2}} \right)^2.$$

We modified a program obtained from J. Algina (University of Florida) that yields a value of T^{*2} and a transformed F for the two-group comparison. The program is labeled YAO2 and is available at the Wiley website. For the data in Table 2.1 the transformed F statistic equals .559 and the P value equals .576. These results happen to be consistent with the previous analysis that assumed the covariance matrices are equal.

3.5 MULTIPLE-GROUP ANALYSES—SINGLE FACTOR

3.5.1 Univariate Analysis

When there is an interest in comparing more than two populations, multiple t tests may be considered, but such analyses can result in an unacceptably high Type I error rate. To avoid this problem, a single hypothesis testing procedure (omnibus test) based on the analysis of variation among group means and variation of units within groups may be conducted. (The simultaneous comparison of two or more group means, however, may not address the specific research questions of interest. If specific research questions have been identified, focused tests or contrast tests may be of greater interest, and the omnibus hypothesis test may be unnecessary. We discuss focused tests in the next chapter.)

If the populations under consideration differ along a single grouping variable (e.g., high school class level—freshmen, sophomore, junior, senior), the univariate analysis is described as a one-way analysis of variance (ANOVA). If the levels of the grouping variable exhaust all of the possible levels, or if the levels are specifically chosen by the researcher for investigation, the design is further described as a fixed-effects design. The null hypotheses that can be tested with this design can be viewed as an extension of the hypothesis tested using the two-group t test. That is, the null hypothesis may be expressed by any one of the following:

$$H_0\colon \alpha_1 = \alpha_2 = \cdots = \alpha_J (= 0),$$

$$H_0\colon \eta^2_{\text{pop}} = 0, \quad \text{or} \quad H_0\colon \mu_1 = \mu_2 = \cdots = \mu_J.$$

In the second statement, η_{pop} denotes the population correlation between the outcome variable and the grouping variable. The research questions behind these hypothesis statements are similar to those addressed by the two-group t test discussed in the previous section.

In Chapter 2, data for two outcome variables from two groups in the Baumann study were presented. We now provide the data for the same outcome variables for the third treated group. The third group of students was presented a combined Directed Reading and Thinking Activity (DRTA) instructional program. The complete data set is presented in Table 3.1. Table 3.2 provides the separate group means and variances for Y_1 and Y_2. (An SPSS data file containing these data, labeled BAUMANN3g2v, is available at the Wiley website.)

The statistic for testing the hypothesis involving multiple populations is computed as the ratio of a weighted variance of group means, the hypothesis mean-square (MSH),

TABLE 3.1 Scores on the Error Detection Task (Y_1) and Degrees of Reading Power (Y_2) for the Think Aloud (TA), Directed Reading Activity (DRA), and Directed Reading and Think Aloud (DRTA) Groups

TA		DRA		DRTA	
Y_1	Y_2	Y_1	Y_2	Y_1	Y_2
4	43	5	34	6	27
4	34	9	36	6	36
4	45	5	42	5	51
3	39	7	37	5	51
8	40	4	44	0	50
1	27	9	49	6	55
7	46	3	38	6	52
7	39	4	38	11	48
9	31	2	38	6	53
6	39	5	50	8	45
4	40	7	31	8	47
12	52	8	49	3	51
14	53	10	54	7	30
12	53	9	52	7	50
7	41	12	50	6	55
5	41	5	35	9	48
9	46	8	36	7	52
13	52	12	46	6	46
11	55	4	42	7	36
5	36	8	47	6	45
11	50	6	39	6	49
15	54	5	38	6	49

to the average variance of observations within the groups, the error mean-square (MSE), and is referred to as the F ratio, $F = \text{MSH}/\text{MSE}$.

The MSH is computed as:

$$\text{MSH} = \frac{\sum_{j=1}^{J} n_j(\overline{Y}_{.j} - \overline{Y}_{..})^2}{J - 1},$$

TABLE 3.2 Means and Variances for Test Scores on the Error Detection Task (Y_1) and Degrees of Reading Power (Y_2) for the Think Aloud (TA), Directed Reading Activity (DRA), and Directed Reading and Think Aloud (DRTA) Groups

	TA		DRA		DRTA	
	Y_1	Y_2	Y_1	Y_2	Y_1	Y_2
Mean ($\overline{Y}_{.j}$)	7.77	43.45	6.68	42.05	6.23	46.64
Variance (s_j^2)	15.422	61.784	7.656	43.760	4.374	56.160

where n_j = number of individuals in Group j ($j = 1, 2, \ldots, J$)

$\overline{Y}_{.j}$ = mean of the scores in Group j $\left(\sum_{u=1}^{n_j} Y_{uj}/n_j\right)$

$\overline{Y}_{..}$ = grand mean of all observations in the study $\left(\sum_{j=1}^{J}\sum_{u=1}^{n_j} Y_{uj}/N\right)$

N = total sample size $\left(N = \sum_{j=1}^{J} n_j\right)$

Y_{uj} = score for unit u in Group j

If sample sizes are equal, $n_1 = n_2 = \cdots = n_J = n$, the numerator of MSH may be written as:

$$n\sum_{j=1}^{J}(\overline{Y}_{.j} - \overline{Y}_{..})^2$$

and is referred to as the hypothesis sum-of-squares, SSH. The denominator, $J - 1$, is the hypothesis degrees of freedom, df_h. In terms of the first expression for the null hypothesis, H_0: $\alpha_1 = \alpha_2 = \cdots = \alpha_J = 0$, where $\alpha_j = \mu_j - \mu_{..}$. Using sample data to estimate the population means, $\hat{\alpha}_j = \overline{Y}_{.j} - \overline{Y}_{..}$, the SSH can also be written as $n\sum_{j=1}^{J}\hat{\alpha}_j^2$ when sample sizes are equal. The error variance is computed as:

$$\text{MSE} = \frac{\sum_{j=1}^{J}\sum_{u=1}^{n_j}(Y_{uj} - \overline{Y}_{.j})^2}{N - J} = \frac{\text{SSE}}{\text{df}_e}.$$

The error sum-of-squares (SSE) equals

$$\sum_{j=1}^{J}\sum_{u=1}^{n_j}(Y_{uj} - \overline{Y}_{.j})^2$$

and the error degrees of freedom, df_e, equals $N - J$.

Using the data for the Y_1 outcome, the F test for the null hypothesis that the three population means are equal, $H_0: \mu_{\text{TA}} = \mu_{\text{DRA}} = \mu_{\text{DRTA}}$ is computed as:

$$F \doteq \frac{27.758/2}{576.502/63} \doteq 1.517.$$

Assuming that the student scores are independent of each other, the populations have normal distributions, and the populations have equal variances, the F statistic has a central F distribution with $\text{df}_h = J - 1$ and $\text{df}_e = N - J$ degrees of freedom. The P value for the observed F ratio with 2 and 63 degrees of freedom is .210. These results indicate that there is little evidence to conclude that one type of reading instruction provides greater reading comprehension as measured by the Error Detection Task than any of the others. An index of association for the relationship between the grouping variable and the outcome variable should also be reported, along with the P value. We delay a discussion on *effect size* in the ANOVA context until the next chapter.

3.5.2 Multivariate Analysis

Although separate F tests could be computed for each outcome variable, multiple hypothesis testing would result in an increased likelihood of at least one Type I error. In addition, because multiple outcome variables are likely to be correlated, the structure underlying the system of outcomes would be ignored with separate tests and statistical power could be decreased. A more appropriate way to examine two or more outcome variables would be to take a multivariate approach. This is especially true when the outcome variables are different indicators of a single common construct, or when the collection of outcome variables is believed to represent different constructs.

The multivariate null hypothesis is analogous to the univariate hypothesis, but the multivariate hypothesis is a statement regarding the vector of outcome variable means (i.e., centroids). The multivariate null hypothesis states that the population centroids do not differ:

$$H_0\colon \boldsymbol{\mu}_1 = \boldsymbol{\mu}_2 = \cdots = \boldsymbol{\mu}_J,$$

where $\boldsymbol{\mu}'_j = [\mu_{1j}, \mu_{2j}, \ldots, \mu_{pj}]$.

For the Baumman study the three centroids are

$$\mathbf{y}_{.1} \doteq \begin{bmatrix} 7.77 \\ 43.45 \end{bmatrix}, \qquad \mathbf{y}_{.2} \doteq \begin{bmatrix} 6.68 \\ 42.05 \end{bmatrix}, \qquad \mathbf{y}_{.3} \doteq \begin{bmatrix} 6.23 \\ 46.64 \end{bmatrix}.$$

To test the null hypothesis of the equality of population centroids, an error sum-of-squares and cross-products matrix, **E**, and a matrix for the hypothesis, **H**, are needed. The error **SSCP** matrix is obtained by summing the separate group $\mathbf{SSCP}_j$ matrices ($\mathbf{E} = \sum_{j=1}^{J} \mathbf{SSCP}_j$). In Chapter 2 we found the **SSCP** for TA and DRA to be

$$\mathbf{SSCP}_1 \doteq \begin{bmatrix} 323.864 & 518.273 \\ 518.273 & 1297.455 \end{bmatrix}$$

$$\mathbf{SSCP}_2 \doteq \begin{bmatrix} 160.773 & 191.318 \\ 191.318 & 918.955 \end{bmatrix}.$$

For the DRTA group, $\mathbf{SSCP}_3$ is

$$\mathbf{SSCP}_3 \doteq \begin{bmatrix} 91.865 & -52.182 \\ -52.182 & 1227.091 \end{bmatrix}.$$

The error **SSCP** matrix is

$$\mathbf{E} \doteq \begin{bmatrix} 576.502 & 657.409 \\ 657.409 & 3443.501 \end{bmatrix}. \tag{3.14}$$

If the elements of this matrix were multiplied by the reciprocal of the error degrees of freedom, the result would be the error covariance matrix, $\mathbf{S}_e$.

For the Baumann data the grand mean centroid is

$$\mathbf{y}_{..} \doteq \begin{bmatrix} 6.894 \\ 44.046 \end{bmatrix},$$

and the separate group centroids minus the grand mean centroid are

$$(\mathbf{y}_{.1} - \mathbf{y}_{..}) \doteq \begin{bmatrix} .88 \\ -.60 \end{bmatrix},$$

$$(\mathbf{y}_{.2} - \mathbf{y}_{..}) \doteq \begin{bmatrix} -.21 \\ -2.00 \end{bmatrix},$$

$$(\mathbf{y}_{.3} - \mathbf{y}_{..}) \doteq \begin{bmatrix} -.66 \\ 2.59 \end{bmatrix}.$$

The **SSCP** for each group is computed as:

$$n_j(\mathbf{y}_{.j} - \mathbf{y}_{..})(\mathbf{y}_{.j} - \mathbf{y}_{..})', \tag{3.15}$$

and for the Baumann data the results are

$$\mathbf{SSCP}_1 \doteq 22\begin{bmatrix} .88 \\ -.60 \end{bmatrix}\begin{bmatrix} .88 & -.60 \end{bmatrix} \doteq \begin{bmatrix} 17.037 & -11.616 \\ -11.616 & 7.920 \end{bmatrix},$$

$$\mathbf{SSCP}_2 \doteq 22\begin{bmatrix} -.21 \\ -2.00 \end{bmatrix}\begin{bmatrix} -.21 & -2.00 \end{bmatrix} \doteq \begin{bmatrix} .970 & 9.240 \\ 9.240 & 88.000 \end{bmatrix},$$

$$\mathbf{SSCP}_3 \doteq 22\begin{bmatrix} -.66 \\ 2.59 \end{bmatrix}\begin{bmatrix} -.66 & 2.59 \end{bmatrix} \doteq \begin{bmatrix} 9.583 & -37.607 \\ -37.607 & 147.578 \end{bmatrix}.$$

The *hypothesis* ***SSCP*** *matrix*, $\mathbf{H} = \sum_{j=1}^{J} \mathbf{SSCP}_j$ is

$$\mathbf{H} \doteq \begin{bmatrix} 27.590 & -39.983 \\ -39.983 & 243.498 \end{bmatrix}.$$

It might be noted that 27.590 equals (within rounding) the hypothesis sum-of-squares, and 576.502 in Eq. (3.14) equals the error sum-of-squares for the EDT outcome reported earlier for the univariate F test. Four criteria have been proposed for using **H** and **E** in testing the hypothesis that the population centroids are identical.

Wilks Criterion The oldest and perhaps the most widely used criterion is that due to S. S. Wilks (1906–1964). This statistic (Wilks, 1932), which is often called the *Wilks lambda criterion*, may be defined as the ratio of two determinants,

$$\Lambda = \frac{|\mathbf{E}|}{|\mathbf{H} + \mathbf{E}|}. \tag{3.16}$$

Using our calculations from the Baumann study, the corresponding determinants are $|\mathbf{E}| \doteq 15.53 \times 10^5$ and $|\mathbf{H}+\mathbf{E}| \doteq 18.46 \times 10^5$. The Wilks Λ value is

$$\Lambda \doteq \frac{15.53 \times 10^5}{18.46 \times 10^5} \doteq .841.$$

In the 1930s it was reported that for certain values of p and df_h, a transformation of Λ has a central F distribution under the null hypothesis, assuming independent units, multivariate normality, and homogeneous covariance matrices. In the two-population case, $\mathrm{df}_h = 1$, the following transformation for Λ was suggested:

$$F = \frac{1-\Lambda}{\Lambda}\,\frac{\mathrm{df}_e - p + 1}{p}. \tag{3.17}$$

This statistic has a central F distribution with $\nu_1 = p$ and $\nu_2 = \mathrm{df}_e - p + 1$. For a three-group design, that is, when $\mathrm{df}_h = J - 1 = 2$, the following transformation of Λ,

$$F = \frac{1-\Lambda^{1/2}}{\Lambda^{1/2}}\,\frac{\mathrm{df}_e - p + 1}{p} \tag{3.18}$$

has a central F distribution with $\nu_1 = 2p$ and $\nu_2 = 2(\mathrm{df}_e - p + 1)$. The Baumann study involved three populations, so this transformation may be used to obtain an F value of

$$F \doteq \frac{1-(.841)^{1/2}}{(.841)^{1/2}}\,\frac{63-2+1}{2} \doteq 2.804$$

With 4 and 124 degrees of freedom, the P value for this statistic is .029. Based on this result, we would conclude that there is sufficient evidence to reject the null hypothesis and conclude that the three population centroids are not identical. Further analyses, which will be discussed in the next chapter, are needed to further understand and describe the nature of the differences among the population centroids. [Rao (1952, pp. 260–262) also gives two other transformations of Λ that have exact F distributions; these are for $p = 1$ and $p = 2$.] More generally, when $r > 2$, where $r = \min(p, \mathrm{df}_h)$, a transformation of Λ provides a statistic that has an *approximate* F distribution with $\nu_1 = p(\mathrm{df}_h)$ and $\nu_2 = m(s) - p(\mathrm{df}_h)/2 + 1$. For this expression, p is the number of outcome variables, and $\mathrm{df}_h = J - 1$;

$$F = \frac{1-\Lambda^{1/s}}{\Lambda^{1/s}}\,\frac{m(s) - \dfrac{p(\mathrm{df}_h)}{2} + 1}{p(\mathrm{df}_h)}, \tag{3.19}$$

where

$$m = \mathrm{df}_e - \frac{p - \mathrm{df}_h + 1}{2}, \tag{3.20}$$

and

$$s = \sqrt{\frac{p^2(\mathrm{df}_h^2) - 4}{p^2 + \mathrm{df}_h^2 - 5}}. \tag{3.21}$$

When $\mathrm{df}_h = 1$, (3.19) simplifies to (3.17), and when $\mathrm{df}_h = 2$, (3.19) simplifies to (3.18).

It can be shown that the value of Λ may be determined in a manner other than as in (3.16). This determination involves finding functions of a matrix called eigenvalues. The matrix of interest for the Wilks Λ is actually a product of two matrices $\mathbf{E}^{-1}\mathbf{H}$, where the $p \times p$ matrices, $\mathbf{E}$ and $\mathbf{H}$, are as described earlier in this chapter. For the Baumann data,

$$\mathbf{E} \doteq \begin{bmatrix} 576.502 & 657.409 \\ 657.409 & 3443.501 \end{bmatrix}, \qquad \mathbf{E}^{-1} \doteq \begin{bmatrix} .0022 & -.0004 \\ -.0004 & .0004 \end{bmatrix},$$

$$\mathbf{H} \doteq \begin{bmatrix} 27.590 & -39.983 \\ -39.983 & 243.498 \end{bmatrix}, \qquad \mathbf{E}^{-1}\mathbf{H} \doteq \begin{bmatrix} .0767 & -.1854 \\ -.0270 & .1134 \end{bmatrix}.$$

The eigenvalues are obtained by solving the *determinantal equation* $|\mathbf{E}^{-1}\mathbf{H} - \lambda\mathbf{I}| = 0$ for λ. For $p = 2$ and $J = 3$ in the Baumann study, we have

$$\left| \begin{bmatrix} .0767 & -.1854 \\ -.0270 & .1134 \end{bmatrix} - \begin{bmatrix} \lambda & 0 \\ 0 & \lambda \end{bmatrix} \right| \doteq 0,$$

$$(.0767 - \lambda)(.1134 - \lambda) - (-.0270)(-.1854) \doteq 0,$$

$$\lambda^2 - .1901\lambda + .0037 \doteq 0.$$

For equations of this form: $a\lambda^2 + b\lambda + c = 0$ with a, b, c being constants and λ an unknown, a solution for λ is provided using the following from Chapter 2:

$$\lambda = \frac{-b \pm \sqrt{b^2 - 4ac}}{2a}.$$

Using our determinant,

$$\lambda \doteq \frac{.1901 \pm \sqrt{-.1901^2 - 4(1)(.0037)}}{2(1)}.$$

Thus, the eigenvalues for the $\mathbf{E}^{-1}\mathbf{H}$ matrix are $\lambda \doteq .1681$ and $.0220$. Using these values, Λ may be computed as:

$$\Lambda = \prod_{v=1}^{r} \left(\frac{1}{1 + \lambda_v} \right) \tag{3.22}$$

Using the Baumann data Λ is

$$\Lambda \doteq \left(\frac{1}{1.1681} \right) \left(\frac{1}{1.0220} \right) \doteq .838,$$

which is, within rounding, that obtained using (3.16).

Bartlett–Pillai Criterion An alternative criterion for testing the null hypothesis $H_0\colon \boldsymbol{\mu}_1 = \boldsymbol{\mu}_2 = \cdots = \boldsymbol{\mu}_J$ was proposed by M. S. Bartlett (1910–2002) and K. C. S. Pillai (1920–1985). This criterion, uses the eigenvalues of the $\mathbf{E}^{-1}\mathbf{H}$ matrix in a different statistic. The Bartlett–Pillai test statistic is

$$U = \sum_{v=1}^{r} \left[\frac{\lambda_v}{1 + \lambda_v} \right], \tag{3.23}$$

where $r = \min(p, \mathrm{df}_e)$. Using the eigenvalues, λ_1 and λ_2, computed in the previous section we get

$$U \doteq \frac{.1681}{1.1681} + \frac{.0220}{1.0220} \doteq .1654.$$

This statistic can be transformed to a statistic having an F distribution with $\nu_1 = br$ and $\nu_2 = r(\mathrm{df}_e - p + r)$ degrees of freedom using the following:

$$F = \frac{U}{r - U} \frac{\mathrm{df}_e - p + r}{b}, \tag{3.24}$$

where $b = \max(p, \mathrm{df}_h)$, and $r = \min(p, \mathrm{df}_h)$. For the Baumann data,

$$F \doteq \frac{.1654}{2 - .1654} \frac{63 - 2 + 2}{2} \doteq 2.840.$$

With $\nu_1 = 4$ and $\nu_2 = 126$ degrees of freedom, $P \doteq .027$. The P value using the Bartlett–Pillai criterion is about the same as that obtained using the Wilks criterion.

Roy Criterion A third MANOVA criterion, often attributed to S. N. Roy (1906–1964), that has been advocated by some writers (e.g., Harris, 2001, p. 231; Morrison, 1990, p. 209; Tatsuoka, 1988, p. 285), is the greatest *characteristic root* or eigenvalue (λ_1) of $\mathbf{E}^{-1}\mathbf{H}$. The test statistic is computed as:

$$\Theta = \frac{\lambda_1}{1 + \lambda_1}. \tag{3.25}$$

Special tables found in the Harris (2001) and Morrison (1990) books are needed when using this criterion. However, the General Linear Model (GLM) program in SPSS uses a transformation of λ_1 that provides an upper bound of the F statistic. This provides a lower bound on the P value. The P value is therefore slightly liberal. If the null hypothesis is not rejected with this approximate test, it would not be rejected with a more exact test. If, however, the null hypothesis is rejected using this F test, complete confidence cannot be given to this conclusion. The transformation used by SPSS is

$$F = \frac{(N - b - 1)\lambda_1}{b}. \tag{3.26}$$

[Rencher (2002, p. 165) suggests that df_e might be used in place of N.]

For the Baumann data,

$$F \doteq \frac{(66-2-1).1681}{2} \doteq 5.295.$$

The degrees of freedom are $\nu_1 = b$ and $\nu_2 = N - b - 1$, where $b = \max(p, \mathrm{df}_h)$. For the Baumann data, the degrees of freedom are 2 and 63, and the corresponding P value is .007.

Hotelling–Lawley Criterion A fourth criterion, known as the *Hotelling–Lawley trace criterion*, is given simply by:

$$V = \sum_{\upsilon=1}^{r} \lambda_\upsilon, \tag{3.27}$$

where λ_υ is the υth eigenvalue of $\mathbf{E}^{-1}\mathbf{H}$. For the Baumann data,

$$V \doteq .1681 + .0220 \doteq .1901.$$

This criterion can be transformed to a statistic having a central F distribution with $\nu_1 = br$ and $\nu_2 = r(\mathrm{df}_e - p - 1) + 2$ degrees of freedom, where $b = \max(p, \mathrm{df}_h)$, and $r = \min(p, \mathrm{df}_h)$:

$$F = V\left(\frac{r(\mathrm{df}_e - p - 1) + 2}{r^2 b}\right). \tag{3.28}$$

For the Baumann data,

$$F \doteq .1901\left(\frac{2(63-2-1)+2}{2^2(2)}\right) \doteq 2.899.$$

With $\nu_1 = 4$ and $\nu_2 = 122$ degrees of freedom, $P \doteq .025$. The P value using the Hotelling–Lawley trace criterion is slightly smaller than that obtained using the Wilks and Bartlett–Pillai criteria.

3.6 COMPUTER APPLICATION

In this section we provide the SPSS syntax that yields basic descriptive statistics, the test covariance matrix equality, and the omnibus multivariate and univariate F tests. These commands are used repeatedly for other analyses presented in this book. A brief explanation of the SPSS commands is provided here. Following the SPSS commands, the output is shown and labeled ***Analysis***. Finally, we provide a brief interpretation of the analysis in the section labeled ***Interpretation***.

SPSS SYNTAX FOR DESCRIPTIVE STATISTICS, MULTIVARIATE, AND UNIVARIATE ANALYSES

```
manova Y1 Y2 by treatmnt(1,3)
/print=cellinfo(means sscp cov) homogeneity(box) error(sscp cor)
signif(multiv ).
```

manova is the system command requesting the multivariate analysis of variance.
Y1 Y2 identifies the outcome variables for the analysis.
treatmnt(1,3) identifies the explanatory variable with the lower and upper levels of the variable in parenthesis.
/print is the system command to print desired intermediary analyses.
cellinfo(means sscp cov) reports the within-cell means, standard deviations, and sum-of-squares, cross-products, and covariance matrices.
homogeneity(box) reports the results of the Box test for covariance matrix equality.
error(sscp cor) reports the error sum-of-squares and cross-products matrix, **E**, and the error correlation matrix.
signif(multiv) reports the multivariate F test statistics, degrees of freedom, and P values.

OUTPUT

Analysis: Descriptive Statistics and the Box Test for Covariance Equality

Cell Means and Standard Deviations
Variable .. Y1

FACTOR	CODE	Mean	Std. Dev.	N
treatmnt	TA	7.773	3.927	22
treatmnt	DRA	6.682	2.767	22
treatmnt	DRTA	6.227	2.092	22
For entire sample		6.894	3.049	66

- -

Variable .. Y2

FACTOR	CODE	Mean	Std. Dev.	N
treatmnt	TA	43.455	7.860	22
treatmnt	DRA	42.045	6.615	22
treatmnt	DRTA	46.636	7.644	22
For entire sample		44.045	7.531	66

- -

Cell Number .. 1 (TA)
Sum of Squares and Cross-Products matrix

	Y1	Y2
Y1	323.864	
Y2	518.273	1297.455

Variance-Covariance matrix

	Y1	Y2
Y1	15.422	
Y2	24.680	61.784

Determinant of Covariance matrix of dependent variables = 343.74541
LOG(Determinant) = 5.83990

- - - - - - - - - - - - - - - -

Cell Number . . 2 (DRA)
Sum of Squares and Cross-Products matrix

	Y1	Y2
Y1	160.773	
Y2	191.318	918.955

Variance-Covariance matrix

	Y1	Y2
Y1	7.656	
Y2	9.110	43.760

Determinant of Covariance matrix of dependent variables = 252.01855
LOG(Determinant) = 5.52950

- - - - - - - - - - - - - - - -

Cell Number . . 3 (DRTA)
Sum of Squares and Cross-Products matrix

	Y1	Y2
Y1	91.864	
Y2	−52.182	1227.091

Variance-Covariance matrix

	Y1	Y2
Y1	4.374	
Y2	−2.485	58.433

Determinant of Covariance matrix of dependent variables = 249.43785
LOG(Determinant) = 5.51921

- - - - - - - - - - - - - - - -

	Y1	Y2
Y1	9.151	
Y2	10.435	54.659

Determinant of pooled Covariance matrix of dependent vars. = 391.28018
LOG(Determinant) = 5.96942

- -

Multivariate test for Homogeneity of Dispersion matrices
Boxs M = 21.41281
F WITH (6,98919) DF = 3.40494, P = .002 (Approx.)
Chi-Square with 6 DF = 20.43092, P = .002 (Approx.)

- -

WITHIN CELLS Correlations with Std. Devs. on Diagonal

	Y1	Y2
Y1	3.025	
Y2	.467	7.393

- -

Interpretation: Descriptive Statistics and the Box Test for Covariance Matrix Equality

Group means and standard deviations are reported at the beginning of the output. These results correspond to those in Table 3.2. The separate group and error covariance matrices are reported next. The separate group covariance matrices appear to differ to some degree with the first group, TA, having a generalized variance greater than either the DRA or DRTA groups. The Box (1949) test using either the F or χ^2 transformation of M provides some evidence to support the conclusion that the three population covariance matrices differ, $M \doteq 21.413$, $F(6, 98919) \doteq 3.405$, $\chi^2(6) \doteq 20.431$, $P \doteq .002$.

Another assessment (albeit fairly subjective) of covariance matrix equality may be made by examining the natural logarithms of the determinants of the $J + 1$ covariance matrices—J separate group matrices plus the error matrix. For the Baumann data these four logarithms are 5.84 (TA), 5.53 (DRA), 5.52 (DRTA), and 5.97 (error). We judge these numbers to be "in the same ballpark." Still another assessment of across-group homogeneity is to consider the four sums of outcome variable variances. These are the *traces* of the four covariance matrices. For example, the trace for the TA group is $15.422 + 61.784 \doteq 77.2$. The other three traces are approximately 51.4 (DRA), 62.8 (DRTA), and 63.8 (error). Although the range of $77.2 - 51.4 \doteq 25.8$ may appear "large" to some, the positive square roots of 8.8 to 7.2 are not judged by us to be "large."

Considering the two assessments in addition to the chi-squared and F statistics, plus the design of 22 students in each of the three groups, we will rely on the robustness of the multivariate tests of equality of population centroids.

The correlation matrix for the outcome variables (at the very end of the above output), obtained by averaging the separate group correlations, is often useful when describing relationships among the outcome variables.

Analysis: Multivariate and Univariate Omnibus Hypothesis Tests

EFFECT .. treatmnt

Multivariate Tests of Significance (S = 2, M = −1/2, N = 30)

Test Name	Value	Approx. F	Hypoth. DF	Error DF	Sig. of F
Pillais	.16185	2.77356	4.00	126.00	.030
Hotellings	.18583	2.83393	4.00	122.00	.027
Wilks	.84094	2.80488	4.00	124.00	.029
Roys	.14225				

Note.. F statistic for WILKS' Lambda is exact.

- -

EFFECT .. treatmnt (Cont.)

Univariate F-tests with (2,63) D. F.

Variable	Hypoth. SS	Error SS	Hypoth. MS	Error MS	F	Sig. of F
Y1	27.75758	576.50000	13.87879	9.15079	1.51668	.227
Y2	243.36364	3443.50000	121.68182	54.65873	2.22621	.116

- -

Interpretation: Multivariate and Univariate Omnibus Hypothesis Tests

The results of the omnibus hypothesis test of the equality of the three population mean centroids indicate sufficient evidence that the centroids differ; Wilks $\Lambda \doteq .841$, $F(4, 124) \doteq 2.805$, $P \doteq .029$. A similar statement could be made using either the Bartlett–Pillai or Hotelling–Lawley multivariate test criterion. Neither of the univariate hypothesis tests, however, provide sufficient evidence to indicate the three populations differ with respect to the two outcomes, Y_1, $F(2, 63) \doteq 1.517$, $P \doteq .227$ and Y_2, $F(2, 63) \doteq 2.226$, $P \doteq .116$. The inconsistency between the multivariate and univariate results reflect the greater statistical power associated with the multivariate hypothesis test.

3.7 SUMMARY

A summary of transformations of the Wilks, Bartlett–Pillai, Roy, and Hotelling–Lawley criteria is given in Table 3.3. The results of all four statistics will be identical if $\mathrm{df}_h = 1$ (two-group comparison), and will only differ slightly for multiple groups in most situations. Rencher (2002, pp. 176–178) points out that the difference in statistical power among the four criteria depends on the mean configuration of the group centroids. If the outcome variables are highly intercorrelated, reflecting a single construct (we will discuss construct identification in Chapter 5), then the Roy test would be most powerful; but for other configurations, the Roy test can be the least sensitive. Rencher suggests the following order of tests with respect to power for a single construct context: Roy > Hotelling–Lawley > Wilks > Bartlett–Pillai. With

TABLE 3.3 Summary of Four MANOVA Test Statistics[a]

Criterion	Symbol	Statistic	ν_1	ν_2
Wilks (3.22)	$\Lambda = \prod\left(\frac{1}{1+\lambda_v}\right)$	$\frac{1-\Lambda^{1/s}}{\Lambda^{1/s}} \frac{m(s) - p(\mathrm{df}_h)/2 + 1}{p(\mathrm{df}_h)}$	$p(\mathrm{df}_h)$	$m(s) - p(\mathrm{df}_h)/2 + 1$
Bartlett–Pillai (3.23)	$U = \sum\left[\frac{\lambda_v}{1+\lambda_v}\right]$	$\frac{U}{r-U} \frac{\mathrm{df}_e - p + r}{b}$	br	$r(\mathrm{df}_e - p + r)$
Roy (3.24)	$\Theta = \frac{\lambda_1}{1+\lambda_1}$	$\frac{(N-b-1)\lambda_1}{b}$	b	$N-b-1$
Hotelling–Lawley (3.27)	$V = \sum \lambda_v$	$V\frac{r(df_e - p - 1) + 2}{r^2 b}$	br	$r(df_e - p - 1) + 2$

[a] p = number of outcome variables; $\mathrm{df}_h = J - 1$; $m = \mathrm{df}_e - (p - \mathrm{df}_h + 1)/2$; $s = \sqrt{\frac{p^2(\mathrm{df}_h^2) - 4}{p^2 + \mathrm{df}_h^2 - 5}}$; $r = \min(p, \mathrm{df}_h)$; $b = \max(p, \mathrm{df}_h)$.

multiple constructs, the order of preferred tests is Bartlett–Pillai > Wilks > Hotelling–Lawley > Roy. As stated above, however, in most situations all four test criteria provide the same conclusion.

Definition *ANOVA*: One egg.

EXERCISES

Exercises 1 to 4 are based on the following context and data:

An educational psychologist was interested in evaluating the effectiveness of an intervention designed to reduce test anxiety. Ten students were randomly assigned to the intervention condition, and 10 students were randomly assigned to a control condition. Following the intervention period, all students completed two measures of anxiety. The first measure, Y_1, was a self-report measure, and the second measure, Y_2, was a well-known standardized measure of test anxiety. Below are some descriptive statistics from the study.

	Intervention Group		Control Group	
	Y_1	Y_2	Y_1	Y_2
Mean	4.40	13.00	5.50	15.20
Sd	1.174	2.582	1.269	2.616
$r_{Y_1 Y_2}$	.147		−.167	

1. Compute the univariate t statistic for each outcome variable. The critical t value is 2.10. Do the data support the researcher's belief that the intervention is successful in reducing test anxiety?

2. For each outcome variable, compute the squared point-biserial correlation, r^2_{pb}, and the standardized mean difference, d, for the Intervention versus Control groups.

3. Analyze the bivariate data using the Hotelling T^2 and transform it to an F statistic.
   (a) Provide $\mathbf{S}_e$, $|\mathbf{S}_e|$, and $\mathbf{S}_e^{-1}$
   (b) State T^2 and F
   (c) D^2

4. If the critical F value for statistical significance equals 3.59, what conclusion would be drawn? How would you explain the difference in the results for Exercises 1 and 3?

Exercises 5 to 10 are based on the following context and data:

A Mathematics educator was interested in evaluating three approaches to teaching fractions to third-grade students. Method 1 used computer software to enhance instruction, Method 2 used peer tutoring, and Method 3 used workbooks. Fifteen students were assigned to each method. At the end of the intervention period two cognitive measures, Y_1, Computation Skills and, Y_2, Application Skills, were obtained. In addition, all students completed an attitudinal measure, Y_3. Descriptive statistics are reported below.

Method 1

	Y_1	Y_2	Y_3
Mean	12.7	14.3	23.0
Sd	3.0	2.9	3.3

Method 2

	Y_1	Y_2	Y_3
Mean	15.5	16.7	18.8
Sd	2.8	3.1	3.7

Method 3

	Y_1	Y_2	Y_3
Mean	14.7	13.6	15.9
Sd	2.6	1.8	3.1

5. Compute **E** and $\mathbf{S}_e$ assuming the following separate group **SSCP** matrices:

$$\mathbf{SSCP}_1 \doteq \begin{bmatrix} 126.0 & 87.7 & 48.0 \\ 87.7 & 117.7 & 52.0 \\ 48.0 & 52.0 & 152.2 \end{bmatrix}$$

$$\mathbf{SSCP}_2 \doteq \begin{bmatrix} 109.8 & 104.3 & 93.4 \\ 104.3 & 134.5 & 106.0 \\ 93.4 & 106.0 & 191.7 \end{bmatrix}$$

$$\mathbf{SSCP}_3 \doteq \begin{bmatrix} 94.6 & 28.4 & 62.5 \\ 28.4 & 45.4 & 48.2 \\ 62.5 & 48.2 & 134.5 \end{bmatrix}$$

6. Compute the **H** matrix for comparing the group centroids.

7. Using only variables Y_1 and Y_2, compute Wilks Λ to test the hypothesis that the centroids do not differ among the three levels of Method.

8. Using the results from Exercise 7, transform Λ to an F statistic. State the appropriate degrees of freedom for this F statistic.

9. When all three outcomes are considered, the eigenvalues for the $\mathbf{E}^{-1}\mathbf{H}$ product matrix are $\lambda_1 \doteq 1.694$ and $\lambda_2 \doteq 0.312$. Use these results to compute:

(a) Λ

(b) Bartlett–Pillai U

(c) Roy Θ

(d) Hotelling–Lawley V

10. Transform each multivariate test criterion computed in Exercise 9 to an F statistic. State the appropriate degrees of freedom associated with each test statistic.

Computer Applications

Exercises 11 to 15 require the analysis of the 5-group Ethington data set (5GED) described in Appendix A. Using the SPSS (or SAS) computer software package, compare the Black, Hispanic, and White student group centroids based on 9 variables (Counselor Interaction, Writing and Speaking Skills, Self-Understanding, Instruction Received, Library Effort, Student–Faculty Effort, Interstudent Effort, Art/Music/Theater Effort, Writing Effort, and Science Effort).

11. What are the sample sizes for the three racial groups in the Ethington data set?

12. Compare the covariance maticies for the three groups.

(a) What is the value of the Box M statistic?

(b) What is the χ^2 statistic value, degrees of freedom, and P value for testing the equality of the covariance matrices?

(c) For these covariance matrices, what are the three log determinants?

(d) Do these results indicate that the statistical validity of an hypothesis test comparing the group centroids using an χ^2 statistic is doubtful?

13. Which two outcome variables have the highest correlation?

14. What are the values of Λ, U, and V?

15. Do the results of the analysis provide sufficient evidence to conclude that the three population centroids differ? Support you answer with appropriate statistics.

CHAPTER 4

Assessing MANOVA Effects

4.1 INTRODUCTION

In Chapter 3 we discussed procedures that are useful for comparing two or more sample centroids to test the null hypothesis that the population centroids are identical. If the null hypothesis is rejected, it is very likely that the researcher would be interested in further analyses of the data to describe the resultant relationship between the grouping variable and the collection of outcome variables, and to make comparisons between or among specific population centroids. It is sometimes suggested that if the multivariate omnibus hypothesis test is statistically significant, a series of univariate omnibus tests be conducted, one such test for each separate outcome variable. We discourage this type of analysis because it ignores correlational structure among the outcomes, it does not adequately control Type I errors, and such analyses can have low statistical power. In this chapter we briefly review two measures of association—eta squared, η^2, and omega squared, ω^2—frequently used in univariate analysis of variance to describe the relationship between the grouping variable and the single outcome variable. For the multivariate context, several indices of effect size have been suggested, and we will present five of the more popular indices to describe the relationship between the grouping variable and the collection of outcome variables. We end this chapter with a review of contrast analyses for the univariate context and present the multivariate analog for focused tests.

The assessment of effects of grouping variables in a univariate or multivariate analysis of variance situation involves more than the reporting of a test statistic value for an omnibus test. If omnibus effects are of interest, the test statistic value should be reported along with a P value and an effect-size value. The very same information is of interest for assessing multivariate contrast effects, effects that are sometimes of more interest than omnibus effects. Following a discussion of a number of effect-size indices, the conduct and assessment of contrast effects are discussed in detail.

Applied MANOVA and Discriminant Analysis, Second Edition, by Carl J. Huberty and Stephen Olejnik
Copyright © 2006 John Wiley & Sons, Inc.

4.2 STRENGTH OF ASSOCIATION

4.2.1 Univariate

In our discussion on the comparison of two population means using the two independent samples t test, we stated that the squared point-biserial correlation—see Eq. (3.6)—provides the proportion of variation in the single outcome variable that is explained by the grouping variable. The point-biserial correlation is one index that can be used to quantify the magnitude of the grouping variable effect.

An analogous statistic that can be used when two or more populations are being compared is eta squared, η^2, and is defined as:

$$\eta^2 = \frac{\text{SSH}}{\text{SSH} + \text{SSE}},$$

or, equivalently, as $1 - (\text{SSE}/\text{SST})$. Using the data for the EDT outcome in Table 3.1, the strength of the relationship between reading program and EDT scores is estimated as:

$$\eta^2 \doteq \frac{27.758}{604.258} \doteq .046.$$

Eta squared, however, is positively biased (Carroll and Nordholm, 1975; Keselman 1975), especially when sample sizes are small. An alternative index of effect size was suggested by Hays (1963), called omega squared, ω^2, based on unbiased estimators of the total variance and the variance associated with the grouping variable. Omega squared is defined as:

$$\omega^2 = \frac{\text{SSH} - \text{df}_h(\text{MS}_{\text{error}})}{\text{SST} + \text{MS}_{\text{error}}}.$$

Using the Baumann data for the EDT outcome, ω^2 is

$$\omega^2 \doteq \frac{27.58 - 2(9.151)}{604.258 + 9.151} \doteq .015.$$

Both estimates of association are very small, which is consistent with the earlier finding of no statistically significant difference in the three population means on the EDT outcome.

4.2.2 Multivariate

Several multivariate indices of effect or association have been suggested. One such index is based on a generalization of η^2 using Λ. This index is defined as:

$$\eta^2_{\text{Mult}} = 1 - \Lambda. \tag{4.1}$$

One approach to the computation of Λ was presented in Eq. (3.16) as the ratio of the determinant of the error sum-of-squares matrix to the determinant of the total sum-of-squares matrix. In the univariate case ($p = 1$), **E** and **H** are scalars equaling SSE and SSH, respectively. Lambda, therefore, would equal

$$\Lambda = \frac{\text{SSE}}{\text{SSH} + \text{SSE}}.$$

So, in the univariate case,

$$\eta^2 = 1 - \frac{\text{SSE}}{\text{SSH} + \text{SSE}} = 1 - \frac{\text{SSE}}{\text{SST}},$$

as given above. In the multivariate case, Λ is an index of the proportion of the total generalized variance of the collection of outcome variables that is *not* associated with the grouping variable. One minus this proportion is the proportion of the total variance that is associated with the grouping variable. Using the Baumann data, $\Lambda \doteq .841$ and $\eta^2_{\text{Mult}} \doteq 1 - .841 \doteq .159$.

An alternative approach for computing Λ was presented in Eq. (3.22) as:

$$\Lambda = \prod_{v=1}^{r} \left(\frac{1}{1 + \lambda_v} \right),$$

or, equivalently, as

$$\prod_{v=1}^{r} \left(1 - \frac{\lambda_v}{1 + \lambda_v} \right).$$

The ratio $\dfrac{\lambda_v}{1 + \lambda_v}$ is the *squared canonical correlation* between the grouping variable and the vth linear discriminant function, or the vth construct underlying the p outcome variables. Cramer and Nicewander (1979) suggested as an index of association, τ^2, tau squared, computed as:

$$\tau^2 = 1 - \left[\prod_{v=1}^{r} \left(1 - \frac{\lambda_v}{1 + \lambda_v} \right) \right]^{1/r}, \qquad (4.2)$$

or, equivalently, as

$$\tau^2 = 1 - \Lambda^{1/r}, \qquad (4.3)$$

where $r = \min(p, \text{df}_h)$. The value $\left[\prod_{v=1}^{r} \left(1 - \dfrac{\lambda_v}{1 + \lambda_v} \right) \right]^{1/r}$ is the geometric mean of the proportion of variation in the constructs underlying the outcome variables that is *not* explained by the grouping variable. One minus this value is the proportion of variation in the underlying constructs that is explained by the grouping variable. The Cramer and Nicewander (1979) effect size measure is associated with the Wilks statistic in SPSS multivariate output. Using the Baumann data, $r = 2$ and τ^2 is estimated as:

$$\tau^2 \doteq 1 - .841^{1/2} \doteq .083.$$

An effect size index related to the Bartlett–Pillai test statistic (U) is ξ^2. Xi squared (ks-eye squared) is written as:

$$\xi^2 = \frac{U}{r}. \tag{4.4}$$

The statistic U is computed as the sum of the squared canonical correlations, $\sum_{v=1}^{r} \frac{\lambda_v}{1+\lambda_v}$. So, ξ^2 is the mean-squared canonical correlation. That is, ξ^2 is the mean proportion of variation in the underlying constructs that is explained by the grouping variable. In the SPSS multivariate output, ξ^2 is reported as the "Pillais" effect size. Using the Baumann data, ξ^2 is computed as:

$$\xi^2 \doteq \frac{.1654}{2} \doteq .083.$$

An effect size index associated with the Hotelling–Lawley test statistic, ζ^2, zeta squared, can be written as:

$$\zeta^2 = \frac{V}{r+V}, \tag{4.5}$$

where $V = \sum_{v=1}^{r} \lambda_v$ and $r = \min(p, \mathrm{df}_h)$. ζ^2 is associated with the Hotelling statistic in SPSS multivariate output. Using the Baumann data,

$$\zeta^2 \doteq \frac{.1901}{2+.1901} \doteq .087.$$

Tatsuoka (1970) suggests a generalization of the univariate ω^2 statistic as an index of association between the collection of outcome variables and the grouping variable. The multivariate ω^2 is defined as:

$$\omega^2_{\mathrm{Mult}} = 1 - \frac{N\Lambda}{(N - \mathrm{df}_h - 1) + \Lambda}. \tag{4.6}$$

Using the Baumann data,

$$\omega^2_{\mathrm{Mult}} \doteq 1 - \frac{66(.841)}{(66-2-1)+.841} \doteq .131.$$

Table 4.1 summarizes the multivariate effect size indices and their values using the Baumann data.

As demonstrated above, the five indices of association may provide considerably different estimates. The difference among these indices of effect size is a function of how these indices of association are conceptualized. The multivariate η^2 and ω^2 define effect size as a function of the total generalized variance that is associated with the grouping variable regardless of the number of constructs underlying the outcome variables. The other three indices of effect size define the strength of association by taking into consideration the number of estimated dimensions or constructs underlying

TABLE 4.1 Five Multivariate Effect Size Indices

Statistic	Expression	Estimate
η^2_{Mult} (4.1)	$1 - \prod\left(1 - \frac{\lambda_v}{1+\lambda_v}\right)$	.159
τ^2 (4.2)	$1 - \left[\prod\left(1 - \frac{\lambda_v}{1+\lambda_v}\right)\right]^{1/r}$	.083
ξ^2 (4.4)	$\frac{\sum_{v=1}^{r} \frac{\lambda_v}{1+\lambda_v}}{r}$	.083
ζ^2 (4.5)	$\frac{\sum_{v=1}^{r} \lambda_v}{r + \sum_{v=1}^{r} \lambda_v}$	.087
ω^2_{Mult} (4.6)	$1 - \frac{N\Lambda}{(N - df_h - 1) + \Lambda}$	.131

the outcome variables. Because there is no single agreed upon definition of "effect" in a multivariate context, no one index is viewed as being preferred over the others. If there are only two populations being compared ($r = 1$), however, all five statistics provide the same estimate. In this case, only one linear discriminant function or construct can be determined regardless of the number of outcome variables examined. When there are three or more levels of the grouping variable, two or more linear discriminant functions or constructs can be estimated. The more dimensions estimated, the larger the difference among the effect-size indices.

SPSS output from the MANOVA program provides the effect-size values using τ^2, ζ^2, and ξ^2. These effect-size indices are associated with the Wilks, Hotelling–Lawley, and Bartlett–Pillai multivariate test criteria, respectively. Typically, if one multivariate test criterion is preferred and reported, the corresponding effect-size index would also be reported. Regardless of which effect-size index is chosen, the assessment of its magnitude is problematic at this time. What is a "large" or "substantial" multivariate effect-size? Such definitions have yet to be suggested. Although researchers have been encouraged (even required by some journal editors) to report effect size values, multivariate effect-size values are rarely, if ever, currently reported by applied researchers. Definitions of effect-size indicies that might span many variable domains, types of analysis units, and substantive areas will, undoubtedly, be difficult to advance. If Wilks Λ values would be routinely reported—as yet they are not!—perhaps some research synthesis could lead to some rough magnitude guides.

4.2.3 Bias

Regardless of which effect-size index a researcher chooses, the sample statistic will overestimate the strength of the association between the collection of outcome variables and the grouping variable. The degree of overestimation, bias, is a function of the number of levels of the grouping variable J, the number of outcome variables p, and the sample size N. The degree of bias increases as J and p increase, and for small group sizes.

Serlin (1982) recommends the following adjustment for ξ^2, which is analogous to the adjustment suggested by M. J. B. Ezekiel (1899–1974) in 1930 for the multiple squared correlation coefficient, R^2:

$$\xi^2_{\text{adj}} = 1 - \frac{N-1}{N-b-1}(1-\xi^2), \tag{4.7}$$

where $b = \max(p, \text{df}_h)$. Applying this adjustment to our numerical estimate of ξ^2 gives the following results:

$$\xi^2_{\text{adj}} \doteq 1 - \frac{66-1}{66-2-1}(1-.083) \doteq .054.$$

A simulation study by Kim and Olejnik (2005) provided support for this adjustment, and it can, in general, be used with τ^2 and ζ^2, as well. For τ^2 and ζ^2, the adjustment works best for two-group studies and for $\text{df}_h \geq 2$ if the sample size is "large."

4.3 COMPUTER APPLICATION I

In this section, the SPSS command to request τ^2, ξ^2, and ζ^2 is given. SPSS does not report η^2_{Mult} or ω^2_{Mult}, nor does SPSS adjust these effect-size indices for their overestimation of the relationship between the grouping variable and the collection of outcome variables.

SPSS SYNTAX FOR COMPUTING EFFECT SIZE

```
manova Y1 Y2 by treatmnt(1, 3)
/print = signif(efsize).
```

signif(efsize) requests three unadjusted multivariate effect-size indices, ξ^2, ζ^2, and τ^2.

OUTPUT

Analysis: Effect-Size Estimates

EFFECT . . treatmnt
Multivariate Tests of Significance (S = 2, M = −1/2, N = 30)

Test Name	Value	Approx. F	Hypoth. DF	Error DF	Sig. of F
Pillais	.16185	2.77356	4.00	126.00	.030

Hotellings	.18583	2.83393	4.00	122.00	.027
Wilks	.84094	2.80488	4.00	124.00	.029
Roys	.14225				

Note . . F statistic for WILKS' Lambda is exact.

- -

Multivariate Effect Size

TEST NAME	Effect Size
Pillais	.081
Hotellings	.085
Wilks	.083

- -

Univariate F-tests with (2,63) D. F.

Variable	Hypoth. SS	Error SS	Hypoth. MS	Error MS	F	Sig. of F
Y1	27.75758	576.50000	13.87879	9.15079	1.51668	.227
Y2	243.36364	3443.50000	121.68182	54.65873	2.22621	.116

Variable	ETA Square
Y1	.04594
Y2	.06601

- -

Interpretation: Effect-Size Estimates

The three effect-size values reported here differ slightly from the results computed in Section 4.2.2. The small difference reflects the increased level of accuracy obtained by using the computer. All three effect-size values are similar, which will generally be the case. While it may be tempting to compare the multivariate with the univariate effect-size indices, such a comparison should be avoided. The univariate index represents the proportion of variance in one outcome variable that is associated with the grouping variable, while the multivariate index represents the proportion of the variation in the construct(s) that is shared with the grouping variable. The magnitude of the multivariate effect size indices will be influenced to some degree by the number of constructs—the number of constructs identified in a variable system is discussed in Chapter 5—which underlie the p variable system of outcomes.

4.4 GROUP CONTRASTS

4.4.1 Univariate

In a univariate context involving more than two levels of a grouping variable, researchers are seldom satisfied with the conclusion from the omnibus F test that populations do not have identical means. Rather, analyses involving comparisons or contrasts of specific groups are generally considered. These are sometimes referred to as *focused tests*. It should be noted that a researcher may very well go directly to an examination of contrasts; that is, the omnibus test is *not* a necessary preliminary test. A contrast, ψ, is defined as a linear combination of means:

$$\psi = \sum_{j=1}^{J} a_j \mu_j = a_1\mu_1 + a_2\mu_2 + \cdots + a_J\mu_J, \tag{4.8}$$

with $\sum_{j=1}^{J} a_j = 0$. Examples are $\psi = \mu_1 - \mu_2$ $(a_1 = 1, a_2 = -1, a_j = 0$ for $j = 3 \ldots J)$, and $\psi = 2\mu_1 - (\mu_2 + \mu_3)$ $(a_1 = 2,\ a_2 = -1, a_3 = -1, a_j = 0$ for $j = 4 \ldots J)$. Sample estimates of contrasts are obtained by substituting sample means for the population means:

$$\hat{\psi} = \sum_{j=1}^{J} a_j \overline{Y}_{.j} = a_1\overline{Y}_{.1} + a_2\overline{Y}_{.2} + \cdots + a_J\overline{Y}_{.J}.$$

The ratio of a contrast value to its estimated standard error results in a t statistic. The estimated standard error of a sample contrast is computed as:

$$s_{\hat{\psi}} = \sqrt{s_e^2 \sum_{j=1}^{J} a_j^2/n_j},$$

and

$$t = \frac{\hat{\psi}}{s_{\hat{\psi}}}. \tag{4.9}$$

It should be recognized that (4.9) has the same form as (3.1) when $J = 2$. Equation (4.9) is more general, however, allowing more than two groups to be compared, and the error variance (s_e^2) is based on all J groups in the study. The error variance across all of the groups in the study provides greater statistical power, assuming variance equality.

4.4.2 Multivariate

In a J-group MANOVA, one can merely conclude that the J population centroids are not equal or that there is a relationship between the grouping variable and the outcome variable composite (i.e., a linear combination of the outcome variables). But, specific vector differences or specific relationships may be of more interest than the omnibus test of centroid differences, or an overall relationship between the grouping variable and the outcome variables. In fact, it may be argued that if contrast-based questions are of interest, the researcher should proceed directly to contrast analyses and bypass the omnibus analysis. We do not want to minimize, however, the usefulness of the omnibus hypothesis test. There are likely to be some research contexts where the omnibus test would be of both practical and/or theoretical interest. For example, a researcher may be interested in combining data obtained by several observers or from several locations. An omnibus test might be used to provide support for such data aggregation if there is no evidence to indicate mean differences among observers or among locations.

In the multivariate context, contrasts are formed using the outcome variable mean vectors rather than individual variable means. For example, a *pairwise* contrast comparing the TA and the DRTA reading conditions from the Baumann study may be of interest, and the null hypothesis H_0: $\boldsymbol{\psi} = \mathbf{0}$, where $\boldsymbol{\psi} = \boldsymbol{\mu}_1 - \boldsymbol{\mu}_3$, may be tested. The sample mean vectors are

$$\mathbf{y}_{.1} \doteq \begin{bmatrix} 7.77 \\ 43.45 \end{bmatrix}, \qquad \mathbf{y}_{.2} \doteq \begin{bmatrix} 6.23 \\ 46.64 \end{bmatrix},$$

and

$$\hat{\boldsymbol{\psi}} \doteq \begin{bmatrix} 1.54 \\ -3.19 \end{bmatrix}.$$

For any contrast, the Hotelling T^2 test statistic can be computed as:

$$T^2 = \hat{\boldsymbol{\psi}}' \left(\sum_{j=1}^{J} \frac{a_j^2}{n_j} \mathbf{S}_\mathrm{e} \right)^{-1} \hat{\boldsymbol{\psi}}, \tag{4.10}$$

where $\hat{\boldsymbol{\psi}}$ is the sample centroid contrast of interest, $\mathbf{S}_e$ is the error covariance matrix, a_j is the weight for Group j, and n_j is the sample size for Group j. For a pairwise contrast this can be simplified to

$$T^2 = \frac{n_j n_{j'}}{n_j + n_{j'}} \left(\hat{\boldsymbol{\psi}}' \mathbf{S}_\mathrm{e}^{-1} \hat{\boldsymbol{\psi}} \right) \tag{4.11}$$

Equation (3.14) provided the error **SSCP** matrix, **E**, for the Baumann data. The error covariance matrix, $\mathbf{S}_e$ and its inverse, are

$$\mathbf{S}_e \doteq \begin{bmatrix} 9.151 & 10.435 \\ 10.435 & 54.659 \end{bmatrix}, \qquad \mathbf{S}_e^{-1} \doteq \begin{bmatrix} .1399 & -.0267 \\ -.0267 & .0234 \end{bmatrix}.$$

The test statistic value, then, is

$$T^2 \doteq \frac{(22)(22)}{22+22} [1.54 \quad -3.19] \begin{bmatrix} .1399 & -.0267 \\ -.0267 & .0234 \end{bmatrix} \begin{bmatrix} 1.54 \\ -3.19 \end{bmatrix} \doteq 9.156.$$

Using Eq. (3.11) T^2 can be transformed to a statistic that has an F distribution with degrees of freedom $\nu_1 = p$ and $\nu_2 = \mathrm{df}_e - p + 1$:

$$F \doteq \frac{63 - 2 + 1}{2(63)} 9.156 \doteq 4.505.$$

Assuming assumptions have been met and the null hypothesis is true, the P value for this statistic, having 2 and 62 degrees of freedom, is .015.

Alternatively, any contrast could be tested for statistical significance using the eigenvalues obtained from the matrix product $\mathbf{E}^{-1}\mathbf{H}$. Because contrast tests are single degree-of-freedom tests, all four test criteria (Wilks, Bartlett–Pillai, Roy, and Hotelling–Lawley) will result in the same F statistic.

The matrix $\mathbf{E}$ is the error **SSCP** matrix and was computed in Section 3.5.2 and found to be

$$\mathbf{E} \doteq \begin{bmatrix} 576.502 & 657.409 \\ 657.409 & 3443.501 \end{bmatrix}.$$

Its inverse is

$$\mathbf{E}^{-1} \doteq \begin{bmatrix} .0022 & -.0004 \\ -.0004 & .0004 \end{bmatrix}.$$

The matrix $\mathbf{H}$ is the **SSCP** matrix for the specified contrast and is computed as:

$$\mathbf{H}_{\hat{\boldsymbol{\psi}}} = \left(\hat{\boldsymbol{\psi}}\hat{\boldsymbol{\psi}}'\right) \Big/ \sum_{j=1}^{J} \frac{a_j^2}{n_j}. \tag{4.12}$$

When sample sizes are equal, (4.12) can be written as:

$$\mathbf{H}_{\hat{\boldsymbol{\psi}}} = \frac{n}{\sum_{j=1}^{J} a_j^2} \left(\hat{\boldsymbol{\psi}}\hat{\boldsymbol{\psi}}'\right). \tag{4.13}$$

For testing $H_0\colon \boldsymbol{\mu}_1 - \boldsymbol{\mu}_3 = \mathbf{0}$ the Baumann data yields

$$\mathbf{H}_{\hat{\boldsymbol{\psi}}} \doteq \frac{22}{(1)^2 + (-1)^2} \begin{bmatrix} 1.54 \\ -3.19 \end{bmatrix} [1.54 - 3.19]$$

$$\doteq \begin{bmatrix} 26.0876 & -54.0386 \\ -54.0386 & 111.9371 \end{bmatrix}.$$

Using the above results for $\mathbf{E}^{-1}$ and $\mathbf{H}_{\hat{\boldsymbol{\psi}}}$, the matrix product is

$$\mathbf{E}^{-1}\mathbf{H}_{\hat{\boldsymbol{\psi}}} \doteq \begin{bmatrix} .0790 & -.1638 \\ -.0321 & .0664 \end{bmatrix}.$$

and the eigenvalues are .145 and 0. There is only one nonzero eigenvalue, as expected, because $df_h = 1$. Based on this one eigenvalue, the four multivariate criteria can be determined as:

Criterion	Statistic	Calculated Value
Wilks	$\Lambda = 1/(1+\lambda)$	$\doteq 1/1.145 \doteq .873$
Bartlett–Pillai	$U = \lambda_1/(1+\lambda_1)$	$\doteq .145/1.145 \doteq .127$
Roy	$\Theta = \lambda_1/(1+\lambda_1)$	$\doteq .145/1.145 \doteq .127$
Hotelling–Lawley	$V = \lambda$	$\doteq .145 \doteq .145$

Using Eq. (3.17) and the Wilks Λ, the resulting F statistic is computed as 4.507, with 2 and 62 degrees of freedom. The P value, assuming the null hypothesis is true, that corresponds with these results, is .015. The small difference between the F statistic obtained from the transformation of Hotelling's T^2 ($F \doteq 4.492$) and the F statistic obtained from the four multivariate test criteria is due to rounding while computing these statistics.

If a comparison between the mean of TA and DRA with DRTA, $\boldsymbol{\psi}_2$, was also of interest, the null hypothesis might be written as H_{02}: $\frac{1}{2}(\mu_{\text{TA}} + \mu_{\text{DRA}}) - \mu_{\text{DRTA}} = 0$. Using the mean vectors for the three groups reported in Chapter 3 the contrast mean vector is

$$\hat{\boldsymbol{\psi}}_2 \doteq \begin{bmatrix} .995 \\ -3.89 \end{bmatrix}.$$

The eigenvalue for the $\mathbf{E}^{-1}\mathbf{H}_{\boldsymbol{\psi}_2}$ matrix product for this complex contrast hypothesis is .1626 and the Wilks Λ is .860. The corresponding F, degrees of freedom, and P value for this result are $F \doteq 5.047$ with 2 and 62 degrees of freedom, and $P \doteq .009$.

When multiple hypotheses are tested, it is generally advisable to adjust the obtained P values. The adjustment procedure we suggest, while very simple, assumes that the identification of the hypotheses to be tested is made a priori to the data collection, or at least before the analysis of the data has begun. Deciding which comparisons to test after examining the sample vector means will invalidate the probability statements made. A sensible adjustment for multiple testing associated with contrasts may be accomplished by multiplying each of the individual P values by the number of hypotheses tested, say c. That is, $P' = c(P)$. For the two hypotheses tested above, the adjusted P values would equal $P_1' \doteq 2(.016) = .032$ and $P_2' \doteq 2(.009) = .018$. These results may be interpreted as providing evidence to conclude that the mean centroids for the Think Aloud strategy and the Directed Reading and Think Aloud strategy are different, and the mean centroid for the Think Aloud and the Directed Reading Activity strategies differ from the centroid of the Directed Reading and Think Aloud strategy. To further understand the nature of these differences still additional analyses are necessary. These additional analyses focus on the determination and characterization of the constructs underlying the centroid differences. These additional analyses are presented in Chapter 5.

Measures of association for contrasts may also be computed using the procedures presented in Section 4.2.2. Because contrasts have only one degree of freedom for the hypothesis, $df_h = 1$, the measures of association provided by η^2_{Mult}, τ^2, ζ^2, and ξ^2 are all the same. For the first contrast, $\Lambda \doteq .873$, so $\eta^2_{\text{mult}} \doteq .127$. Tatsuoka's (1970) multivariate generalization of ω^2, ω^2_{Mult} is .112. Using the Serlin (1982) adjustment for these data, the adjusted effect size measure is

$$\xi^2_{\text{adj}} \doteq 1 - \frac{66-1}{66-2-1}(1 - .127) \doteq .099.$$

4.5 COMPUTER APPLICATION II

In this section we present the SPSS syntax to conduct focused tests. We demonstrate the procedure using the same contrasts that were used in Section 4.4. That is, we analyze the pairwise contrast comparing the TA group with the DRTA group and the *complex contrast* comparing the mean centroid for the TA and the DRA groups with the DRTA group. The degrees of freedom for the grouping variable determines the number of contrasts that may be requested on any one contrast command. But multiple contrast commands may be used for any MANOVA run but multiple **/design** statements must also be provided.

SPSS SYNTAX FOR MANOVA CONTRAST ANALYSES

```
manova Y1 Y2 by treatmnt(1,3)
/print = signif(efsize)
/contrast(treatmnt) = special(1 1 1, .5 .5 −1, 1 0 −1)
/design = treatmnt(1) treatmnt(2).
```

/contrast is a SPSS system command requesting a contrast.
(treatmnt) identifies the explanatory variable.
special introduces the researcher generated contrast weights.
1 1 1 identifies the grand mean, the number of 1's is determined by the number treatmnt levels.
1 0 −1 compares level 1 of treatment with level 3 of treatment.
.5 .5 −1 compares the average of levels 1 and 2 with level 3 of treatment.
/design is an SPSS system command to identify subhypotheses for analysis.
treatmnt(1) labels the first contrast on the explanatory variable.
treatmnt(2) labels the second contrast on the explanatory variable.

OUTPUT

Analysis: Focused Tests

EFFECT . . TREATMNT(2)

Multivariate Tests of Significance (S = 1, M = 0, N = 30)

Test Name	Value	Exact F	Hypoth. DF	Error DF	Sig. of F
Pillais	.12694	4.50712	2.00	62.00	.015
Hotellings	.14539	4.50712	2.00	62.00	.015
Wilks	.87306	4.50712	2.00	62.00	.015
Roys	.12694				

Note . . F statistics are exact.

- -

Multivariate Effect Size

TEST NAME	Effect Size
(All)	.127

- -

EFFECT . . TREATMNT(1)

Multivariate Tests of Significance (S = 1, M = 0, N = 30)

Test Name	Value	Exact F	Hypoth. DF	Error DF	Sig. of F
Pillais	.14016	5.05339	2.00	62.00	.009
Hotellings	.16301	5.05339	2.00	62.00	.009
Wilks	.85984	5.05339	2.00	62.00	.009
Roys	.14016				

Note . . F statistics are exact.

- -

Multivariate Effect Size

TEST NAME	Effect Size
(All)	.140

Interpretation: Focused Tests

The computer results reported here differ slightly from those reported in Section 4.4.2, reflecting the increased precision available through computer analysis. The conclusions are, however, the same. The second contrast, which compared the centroids for the TA with the DRTA groups, is reported first. The complex contrast, comparing the mean of the TA and DRA group centroids with the DRTA group centroid, is reported second. The results provide considerable evidence, adjusted (Bonferroni method) for the two hypotheses, to conclude that the observed difference between the TA and DRTA centroids is generalizable to the populations that the samples represent [$\Lambda \doteq .873$, $F(2, 62) \doteq 4.507$, $P' \doteq .030$, $\xi^2_{\text{adj}} \doteq .099$]. And there is evidence to conclude that the difference between the mean of the TA and DRA centroids and the DRTA centroid is generalizable to the populations that the groups represent [$\Lambda \doteq .860$, $F(2, 62) \doteq 5.053$, $P' \doteq .018$, $\xi^2_{\text{adj}} \doteq .113$].

4.6 COVARIANCE MATRIX HETEROGENEITY

As pointed out earlier, the hypothesis tests presented here on contrasts assume J-population multivariate normality and covariance matrix homogeneity. While multivariate nonnormality typically does not affect the statistical validity of the analysis, heterogeneous covariance matrices can result in reported P values that are either too small (liberal) or too large (conservative). We discussed this issue in Section 3.3 and refer the reader to the guidelines provided by Hakstain et al. (1979) when covariance matrix homogeneity is not tenable.

If sample sizes are not equal and covariance matrix homogeneity is not tenable, the test criterion to be used for the contrast $\boldsymbol{\mu}_j - \boldsymbol{\mu}_{j'}$ is the procedure suggested by Yao (1965) and discussed in Section 3.2.2. A computer program written by J. Algina (University of Florida) that may be used for pairwise contrasts in a single-factor MANOVA context is available at the Wiley website. The program is labeled YAOPC. In the Technical Notes at the end of this chapter we provide a generalization of the Yao test for complex contrasts.

Suppose, in a three-group situation, it is concluded that

$$\boldsymbol{\Sigma}_1 = \boldsymbol{\Sigma}_2 \neq \boldsymbol{\Sigma}_3;$$

that is, the first two sample covariance matrices are reasonably close but different from the third. In this case it seems reasonable to use a sample covariance matrix pooled across Groups 1 and 2 in the Hotelling T^2—see Eq. (3.10)—for the contrast $\boldsymbol{\mu}_1 - \boldsymbol{\mu}_2$. On the other hand, for the contrast, say, $\boldsymbol{\mu}_1 - \boldsymbol{\mu}_3$ one could use the Yao test (3.12) and (3.13).

Usually, in multiple group situations, if specific group comparison questions are of interest, more than one such type of question will be posed. Therefore, a multiple testing situation arises. As suggested at the end of Section 4.4.2, a multiple testing situation calls for an adjustment of the probability distribution tail areas for the individual tests. Simply multiplying each tail area by the number of tests conducted seems to be a reasonable adjustment in most situations. These products, then, can be assessed in judging statistical significance.

4.7 SAMPLE SIZE

In the ANOVA context, sample size tables have been proposed by Cohen (1988) and Kraemer and Thiemann (1987). Input for the use of these tables consists of α and β (probabilities of Type I and Type II errors, respectively, with which we are willing to live) and effect size. The idea of the use of these tables is that specifications of these three numerical indicators—all researcher judgments—will suggest a sample size that for the given α value and effect size value, the sample size used will yield adequate (for the specified effect size value) statistical power $(1 - \beta)$. Such tables should only be considered as guides to go along with judgments.

Similarly, in the MANOVA context, given some α and β values, and some indication of effect size, we would like to have a guide to a desirable sample size. Some sample size tables have been developed by Lauter (1978); Lauter's tables are also given in Kres (1983, pp. 432–451). These sample size tables are based on use of the Hotelling–Lawley trace criterion (see Table 3.3) for a one-factor design. As one might expect, the multivariate interpretation of deviations from the null hypothesis (H_0: $\boldsymbol{\mu}_1 = \boldsymbol{\mu}_2 = \cdots = \boldsymbol{\mu}_J$) gets messy because of the many ways (with p variables) that H_0 may be false. For $J = 2$ and $J = 4$, $p = 5$ and $p = 15$, $\alpha = .05$ and $\alpha = .01$, and $\beta = .30$ and $\beta = .20$ the use of the Lauter tables suggests that a rule of thumb like $4p$ or $5p$ for a "moderate" effect size may provide an adequate sample size per group. Rencher (1998), however, states that the Lauter tables "have limited usefulness because they are based on an overly simplified alternative hypothesis" (pp. 140–141).

4.8 SUMMARY

In this chapter we presented procedures that are useful to further understand multiple population separation. Specifically, effect-size indices associated with each multivariate test criterion and methods for focused hypothesis tests are presented and demonstrated. Although the interpretation of multivariate effect size indices at the present time is as best questionable, we nevertheless believe that the reporting of such indices will ultimately be useful for the advancement of our understanding of population separation. We also caution that the currently available effect size indices are biased to some degree, and the degree of bias is determined by four factors: sample size, the number of outcome variables, the number of populations, and the degree of population separation. The adjustment procedure suggested by Serlin (1982) is strongly recommended.

Focused tests or contrast analyses examine specific population separation. There may very well be research situations where the overall (i.e., omnibus) comparison of J population mean vectors would be of interest. For example, Smart (1982) was interested in studying a typology of six college types of academic behavior (using scores on three factors of faculty teaching goals as outcome measures). However, many times in population comparison studies specific questions are of more interest (i.e., questions associated with multivariate group contrasts). The omnibus test examines the linear composite of outcome variables that best separates all groups simultaneously. Contrast analyses, on the other hand, examine the linear composite that best separates specific populations, and it may be considerably different than the composites provided by the omnibus test. Consequently, if specific comparisons can be identified at the beginning of the study, we recommend ignoring the omnibus test and go directly to the focused tests that address the primary interest. For such analyses, a P value along with an effect-size index (e.g., η^2) value may be used to assess the strength of empirical evidence for the hypotheses under consideration. Of course, P values for individual tests should be adjusted because of multiple testing.

In the next chapter we will present procedures that will help determine the number and identification of the underlying constructs that separate the populations being compared.

Technical Notes

1. Very little, if anything, has been written about the assessment of the magnitude of the various effect-size indices presented in this chapter. These are very difficult judgments to make at this point in time, with the paucity of MANOVAs conducted in a given area of study. There is another index that may be attractive to some researchers. This index involves the use of a predictive discriminant analysis (PDA). The index itself is the classification hit rate—either across groups or for a particular group. (*Note*: The use of a PDA in a group separation context implies a reversal of the roles of the variables involved; the MANOVA outcome variables are now serving as predictors. This may be bothersome to some.) The use of PDA results in assessing effect is discussed by Huberty and Lowman (1997).

2. A statistical test for a univariate pairwise contrast in the context of unequal variances was mentioned in Section 3.2 [see Eqs. (3.4) and (3.5)]. The extension of this test to one for a univariate *complex* contrast is fairly straightforward (Maxwell and Delaney, 2000, p. 180). Let us discuss the test of a *multivariate* complex contrast (when the $\boldsymbol{\Sigma}$'s are unequal):

$$H_0\colon \boldsymbol{\psi} = \sum_{j=1}^{J} a_j \boldsymbol{\mu}_j = 0.$$

One might be interested in, for example, $\boldsymbol{\psi} = \boldsymbol{\mu}_2 + \boldsymbol{\mu}_3 - 2\boldsymbol{\mu}_4$; here $a_1 = 0$, $a_2 = 1$, $a_3 = 1$, and $a_4 = -2$.

Now let

$$W = \sum_{j=1}^{J} \frac{a_j^2}{n_j} \mathbf{S}_j,$$

where $\mathbf{S}_j$ is the Group j covariance matrix. Then our test statistic is a direct generalization of T^{*2} in (3.12):

$$T^{*2} = \hat{\boldsymbol{\psi}}' W^{-1} \hat{\boldsymbol{\psi}},$$

where $\hat{\boldsymbol{\psi}}$ is the $p \times 1$ vector of sample contrast "scores" for each Y. The following approximate transformation may be utilized:

$$\frac{f - p + 1}{pf} T^{*2} \longrightarrow F(p, f - p + 1), \tag{4.14}$$

where p is the number of outcome variables, and

$$f = \frac{\operatorname{tr}\left(\sum_{j=1}^{J} \frac{a_j^2}{n_j} \mathbf{S}_j\right)^2 + \left[\operatorname{tr}\left(\sum_{j=1}^{J} \frac{a_j^2}{n_j} \mathbf{S}_j\right)\right]^2}{\sum_{j=1}^{J} \frac{1}{n_j - 1} \left\{\operatorname{tr}\left(\frac{a_j^2}{n_j} \mathbf{S}_j\right)^2 + \left[\operatorname{tr}\left(\frac{a_j^2}{n_j} \mathbf{S}_j\right)\right]^2\right\}},$$

where "tr" denotes the trace function. [The expression for f is based on the work of Nel and van der Merwe (1986, p. 771), as suggested by J. Algina (University of Florida).]

Further Reading

Cramer and Nicewander (1979) suggest six different indices of multivariate association, all of which are functions of squared canonical correlation coefficients and two of which (η^2, τ^2) are discussed in Section 4.2.

Haase (1991) presents formulas for computing multivariate effect size indices from reported F and chi-squared values.

Tate (1981) discusses multivariate aptitude–treatment–interaction analyses, including interaction contrasts.

Definition ***Biserial correlation:*** Relationship between Wheaties and Rice Crispies.

EXERCISES

Use the following context to answer Exercises 1 to 10.

A University administrator was interested in comparing attributions for success in college held by students majoring in Mathematics (Math), English (Eng), Psychology (Psy), and Business (Bus). A random sample of 12 students from each Major was selected. Participants completed a battery of four instruments assessing attributions for success to Effort, Ability, Luck, and their Teachers. The data were analyzed using a multivariate analysis of variance program. The results of the analysis reported the following eigenvalues: 4.764, 1.237, and .044.

1. Given the research design, should more eigenvalues be expected? Briefly explain.

2. Given the above results compute:

(a) Λ

(b) Bartlett–Pillai U

(c) Roy Θ

(d) Hotelling–Lawley V

3. Convert each of the multivariate test criteria in Exercise 2 to an an F statistic and report the appropriate degrees of freedom.

4. Compute the following effect size measures:

(a) η^2_{Mult}

(b) τ^2

(c) ξ^2

(d) ζ^2

(e) ω^2_{Mult}

5. Using the Serlin procedure, provide the adjusted effect-size measures for:
   (a) τ^2
   (b) ζ^2
   (c) ξ^2

6. The administrator was interested in three specific contrasts and provided the following SPSS command **/contrast = special(1 1 1 1, 1 0 0 −1, 0 1 0 −1, 0 0 1 −1)**. What comparisons is she requesting?

7. How many eigenvalues are associated with each contrast? Briefly explain.

8. Suppose the following are the group centroids:

$$\text{Math} = \begin{bmatrix} 11.5 \\ 25.3 \\ 22.7 \\ 32.3 \end{bmatrix}, \quad \text{Eng} = \begin{bmatrix} 22.2 \\ 31.5 \\ 30.5 \\ 19.0 \end{bmatrix}, \quad \text{Psy} = \begin{bmatrix} 17.9 \\ 28.6 \\ 30.3 \\ 22.4 \end{bmatrix}, \quad \text{Bus} = \begin{bmatrix} 14.3 \\ 26.4 \\ 34.1 \\ 31.3 \end{bmatrix}$$

   For the last contrast in Exercise 6 what would $\mathbf{H}_{\hat{\psi}}$ equal?

9. Suppose the eigenvalue for the last contrast equals 3.248.
   (a) Compute Wilks Λ.
   (b) Transform Λ to an F statistic.
   (c) State the degrees of freedom for the transformed F statistic.

10. For the last contrast in Exercise 6 provide
   (a) ξ^2
   (b) ξ^2_{adj}

Computer Applications

Exercises 11 to 15 require a continuation of the analyses begun in the Exercises at the end of Chapter 3, using the Ethington 5-group data set (5GED) described in Appendix A. Use the SPSS (or SAS) computer software package to compare the Black, Hispanic, and White student group centroids based on 9 variables (Counselor Interaction, Writing and speaking skills, Self-understanding, Instruction received, Library effort, Student–faculty effort, Interstudent effort, Art/music/theater effort, Writing effort, and Science effort).

11. What are the numerical values of the following:
   (a) η^2_{Mult}
   (b) τ^2
   (c) ξ^2
   (d) ζ^2

12. Using the Serlin procedure, adjust

(a) τ^2

(b) ξ^2

(c) ζ^2

13. Contrast the mean centroids for the Black and Hispanic groups.

(a) What are the values for Λ, U, and V?

(b) State the values of F, degrees of freedom, and P.

(c) Do the results support the conclusion that the population centroids differ?

14. Contrast the mean centroid for the minority students with the mean centroid for the White students.

(a) What are the values for Λ, U, and V?

(b) State the values of F, degrees of freedom, and P.

(c) Do the results support the conclusion that the population centroids differ?

15. For the two contrasts examined in Exercises 13 and 14:

(a) Provide ξ^2 as an index of effect size.

(b) Adjust the ξ^2 using the Serlin procedure.

CHAPTER 5

Describing MANOVA Effects

5.1 INTRODUCTION

Once it has been determined that the resultant MANOVA omnibus or contrast effects are generalizable, we generally are interested in further describing or characterizing the nature of these effects. Reporting test statistic, P, and effect size values is only a start. For example, in analyzing the Baumann data, we found in Chapter 3 that the three group centroids differed beyond what we might reasonably expect due to chance or sampling error [$\Lambda \doteq .841$, $F(4, 124) \doteq 2.804$, $P \doteq .029$]. In assessing the nature of the group differences in Chapter 4, we first characterized the relationship between the grouping variable (Reading Program) and the outcome variables (Y_1, Y_2) to be fairly strong ($\xi^2_{\text{adj}} \doteq .054$), and in an examination of two contrasts (TA vs. DRTA and TA with DRA vs. DRTA) both focused tests had small adjusted P values (.032 and .019, respectively) indicating that the differences between these groups are generalizable to their respective populations with effect sizes of $\xi^2_{\text{adj}} \doteq .092$ and $\xi^2_{\text{adj}} \doteq .113$, respectively.

These analyses have only convinced us that the observed differences are real, but it is not clear up to this point in what way the populations differ. It is on the description of the nature of the differences that we focus in this chapter. To describe the group differences we closely examine linear composites of the outcome variables. These linear composites are useful in identifying outcome variable *constructs* (or latent variables) that underlie the group differences; that is, constructs that underlie the grouping variable effect. The number of constructs that evolve and are useful in defining a variable structure associated with the resultant effect refers to the *structure dimension*. Construct definition and structure dimension constitute the focus of a *descriptive discriminant analysis* (DDA).

One DDA question, then, is: In how many dimensions can this effect be represented? Another question: The variability in what underlying construct(s) is accounted for by the grouping variable (represented by the J levels)? Still another

Applied MANOVA and Discriminant Analysis, Second Edition, by Carl J. Huberty and Stephen Olejnik
Copyright © 2006 John Wiley & Sons, Inc.

question: Given that the J groups are "significantly" separated, what group separation configuration is yielded by the group centroids? Yet another question might be: What configuration of group separation is associated with each underlying construct? If contrasts are examined and found to be "significant," another question might be raised: Are different constructs associated with different contrasts?

Attempts to answer the above questions may be made through the study of some linear composites of the outcome variables. These linear composites are called *linear discriminant functions* (LDFs). LDFs are the focus of this chapter. We begin by providing some background on how to derive the LDFs although we depend on SPSS for the actual calculation. Next, we discuss their interpretation. We conclude this chapter with a discussion on the determination of the number of LDFs to interpret. Because the description of effects are different when examining the omnibus findings than those obtained from contrast analyses, we discuss the LDFs from these two analyses separately.

5.2 OMNIBUS EFFECTS

5.2.1 An Eigenanalysis

You might recall the multiple correlation situation in which there is a single response variable on the one "side" and multiple response variables on the other side. A linear composite of the multiple response variables—label it Z—is determined so that the simple correlation between the single response variable and Z is maximized. That is, a set of weights (b values) for the set of multiple response variables is determined so that the correlation (for the data on hand) is higher than if any other set of b values is used in determining Z.

The idea of determining a set of weights for some response variables so as to maximize a correlation is also basic to DDA. The correlation that is being maximized in DDA, for a one-factor layout with J groups, is that between the grouping variable (represented by a set of $J - 1$ *indicator variables*) on the one hand and a composite, Z, of the outcome variables on the other hand. Maximizing this (canonical) correlation is equivalent to maximizing the ratio of the hypothesis mean square to the error mean square, where the mean squares are found with respect to the composite, or Z, scores. Of course, maximizing this mean-square ratio is equivalent to maximizing the sum-of-squares ratio:

$$\frac{\text{SSH}_Z}{\text{SSE}_Z}. \tag{5.1}$$

So, the task is to find a set of weights (b values) for the outcome variables to determine a linear composite;

$$Z = b_1 Y_1 + b_2 Y_2 + \cdots + b_p Y_p \tag{5.2}$$

so that the ratio in (5.1) is maximized. That is, a set of weights to determine Z is found so that the separation among the groups is maximized (with respect to the Z scores).

[Stated yet another way, a univariate analysis of variance using Z scores determined by the (optimal) set of b values will yield a larger univariate F ratio than if any other set of weights is employed.] This maximization problem is solved, as are many other scientific maximization problems, through the use of partial derivatives. It turns out that to obtain the b values in (5.2), one first finds the largest eigenvalue, λ_1, of the matrix product,

$$\mathbf{E}^{-1}\mathbf{H}.$$

(Conceptually, this product may be viewed as a generalization of SSH/SSE.)

The value of λ_1 is obtained by solving the eigenequation,

$$|\mathbf{E}^{-1}\mathbf{H} - \lambda_1\mathbf{I}| = 0. \tag{5.3}$$

The $p \times p$ matrices, **E** and **H**, are the error and hypothesis **SSCP** matrices, respectively. The identity matrix, **I**, is also $p \times p$. Solving for λ_1 involves finding the largest root of a polynomial of degree $r = \min(p, \mathrm{df}_h)$, where $\mathrm{df}_h = J - 1$. Of course, finding such a root is best done via a computer package such as SPSS. Earlier (in Section 3.5.2), we found the largest eigenvalue, λ_1, to equal .1681 for the Baumann data. After finding λ_1, a $p \times 1$ *eigenvector*, **b**, associated with λ_1 is found by solving the set of p equations,

$$(\mathbf{E}^{-1}\mathbf{H} - \lambda_1\mathbf{I})\mathbf{b} = \mathbf{0}. \tag{5.4}$$

The p elements of **b** are, within a constant of proportionality, the weights of the linear composite in (5.2). These weights, then, define the first (or leading) LDF. [See Tatsuoka (1988, pp. 313–314) for a more formal discussion of the relationship between (5.1) and (5.4).]

Recall that solving Eq. (5.3) involves solving a polynomial equation of degree $r = \min(p, \mathrm{df}_h)$—typically, $r = \mathrm{df}_h$, the *rank* of the matrix **H**. This solution, then, results in r eigenvalues (that are necessarily decreasing in value, i.e., $\lambda_1 > \lambda_2 > \cdots > \lambda_r$) and, therefore, r eigenvectors. Hence, there is a potential for obtaining r LDFs. These functions are mutually uncorrelated; the successive functions are determined so as to maximize relative group separation after preceding functions are "partialed out."

It should be recognized that the analysis in this section is done under the condition of covariance matrix homogeneity. It is only under this condition that finding eigenvalues of $\mathbf{E}^{-1}\mathbf{H}$ makes sense; **E** is the error **SSCP** matrix obtained by summing the separate group $\mathbf{SSCP}_j$ across the groups, which is only sensible when the covariance matrices across groups are "in the same ballpark."

5.2.2 Linear Discriminant Functions

An aspect, then, of DDA involves a set of $r = \min(p, \mathrm{df}_h)$ linear composites of the form (5.2). As indicated earlier, these composites are called linear discriminant functions. Therefore, each analysis unit may be assigned r LDF "scores." Consider for a moment the score on Z_1, the first or leading LDF, for each unit. A question arises: What do these scores represent? That is, these are scores on what? If these are scores on some underlying construct (a latent, unobservable variable), how is this construct defined or described? What is this latent characteristic of our analysis units?

The predominant method of identifying latent constructs in multivariate analyses—this includes factor analysis and canonical correlation—is to examine correlations between linear composite scores and scores on the individual variables in the composite. These LDF–variable correlations are often called *structure* r's. The corresponding term sometimes used by factor analysts is "loadings." (Some methodologists espouse the examination of standardized LDF weights to identify underlying constructs. The weight versus structure r issue is discussed briefly in Section 5.4.)

Conceptually, there are three ways of finding correlations between variables and LDFs. There are total-group structure correlations, between-group structure correlations, and within-group (or error) structure correlations. Which type of correlation is most appropriate in the current context? The total-group correlations ignore intergroup mean differences (which were judged to be real in the first place); R. E. Bargmann (1921–2004) provides an example (in 1969) where total-group correlations are misleading. The between-group correlations would involve means and not individual variable/variate scores and thus would not be useful for construct identification. Although the use of total-group correlations has been advocated (e.g., Cooley and Lohnes, 1985, p. 248), it seems more reasonable to focus on the error structure r's. The latter correlations take into consideration group differences of mean vectors.

So, then, how are these structure r's used to interpret constructs that underlie the resultant group differences? Remember, we are attempting to "get a handle" on what attributes of the analysis units under study are being measured when scores on the LDFs are obtained. The idea behind the use of structure r's is that the variables that share the most variation with a given LDF should define what attribute the LDF represents. For example, consider the following hypothetical sets of structure r's:

	LDF_1	**LDF_2**
Y_1	.72	−.23
Y_2	−.60	.00
Y_3	.19	.49
Y_4	.12	−.42
Y_5	−.03	.12

These results indicate that the scores on LDF_1 are scores on an attribute that is (basically) comprised of whatever is embodied in Y_1 and Y_2 (r's of .72 and −.60, respectively). The descriptive label assigned to LDF_1 is a substantive, rather than a statistical, concern. The label assigned is dependent on how Y_1 and Y_2 are defined and how these two attributes are "put together" in the mind of the researcher—another judgment call in the research process. LDF_2 would be defined, basically, by Y_3 and Y_4 (r's of .49 and $-.42$, respectively).

If a factor analyst were attempting to "interpret" the two LDFs above, he or she may want to rotate them first. The idea of factor rotation—however it is accomplished—is to facilitate factor interpretation. The rotation idea may also be applied in DDA—see Bray and Maxwell (1982, p. 345) for references. The lone computer package that may be used to rotate LDFs is SPSS. If the LDFs are rotated and identified via structure r's—or via standardized LDF weights, for that matter—a question arises: Are we interpreting the same outcome variable composites that were determined so as to maximize group differences (even if the rotations are "rigid")? Because of the negative answer to this question, many methodologists are somewhat skeptical about the practice of interpreting rotated LDFs as constructs or latent variables associated with grouping variable effects.

Because three instructional groups were considered in the Baumann study ($df_h = 2$), two LDFs are determined. The eigenvalues were computed earlier as .1681 and .0220 and the canonical correlation $\sqrt{\lambda_1/(1+\lambda_1)}$ for the first LDF equals .379, and the second canonical correlation equals .147. Because the second squared canonical correlation, .022, indicates a minimal amount of shared variation between the grouping variable and the second LDF, only the first LDF is likely to be meaningful. Alternative approaches for determining the number of meaningful dimensions present in the outcome variables studied is discussed in Section 5.5.

5.3 COMPUTER APPLICATION I

In this section we provide the SPSS syntax to obtain eigenvalues, the raw and the standardized discriminant function weights, and the structure r's. As discussed above, examining the eigenvalues and the squared canonical correlations can be helpful in determining how many functions are likely to be meaningful. The raw discriminant function weights are useful to determine group centroids in LDF space. Finally, standardized discriminant function weights and structure r's are useful for defining the identified constructs.

SPSS SYNTAX FOR COMPUTING EIGENVALUES AND STRUCTURE *r*'s

```
manova Y1 Y2 by treatmnt(1, 3)
/print=signif(eigen)
/discrim=raw stan cor alpha(1).
```

signif(eigen) requests the eigenvalues for the $\mathbf{E}^{-1}\mathbf{H}$ matrix be reported.
/discrim is the SPSS system command requesting the discriminant functions.

raw requests the raw discriminant function weights be reported.
stan requests the standardized discrimninant function weights be reported.
cor requests the structure r's to be reported.
alpha(1) requests that all (both statistically "significant" and "nonsignificant") functions be reported. If alpha is not included, only statistically significant, at the .15 level, functions are reported.

OUTPUT

Analysis: Eigenvalues and Canonical Correlations

EFFECT . . TREATMNT
Multivariate Tests of Significance (S = 2, M = −1/2, N = 30)

Test Name	Value	Approx. F	Hypoth. DF	Error DF	Sig. of F
Pillais	.16185	2.77356	4.00	126.00	.030
Hotellings	.18583	2.83393	4.00	122.00	.027
Wilks	.84094	2.80488	4.00	124.00	.029
Roys	.14225				

Note . . F statistic for WILKS' Lambda is exact.

Eigenvalues and Canonical Correlations

Root No.	Eigenvalue	Pct.	Cum. Pct.	Canon Cor.
1	.166	89.244	89.244	.377
2	.020	10.756	100.000	.140

Raw discriminant function coefficients
Function No.

Variable	1	2
Y1	−.302	.220
Y2	.137	.067

Standardized discriminant function coefficients
Function No.

Variable	1	2
Y1	−.913	.666
Y2	1.016	.497

Correlations between DEPENDENT and canonical variables
Canonical Variable

Variable	1	2
Y1	−.439	.898
Y2	.589	.808

Interpretation: Eigenvalues, Canonical Correlations, and Structure Correlations

The eigenvalues reported by SPSS are slightly smaller than those we calculated in Chapter 3, but the difference is very small. The output includes a column labeled "Pct." and "Cum. Pct."; we will discuss these results in Section 5.5.2. The last column presents the canonical correlations, each of which is an index of the relationship between the grouping variable and each construct identified by our analysis. A squared canonical correlation indicates the proportion of variation in the construct that is explained by the grouping variable. For example, for the first construct the canonical correlation is .377 and the squared canonical correlation is .142. These results indicate that 14.2 percent of the variation in the first construct is explained by the grouping variable.

The raw and standardized discriminant function coefficients (i.e., weights) are reported next. The raw discriminant function coefficients are useful for computing mean composite scores for each group, and we will discuss these composite scores to describe group differences in Section 5.5.3. The standardized discriminant function weights are sometimes used for identifying the constructs that separate the groups. We discuss these weights in Section 5.4.

The error structure r's are reported in the output section labeled: "Correlations between DEPENDENT and Canonical Variables." The structure r's for the two LDF's are:

	$\mathbf{LDF_1}$	$\mathbf{LDF_2}$
EDT	−.439	.898
DRP	.589	.808

The absolute values of the reported correlations between each of the outcome variables and the first LDF are similar. Because both variables are "moderately" correlated with the first LDF, it might be concluded that both variables contribute equally to the definition of the first underlying construct. The two variables, EDT and DRP, are two different indicators of reading comprehension, so we might label the first composite as "reading comprehension." The structure r's for the second LDF are also similar, and, therefore, again both variables contribute to definition of the second construct. The number of constructs that we judge to be meaningful is discussed in Section 5.5.

5.4 STANDARDIZED LDF WEIGHTS

When researchers report results of a multiple correlation analysis, it is fairly common to find weights reported for the set of multiple response variables. These weights reported may be either of two types: (1) those applicable to "raw" response variable scores and (2) those applicable to standardized response variable scores (usually z scores).

Similarly, in reporting results of a DDA, researchers sometimes report LDF weights applicable to raw outcome variable (or Y) scores [see (5.2)] and sometimes "standardized" LDF weights. In general, the standardized weight for variable i, b_i^*, is calculated from the raw weight—SPSS seems to reverse the order of calculation—using

$$b_i^* = b_i \sqrt{d_{ii}}. \tag{5.5}$$

The radicand in (5.5) is a main diagonal element of a covariance matrix, that is, a variance. Two matrices have been considered as a source of the d_{ii} term, a total-group or an error covariance matrix. The two alternative approaches are discussed by Mueller and Cozad (1988, 1993) and Nordlund and Nagel (1991). For reasons given below, issues of computation and use of such LDF weights will not be discussed here.

Why do empirical researchers concern themselves with the standardized LDF weights? There is an apparent fairly common belief that "standardized weights" are useful for an interpretation purpose. [The reason for the quotes is that there are two approaches to standardization: (1) determining weights for Y variables that have zero means and unit standard deviations, and (2) determining weights so that the composite has zero mean and unit variance (actually, a normalization).] This purpose pertains to using these weights to assess the relative "importance" of response variables. The tradition of using standardized weights in a multiple correlation context to order response variables has probably been carried over to a DDA context for the purpose of using such LDF weights again for variable ordering purposes. As discussed by Huberty and Wisenbaker (1992b), such an approach to ordering variables in a DDA, and in a multiple correlation analysis, is questionable (see, also, Joy and Tollefson, 1975, p. 729). [Rencher (2002, pp. 282–284), however, disagrees and advocates the use of standardized LDF weights to order variables.] It was mentioned above that some methodologists/statisticians (e.g., Rencher, 2002, p. 289) favor standardized LDF weights over structure r's for the purpose of LDF identification. Because the use of standardized LDF weights is not herein recommended for variable ordering, for construct definition, or for variable selection, standardized LDF weights are not discussed further.

5.5 LDF SPACE DIMENSION

Given that $r = \min(p, \mathrm{df}_h)$ LDFs may be extracted, how many should be considered in the interpretation of resultant group differences? The conclusion of generalizable group differences implies that at least one LDF should be considered. In Section 5.2.2 we suggested that the squared canonical correlation associated with each eigenvalue might be used to identify the meaningfulness of LDFs. In this section we discuss three additional methods that one may consider in determining the final number of LDFs to retain for interpretation purposes, that is, in determining the dimensionality of the LDF space: statistical tests, proportion of variance, and LDF plots.

5.5.1 Statistical Tests

The testing may be described as follows. Recall from Section 3.5.2 that one MANOVA test statistic is based on the Wilks Λ and can be computed as $\prod_{v=1}^{r}[1/(1+\lambda_v)]$. This statistic may be used to test H_0: $\boldsymbol{\mu}_1 = \boldsymbol{\mu}_2 = \cdots = \boldsymbol{\mu}_J$ or, equivalently, H_0: all r population counterparts to eigenvalues λ_v are equal to zero. The alternative hypothesis may be stated as H_a: at least one dimension is needed to interpret group differences.

Because Λ can be expressed as $\prod_{v=1}^{r}[1/(1+\lambda_v)]$, H_0 can be tested using the transformation of Λ to an F statistic:

$$F = \frac{1-\Lambda^{1/s}}{\Lambda^{1/s}} \frac{m(s) - p(\mathrm{df}_h)/2 + 1}{p(\mathrm{df}_h)},$$

where

$$m = \mathrm{df}_e - \frac{p - \mathrm{df}_h + 1}{2},$$

and

$$s = \sqrt{\frac{p^2(\mathrm{df}_h^2) - 4}{p^2 + \mathrm{df}_h^2 - 5}}.$$

This statistic has an approximate F distribution with degrees of freedom $\nu_1 = p(\mathrm{df}_h)$ and $\nu_2 = m(s) - p\,(\mathrm{df}_h)/2 + 1$. Rejection of H_0 tells us that at least one dimension is needed to interpret group differences. Failure to reject this null hypothesis tells us that there is insufficient evidence that the groups are separate. Now, to determine if more dimensions are statistically involved in the group differences, the Λ statistic is (repeatedly) partitioned. To start the partitioning, let

$$\Lambda_1 = \prod_{v=2}^{r}\left(\frac{1}{1+\lambda_v}\right).$$

This statistic can be transformed to have an F distribution with $\nu_1 = (p-1)(\mathrm{df}_h - 1)$ and $\nu_2 = m(s) - (p-1)(\mathrm{df}_h - 1)/2 + 1$ degrees of freedom, and may be used to test H_{01}: at most one dimension is needed to interpret group differences versus H_{a1}: at least two dimensions are needed to interpret group differences. The values of p and df_h used to obtain the value of s are sequentially decreased by 1. Proceeding, let

$$\Lambda_2 = \prod_{v=3}^{r}\left(\frac{1}{1+\lambda_v}\right),$$

which can be transformed to a statistic having an F distribution with $\nu_1 = (p-2)(\mathrm{df}_h - 2)$ and $\nu_2 = ms - (p-2)(\mathrm{df}_h - 2)/2 + 1$ degrees of freedom. This statistic may be used to test H_{02}: at most two dimensions are needed to interpret group differences versus H_{a2}: at least three dimensions are needed to interpret group differences. If necessary, Λ can be further portioned following the same procedure.

This sequential test procedure, then, may be employed to statistically determine the desired dimensionality of the LDF space for purposes of interpreting the resultant MANOVA effects. Each succeeding statistic is used to test the significance of residual effects after removing the effects of the preceding dimensions. Collectively, the procedure should be considered a *test of dimensionality*.

Applying this procedure using the Baumann data with the computed generated $\lambda_1 \doteq .166$ and $\lambda_2 \doteq .020$,

$$\Lambda \doteq \frac{1}{1+.166}\frac{1}{1+.020} \doteq .841.$$

So, $F \doteq 2.804$ with $\nu_1 = 4$ and $\nu_2 = 126$, $P \doteq .029$. These results indicate that the populations differ on at least one dimension. Partitioning Λ to determine whether at least two dimensions are needed to describe group differences

$$\Lambda_1 = \prod_{v=2}^{r}\left(\frac{1}{1+\lambda_v}\right) \doteq \frac{1}{1+.020} \doteq .980.$$

Transforming Λ_1 with $s = 1$ and $m = 62.5$,

$$F \doteq \frac{1-.980}{.980}\frac{62.5-(2-1)(2-1)/2+1}{(2-1)(2-1)} \doteq 1.286.$$

With degrees of freedom $\nu_1 = 1$ and $\nu_2 = 63$, $P \doteq .261$. These results indicate that one dimension is sufficient to describe group differences.

A summary of the sequential test procedure for a five-group design is given in Table 5.1. The purpose of the testing sequence above is to determine the number

TABLE 5.1 Summary of Dimensionality Tests[a]

Null Hypothesis	Test Statistic	F df Values: ν_1	ν_2
No separation on any dimension	$\Lambda = \prod_{v=1}^{r} \frac{1}{1+\lambda_v}$	pq	$m(s) - pq/2 + 1$
Separation on at most one dimension	$\Lambda_1 = \prod_{v=2}^{r} \frac{1}{1+\lambda_v}$	$(p-1)(q-1)$	$m(s) - (p-1)(q-1)/2 + 1$
Separation on at most two dimensions	$\Lambda_2 = \prod_{v=3}^{r} \frac{1}{1+\lambda_v}$	$(p-2)(q-2)$	$m(s) - (p-2)(q-2)/2 + 1$
Separation on at most three dimensions	$\Lambda_3 = \prod_{v=4}^{r} \frac{1}{1+\lambda_v}$	$(p-3)(q-3)$	$m(s) - (p-3)(q-3)/2 + 1$

[a]The values for p and q used to obtain the value of s in ν_2 are sequentially decreased by 1 just as in λ_1; for the first test, $s = \sqrt{(p^2q^2-4)/(p^2+q^2-5)}$.

of LDFs to retain for interpretation in explaining group differences assessed via MANOVA. The Λ_v statistics in Table 5.1 *cannot* be used to test the significance of the individual eigenvalues or of individual LDFs. That is, results of the testing sequence above cannot be used to conclude that the "vth LDF is significant." A reason for the difficulty of testing individual eigenvalues is the possible mismatch between population and sample eigenvalues. As Harris (2001, p. 235) states it: "... there is no way of knowing which of the r population roots has generated the vth largest sample value of λ_v."

5.5.2 Proportion of Variance

A second way of considering the dimensionality problem that may appeal to some researchers is a proportion-of-variance approach. As discussed by Tatsuoka (1988, p. 213) the vth eigenvalue of $\mathbf{E}^{-1}\mathbf{H}$ reflects the ratio of between-group to within-group variability with respect to scores on the vth LDF. That is, λ_v reflects a proportion of variation in a p-variable system accounted for by the vth LDF. Thus, to determine the desired dimensionality, one might consider, in turn,

$$\frac{\lambda_1}{\sum_{v=1}^{r} \lambda_v},$$

$$\frac{\lambda_1 + \lambda_2}{\sum_{v=1}^{r} \lambda_v},$$

$$\frac{\lambda_1 + \lambda_2 + \lambda_3}{\sum_{v=1}^{r} \lambda_v},$$

and so on, until a "substantial" proportion of variance is accumulated.

For the Baumann data, the two eigenvalues were computed as $\lambda_1 = .166$ and $\lambda_2 = .020$. The sum of the eigenvalues is .186. The first LDF, therefore, reflects $.166/.186 \doteq 89.2$ percent of the variance in the two-variable system. The second LDF reflects only 10.7 percent of the variance in the two variable system. (The percent of variation in the p-variable system reflected in each λ_v is reported in SPSS, under the column labeled "Pct" and the cumulative percent under the column labeled "Cum Pct," as part of the output when eigenvalues are requested—see Section 5.3.) It may be judged that .893 is "large" enough and, therefore, one dimension would suffice. It is, as often is the case, a judgment call.

5.5.3 LDF Plots

Third, and finally, one could examine a plot of the group centroids in the space of the LDFs to determine the number of LDFs to retain for interpretation. Let us assume, for the moment at least, in a J-group one-factor design that there is some interest in assessing and describing the omnibus effects (i.e., overall group differences). Suppose further that using either of the foregoing two approaches to dimensionality determination, it is reasonable to retain two LDFs for description purposes. That being the case, it might be instructive to view a plot of the LDF mean vectors for each group. Let us take a sidestep for a moment. An LDF is a linear composite as indicated in (5.2) with the weights being elements of an eigenvector. There are $r = \min(p, \mathrm{df}_h)$ LDFs; that is, there are r linear composites. To determine the mean on the vth LDF for Group j, one can merely substitute the p-variable means for Group j into the v composite. This process will yield r LDF mean vectors for each group; that is, for each group you have a centroid comprised of r LDF means. The raw discriminant function weights can be obtained from SPSS by requesting raw on the **/discrim** line [see Section 5.3 **/discrim = raw alpha(1)**]. For the Baumann data these weights are reported as:

	LDF_1	LDF_2
EDT	−302	.220
DRP	.137	.067

Group means for the three reading programs were reported in Table 3.2:

	TA	DRA	DRTA
EDT	7.77	6.68	6.23
DRP	43.45	42.05	46.64

So, the group mean composites for the two LDFs are:

	TA	DRA	DRTA
LDF_1	3.60	3.74	5.41
LDF_2	4.62	4.29	4.49

Typically, a rectangular coordinate system—where the two LDF axes intersect at right angles—is used in plotting the LDF centroids. This is a convenient system and one that generally presents a reasonably correct configuration of the groups in the space of the two LDFs. A plot of the group centroids for the Baumann data is presented in Figure 5.1.

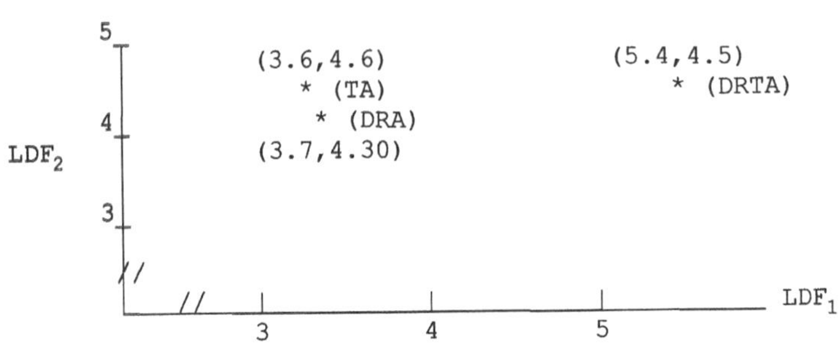

Figure 5.1 LDF plot of group centroids for Baumann study.

By projecting the centroid points onto the two respective axes, one gets a general idea of group separation that may be attributed to each LDF. From Figure 5.1 it looks like LDF_1 may account for separation of TA and DRA on one hand versus DRTA on the other. From Figure 5.1, too, it might be concluded that LDF_2 is not related to any specific separation of the three reading groups. This interpretation is consistent with our previous interpretations based on the squared canonical correlations, the statistical tests, and the proportion of shared variance. Such consistency in results may not always be present. Of course, we must be careful of too much reliance on visual inspection because of human differences in visual perception, and because perceptions may depend on the numerical scales used for the LDF axes.

From Figure 5.1, it is reasonable to conclude that separation of DRTA versus TA and DRA may be attributed to LDF_1. The next obvious question is: What does LDF_1 represent? This is the structure identification question, the question addressed in Section 5.2.2.

[A graphical representation of the J groups of units in the resultant LDF space may be obtained using the S-PLUS function CLUSPLOT—see Pison et al. (1999).]

5.6 COMPUTER APPLICATION II

In this section we provide the SPSS syntax for obtaining the tests of dimensionality. These tests are labeled by SPSS as Dimension Reduction Analysis.

SPSS SYNTAX FOR DIMENSIONALITY TESTS

```
manova Y1 Y2 by treatmnt(1,3)
/print signif(dimenr).
```

/print signif(dimenr) requests the tests of dimensionality be reported.

OUTPUT

Analysis: Dimensionality Tests

EFFECT .. TREATMNT

Multivariate Tests of Significance (S = 2, M = −1/2, N = 30)

Test Name	Value	Approx. F	Hypoth. DF	Error DF	Sig. of F
Pillais	.16185	2.77356	4.00	126.00	.030
Hotellings	.18583	2.83393	4.00	122.00	.027
Wilks	.84094	2.80488	4.00	124.00	.029
Roys	.14225				

Note.. F statistic for WILKS' Lambda is exact.

------------ Dimension Reduction Analysis

Roots	Wilks L.	F Hypoth.	DF	Error DF	Sig. of F
1 TO 2	.84094	2.80488	4.00	124.00	.029
2 TO 2	.98040	1.25922	1.00	63.00	.266

Interpretation: Dimensionality Tests

The first test in the dimension reduction analysis is identical to the omnibus MANOVA F test comparing the group centroids, $\Lambda \doteq .841$, $F(4, 124) \doteq 2.805$, $P \doteq .029$. If the omnibus test indicates that the population centroids differ, then the populations differ on at least one dimension. The second test, $\Lambda_1 \doteq .980$, $F(1, 63) \doteq 1.259$, $P \doteq .266$, provides little evidence to indicate that the populations differ on two dimensions. The results of these analyses indicate that only the first LDF needs to be interpreted to explain the group differences. (It should be noted that testing the first null hypothesis in Table 5.1 is equivalent to testing the null hypothesis, H_0: $\boldsymbol{\mu}_1 = \boldsymbol{\mu}_2 = \cdots = \boldsymbol{\mu}_J$. This is evidenced by the Wilks Λ value of 2.80488 reported twice in the above output.)

5.7 COMPUTER APPLICATION III

In this section we provide the SPSS syntax to obtain a plot of the mean group centroids in LDF space. This program, however, is of limited usefulness for obtaining LDF plots if only a single grouping variable can be identified. Still, the program does report the group centroids in LDF space along with the plot, and they can be helpful when interpreting the dimensionality of the variable space.

SPSS SYNTAX TO OBTAIN AN LDF PLOT

```
discriminant groups=treatmnt(1,3)
/variables Y1 Y2
/plot=combined.
```

discriminant groups = treatmnt(1.3) is the SPSS system command to obtain discriminant functions and a plot of the group centroids.

/variables Y1 Y2 identifies the outcome variables.

/plot=combined requests that all group centroids be presented on a single plot.

OUTPUT

Analysis: LDF Plot

	Function	
Treatment	1	2
TA	−.346	.154
DRA	−.211	−.181
DRTA	.557	.027

Unstandardized canonical discriminant functions evaluated at group means

Interpretation: LDF Plot

In addition to providing the eigenvalues, dimension reduction analysis, and the structure *r*'s, the DISCRIMINANT program provides the LDF group centroids. These centroids are obtained by multiplying the group means by the raw discriminant function weights and adding the constant, b_0 (see Technical Note). The interpretation

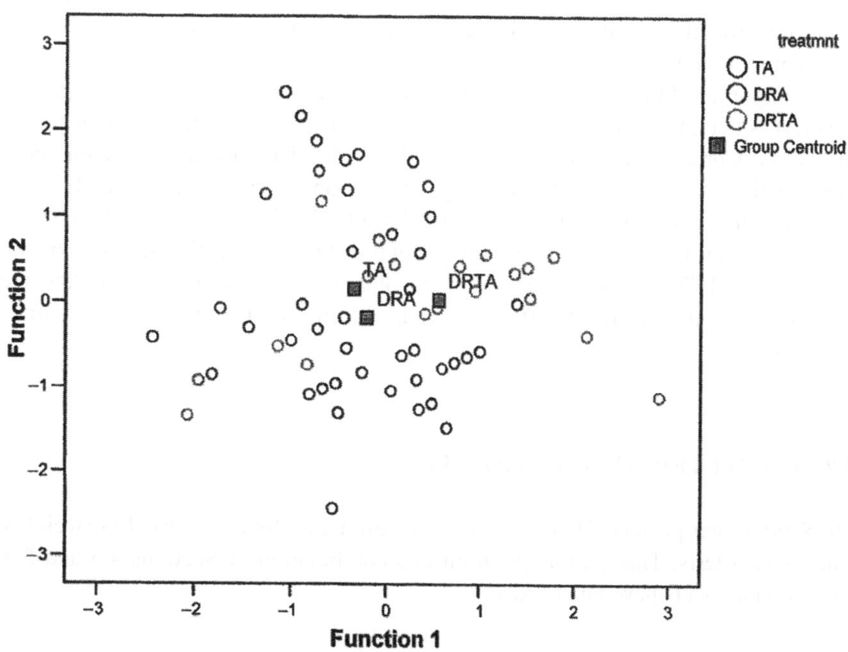

Figure 5.2 Plot of group centroids in LDF space.

would be the same as presented earlier, the DRTA group differs from the TA and DRA groups with respect to the first function.

The program also generates a plot of the group centroids. The computer-generated plot (Fig. 5.2) differs from Figure 5.1 only in terms of the scaling of the axes. Regardless of the scaling, the interpretation remains the same. Only LDF_1 provides clear separation among the groups, with the DRTA group being noticeably distinct from the TA and DRA groups.

5.8 CONTRAST EFFECTS

In Section 5.2, procedures for describing group differences via the omnibus hypothesis test were discussed and demonstrated. The omnibus test examines all group differences simultaneously and determines the appropriate LDFs that maximize the separation of all groups considered. Consequently, the structure r's used to define the constructs on which all groups differ are relevant to the situation where all groups are considered as a set. Contrasts, or focused tests, address different questions and examine separation between pairs of groups or between different subsets of groups (i.e., complex contrasts). For example, in Chapter 4 the difference between the TA and DRTA instructional methods was examined. Maximizing the difference between just two groups is considerably different than maximizing differences among all three instructional programs. Because the questions addressed in the omnibus test and the focused tests are different, the LDFs and structure r's obtained for a focused test will differ from those computed for the omnibus test, and from those computed for a different focused test.

An important difference in an eigenanalysis for an omnibus test and for a contrast test is that all contrasts have a single degree of freedom for the hypothesis test ($df_h = 1$). Consequently, only a single LDF can be determined for any one contrast. Statistical tests for dimensionality, proportion of variance, and LDF plots, discussed in Sections 5.2 and 5.5, are not relevant. The discriminant function structure r's, however, are useful to describe group differences. In the next section we present the SPSS computer application to obtain the structure r's to define the construct that appears to underlie the group differences reflected in the contrast statement.

5.9 COMPUTER APPLICATION IV

The SPSS syntax presented here combines the commands for effect size, LDF weights, and focused tests. That is, this program is a combination of Sections 4.5 and 5.3. No new statements have been added.

SPSS SYNTAX FOR CONTRAST RAW DISCRIMINANT FUNCTION WEIGHTS AND STRUCTURE *R*'s

```
manova Y1 Y2 by treatmnt(1, 3)
/print signif(efsize)
/discrim=raw cor alpha(1)
/contrast(treatmnt)=special(1 1 1, .5 .5 −1, 1 0 −1)
/design=treatmnt(1) treatmnt(2).
```

OUTPUT

Analysis: Contrast Raw Discriminant Function Weights and Structure r's

EFFECT . . TREATMNT(2)
Multivariate Tests of Significance (S = 1, M = 0, N = 30)

Test Name	Value	Exact F	Hypoth. DF	Error DF	Sig. of F
Pillais	.12694	4.50712	2.00	62.00	.015
Hotellings	.14539	4.50712	2.00	62.00	.015
Wilks	.87306	4.50712	2.00	62.00	.015
Roys	.12694				

Note . . F statistics are exact.

- -

Multivariate Effect Size

TEST NAME	Effect Size
(All)	.127

- -

Raw discriminant function coefficients

	Function No.
Variable	1
Y1	−.330
Y2	.127

Correlations between DEPENDENT and canonical variables

	Canonical Variable
Variable	1
Y1	−.560
Y2	.472

- -

EFFECT . . TREATMNT(1)
Multivariate Tests of Significance (S = 1, M = 0, N = 30)

Test Name	Value	Exact F	Hypoth. DF	Error DF	Sig. of F
Pillais	.14016	5.05339	2.00	62.00	.009
Hotellings	.16301	5.05339	2.00	62.00	.009

```
Wilks           .85984       5.05339        2.00         62.00        .009
Roys            .14016
Note . . F statistics are exact.
- - - - - - - - - - - - - - - - - - - - - - - - - - - - - - - - - - - - - -
Multivariate Effect Size
TEST NAME Effect Size
 (All)                .140
- - - - - - - - - - - - - - - - - - - - - - - - - - - - - - - - - - - - - -
 Raw discriminant function coefficients
             Function No.
Variable                1
Y1                  -.291
Y2                   .140
Correlations between DEPENDENT and canonical variables
             Canonical Variable
Variable                1
Y1                  -.395
Y2                   .628
- - - - - - - - - - - - - - - - - - - - - - - - - - - - - - - - - - - - - -
```

Interpretation: Contrast Structure r's

As discussed in Section 5.2.2, structure r's are correlations between each outcome variable and an LDF. Variables having a high correlation with a composite score are used to define a construct underlying the p-variable system. Because only one LDF is determined for each contrast, only one construct can be defined for each contrast.

For the pairwise contrast between the TA and DRTA reading approaches, the correlations between the EDT and DRP variables and the LDF are −.560 and .472, respectively. Because the absolute values of the two correlations are judged to be similar in magnitude, the results might be interpreted as indicating that both variables contribute equally to the underlining construct that separates the two groups.

For the complex comparison of the average of the TA and DRA reading approaches with the DRTA approach, the correlations are −.395 and .628 for the EDT and DRP, respectively. Because the absolute values of the correlation are judged to differ noticeably, these results indicate that the construct being assessed is primarily determined by DRP.

Comparing the structure r's obtained following the omnibus test (−.439 and .589) with those obtained from the pairwise contrast (−.560 and .472) and the complex contrast (−.395 and .628) demonstrate that the constructs that define group differences can be different depending on the research question addressed.

5.10 SUMMARY

In this chapter we introduced the use of linear discriminant functions (LDFs) as a means of describing the nature of group differences. This use of LDFs is basic to descriptive discriminant analysis (DDA). What one is doing is searching for hypothetical constructs or latent variables, determined by the data on hand, that underlie the grouping variable effects that were already judged to be generalizable. A search for data-based constructs is a common focus for some other multivariate analysis methods (e.g., common factor analysis, canonical correlation analysis, and multidimensional scaling). The interpretation of resultant LDFs, which provide potential explanations, if you will, of the grouping variable effects, is a substantive concern rather than a statistical concern. Also, such an interpretation requires the use of judgment and common sense on the part of the researcher. Explicit and implicit reference to the considerable interpretation potential associated with LDFs has been made in this chapter. Both the derivation and interpretation of LDFs—or canonical variates—are dependent on the tenability of the homogeneity of the group covariance matrices. Discussion of the study of canonical variates under heteroscedasticity has been virtually ignored by methodologists. Campbell (1984) and Krzanowski (1990) have, however, generalized the determination of canonical variates to the heteroscedastic case (but these statisticians do not discuss the "interpretation" of the variates).

There is a tie-in between DDA and a multivariate analysis technique other than MANOVA that may be helpful. This other technique is cluster analysis. When a cluster analysis is completed with a decided-upon number of clusters, it may be helpful to describe the cluster structure in some way. One might use DDA techniques to identify dimensions or constructs that distinguish the clusters; that is, to describe and interpret the cluster solution. Examples where this has been attempted are rare, perhaps because of how cluster analysts (e.g., Blashfield and Aldenderfer, 1988) admonish the use of MANOVA in validating cluster analysis (CA) results. Attempts at such a DDA-CA connection are given by Aversa (1985), Hornsby-Smith et al. (1987), and Huberty et al. (2005).

There may be some multiple-outcome variable group comparison research situations when outcome variable structure would be of limited interest. An example of such a situation may present itself when a researcher is interested in a small number of outcome variables (say, four or fewer). It may be that the outcome variables are from substantively different domains. A context of such a situation is the evaluation of some "treatments" in a factory, school, or laboratory. Interest may basically be on group differences with respect to each outcome variable; this would call for multiple univariate analyses—ANOVAs or contrast analyses. (Multiple testing should not be ignored.) In such a situation, relative outcome variable importance may also be of interest. If so, this is when a multivariate analysis would be imperative.

Technical Note

Computation of LDF weights is accomplished via an eigenanalysis (see Section 5.2.1). The SPSS computer package includes in its LDFs a "constant." That is, rather than the Z in (5.2), the SPSS algorithm yields

$$Z_c = b_0 + b_1Y_1 + b_2Y_2 + \cdots + b_pY_p$$

The b_0 value is a constant determined so that the means Z_c is zero across all N units:

$$b_0 = -\sum_i b_i \overline{Y}_i,$$

where $\overline{Y}_i$ is the mean on variable Y_i across all N units.

Further Reading

Campbell (1980) discusses the idea and use of ridge-based estimators for LDF weights so as to reduce the bias of such estimators.

Hwang (1994) simulates a comparison of a number of methods of determining the number of LDFs to retain for interpretation purposes; normal and elliptical distributions were considered along with varied group sizes and eigenvalues.

Overall and Klett (1972, pp. 292–295) propose the representation of "measurement vectors" in the discriminant space for the purpose of providing information concerning how the groups differ with respect to the original p outcome variables. Such a representation purportedly may be useful in assessing the relative importance of each outcome variable for group separation.

Seo et al. (1995) discuss the effects of nonnormality on dimensionality tests in a DDA context.

von Eye (2002, pp. 342–346) briefly mentions (descriptive) discriminant analysis in the context of configural frequency analysis (CFA).

Definition ***Discriminant function***: Opposite of datcriminant function.

EXERCISES

1. The context provided for Exercises 1 to 10 in Chapter 4 reported three eigenvalues of 4.764, 1.237, and .044. Given these results, what are the values of the three squared canonical correlations (see Section 5.2.2) between the grouping variable and each of the three constructs identified for the data set.

2. Using the eigenvalues in Exercise 1, what proportion of variation in the four variable system is shared with each construct identified? Based on these results, how many "meaningful" constructs appear to be present in this variable system?

3. Partition Wilks Λ to statistically determine the number of dimensions that separate the four groups. Assume that any computed F statistic less than 1.00 is not statistically significant.

4. What is a linear discriminant function (LDF)?

5. What role do linear discriminant functions play in "interpreting" results of a MANOVA?

6. Assume the following are the structure r's for the discriminant function analysis. Use these results to label the constructs that separate the four groups.

	Function Number		
Variable	1	2	3
Effort	−.501	.167	−.492
Ability	−.239	.043	−.405
Luck	−.140	.818	−.482
Teachers	.719	.029	−.568

7. Assume the following are the raw discriminant function weights. Use these results along with the group raw mean centroids provided in Exercise 8 in Chapter 4 to compute the group mean centroids in LDF space.

	Function Number		
Variable	1	2	3
Effort	−.187	.096	.146
Ability	−.044	.035	.163
Luck	.027	.249	−.038
Teachers	.227	.139	.159

8. Compare the group centroids in the LDF space and describe the group differences with respect to the meaningful constructs identified in Exercises 2 and 3.

9. The eigenvalue for the contrast between Mathematics and Business majors equals 1.257. What proportion of variation in the variable system is explained by this contrast?

10. Given the eigenvalue in Exercise 9, compute:
   (a) Wilks Λ
   (b) τ^2

11. For the contrast between Mathematics and Business majors, the structure r's are −.242, −.077, −.836, and .095 for Effort, Ability, Luck, and Teachers, respectively. How would you label the construct that separates the Mathematics and Business majors?

Computer Applications

Exercises 12 to 18 require a continuation of the analyses begun in the Exercises at the end of Chapters 3 and 4, using the 5-group Ethington data set (5GED) described in Appendix A. Use the SPSS (or SAS) computer software package to conduct the analyses necessary to complete these Exercises focusing on 9 outcome variables (Counselor Interaction, Writing and Speaking Skills, Self-Understanding, Instruction Received, Library Effort, Student–Faculty Effort, Interstudent Effort, Art/Music/Theater Effort, Writing Effort, and Science Effort) for Black, Hispanic, and White students.

12. What proportion of variation in the variable system is explained by each of the constructs identified?

13. Based on the results of the Dimension Reduction Analysis, how many LDFs are to be considered in the three-group separation?

14. What proportion of variation in each construct identified is explained by the Race variable?

15. Using the structure r's, identify the variables you believe define the constructs identified.

16. What are the LDF group centroids?

17. For the contrast comparing Black and Hispanic students what variable(s) appear to define group separation based on the structure r's.

18. Using structure r's, what variables appear to define the construct separating the mean centroids of Black and White students from the mean centroid for Hispanic students?

CHAPTER 6

Deleting and Ordering Variables

6.1 INTRODUCTION

Applied researchers appear to have considerable interest in determining a subset of the original set of outcome variables and in assessing relative variable "importance"; this interest is perhaps due to accessibility of discriminant analysis computer programs in the two popular packages. That deleting and ordering variables "appear" to be of interest may be evidenced by analysis interpretation statements in journal articles in which descriptive discriminant analysis (DDA) techniques are used. Very often researchers seem to want to eliminate or discard "worthless" variables; and they often want to identify the "most important" variable(s) and the "least important" variables(s).

An analysis approach very often used for the variable deletion and variable ordering problems is a *stepwise discriminant analysis*. The same general analysis strategy is often used in a multiple regression/correlation situation to delete and/or order variables. It has been argued elsewhere (e.g., Huberty, 1989) that stepwise methods should *not* generally be used for variable deletion and variable ordering purposes. Therefore, a stepwise discriminant analysis will not be advocated in this book. (Partial stepwise output will, however, be used, as discussed in Section 6.3.)

Methods for deleting subsets of outcome variables—assuming that it is desirable to delete some variables—and for determining relative variable importance are suggested in this chapter. Suggestions are made with respect to contrast effects as well as omnibus effects.

6.2 VARIABLE DELETION

6.2.1 Purposes of Deletion

There is a theoretical reason for retaining only "worthwhile" variables or deleting "worthless" variables in a DDA context. This reason pertains to parameter estimation

Applied MANOVA and Discriminant Analysis, Second Edition, by Carl J. Huberty and Stephen Olejnik
Copyright © 2006 John Wiley & Sons, Inc.

(e.g., LDF weights, structure r's). For a fixed total sample size, fewer outcome variables will lead to more precise estimates. Suppose that a system of outcome variables is initially chosen but upon data collection it is concluded that a limited total sample size is feasible. It may then be desirable to delete some variables; in doing so it must be recognized that the variable system initially chosen is no longer intact. There may also be what is termed a practical reason for deleting some variables in a MANOVA/DDA context. This reason pertains to parsimony; that is, economy or simplicity of description. Fewer outcome variables may make explanations and interpretations substantively more simple.

If one were to scan the many journal articles in which results of a "discriminant analysis" are reported, very often some attempt at variable deletion will be reported. Again, it is surmised that the predominant reason for finding such results is the availability of a stepwise discriminant analysis in the popular computer packages. But should variable deletion be "automatic" in a MANOVA/DDA context? No! In PDA (see Section 17.2), fewer variables can yield greater classification accuracy, whereas in DDA, fewer variables cannot yield greater separation—as assessed by, say, the Wilks Λ criterion. For a given design, separation generally decreases (i.e., the Wilks Λ increases) as the number of variables decreases.[1] True, parsimony in explanation may be a reasonable goal. (Note the use of "may" here and in the preceding paragraph.) It was noted in Section 1.5 that emphasis should be placed on thoroughness and thoughtfulness in the initial choice of response variables for a multivariate study. Careful consideration should be given to choosing individual variables and variable domains that are particularly relevant to the study to be conducted. If it is also the case that measures on the chosen variables are substantively appropriate, why might there be subsequent interest in data-based variable deletion? It is recognized that regardless of the care taken in the initial variable choice, some relatively worthless variables may be chosen for inclusion. Thus, it may be desirable to determine whether the data suggest that some variables may be deleted without substantial decrease in group separation.

6.2.2 McCabe Analysis

If it is desirable to determine whether some variables may be deleted on the basis of the data on hand, how might the deletion be accomplished? The analysis approach favored is an all-possible variable subset approach. Such an analysis may be accomplished via an algorithm developed by G. P. McCabe (Purdue University). The McCabe (1975) FORTRAN program is available at the Wiley website labeled MCCABEPC. This program searches all possible subsets of a given size and outputs the 10 (at most) best—in the sense of smallest U ratios, Λ values—subsets of that size. (The Wilks Λ statistic is labeled the U ratio by McCabe.) For example, for a 15-variable problem, McCabe's program outputs the 10 variables that yield the 10 lowest U ratios, each for a univariate analysis; the 2-variable subsets that yield the 10 lowest U ratios, each for

[1]This is similar to multiple regression/correlation whether the R^2 value cannot increase as the number of response variables decreases. (An adjusted R^2 value can, however, increase with a decrease in the number of variables.)

a bivariate analysis; the 3-variable subsets that yield the 10 lowest U ratios, each for a trivariate analysis; and so on to the U ratios for the 10 best 14-variable subsets, and the U ratio for the complete set of 15 variables.

6.2.3 Computer Application I

To run the McCabe program a file must be prepared using Microsoft NOTEPAD. An example of such a file is presented here using the 3-group Ethington data set, 3GED (as described in Appendix A). When executing the McCabe program the file name will be requested. For example, the program below was saved as MCC.dat.

MCCABE SYNTAX FOR FINDING THE BEST SUBSET OF OUTCOME VARIABLES

```
MCCABE.OUT
9 3
66 122 76
(2X,F2.0,1X,8F3.0)
Enter the data file here. Be sure the data are sorted by group level.
```

MCCABE.OUT defines the output file
9 3 Columns 1–5 define the number of variables; Columns 6–10 state the number of groups.
66 122 76 Columns 1–5 define Group 1 size; Columns 6–10 define Group 2 size; Columns 11–15 define Group 3 size.
(2X,F2.0,1X,8F3.0) FORTRAN format statement for the data set.

OUTPUT

Analysis: Partial McCabe Output for the 3-Group Ethington Data (Table 6.1)

Interpretation: McCabe Best Subset of Outcomes for the 3-Group Ethington Data

From this output, the best subset of size 5 is comprised of Y_1, Y_2, Y_3, Y_5, and Y_9; add Y_7 to this subset for the best subset of size 6; and then add Y_8 to get the best subset of size 7.[2]

Note that not much separation is lost when using the, say, third best subset of size 6 ($\Lambda = .862$) rather than the best subset ($\Lambda = .851$). It turns out that the tenth best subset of size 5 yields "close" to the same separation ($\Lambda = .881$) as that yielded by the best subset of size 5. Separation information about multiple subsets of a given size may

[2]It should be noted that McCabe results obtained with the Ethington data set are *not* to be expected with all data sets, in that the best subset of size q will *not* always be contained in the best subset of size $q + 1$.

TABLE 6.1 Partial McCabe Output for the 3-Group Ethington Data

			U Ratio for Five Variables				
.865	1	2	3	5	9		
.866	1	3	5	7	9		
.870	1	2	3	7	9		
.874	2	3	5	7	9		
			U Ratio for Six Variables				
.851	1	2	3	5	7	9	
.859	1	3	5	7	8	9	
.862	1	2	3	5	6	9	
.863	1	2	3	4	5	9	
			U Ratio for Seven Variables				
.848	1	2	3	5	7	8	9
.849	1	2	3	4	5	7	9
.850	1	2	3	5	6	7	9
.858	1	3	5	6	7	8	9

be of potential interest to substantive researchers. This information gives a researcher some choices—an opportunity to make judgments. Perhaps a third- or fourth-best subset of a given size is more attractive than the statistically best subset for reasons related to some variable characteristics, variable measures, previous research, or to the composition of a particular subset, goodness of which is assessed in whatever manner.

In the variable deletion process, one needs to decide on the number of variables to retain for the final analysis. Of course, the judgment may be made on nonstatistical grounds; or one may let the data themselves determine the choice of subset size. Consider Figure 6.1. The Wilks Λ value for the best subset of size 5 is .865, for size 6 is .851, for size 7 is .848, and for size 8 is .847. From the plot it may appear (to some eyes, at least) that a subset of size 6 will not be improved upon very much by increasing the number of variables. Of course, there is no reason why a researcher could not look at subsets across subset sizes. It may just be sensible to do so!

It is important that *if* variables are deleted by some process, statistical or otherwise, the final analysis and all interpretations should be based only on the final subset retained. This includes omnibus analyses, contrast analyses, structure identification, and variable ordering, the topic to which we now turn.

6.3 VARIABLE ORDERING

6.3.1 Meaning of Importance

What does it mean to say that one variable is more "important" than another in the context of MANOVA/DDA? Important with respect to what? Applied researchers

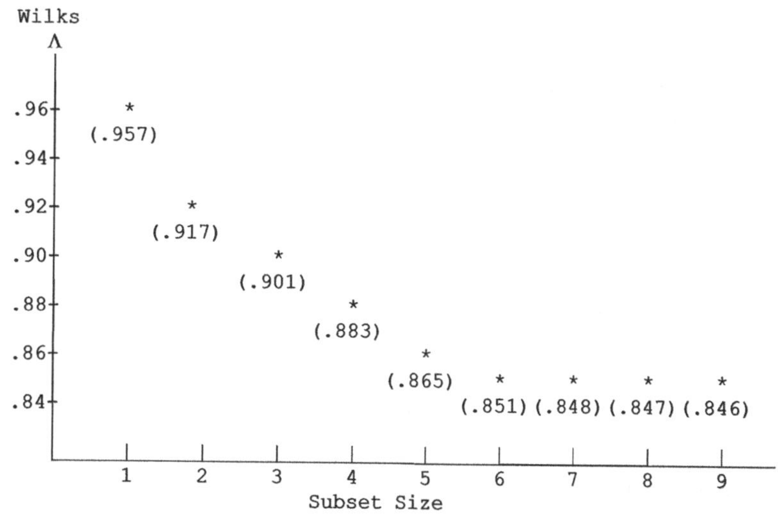

Figure 6.1 Plot of Wilks Λ values versus best subset size for the 3-group Ethington data.

usually view relative variable importance in either of two ways. The *first* way, perhaps the more popular way, is to consider the contribution of a variable to linear discriminant function (LDF) scores. The index considered in this case is (the absolute value of) of standardized LDF weight (of, usually, the leading LDF). The consideration, for a given variable, of weights of multiple LDFs by applied researchers has not been observed. The *second* way in which relative variable importance has been viewed is to consider the contribution of a variable to construct definition (usually associated with the leading LDF). The index considered here is the variable–LDF correlation, that is, the structure r. Again, consideration of multiple index values (for the multiple LDFs) for each variable has been virtually ignored by methodologists and applied researchers.

Conceptual and practical problems with both of these views of relative variable importance are discussed in some detail by Huberty and Wisenbaker (1992b) and Thomas (1997). The preferred point of view is to assess relative contribution of the variables to group separation. To make this assessment, the following question is to be addressed (just as in PDA; see Section 17.4): How well can we do without a variable? The "do" in this context pertains to group separation (i.e., to grouping variable effects). One analysis approach in answering this question is to conduct p MANOVAs, each involving $p - 1$ outcome variables. That is, delete each variable in turn and conduct a MANOVA using the remaining $p - 1$ outcome variables. The most "important" variable, then, is the one for which the MANOVA on the remaining variables yields the *largest* Wilks Λ value. For the ith variable, Y_i, deleted, let $\Lambda_{(i)}$ denote the associated Wilks Λ value. (Recall that a large Wilks Λ value implies little separation.) The most important variable is the one for which the remaining $p - 1$ variables yield the least separation, that is, the largest Wilks Λ value.

So, the variables may be ordered according to the $\Lambda_{(i)}$ values, $i = 1, 2, \ldots, p$. These values may be obtained using the SPSS DISCRIMINANT program (with the METHOD = WILKS command).[3]

An index equivalent—in the sense of variable ordering—to $\Lambda_{(i)}$ is the *partial lambda*, $\Lambda(Y_i|Y_{(i)})$. This is the Λ value for Y_i when the remaining Y variables are "partialed out." This index may be used in the same spirit as the index $\Lambda_{(i)}$; namely, $\Lambda(Y_i|Y_{(i)})$ reflects the separation yielded by Y_i in addition to that yielded by the remaining $p - 1$ variables. The partial Λ values are not given as output by either of the two package stepwise programs. *One* transformation of $\Lambda(Y_i|Y_{(i)})$ is, however, reported by both package programs. The transformation is

$$F_{(i)} = \frac{1 - \Lambda(Y_i|Y_{(i)})}{\Lambda(Y_i|Y_{(i)})} \frac{\mathrm{df}_e - p}{J - 1}.$$

This statistic may also be expressed as:

$$F_{(i)} = \frac{1 - \Lambda/\Lambda_{(i)}}{\Lambda/\Lambda_{(i)}} \frac{\mathrm{df}_e - p}{J - 1},$$

where Λ is the Wilks Λ value based on all p variables. The value of this statistic is labeled F TO REMOVE by SPSS DISCRIMINANT and simply F by SAS STEPDISC. This F statistic may be used to test the null hypothesis of the equality of the J means on Y_i when the remaining $p - 1$ variables are partialed out of Y_i. A high $F_{(i)}$ value indicates a large loss in group separation if Y_i were deleted. A *second* transformation of $\Lambda(Y_i|Y_{(i)})$ is given by SAS STEPDISC and labeled "Partial R**2":

$$R_i^2 = 1 - \Lambda(Y_i \mid Y_{(i)}).$$

This index may be thought of as a squared partial correlation; it is an η^2-like index. It reflects the relationship between the grouping variable and Y_i when the other $p - 1$ variables are partialed out of Y_i.

6.3.2 Computer Application II

The computer package commands to use with the 3-group Ethington data set (3GED) are given below. The DISCRIMINANT program in SPSS was discussed earlier (see Section 5.7) and only new commands are defined below. The analysis done with these commands is a "forward stepwise analysis." The output information (F-to-Remove and $\Lambda_{(i)}$) useful for outcome variable ordering is found under the heading Variables in the Analysis after Step 9 (i.e., the last step).

[3]The analysis must be done so that all p variables are entered; this may be accomplished by using FIN = .00001 and FOUT = .00001, or by using the DIRECT command.

SPSS SYNTAX FOR STEPWISE DESCRIPTIVE DISCRIMINANT ANALYSIS

```
discriminant groups = grade (1,3)
/variables = counsum gainsum learnsum qelib qefac qestacq qeamt qewrite qesci
/method = wilks
/fin = .00001
/fout = .00001
/history  = nostep
/statistics = all.
```

/method = wilks selects variables based on the smallest Wilks Λ value.
/fin = .00001 F value to enter into the model, default is $F = 1.0$.
/fout = .00001 F value to remove, default is $F = 1.0$.
/history = nostep supresses the step by step output.
/statistics = all requests a variety of statistics from means, standard deviations, and covariance matrices.

The following are the SAS STEPDISC commands that also may be used to order the outcome variables. The analysis completed with these commands is a "backward stepwise analysis"—thus the inclusion of the **bw** command. The output information ("F" and "Partial R**2") useful for variable ordering may be found in the last analysis step under the heading Statistics for Removal.

SAS SYNTAX FOR STEPWISE DESCRIPTIVE DISCRIMINANT ANALYSIS

```
proc stepdisc sle = .99999 sls = .99999 bw maxstep = 1;
class grade;
var counsum gainsum learnsum qelib qefac qestacq qeamt qewrite qesci;
run;
```

proc stepdisc SAS command requesting the stepwise discriminant analysis procedure.
sle = .99999 significance level to enter all variables.
sls = .99999 significance level to retain variables.
bw requests the backward elimination of variables.
maxstep = 1 prints the last step.
class grade identifies the grouping variable.
consum–qesci identifies the potential predictor variables.

TABLE 6.2 Results Used to Order Outcome Variables for the 3-Group Ethington Data

Deleted	$F_{(i)}$	$\Lambda_{(i)}$	R_i^2	Rank
Instruction Received	4.91	.879	.037	1.5
Science Effort	4.76	.878	.036	1.5
Counselor Interaction	3.29	.868	.025	3.5
Student–Faculty Effort	3.29	.868	.025	3.5
Art/Music/Theater Effort	1.88	.859	.015	5.5
Self-Understanding	1.61	.857	.013	5.5
Writing Effort	0.28	.848	.002	8
Interstudent Effort	0.09	.847	.001	8
Library Effort	0.07	.847	.001	8

OUTPUT

Results of either analysis that may be used to order the nine outcome variables with respect to omnibus group separation are given in Table 6.2. A rank ordering of the nine variables is discussed in the next subsection.

6.3.3 Variable Ranking

As is obvious, the naive rankings are identical for the two indices. The reason "naive" is used is that no matter which index is considered, it would be stretching matters a bit to conclude that Student–Faculty Effort and Art/Music/Theater Effort (with $\Lambda_{(i)}$ values of .868 and .859, respectively), for example, should be judged as being differentially important. The bottom line suggested herein is: Use common sense and judgment!

6.4 CONTRAST ANALYSIS

The discussion in this chapter so far has centered around variable deletion and variable ordering for omnibus effects. The same ideas, however, may be applied to deleting and ordering variables with respect to contrast effects. When contrast effects are of interest along with, or without, omnibus effects (e.g., interaction effects in a two-factor design), a variable deletion strategy may proceed as follows. Suppose that there are four statistical tests of interest; say, one omnibus test and three contrast tests. "Good" variable subsets associated with *each* of the four tests may be determined. Using some judgment, then, a "good" subset for *all* tests may be determined. One approach might be to consider the union of the four subsets as the subset to be used in the final analysis.

But how can one conduct a variable deletion analysis for contrast effects or for, say, omnibus B effects in an $A \times B$ design? The McCabe program discussed earlier in this chapter uses raw data as input; therefore, it cannot be used here. The reason it cannot be used is because the McCabe program bases the error **SSCP** matrix, **E**, on the data entered. Suppose, for example, that one has a 2×4 design with interest in the main B effects. Collapsing across the two levels of A, a McCabe analysis would yield an **E** matrix based on four "cells" rather than on the original eight cells.

A computer program by C. E. McHenry (Tennessee Tech University), however, can well serve our needs. Like the McCabe program, the McHenry (1978) program yields an all-subsets analysis. Unlike the McCabe program, the McHenry program uses matrix input—an **E** matrix and an **H** matrix. (See Section 3.3 for a discussion regarding elements of these matrices.) The appropriate ($p \times p$) **E** matrix can be determined from an "overall analysis" with the data from the original $A \times B$ design. The **E** matrix is available via SAS GLM (with the PRINTE option), SAS DISCM or CANDISC (with the PSSCP option), and SPSS MANOVA [with PRINT = ERROR (SSCP)]. To get an **E** matrix for a 2×4 design, an eight-group one-way MANOVA can be conducted with either of the two packages. The $p \times p$ hypothesis **SSCP** matrix, **H**, is also available via the SAS GLM program (with the PRINTH option), the SAS DISCRIM and CANDISC program (with the BBSCP option), and the SPSS MANOVA program [with PRINT = SIGNIF (HYPOTH)]. An **H** matrix for either type of test, omnibus or contrast, may be obtained. Thus, one has the necessary input information for a McHenry analysis. (An example of outcome variable ordering using the McHenry program in a two-factor design context will be given in Section 8.8.)

McHenry output consists of simply the best—in the sense of the smallest Wilks Λ value—subset of a given size rather than the 10 best subsets of a given size as one gets with a McCabe analysis. The McHenry program is available at the Wiley website labeled MCHENPC.

6.5 COMPUTER APPLICATION III

For an example of an application of the McHenry program, the 3-group Ethington data set (3GED) is used. A multivariate contrast of Group 1 versus Group 2—$H_0\colon \boldsymbol{\mu}_1 - \boldsymbol{\mu}_2 = \mathbf{0}$—is considered. Following are the steps used to retrieve the **E** and **H** matrices from SAS GLM output, store them, and use them as input for the McHenry program:

1. Store the output in an external file.
2. Edit the stored output file so that it contains the **E** and **H** matrices.
3. Modify the original program to run it with the matrix data file being separated from the main program.

To run the McHenry program, a file must be prepared using Microsoft NOTEPAD. An example of such a file is presented here using the 3-group Ethington data set (3GED). When executing the McHenry program, the file name will be requested. For example, the program below is saved as MCH.dat.

MCHENRY SYNTAX FOR CONTRAST ANALYSIS

```
MCH.OUT
9261 2 1 9

ETHINGTON (3F13.0) Enter the error matrix followed by the hypothesis matrix without
blank lines separating the matrices.
```

mch.out defines the output file.
9261 2 1 9 Columns 1–2 define the number of variables; columns 3–5 state the error degrees of freedom; columns 6–8 state the hypothesis degrees of freedom; columns 9–10 define the minimum subset size; columns 11–12 define the maximum subset size.
Prior ordering can be specified on the blank line.
Ethington is the output title.
(3F13.0) is a FORTRAN format statement for the **E** and **H** matrices. Here the matrices are read in three column sets.

OUTPUT

Analysis: Partial McHenry Output for $\mathbf{H}_0$*:* $\mu_1 - \mu_2 = 0$ *Using the 3-Group Ethington Data*

CONTRAST OUT

Modified from the mainframe program described in: McHenry, C.E. (1978). Computation of a best subset in multivariate analysis. Applied Statistics, 27, 291-296.

- - - - - VARIABLE SELECTION - - - - -

NO. OF VARIABLES = 9 D.F. FOR ERROR = 261 D.F. FOR HYPOTHESIS = 1
MINIMUM SUBSET SIZE = 1 MAXIMUM SUBSET SIZE = 9 VARIABLE FORMAT: (3F12.0)

COMMENT: ETHINGTON MU1 VS MU2

SUBSET OF SIZE 1 x(7) WILKS' LAMBDA - .97122

MAXIMUM ADJUSTED F-RATIO = 6.986 WITH 1 AND 261 D.F. PROBABILITY A GREATER F-RATIO = .009

SUBSET OF SIZE 2

x(5) x(7) WILKS' LAMBDA = .97344 MAXIMUM ADJUSTED F-RATIO 3.391 WITH 1 AND 260 D.F.

PROBABILTY OF A GREATER F-RATIO = .067

SUBSET OF SIZE 3

x(9) x(5) x(7) WILKS' LAMBDA - .95507 MAXIMUM ADJUSTED F-RATIO = 2676 WITH 1 AND 259 D.F.
PROBABILITY OF A GREATER F-RATIO = .103

SUBSET OF SIZE 4

x(2) x(9) x(5) x(7)

CONTRAST OUT

WILKS' LAMBDA = .94766

MAXIMUM ADJUSTED F-RATIO = 1.613 WITH 1 AND 258 D.F.
PROBABILITY OF A GREATER F-RATIO = .205

Interpretation: McHenry Best Subset of Outcomes for H_0: $\boldsymbol{\mu}_1 - \boldsymbol{\mu}_2 = \mathbf{0}$ *Using the 3-Group Ethington Data*

From the output, we see that the best subset of size 2 consists of $X5$ and $X7$, and the best subset of size 4 consists of $X2$, $X5$, $X7$, and $X9$. (Note: McHenry uses X as an outcome variable, while we use Y.)

It turns out that for H_0: $\boldsymbol{\mu}_1 = \boldsymbol{\mu}_2 = \boldsymbol{\mu}_3$ for the 3-group Ethington data, the best subset of size 2 consists of $X3$ and $X9$, and the best subset of size 4 consists of $X1$, $X3$, $X7$, and $X9$. It is not too surprising that these two best subsets for the omnibus test are not the same as for the two-group contrast.

All analyses discussed to this point in this chapter are based on the condition of equal population covariance matrices. If this condition is not tenable, the appropriateness of the discussed analyses may be questioned. Recall from Section 3.4 that a pairwise contrast may be tested without assuming covariance matrix equality by using the Yao statistic [see Eq. (3.12)]. Deleting and ordering variables with respect to a contrast when covariance equality is not tenable may be approached by "taking the bull by the horns." That is, for variable deletion, an all-subset ("weekend") analysis may be conducted. Just as in predictive discriminant analysis (see Section 17.2), the number of analyses may be reduced considerably if the researcher judges ahead of time that, say, q of the p variables should definitely be retained for the final analysis. If so, the weekend analysis would consist of $2^{p-q} - 1$ analyses rather than $2^p - 1$ analysis.

Variable ordering when covariance equality is not tenable may be accomplished simply but in a time-consuming manner. This amounts to conducting $p(p-1)$-variable Yao-type analyses for each contrast of interest.

6.6 COMMENTS

Even though both variable deletion and variable ordering appear to be very common concerns of applied researchers in the context of MANOVA and descriptive discriminant analysis, for reasons presented earlier in this chapter, variable deletion should be of less concern. If variable deletion is of concern, the deletion approach suggested is an all-subsets analysis. The ordering of variables in MANOVA/DDA, on the other hand, deals with the question: How well can we do without a variable? It was suggested that a relevant interpretation of variable contribution in a MANOVA/DDA context pertains to group separation (i.e., to grouping variable effects).

When determining which variables are to be deleted and when determining how variables are to be ordered, the *substantive effects of interest should be considered!* If omnibus effects are not of particular interest, whereas some contrast effects are of particular interest, the latter effects should be those of concern in the variable deletion and ordering processes.

A handful of deletion and ordering methods in the MANOVA/DDA context have been used by applied researchers and suggested by methodologists. The variable deletion approach suggested in this book is the all-possible-subset method. With group separation as a basis, an F-to-remove, partial Λ, or partial R^2 index is suggested for variable ordering. No matter what analysis approaches are employed to delete variables and to order variables, there are two potential analysis interpretation problems. These problems—model specificity and sample specificity—pertain to generalizability and are discussed in Section 17.6. This discussion is given in the context of PDA; therefore, to apply it to DDA, one should substitute "outcome variable(s)" for "predictor(s)."

Another comment about variable deletion follows. In a study that calls for a DDA, the original selection of the set of outcome variables is based on substantive considerations. The originally determined collection of outcome variables should "hang together" in a substantive sense. Recall that the basic purpose of conducting a DDA is to *describe* grouping variable effects (determined by MANOVA). Such descriptions are determined by linear composites (i.e., LDFs) of the outcome variables that are interpreted as variable *constructs*. If an outcome variable has a "low" structure r, then it may be decided that this variable is "unimportant." But this result does *not* indicate that the variable should be deleted and a reanalysis be conducted with $p - 1$ outcome variables. That such a variable (in the company of the other variables) does not contribute to construct definition may be informative. (It may be that this outcome variable contributes to the definition of another construct.) So, assuming some serious thought has been given to the initial set of outcome variables to consider in a MANOVA/DDA, variable deletion may not be an issue at all.

Finally, it should be noted that the McCabe analysis is only applicable in a one-factor design context. The McHenry analysis is, however, applicable in a one-factor context (including contrast analyses) and in a multiple-factor design—see Section 8.8.

Further Reading

Ehrenberg (1990) rebukes the wisdom of attempting to order variables.

Huberty and Wisenbaker (1992b) review the issue of assessing variable importance, express three views of importance, and present an application of bootstrapping to the problem of ranking variables.

Meulepas (1990) suggests, in a two-group context, a modified estimate of the decrease in group separation when a variable is deleted for the purpose of statistically testing the decrease.

Rencher and Scott (1990) advocate the use of standardized LDF weights for assessing the contribution of each variable in the presence of the other variables.

Tardif and Hardy (1995) assess relative outcome variable contribution in a DDA context.

Thomas (1992) proposes a new index for variable ordering, a discriminant ratio coefficient (DRC), which is a product of a standardized discriminant function coefficient (SDFC) and a structure coefficient (SC), each for a given variable.

Thomas (1997) suggests a single criterion for ordering MANDOVA/DDA outcome variables.

Quotation "There are three kinds of lies: lies, damned lies, and statistics"—B. Disraeli (1804–1881)

EXERCISES

1. Briefly describe a McCabe analysis.

2. Briefly, what information can one obtain from conducting a McCabe analysis?

3. **(a)** How are the McCabe and McHenry analyses alike?
   (b) How are they different?

4. **(a)** Why may variable deletion be of questionable value in a DDA context?
   (b) In what other multivariate analysis context is variable deletion of questionable value?

5. When carrying out an analysis using SAS STEPDISC or DISCRIMINANT, why should one not simply rely on the program/procedure defaults?

Computer Applications

6. Conduct a McCabe analysis on the 5-group Ethington data set (5GED):
   (a) What is the best subset of size 2? Of size 6?
   (b) Is the best subset of size 2 (or 6) contained in the best subset of size 3 (or 7)? (See footnote 2.)
   (c) What subset of variables would you retain?
   (d) What are the two variables that contribute most to total-group separation?
   (e) What two variables contribute least?
   (f) Consider (d) and (e) in terms of contribution to the definition of the first LDF.

7. Consider your personal data set.

(a) Which effects, omnibus or contrast, are of interest to you?

(b) Is potential variable deletion of interest to you? If so, are there some variables you can specify in advance to retain in your final subset? If so, which one(s)?

(c) If variable deletion is of interest, conduct a variable deletion analysis (with respect to effects of interest).

(d) Order your variables—total set or subset—with respect to effects of interest.

CHAPTER 7

Reporting DDA Results

7.1 INTRODUCTION

Just what is included in a descriptive discriminant analysis (DDA) is not agreed upon by all methodologists. For example, does DDA include MANOVA? How about variable deletion? Variable ordering? Some writers refer to a *discriminant function analysis*. To those, would a DDA be restricted to analyses directly involving composites of variables called linear discriminant functions (LDFs)? As far as this chapter goes, all such results will be considered. An excellent discussion pertaining to the general reporting of quantitative results is given by Bailar and Mosteller (1992). The reporting of discriminat analysis results, in particular, is reviewed by Huberty and Hussein (2003).

7.2 EXAMPLE OF REPORTING DDA RESULTS

Introduction (This initial section of a group comparison study write-up would be a discussion of the research context. This would include a rationale of the study and a review of related research. This being done, an *explicit research purpose statement* should be given. This statement should relate directly to the analysis procedure, given later, used.)

The current research purpose is that of examining differences among three grade levels (A, B, C or C−) earned by community college students—see Data Set A2(3GED) in Appendix A.

Study Design For our data set (3GED), the three grouping-variable levels are defined as A students ($n_1 = 76$), B students ($n_2 = 122$), and students who earned C or C− ($n_3 = 66$). Of the large number of "variables" obtained on the basis of responses to items on the CCSEQ, nine outcome variables were identified. Outcome variable identification was based on a minimum of the sum of six item scores. Reliability

Applied MANOVA and Discriminant Analysis, Second Edition, by Carl J. Huberty and Stephen Olejnik
Copyright © 2006 John Wiley & Sons, Inc.

and validity information for the last six variables is provided by Ethington and Polizzi (1996). The nine outcome variables were selected on the basis of professional judgment of their relevance to the study of grade-group differences. [A good example of outcome variable choice and how each variable is measured is given by Ellis and Armstrong (1989).]

No missing data were found in the 264×9 data matrix; also no aberrant outcome variable scores were present.

[To this point in the reporting, we have considered research context, purpose statement, and (grouping and outcome) variables considered.]

Descriptives The data were analyzed using both SPSS Version 12.0 and SAS Version 8. Descriptive information for our nine outcome variables is given in Table 7.1. The results of univariate hypothesis tests ($\text{df}_1 = 2$, $\text{df}_2 = 261$) indicate that the three populations differed on only two of the nine variables (Y_3 and Y_9). Also, the error correlations among the nine outcome variables are in the "small-to-moderate" range.

Requisite Data Conditions Justification of reporting the error correlations is based on the tenability of group-covariance-matrix homogeneity. The Box test for covariance homogeneity provided little support to indicate that the population covariance matrices differ [$F(90, 125132) \doteq 1.19$, $P \doteq .105$ and $\chi^2(90) \doteq 107.25$, $P \doteq .104$]. Further support for this conclusion is provided by the (natural) logarithms of the determinants of the four covariance matrices: Group 1, 24.6; Group 2, 23.1; Group 3, 22.5; and Error, 23.1. These four log determinants are clearly "in the same ballpark" [see Huberty (2002, pp. 587–588)]. Based on theoretical considerations and the nature of the variables studied, we assumed that the joint distribution of the nine vaiables within each population is approximately multivariate normal.

Group Comparison Because there is support for the equality of the three covariance matrices, we proceed with a multivariate analysis of variance (MANOVA). For our data, we have $\Lambda \doteq .846$, $F(18, 506) \doteq 2.45$, $P \doteq .001$, $\tau^2 = 1 - \Lambda^{1/2} \doteq .080$, and $\tau^2_{\text{adj}} \doteq .066$. We thus conclude that the observed differences among the three Grade groups are generalizable to the populations they represent with respect to the nine outcome variables.

The nine F-to-remove values (obtained via SPSS DISCRIMINANT) reported in Table 7.2 indicate that Instruction Received (Y_3) and Science Effort (Y_9) are the two variables contributing most to overall group differences, followed by Counselor Interaction (Y_1) and Student–Faculty Effort (Y_5). Writing Effort (Y_8), Interstudent Effort (Y_6), and Library Effort (Y_4) contributed the least to group differences.

Dimensionality To further study the resulting group differences, we considered the linear discriminant functions (LDFs). With our three groups, two LDFs may be obtained. To determine if the group differences are to be described in one or two dimensions, statistical test results and an LDF plot were examined. The statistical test results (obtained via SPSS DISCRIMINANT) are given in Table 7.3.

TABLE 7.1 Descriptive Information for the 3-Group Ethington Data

Variable		Group Means/(SDs)			Univariate	Error Correlations							
		1	2	3	F^a	Y_2	Y_3	Y_4	Y_5	Y_6	Y_7	Y_8	Y_9
Counselor Interaction	(Y_1)	4.1 (2.29)	4.7 (2.06)	4.6 (1.72)	2.33	.44	.27	.38	.40	.27	.28	.36	.24
Self-Under standing	(Y_2)	17.4 (5.72)	18.2 (4.86)	16.6 (4.78)	2.24		.41	.38	.40	.35	.28	.47	.39
Instruction Received	(Y_3)	16.9 (5.20)	18.3 (4.81)	19.3 (4.79)	4.30			.27	.20	.14	.16	.28	.14
Library Effort	(Y_4)	13.7 (5.26)	14.5 (4.58)	13.7 (4.37)	0.80				.35	.37	.27	.48	.22
Student–Faculty Effort	(Y_5)	15.9 (4.68)	15.1 (3.86)	15.2 (4.55)	1.02					.49	.36	.46	.24
Interstudent Effort	(Y_6)	11.9 (4.29)	11.7 (3.94)	10.8 (3.67)	1.62						.36	.30	.33
Arts/Music Theater Effort	(Y_7)	8.8 (3.21)	9.0 (3.15)	8.1 (2.11)	2.20							.18	.11
Writing Effort	(Y_8)	20.8 (6.05)	21.7 (5.47)	21.0 (5.12)	0.72								.16
Science Effort	(Y_9)	16.4 (7.43)	15.8 (6.51)	13.6 (5.77)	5.58								
n		76	122	66									

[a] $df_1 = 2, \quad df_2 = 261.$

TABLE 7.2 Variable Ordering for the 3-Group Ethington Data

Variable	F to Remove
Instruction Received (Y_3)	4.91
Science Effort (Y_9)	4.76
Counselor Interaction (Y_1)	3.29
Student–Faculty Effort (Y_5)	3.29
Arts/Music Theater Effort (Y_7)	1.88
Self–Understanding (Y_2)	1.61
Writing Effort (Y_8)	0.28
Interstudent Effort (Y_6)	0.09
Library Effort (Y_4)	0.07

TABLE 7.3 Test of Dimensionality for the 3-Group Ethhington Data

Number of Dimensions	Λ	χ^2	df	P
1	.85	42.88	18	.001
2	.96	11.84	8	.158

From these results, we conclude that it may be reasonable to consider two dimensions in describing Grade-group differences. The group centroids are reported in Table 7.4. [Typically, one would not consider the second dimension (with $P \doteq .158$); it is considered here merely for illustrative purposes.]

A plot of the group centroids (nine-element means) is given in Figure 7.1. From this plot, it appears that, with respect to LDF_1, there is a general "separation" among the three Grade groups. With respect to LDF_2, it appears that Group 2 (B students) is "separated" from, collectively, Group 1 and Group 3 (A and C or C− students, respectively).

Group Difference Structure An interpretation of the resulting Grade group differences is based on correlations between each of the nine outcome variable scores and the two respective LDF scores—these are the *structure r's*. The two sets of structure

TABLE 7.4 LDFs at Group Centroids

	LDF	
Group	1	2
A	.464	−.190
B	−.014	.233
C or C−	−.509	−.212

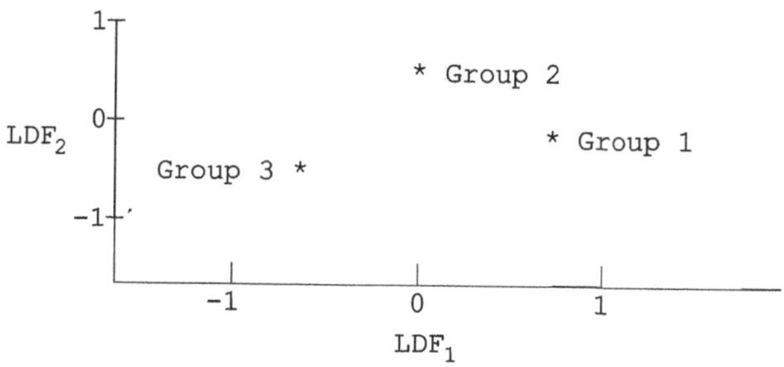

Figure 7.1 LDF plot of group centroids.

r's for our data are given in Table 7.5. From these results, the first *construct* is defined primarily by Science Effort and Instruction Received. A possible definition for the first construct is, simply, "Science Effort and Instruction Received." And the second construct may be labeled "Self-Understanding and Art/Music/Theater Effort."

Therefore, the separation among all three Grade groups may be attributed to "Science Effort and Instruction Received." And, the separation of B students from A and C or C– students may be attributed to "Self-Understanding and Art/Music/Theater Effort."

(The obtained information would then be discussed in the context of the stated research purpose, with some substantive conclusions. The results of the study would also be discussed and compared with results of previous related studies—that were referenced in the study Introduction.)

Group Contrast Of particular interest in this study is a comparison of A Students (Group 1) with the C or C– Students (Group 3). The results of a contrast analysis indicated sufficient evidence to conclude that the two populations represented by these groups differed [$\Lambda \doteq .886$, $F(9, 253) \doteq 3.60$, $P \doteq .000$, $\tau^2_{\text{adj}} \doteq .083$].

TABLE 7.5 Structure r's for the 3-Group Ethington Data[a]

Variable	LDF_1	LDF_2
Counselor Interaction	–.30	.36
Self-Understanding	.15	**.55**
Instruction Received	–**.50**	.11
Library Effort	–.02	.36
Student–Faculty Effort	.19	–.25
Interstudent Effort	.29	.19
Art, Music, Theater Effort	.23	**.46**
Writing Effort	–.04	.34
Science Effort	**.59**	.08

[a]Construct identifiers in bold.

TABLE 7.6 Structure r's for Group 1 versus Group 3[a]

Variable	r
Counselor Interaction	−.29
Self-Understanding	.16
Instruction Received	**−.50**
Library Effort	−.01
Student–Faculty Effort	.19
Interstudent Effort	.29
Art, Music, Theater Effort	.24
Writing Effort	−.03
Science Effort	**.59**

[a]Construct identifiers in bold.

With these two groups, there is only one LDF to consider. The group-difference structure is given in Table 7.6. From these structure r's, we conclude that the construct underlying the difference between the A group and the C or C− group is the same as for the first construct underlying the three group differences. (The relative contributions of the nine outcome variables would be given here, just as discussed for the omnibus effects.)

Conclusions/Discussion The purpose of this study was to examine and assess differences among community college students achieving at three academic levels, A students, B students, and C or C− students. We found, as might be expected, that Grade differences existed. Two of the nine outcome variables, Instruction Received and Science Effort, contributed most to the resulting Grade differences. On the other hand, Library Effort, Interstudent Effort, and Writing Effort contributed least to Grade differences.

We also found that the resulting three Grade differences may be attributed to a construct defined by Instruction Received and Science Effort. This same construct was found to describe the differences of A students from the C or C− students.

[The results of this study would be related to results of previous related research, which was discussed in the Introduction. Also, when describing results or some specific analysis results, such as the Box test, MANOVA, test of LDF dimensionality, and structure r's, it is *strongly* suggested that references be given. Book references for such analyses should include the page number(s).]

7.3 COMPUTER PACKAGE INFORMATION

Most computational results reported in applications of DDA have been obtained via the use of some statistical computer package(s). The "big two" (SAS and SPSS) are very popular. Each of these two has programs that deal with DDA. Specific output information for the two packages is indicated in Table 7.7. It is recognized that some additional packages exist that relate to DDA procedures.

TABLE 7.7 DDA Printout Information

	SAS			SPSS	
	DISCRIM	CANDISC	GLM	DISCRIMINANT	MANOVA
Preliminary					
Uni. F's	Yes	Yes	Yes	Yes	Yes
Wilks	Yes	Yes	Yes	Yes	Yes
$=\Sigma$s test	Yes[a,b]	No	No	Yes[c]	Yes[a,c]
Effect size	No	No	No	No	Yes
Error correlation matrix	Yes	Yes	No	Yes	Yes
DDA					
LDFs					
Raw	No	Yes	Yes	Yes	Yes
Standardized	No	Yes[d]	Yes[d]	Yes	Yes
Rotated	No	No	No	Yes	Yes
No. LDFs test	Yes	Yes	No	Yes	Yes
Structure correlations	No	Yes	Yes	Yes	Yes
Partial F's	No	No	No	Yes	No
Intergroup distances					
D^2	Yes	Yes	No	No	No
F	No	Yes	No	Yes	No
P	No	Yes	No	Yes	No
Discriminant space plots	No	Yes	No	Yes	No
LDF scores	No	No	No	Yes	No
Contrasts	No	No	Yes	No	Yes

[a]Chi-squared test.
[b]Not usual statistic.
[c]F test.
[d]Based on total group variances.

7.4 REPORTING TERMS

Following is a list of terms that might be utilized in writing up a report of a multivariate study in which group separation is of interest. Of course, particular research situations may call for more specific terms.

- Grouping variable
- Outcome variable
- Multivariate analysis of variance
- Covariance matrix homogeneity
- Wilks Λ $F(\nu_1, \nu_2)$, P value
- Effect size
- Multivariate contrast (?)
- Linear discriminant function

- Dimensionality
- LDF plot
- Outcome variable deletion (?)
- Outcome variable ordering
- F-to-remove/partial Λ
- Structure/construct
- Judgment

There is also a list of "expressions" we suggest that a writer of a manuscript involving a DDA *not* use. These terms follow.

- Chi-square (vs. chi-squared)
- Data was (vs. data were)
- Discriminant function analysis (vs. DDA)
- Discriminate (vs. separate)
- Discrimination (vs. separation)
- Independent, dependent variable (vs. grouping, outcome)
- Methodology (vs. methods)
- $P < .05$ (vs., e.g., $P \doteq .023$)

[The statistically inclined reader might refer to David (1995, 1998) for first users of many statistical terms.]

7.5 MANOVA/DDA APPLICATIONS

A number of applications of MANOVA/DDA in the user field of study may be found using a Web search with the key term "discriminant analysis." Two cautions need be taken with such a search: (1) a located study may involve both a DDA and a PDA; and (2) the reporting of DDA/PDA results may not be very clear (and confusing?). Thus, a researcher wanting to write up an application of a MANOVA and a DDA may *not* want to use a published study as a "model." It is recognized that some researchers look at published articles as a guide to reporting: "They got their manuscript published, so I'll do a similar report." As implied by Huberty and Hussein (2003), this is a principle that should *not* necessarily be followed. Some applications of DDA in horticultural research are given by Cruz-Castillo et al. (1994).

7.6 CONCERNS

With the current availability and popularity of statistical computer packages it is fairly straightforward to conduct a discriminant analysis (and many other analyses, for that matter). But as pointed out in a number of places in the current book, computer output can sometimes be misleading and, in a few isolated instances, incorrect. In applying discriminant analysis in a given research situation, then, one must take care in using and reporting numerical information from computer printouts. There are four concerns with the write-ups of discriminant analysis applications in professional journals that will now be mentioned.

The *first* concern is the apparent mixup of DDA and PDA. As emphasized earlier, the former pertains to group differences and the characterizations of such, while the latter pertains to group membership prediction. In some journal articles DDA results *as well as* PDA results are reported for no apparent good reason—usually in the article introduction it is clear that the research pertains to either group differences or prediction of group membership. In one journal article, it was written, for example, in a three-group study that the two resulting LDFs were used to classify units into the three groups. [The LDF scores may, however, be used as input into classification functions, values of which are used for group assignment (as SPSS DISCRIMINANT does)—this is quite different from basing group assignment decisions directly on LDF scores.] Some distinctions between PDA and DDA are spelled out in Table 7.8.

A *second* concern pertains to the covariance matrix homogeneity condition that is required to employ the usual MANOVA test criteria. Admittedly, the assessment of this

TABLE 7.8 DDA versus PDA; Context: J Groups of Units, p Response Variables

	DDA	PDA
1. Research concern	Description of group separation	Prediction of group membership
2. Variable roles		
Predictor(s)	Grouping variables	Response variable
Criterion(ia)	Response variables	Grouping variable
3. Response variable composite	LDF	LCF/QCF
4. Number of composites	$\min(p, J - 1)$	J
5. Equality of		
covariance matrices	Yes	Yes
MANOVA	Yes	No
6. Analysis aspects		
Variable construct(s)	Yes (!)	No
Response variable deletion	Maybe	Yes (!)
Response variable ordering	Yes	Yes
7. Criterion for deletion/ordering	Group separation	Classification accuracy
8. Interest in generalizability?	Yes	Yes

condition is difficult, but if the matrices are clearly(?) heterogeneous, the calculated outcome variable linear composites (LDFs) may have very limited interpretability. It is suggested that it should be made clear that the author is cognizant of the condition.

A *third* concern is with the interpretation of *relative variable importance*. An impression one can get from reading some applications of discriminant analysis is that an undefined interpretation of relative variable importance is used, an overreliance on numerical values of indices is taken, and step-by-step results of stepwise analyses are still being used (for variable ordering as well as for variable deletion).

The latter comment indicates the *fourth* concern, an overuse of stepwise discriminant analysis computer programs. Reasons for this concern are expressed in some detail by Huberty (1989).

7.7 OVERVIEW

An overview of multivariate analysis of variance, descriptive discriminant analysis and related techniques is given in Figure 7.2.

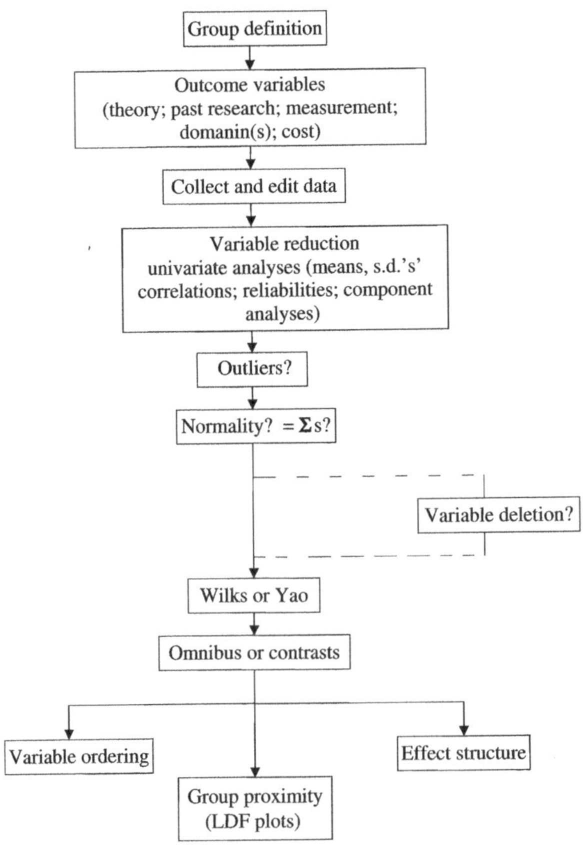

Figure 7.2 MANOVA and descriptive discriminant analysis.

Further Reading

Maxwell and Cole (1995) make some good suggestions about writing journal articles.

Definition ***Content validity***: Seeing if the text developer is happy.

EXERCISES

1. Locate a journal article in your area of study that reports an application of DDA. Consider the following:
   (a) The groups/factor(s)
   (b) The outcome variables; how were they measured?
   (c) Variable deletion
   (d) Variable ordering
   (e) Effects interpretation; structure r's, standardized LDF weights
   (f) LDF plot
   (g) Other evaluative factors pertaining to design, analysis, and interpretation

2. Assess the completeness of reporting of information: descriptive information in your located study:, statistical test information, and information related to LDFs. Are variable measures clear? What computer software was used?

3. Propose a study in your discipline that would call for a DDA. Indicate some design and analysis specifics for your research study.

4. Consider your personal data set. From your printout(s), prepare tables as suggested in this chapter. How about an LDF plot (if $J > 2$)?

PART III

Factorial MANOVA, MANCOVA, and Repeated Measures

In Part III we generalize the procedures introduced in Part II to more complex research designs. We begin by presenting the application of MANOVA to a factorial design. Our presentation is limited to a two-factor design but the procedures easily generalize to multiple grouping variables. We then introduce multivariate analysis of covariance. In this chapter we limit our presentation to a single covariate and a single grouping variable. The application of multiple covariates is briefly presented in a Technical Note at the end of Chapter 9. We conclude Part III by presenting the repeated measures design with a single group (Chapter 10) and multiple groups (Chapter 11). Only a single repeated measures variable is considered here, but we briefly provide an example of a design with two repeated measures variables in the Technical Notes at the end of Chapter 10.

Applied MANOVA and Discriminant Analysis, Second Edition, by Carl J. Huberty and Stephen Olejnik
Copyright © 2006 John Wiley & Sons, Inc.

CHAPTER 8

Factorial MANOVA

8.1 INTRODUCTION

All of the discussion in Chapters 3 to 5 dealt with a one-factor design. Now the discussion is extended to a two-factor layout with p outcome variables. The generalization to studies examining multiple grouping variables is straightforward, but the interpretation of the results can become considerably more complex. With additional grouping variables there are more hypotheses to test, more contrasts to examine, more effect sizes to estimate, and more LDFs to consider. Whereas in the previous discussion it was appropriate to refer to "separation" of groups or to grouping variable "effect," the latter term is more generally applicable and more appropriate in connection with multifactor MANOVA.

We begin this chapter by introducing a new research context involving two grouping variables. Then, following a brief review of the univariate two-way ANOVA, we apply the multivariate procedures introduced in Chapters 3 to 5 to the two-factor layout with p outcome variables. We begin with the omnibus test followed by effect size estimation and construct definition. We conclude with multivariate focused tests and their interpretations. It should be recalled that omnibus tests are not necessary if specific contrasts are of primary interest. In this chapter we present some of the matrix formulations used to test hypotheses and estimate effects, but the emphasis is given to the computer applications and interpretations because we do not believe additional conceptual understanding is gained by detailed calculations.

8.2 RESEARCH CONTEXT

Suppose a researcher is interested in the degree of stress experienced by teachers. Table 8.1 provides hypothetical data for such a study. (An SPSS data file containg these data, labeled STRESS, is available at the Wiley website.) The Wilson Stress Profile for Teachers (WSPT) contains nine scales, four of which are designed to assess teacher

---

Applied MANOVA and Discriminant Analysis, Second Edition, by Carl J. Huberty and Stephen Olejnik
Copyright © 2006 John Wiley & Sons, Inc.

TABLE 8.1 Test Scores on Four Measures of Stress[a] for Three School Levels and Two Levels of Gender

	Gender[b]							
	j_1				j_2			
School Level[c]	Y_1	Y_2	Y_3	Y_4	Y_1	Y_2	Y_3	Y_4
k_1	7	10	12	15	9	9	10	12
	7	9	12	15	9	8	9	10
	8	10	13	16	11	12	11	11
	6	9	11	15	11	10	7	9
	6	9	11	14	10	10	9	9
	7	9	12	15	13	12	11	12
	7	9	12	14	10	9	10	11
	9	10	13	16	9	9	9	10
	6	8	11	14	10	9	10	11
	6	9	13	17	10	9	9	11
k_2	11	12	12	15	14	11	8	9
	11	13	12	14	15	13	7	7
	11	12	11	15	14	12	10	10
	11	12	12	14	14	14	9	11
	11	11	13	15	16	14	8	9
	12	13	11	13	13	12	9	10
	11	12	11	14	17	14	9	10
	10	11	12	16	15	13	9	10
	9	10	12	15	14	13	9	10
	10	11	11	15	14	13	8	10
k_3	9	11	15	19	14	13	13	14
	11	12	17	19	14	14	13	13
	8	11	16	18	14	13	12	14
	9	11	14	17	14	13	13	13
	9	12	16	19	14	12	14	14
	12	13	18	20	15	14	12	12
	8	10	16	20	14	13	12	13
	10	13	14	17	13	12	13	15
	10	11	15	18	12	11	13	14
	10	13	15	19	13	12	12	14

[a] Stress: Y_1 = Administration; Y_2 = Colleagues; Y_3 = Parents; Y_4 = Students.
[b] Gender: j_1 = Males; j_2 = Females.
[c] School Level: k_1 = Elementary School; k_2 = Middle School; k_3 = High School.

stress in relationships between a teacher and (1) Administrators, (2) Colleagues, (3) Parents, and (4) Students. The researcher might distribute these four scales to a sample of Male and Female teachers at Elementary, Middle, and High School levels to determine if the degrees of stress in these four sources differ by School Level (L)

TABLE 8.2 Means and Standard Deviations for Four Measures of Stress[a] from Three School Levels and Two Levels of Gender

School Level[c]		Gender[b] j_1 Y_1	Y_2	Y_3	Y_4	j_2 Y_1	Y_2	Y_3	Y_4	Mean
k_1	Mean	6.9	9.2	12.0	15.1	10.2	9.5	9.5	10.6	$\overline{Y}_1 = 8.55$ $\overline{Y}_2 = 9.35$
	Sd	.994	.632	.816	.994	1.229	1.080	1.179	1.075	$\overline{Y}_3 = 10.75$ $\overline{Y}_4 = 12.85$
k_2	Mean	10.7	11.7	11.7	14.6	14.6	12.9	8.6	9.6	$\overline{Y}_1 = 12.65$ $\overline{Y}_2 = 12.30$
	Sd	.823	.949	.675	.843	1.174	.994	.843	1.075	$\overline{Y}_3 = 10.15$ $\overline{Y}_4 = 12.10$
k_3	Mean	9.6	11.7	15.6	18.6	13.8	12.7	12.7	13.6	$\overline{Y}_1 = 11.70$ $\overline{Y}_2 = 12.20$
	Sd	1.265	1.059	1.265	1.075	.919	.949	.675	.843	$\overline{Y}_3 = 14.15$ $\overline{Y}_4 = 16.10$
	Mean	9.07	10.87	13.10	16.10	12.87	11.70	10.27	11.27	$\overline{Y}_1 = 10.967$ $\overline{Y}_2 = 11.283$ $\overline{Y}_3 = 11.683$ $\overline{Y}_4 = 13.683$

[a]Stress: Y_1 = Administration; Y_2 = Colleagues; Y_3 = Parents; Y_4 = Students.
[b]Gender: j_1 = Males; j_2 = Females.
[c]School Level: k_1 = Elementary School; k_2 = Middle School; k_3 = High School.

and by Gender (G). Differences between Male and Female teachers may also differ depending on the School Level (interaction). The researcher may also be interested in determining whether Gender and School Level effects are reflected in a single dimension or whether grouping variable effects are better represented by multiple dimensions. Group means for each School Level and Gender, as well as all combinations of School Level and Gender, are presented in Table 8.2, along with cell standard deviations.

With this 3×2 nonexperimental two-way design, three questions may by examined:

1. Do differences in stress between Male and Female teachers change depend on the Level of the School? Or, do differences among School Levels vary with teacher Gender with respect to four stress variables?
2. Across all three School Levels, is there a difference between Male and Female teachers with respect to stress scales?

3. For all teachers (Male and Female) do the four measures of stress differ among Elementary, Middle, and High School levels?

The first question addresses the issue of an interaction between Gender and School Level. The second question addresses a main effect for Gender. And the third question reflects a School Level main effect.

8.3 UNIVARIATE ANALYSIS

A univariate analysis would answer each of the three questions raised in the previous section for each measure of stress ($Y_1 - Y_4$) separately. The sum-of-squares for Interaction, Gender, and School Level would be computed using the formulas in Table 8.3 assuming a balanced design.

Using the data in Tables 8.1 and 8.2, Table 8.4 provides an ANOVA summary table reporting the sources of variation, degrees of freedom, mean squares, F ratios, P values, and generalized η^2 (Olejnik and Algina, 2003) for variable Y_1.

The results do not support an interaction between Gender and School Level, $F(2, 54) \doteq .900$, $P \doteq .413$, $\eta_G^2 \doteq .005$. But there is evidence to conclude that Male and Female teachers across all School Levels differ with respect to their reported levels of stress with Administrators (Y_1), $F(1, 54) \doteq 185.604$, $P \doteq .000$, $\eta_G^2 \doteq .465$, and teachers (both Male and Female) report different levels of stress with Administrators at different School Levels, $F(2, 54) \doteq 78.935$, $P \doteq .000$, $\eta_G^2 \doteq .395$. Females

TABLE 8.3 Univariate Sum-of-Squares

Source of Variation	Sum-of-Squares[a]
Gender (G)	$n_{.j} \sum_{j=1}^{J} (\overline{Y}_{..j} - \overline{Y}_{...})^2$
School Level (L)	$n_{k.} \sum_{k=1}^{K} (\overline{Y}_{.k.} - \overline{Y}_{...})^2$
Interaction ($G \times L$)	$n_{kj} \sum_{j=1}^{J} \sum_{k=1}^{K} (\overline{Y}_{.kj} - \overline{Y}_{..j} - \overline{Y}_{.k.} + \overline{Y}_{...})^2$
Error (E)	$\sum_{j=1}^{J} \sum_{k=1}^{K} \sum_{u=1}^{n} (Y_{ukj} - \overline{Y}_{.kj})^2$

[a] $n_{.j}$ is the number of units at each level of variable G; $n_{k.}$ is the number of units at each level of variable L; n_{kj} is the number of units at each combination of variables G and L; $\overline{Y}_{..j}$ is the group mean of units at the jth level of factor G; $\overline{Y}_{.k.}$ is the group mean of units at the kth level of factor L; $\overline{Y}_{.kj}$ is the group mean of units at the kth of variable L and the jth level of factor G; $\overline{Y}_{...}$ is the mean of all units in the study (Grand Mean); Y_{ukj} is observation for unit u in the kth level of variable L and the jth level of variable G.

TABLE 8.4 ANOVA Summary for Variable Y_1

Source of Variation	Sum-of-Squares	Degrees of Freedom	Mean Square	F	P	η_G^2
Gender (G)	216.60	1	216.600	185.604	.000	.465
School Level (L)	184.23	2	92.117	78.935	.000	.395
Interaction ($G \times L$)	2.10	2	1.050	.900	.413	.005
Error (E)	63.00	54	1.167			
Total	465.93					

reported higher stress mean with Administration than Males, 12.87 vs. 9.07. The standardized mean difference between Male and Female teachers is 1.83 ($\doteq 3.80/\sqrt{4.299}$). [Standardized mean difference is computed using the Gender standard deviation, ignoring the School Level factor (see Olejnik and Algina, 2000).] Examining all pairwise contrasts among School Levels may be of interest. Marginal means for Elementary, Middle, and High Schools are 8.55, 12.65, and 11.70, respectively. The Bonferroni adjusted P' value for the difference between Elementary and Middle School teachers equals .000 and a standardized mean difference of -1.84 ($\doteq -4.1/\sqrt{4.942}$). (Standardized mean difference is computed using School Level standard deviation, ignoring Gender.) For the difference between Elementary and High School teachers, the Bonferroni adjusted P' value equals .000 and the standardized mean difference equals -1.42. Finally, the difference between Middle School and High School teachers has a Bonferroni adjusted P' value of .022 and a standardized mean difference of .43.

The results of this hypothetical study indicate that with respect to Administrators (Y_1), there is no evidence of an interaction between Gender and School Level, but Female teachers do report greater stress mean levels than Male teachers, and Middle School teachers report greater stress than either Elementary or High School teachers. High School teachers report greater stress with Administration than Elementary teachers—"on the average."

The results of the three omnibus tests using the stress scales for Colleagues, Y_2, Parents, Y_3, and Students, Y_4, are summarized in Table 8.5. These results parallel those

TABLE 8.5 Summary of Omnibus Univariate Results for Outcome Variables Y_2, Y_3, and Y_4

	Variable								
	Y_2			Y_3			Y_4		
Factor	F	P'	η_G^2	F	P'	η_G^2	F	P'	η_G^2
$G \times L$	1.223	.302	.013	.531	.591	.003	.43	.656	.001
G	11.410	.001	.060	136.895	.000	.339	357.703	.000	.599
L	61.467	.000	.644	105.802	.000	.524	92.297	.000	.309

TABLE 8.6 Summary of Pairwise Contrasts Among School Levels with Bonferroni Adjusted P' Values for Outcome Variables Y_2, Y_3, and Y_4[a]

	Variable								
	Y_2			Y_3			Y_4		
Contrast	F	P'	d	F	P'	d	F	P'	d
EvsM	95.418	.000	−2.84	4.081	.144	.35	5.742	.060	.28
EvsH	89.058	.000	−2.73	131.052	.000	−1.98	107.815	.000	−1.22
MvsH	.110	1.000	.10	181.387	.000	−2.32	163.317	.000	−1.50

[a]The standardized mean difference, d, is computed using the School Level standard deviation when Gender is ignored (see Olejnik and Algina, 2000).

reported above for Administrators, Y_1. That is, there appear to be no interactions, and both main effects appear to be generalizable to the respective populations.

The results for all pairwise contrasts for levels of School Level are summarized in Table 8.6. The pattern of differences for variables Y_2, Y_3, and Y_4 are somewhat different than those reported for Y_1. For variable Y_2, the difference between Middle School and High School teachers, and for variables Y_3 and Y_4, the differences between Elementary and Middle School teachers, do not appear to be generalizable to their respective populations. The remaining contrasts are generalizable with Elementary teachers reporting less stress with Colleagues (Y_2) than Middle or High School teachers. High School teachers report greater stress then either Elementary or Middle School teachers with Parents (Y_3) and Students (Y_4).

8.4 MULTIVARIATE ANALYSIS

8.4.1 Omnibus Tests

Rather than examining each individual outcome variable, the multivariate approach addresses the same questions as presented above, with a collection of outcome variables being simultaneously analyzed. Following an analogous procedure as presented in Chapter 3, an eigenanalysis of the $\mathbf{E}^{-1}\mathbf{H}_\theta$ product matrix is conducted, where $\mathbf{E}^{-1}$ is the inverse of the error sum-of-squares and cross-products matrix, and $\mathbf{H}_\theta$ represents the sum-of-squares and cross-products for the θ hypothesis. The error matrix, **E**, for the factorial design, is computed by summing the **SSCP** matrices across all cells in the design (see Section 3.2.2). That is,

$$\mathbf{E} = \sum_{k=1}^{K}\sum_{j=1}^{J}(\mathbf{Y}_{ukj} - \mathbf{Y}_{.kj})'(\mathbf{Y}_{ukj} - \mathbf{Y}_{.kj}), \tag{8.1}$$

where $\mathbf{Y}_{ukj} = n_{kj} \times p$ matrix of observations from n_{kj} units on p outcome variables in the kth row and the jth column of the data matrix, (e.g., Table 8.1)

$\mathbf{Y}_{.kj} = n_{kj} \times p$ matrix of p cell means for the kth row and jth column of the data matrix

(Note the two summation signs indicate that the separate cell **SSCP** matrices are summed across the J columns and then across the K rows in the row-by-column factorial design.)

For the Gender-by-School Level interaction ($\theta = G \times L$), assuming that the sample sizes are equal across all KJ cells, the hypothesis sum-of-squares and cross products is computed as:

$$\mathbf{H}_{G\times L} = n_{kj} \sum_{k=1}^{K} \sum_{j=1}^{J} (\mathbf{y}_{.kj} - \mathbf{y}_{..j} - \mathbf{y}_{.k.} + \mathbf{y}_{...})(\mathbf{y}_{.kj} - \mathbf{y}_{..j} - \mathbf{y}_{.k.} + \mathbf{y}_{...})', \quad (8.2)$$

where n_{kj} = number of analysis units per cell

$\mathbf{y}_{.kj} = p \times 1$ vector of p cell means in the kth row and jth column of the data matrix

$\mathbf{y}_{..j} = p \times 1$ vector of p column means in the jth column of the data matrix

$\mathbf{y}_{.k.} = p \times 1$ vector of p row means in the kth row

$\mathbf{y}_{...} = p \times 1$ vector of p grand means

For the Gender main effect ($\theta = G$) the hypothesis sum-of-squares and cross-products is computed as:

$$\mathbf{H}_G = n_{.j} \sum_{j=1}^{J} (\mathbf{y}_{..j} - \mathbf{y}_{...})(\mathbf{y}_{..j} - \mathbf{y}_{...})', \quad (8.3)$$

where $n_{.j}$ = number of units for each level of factor G

$\mathbf{y}_{..j} = p \times 1$ vector of p means for the jth column

$\mathbf{y}_{...} = p \times 1$ vector of p grand means

For the School Level main effect ($\theta = L$) the hypothesis sum-of-squares and cross-products matrix is computed:

$$\mathbf{H}_L = n_{k.} \sum_{k=1}^{K} (\mathbf{y}_{.k.} - \mathbf{y}_{...})(\mathbf{y}_{.k.} - \mathbf{y}_{...})', \quad (8.4)$$

where $n_{k.}$ = number of units for each level of factor L

$\mathbf{y}_{.k.} = p \times 1$ vector of p means for the kth row

$\mathbf{y}_{...} = p \times 1$ vector of p grand means

These calculations are tedious and better left to the computer. However, the similarity of Eqs. (8.1) to (8.4) with those for univariate sum-of-squares in Table 8.3 should be noted. The univariate analysis involves a single outcome variable, while the multivariate analysis involves a vector of outcome variables. The eigenvalues for each matrix product, $\mathbf{E}^{-1}\mathbf{H}_{L\times G}$, $\mathbf{E}^{-1}\mathbf{H}_G$, and $\mathbf{E}^{-1}\mathbf{H}_L$, can be computed, and the same four multivariate test criteria, Wilks, Bartlett–Pillai, Roy, and Hotelling–Lawley, can be used to test each of the three hypotheses (the interaction between the factors, and the two main effects). Using the eigenvalues, the effect-size indices can also be computed for each effect using the equations for the effect-size statistics presented in Table 4.1, or the bias adjustments in Eq. (4.7). If the effects are judged to be generalizable and $r = \min(p, \mathrm{df}_\theta)$ is greater than 2, an examination of the linear discriminant functions would be appropriate to determine the number of meaningful dimensions and the identification of the constructs that define the differences among the groups. In Section 8.5 we present the SPSS program and output for the three omnibus tests, the linear discriminant functions, dimensionality analysis, the structure r's, and an LDF plot, followed by an interpretation of the findings. In Section 8.6, a contrast analysis using SPSS is presented.

8.4.2 Distribution Assumptions

As was discussed in the single grouping variable analysis, a valid interpretation of the inferential statistics and the linear discriminant functions is dependent on the analysis units being independent of each other, and within-population distribution of response variables being multivariate normal and having equal covariance matrices. If it is reasonable to believe that individual units have little influence on one another with respect to the response variables, the independence assumption is likely met. A violation of the multivariate normality assumption typically has little effect on the P values reported for MANOVA hypothesis tests. A violation of the assumption of equal covariance matrices, however, can affect P values and the interpretation of the LDFs. Under covariance heterogeneity, the LDFs are not interpretable. When group sizes are equal or at least similar, P values are minimally affected. But when group sizes differ substantially—by at least a factor of 2—P values can be overestimated or underestimated depending on the relationship between group size and the generalized variance. If group size and the generalized variance is positively related, P values will be overestimated. That is, the reported P value will be too large, and the hypothesis test is said to be *conservative*. If the group generalized variance and group size is negatively related, the reported P value will be too small and the hypothesis test is said to be *liberal*. If the P value is small in a conservative test, there would be sufficient evidence to reject the null hypothesis regardless of covariance heterogeneity. Similarly, if a large P value is reported in a liberal test, there would be insufficient evidence to reject the null hypothesis. If neither of these patterns occurs, a close examination of the variables may reveal the source or sources of the covariance heterogeneity, and variables or groups may be dropped from the analyses to achieve covariance homogeneity. Alternatively, the Yao test, which was discussed in Section 3.4, might be used to examine specific contrasts rather than omnibus tests.

The Box (1949) F, or chi-squared test for covariance equality, discussed in Section 3.3 and reported in the SPSS MANOVA program, can be used to help guide the data analysis process. However, we want to remind the reader that, because the Box test examines the equality of the KJ group variances and covariances simultaneously, it is an extremely powerful test and it is sensitive to nonnormality. Consequently, if the result of the Box test indicates covariance heterogeneity, the determinants of the group covariance matrices should be examined. If the natural logarithms of the $KJ + 1$ determinants are judged to be approximately equal, one can proceed with the usual MANOVA tests. Interpretation of the hypothesis tests should be guided by the discussion above with respect to the test being liberal or conservative.

8.5 COMPUTER APPLICATION I

In this section we present the SPSS syntax to conduct a MANOVA for a factorial design with two grouping variables. Generalization to situations involving more than two grouping variables is straightforward. The program presented here includes the same commands as in Sections 3.6, 4.3, and 5.3. Explanations of the commands can be found in those sections.

SPSS SYNTAX FOR COVARIANCE EQUALITY, OMNIBUS TEST STATISTICS, DIMENSIONALITY ANALYSIS, AND LINEAR DISCRIMINANT FUNCTIONS

```
manova Y1 Y2 Y3 Y4 by L(1,3) G(1,2)
/print=cellinfo(cov) homogeneity(box) error(cor)
signif(multiv eigen dimenr efsize)
/discrim=raw cor.
```

OUTPUT

Analysis: Homogeneity of Covariance Matrices

Cell Number .. 1
Variance–Covariance matrix

	Y1	Y2	Y3	Y4
Y1	.989			
Y2	.467	.400		
Y3	.556	.333	.667	
Y4	.344	.311	.667	.989

Determinant of Covariance matrix of dependent variables = .01277
LOG(Determinant) = 4.36047

- - - - - - - - - - - - - - - -

Cell Number .. 2
Variance–Covariance matrix

	Y1	Y2	Y3	Y4
Y1	1.511			
Y2	1.222	1.167		
Y3	.444	.389	1.389	
Y4	.311	.222	1.000	1.156

Determinant of Covariance matrix of dependent variables = .14202
LOG(Determinant) = −1.95178

- - - - - - - - - - - - - - - - -

Cell Number .. 3
Variance–Covariance matrix

	Y1	Y2	Y3	Y4
Y1	.678			
Y2	.678	.900		
Y3	−.100	−.211	.456	
Y4	−.467	−.578	.200	.711

Determinant of Covariance matrix of dependent variables = .01853
LOG(Determinant) = −3.98816

- - - - - - - - - - - - - - - - -

Cell Number .. 4
Variance–Covariance matrix

	Y1	Y2	Y3	Y4
Y1	1.378			
Y2	.733	.989		
Y3	−.178	−.044	.711	
Y4	−.289	.178	.711	1.156

Determinant of Covariance matrix of dependent variables =.16985
LOG(Determinant) = −1.77283

- - - - - - - - - - - - - - - - -

Cell Number .. 5
Variance–Covariance matrix

	Y1	Y2	Y3	Y4
Y1	1.600			
Y2	.978	1.122		
Y3	.711	.200	1.600	
Y4	.267	−.022	1.044	1.156

Determinant of Covariance matrix of dependent variables = .43835
LOG(Determinant) = −.82474

- - - - - - - - - - - - - - - - -

Cell Number .. 6
Variance–Covariance matrix

	Y1	Y2	Y3	Y4
Y1	.844			
Y2	.600	.900		
Y3	.044	−.211	.456	
Y4	−.422	−.578	.200	.711

Determinant of Covariance matrix of dependent variables = .04179

LOG(Determinant) = −3.17504

Pooled within-cells Variance–Covariance matrix

	Y1	Y2	Y3	Y4
Y1	1.167			
Y2	.780	.913		
Y3	.246	.076	.880	
Y4	−.043	−.078	.637	.980

Determinant of pooled Covariance matrix of dependent vars. = .17460
LOG(Determinant) = −1.74528

Multivariate test for Homogeneity of Dispersion matrices

Boxs M = 50.41202
F WITH (50,5351) DF = .81149, P = .826 (Approx.)
Chi-Square with 50 DF = 41.04534, P = .813 (Approx.)

-- WITHIN CELLS Correlations with Std. Devs. on Diagonal

	y1	y2	y3	y4
y1	1.080			
y2	.755	.955		
y3	.243	.085	.938	
y4	−.040	−.082	.686	.990

Interpretation: Homogeneity of Covariance Matrices

The current study involves an equal number of teachers for each School Level-by-Gender combination ($n_{kj} = 10$). Consequently, a violation of the covariance homogeneity assumption is not likely to invalidate the P values reported below for the omnibus and contrast hypothesis tests. We present the separate covariance matrices here to provide a comprehensive example of a multivariate analysis for a factorial design. The results of the Box test provides little evidence to indicate that the population covariance matrices differ, $\chi^2(50) \doteq 41.045$, $P \doteq .813$. An examination of the within-cells correlations indicates that variables Y_1 and Y_2 and variables Y_3 and Y_4 are correlated .755 and .686, respectively, but no other pair of variables appear to have a substantial correlation in this data set.

Analysis: Interaction Test

EFFECT .. L BY G

Multivariate Tests of Significance (S = 2, M = 1/2, N = 24 1/2)

Test Name	Value	Approx. F	Hypoth. DF	Error DF	Sig. of F
Pillais	.08851	.60196	8.00	104.00	.774
Hotellings	.09381	.58629	8.00	100.00	.787
Wilks	.91293	.59420	8.00	102.00	.781
Roys	.06712				

Note.. F statistic for WILKS' Lambda is exact.

Multivariate Effect Size

TEST NAME	Effect Size
Pillais	.044
Hotellings	.045
Wilks	.045

Eigenvalues and Canonical Correlations

Root No.	Eigenvalue	Pct.	Cum. Pct.	Canon Cor.
1	.072	76.694	76.694	.259
2	.022	23.306	100.000	.146

Dimension Reduction Analysis

Roots	Wilks L.	F	Hypoth. DF	Error DF	Sig. of F
1 TO 2	.91293	.59420	8.00	102.00	.781
2 TO 2	.97861	.37895	3.00	52.00	.769

Interpretation: Interaction Test

The statistics pertaining to an investigation of the interaction effects are intentionally specified first. Most methodologists suggest that interaction effects are to be examined first. If it is concluded, by judging the size of the associated P value and an effect-size index, that no interaction effects exist, the two sets of marginal means may be examined in terms of an omnibus test or, if appropriate, in terms of contrasts. If it is concluded that interaction effects do exist, one would typically examine simple effects or differences among cell group centroids within each row (simple column effect) or differences among cell group centroids within each column (simple row effect). Alternatively, the omnibus simple row or simple column effects could be ignored and contrasts involving cell centroids could be examined directly. See the Technical Notes for computer applications for simple effects and cell contrasts.

In the present context, SPSS reports three multivariate test criteria, Bartlett–Pillai (labeled Pillais by SPSS), Hotelling–Lawley (labeled Hotellings by SPSS), and Wilks. Using the two eigenvalues reported in the middle of the output section and the formulas presented in Table 3.3, the three statistics could be computed. For example, Wilks Λ is computed as $\prod_{v=1}^{r}[1/(1+\lambda_v)] \doteq [(1/1.072)(1/1.022)] \doteq .913$. The statistical test results are $F(8, 102) \doteq .594$, $P \doteq .781$, $\tau^2_{\text{adj}} \doteq .000$ ($\tau^2_{\text{adj}} = 1 - [(60-1)/(60-4-1)](1-.045)$). Note that because τ^2_{adj} is actually computed to be negative, we report it as 0. These results do not provide sufficient evidence to indicate an interaction between Gender and School Level. We may conclude that any difference (if there are any) among levels of one grouping variable are consistent across all levels of the second grouping variable. Because the omnibus test was not statistically significant, the interaction LDFs are not reported and the Dimension Reduction Analysis at the end of the output is irrelevant. Furthermore, because none of the dimension test results are significant at the default .15 level, the MANOVA program will not report any canonical discriminant functions or correlation analysis for this effect. (Note that the Wilks Λ of .9123 for the interaction effect is identical to the Wilks Λ in the first dimension test.) Given these findings, an examination of the Gender main effect and the School Level main effect would be appropriate, the results of which we examine next.

Analysis: Gender Test

EFFECT .. G

Multivariate Tests of Significance (S = 1, M = 1 , N = 24 1/2)

Test Name	Value	Exact F	Hypoth. DF	Error DF	Sig. of F
Pillais	.92784	163.94181	4.00	51.00	.000
Hotellings	12.85818	163.94181	4.00	51.00	.000
Wilks	.07216	163.94181	4.00	51.00	.000
Roys	.92784				

Note.. F statistics are exact.

- -

Multivariate Effect Size

TEST NAME	Effect Size
(All)	.928

- -

Eigenvalues and Canonical Correlations

Root No.	Eigenvalue	Pct.	Cum. Pct.	Canon Cor.
1	12.858	100.000	100.000	.963

- -

Raw discriminant function coefficients

Function No.

Variable	1
Y1	1.042
Y2	−.777
Y3	−.305
Y4	−.543

- -

Correlations between DEPENDENT and canonical variables

Canonical Variable

Variable	1
Y1	.517
Y2	.128
Y3	−.444
Y4	−.718

- -

Interpretation: Gender Test

Because the Gender variable has only two levels and $r = \min(p, \text{df}_G) = 1$, a single LDF can be determined, and all three multivariate criteria provide the exact same F statistic and P value. Using the Wilks criterion, $\Lambda = .072$, $F(4, 51) \doteq 163.942$,

$P \doteq .000$, $\tau^2_{adj} \doteq .923$; we would conclude that there is sufficient evidence that Male and Female teachers differ in mean reported stress levels. An examination of the structure r's indicates that Y_4 (Students) is the primary scale to define the stress construct that differentiates Male and Female teachers.

Analysis: School Level Test

EFFECT .. L

Multivariate Tests of Significance (S = 2, M = 1/2, N = 24 1/2)

Test Name	Value	Approx. F	Hypoth. DF	Error DF	Sig. of F
Pillais	1.59608	51.36928	8.00	104.00	.000
Hotellings	8.19735	51.23342	8.00	100.00	.000
Wilks	.03961	51.31288	8.00	102.00	.000
Roys	.83235				

Note.. F statistic for WILKS' Lambda is exact.

- -

Multivariate Effect Size

TEST NAME	Effect Size
Pillais	.798
Hotellings	.804
Wilks	.801

- -

Eigenvalues and Canonical Correlations

Root No.	Eigenvalue	Pct.	Cum. Pct.	Canon Cor.
1	4.965	60.568	60.568	.912
2	3.232	39.432	100.000	.874

- -

Dimension Reduction Analysis

Roots	Wilks L.	F	Hypoth. DF	Error DF	Sig. of F
1 TO 2	.03961	51.31288	8.00	102.00	.000
2 TO 2	.23627	56.02782	3.00	52.00	.000

- -

Raw discriminant function coefficients

Function No.

Variable	1	2
Y1	.398	−.914
Y2	−.560	−.115
Y3	−.726	.517
Y4	−.393	−.344

- -

Correlations between DEPENDENT and canonical variables

Canonical Variable

Variable	1	2
Y1	−.125	−.939
Y2	−.236	−.786
Y3	−.888	.003
Y4	−.829	.041

Interpretation: School Level Test

Each of the three multivariate criteria indicate the same conclusion: Teachers at the Elementary, Middle, and High School levels differ with regard to the four self-reported stress levels: $\Lambda \doteq .0396$, $F(8, 102) \doteq 51.313$, $P \doteq .000$, $\tau^2_{\text{adj}} \doteq .787$. An examination of the Eigenvalues and Canonical Correlations and the Dimension Reduction Analysis indicates that two dimensions may be used to describe School Level group differences, $\Lambda \doteq .236$, $F(3, 52) \doteq 56.028$, $P \doteq .000$. The first dimension, with $\lambda_1 \doteq 4.965$, accounts for 60.6 percent of the variation in the four-variable system, while the second dimension with $\lambda_2 \doteq 3.232$ accounts for 39.4 percent of the total variation in the four-variable system. The LDF mean vectors for the Elementary, Middle, and High School levels, obtained by multiplying the LDF weights by the group means for each outcome variable, are given below. Figure 8.1 provides a plot of the LDF mean vectors.

	Elementary School	**Middle School**	**High School**
LDF_1	−14.7	−14.0	−18.8
LDF_2	−7.8	−11.9	−10.3

From Figure 8.1 it appears that the High School teachers differ from both the Elementary and Middle School teachers on the first dimension, while Elementary teachers differ from both the Middle School and High School teachers on the second dimension.

To define the dimensions on which the groups differ, the structure r's (reported in the output by Correlations between DEPENDENT and canonical variables Canonical Variable) may be examined. The first dimension appears to be defined by variables Y_3 and Y_4, and the second dimension is defined by variables Y_1 and Y_2. Because variables Y_3 and Y_4 assess stress associated with working with parents and students, respectively, we might label this dimension Nonprofessional Source of Stress. Variables Y_1 and Y_2 assess stress associated with working with administrators and teachers, respectively,

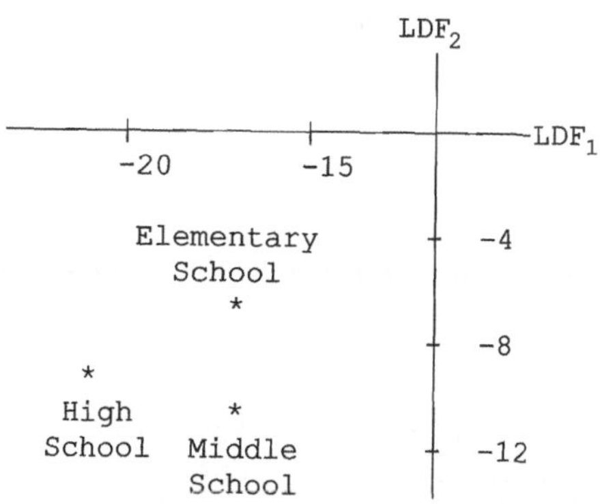

Figure 8.1 LDF plot for the three school levels.

so we might label the second dimension Professional Source of Stress. To make statistical comparisons among the three School Levels, focus tests are needed, and we turn to those analyses next.

8.6 COMPUTER APPLICATION II

The following SPSS syntax for contrasts in a factorial design yields two sets of contrasts. The first contrast statement for grouping variable L requests two contrasts, one pairwise and the other complex. The pairwise contrast compares the centroids of Middle School teachers with the centroid of High School teachers. The complex contrast compares the centroid of Elementary School teachers with the mean centroid of Middle School and High School teachers combined. The second contrast statement, again for L, requests two additional pairwise contrasts comparing the centroids of Elementary School with Middle School teachers, and comparing the centroids of Elementary School with High School teachers. Because the number of contrasts that can be requested on each contrast statement is limited to the degrees of freedom for the factor, all pairwise contrasts among levels of a grouping variable cannot be requested on a single contrast statement. In our example we chose to request one complex contrast and all possible pairwise contrasts among levels of the L grouping variable. The SPSS commands were defined in Section 4.5.

SPSS SYNTAX FOR COMPUTING MAIN EFFECT CONTRASTS IN A FACTORIAL DESIGN

```
manova
Y1 Y2 Y3 Y4 by L(1,3) G(1,2)
/print=signif(multiv eigen dimenr efsize)
/discrim=raw cor
/contrast(L)=special(1 1 1, 0 1 −1, 1 −.5 −.5)
/design G by L G L(1) L(2)
/contrast(L)=special(1 1 1, 1 −1 0, 1 0 −1)
/design G byL G L(1) L(2).
```

OUTPUT

Analysis: Complex Contrast Elementary School vs. Middle School and High School

EFFECT .. L(2)
Multivariate Tests of Significance (S = 1, M = 1 , N = 24 1/2)

Test Name	Value	Exact F	Hypoth. DF	Error DF	Sig. of F
Pillais	.77670	44.34929	4.00	51.00	.000
Hotellings	3.47838	44.34929	4.00	51.00	.000

Wilks	.22330	44.34929	4.00	51.00	.000
Roys	.77670				

Note.. F statistics are exact.

- -

Multivariate Effect Size

TEST NAME	Effect Size
(All)	.777

- -

Eigenvalues and Canonical Correlations

Root No.	Eigenvalue	Pct.	Cum. Pct.	Canon Cor.
1	3.478	100.000	100.000	.881

- -

Raw discriminant function coefficients

	Function No.
Variable	1
Y1	−.637
Y2	−.355
Y3	.135
Y4	−.484

- -

Correlations between DEPENDENT and canonical variables

	Canonical Variable
Variable	1
Y1	−.894
Y2	−.809
Y3	−.398
Y4	−.336

- -

Interpretation: Complex Contrast—Elementary School vs. Middle School and High School

The contrast results provide evidence to indicate a generalizable difference in reported stress of Elementary teachers versus Middle and High School teachers, $\Lambda = .223$, $F(4, 51) \doteq 44.349$, $P' \doteq .000$, $\tau^2_{\text{adj}} \doteq .760$. An examination of the structure r's indicates that the construct is defined primarily by the Administrator (Y_1) and Colleague (Y_2) scales.

Analysis: Pairwise Contrast—Middle School vs High School

EFFECT .. L(1)

Multivariate Tests of Significance (S = 1, M = 1 , N = 24 1/2)

Test Name	Value	Exact F	Hypoth. DF	Error DF	Sig. of F
Pillais	.82514	60.16688	4.00	51.00	.000
Hotellings	4.71897	60.16688	4.00	51.00	.000
Wilks	.17486	60.16688	4.00	51.00	.000
Roys	.82514				

Note.. F statistics are exact.

Multivariate Effect Size

TEST NAME	Effect Size
(All)	.825

Eigenvalues and Canonical Correlations

Root No.	Eigenvalue	Pct.	Cum. Pct.	Canon Cor.
1	4.719	100.000	100.000	.908

Raw discriminant function coefficients

	Function No.
Variable	1
Y1	.663
Y2	−.496
Y3	−.851
Y4	−.266

Correlations between DEPENDENT and canonical variables

	Canonical Variable
Variable	1
Y1	.174
Y2	.021
Y3	−.845
Y4	−.801

Interpretation: Pairwise Contrast—Middle School vs. High School

The results of the analysis contrasting Middle School and High School teachers indicate these school levels differ in their reported levels of stress, $\Lambda \doteq .175$, $F(4, 51) \doteq 60.167$, $P' \doteq .000$, $\tau^2_{\text{adj}} \doteq .812$. Examining the structure r's indicates that Parents (Y_3) and Students (Y_4) with structure r's equaling $-.845$ and $-.801$, respectively, define the construct that separates these two School Levels. Thus, these results indicate that stress from nonprofessional interactions separate teachers at the Middle School and High School levels.

Analysis: Pairwise Contrast—Elementary School vs. High School

EFFECT .. L(2)

Multivariate Tests of Significance (S = 1, M = 1 , N = 24 1/2)

Test Name	Value	Exact F	Hypoth. DF	Error DF	Sig. of F
Pillais	.81176	54.98106	4.00	51.00	.000
Hotellings	4.31224	54.98106	4.00	51.00	.000
Wilks	.18824	54.98106	4.00	51.00	.000
Roys	.81176				

Note.. F statistics are exact.

- -

Multivariate Effect Size

TEST NAME	Effect Size
(All)	.812

- -

Eigenvalues and Canonical Correlations

Root No.	Eigenvalue	Pct.	Cum. Pct.	Canon Cor.
1	4.312	100.000	100.000	.901

- -

Raw discriminant function coefficients

	Function No.
Variable	1
y1	−.149
y2	−.536
y3	−.340
y4	−.516

- -

Correlations between DEPENDENT and canonical variables

	Canonical Variable
Variable	1
y1	−.604
y2	−.618
y3	−.751
y4	−.680

- -

Interpretation: Pairwise Contrast—Elementary vs. High School

The contrast analysis comparing Elementary School with High School teachers indicates the observed difference in LDF centroids is generalizable to the populations they represent, $\Lambda \doteq .188$, $F(4, 51) \doteq 54.981$, $P' \doteq .000$, $\tau^2_{\text{adj}} \doteq .798$. The structure r's for the LDF has a narrow range between −.604 and −.751, indicating that all four areas of stress contribute to the definition of the stress construct separating Elementary School and High School teachers.

Analysis: Pairwise Contrast—Elementary School vs. Middle School

EFFECT .. L(1)

Multivariate Tests of Significance (S = 1, M = 1 , N = 24 1/2)

Test Name	Value	Exact F	Hypoth. DF	Error DF	Sig. of F
Pillais	.76552	41.62631	4.00	51.00	.000
Hotellings	3.26481	41.62631	4.00	51.00	.000
Wilks	.23448	41.62631	4.00	51.00	.000
Roys	.76552				

Note.. F statistics are exact.

- -

Multivariate Effect Size

TEST NAME	Effect Size
(All)	.766

- -

Eigenvalues and Canonical Correlations

Root No.	Eigenvalue	Pct.	Cum. Pct.	Canon Cor.
1	3.265	100.000	100.000	.875

- -

Raw discriminant function coefficients

	Function No.
Variable	1
Y1	.968
Y2	.019
Y3	−.632
Y4	.272

- -

Correlations between DEPENDENT and canonical variables

	Canonical Variable
Variable	1
Y1	.904
Y2	.735
Y3	−.152
Y4	−.180

- -

Interpretation: Pairwise Contrast—Elementary vs. Middle School

Self-reported levels of stress are different between Elementary and Middle School level teachers, $\Lambda \doteq .234$, $F(4, 51) \doteq 41.626$, $P' \doteq .000$, $\tau^2_{\text{adj}} \doteq .749$. The structure r's indicate that the construct separating the two school levels is defined primarily on the basis of Administrators (Y_1) and Colleagues (Y_2).

8.7 NONORTHOGONAL DESIGN

With two or more grouping variables, a researcher may have what is termed a *nonorthogonal* design, either as part of a research plan or because of complexities associated with data collection. For a two-factor A-by-B design, let n_{kj} denote the frequency (i.e., number of analysis units) for cell (k, j), let $n_{k.}$ denote the frequency for row k, let $n_{.j}$ denote the frequency for column j, and let N denote the total sample size. Then, a two-factor design is said to be *orthogonal* if for any row–column combination,

$$n_{kj} = \frac{n_{k.}n_{.j}}{N},$$

and otherwise nonorthogonal. An implication of nonorthogonality is that the test statistics (univariate or multivariate) for the $A \times B$, A, and B effects are not independent. Thus, statistics other than the usual test statistics are used, depending on the type of hypotheses (involving weighted or unweighted means or mean vectors) one chooses to test. This testing issue has been reviewed extensively for the univariate situation (see, e.g., Maxwell and Delaney, 2000, pp. 271–297) and will not be covered here. For a multivariate analysis, the two more relevant test strategies may be implemented via the SPSS MANOVA and SAS GLM programs.

8.8 OUTCOME VARIABLE ORDERING AND DELETION

Variable ordering may be particularly informative in the context of a two-factor design. In a two-factor design there are three sets of effects that are of poterntial interest: the effect associated with the two-factor interaction, plus the effects associated with each of the two factors. Finding that one or two variable are relatively more important for one set of effects, whereas one or two *other* variables are important of a different set of effects may be substantitively revealing. But how does one go about assessing relative variable importance in a two-factor setting? The most straightforward approach may be to conduct $p\ (p-1)$ variable analyses.

For example, consider the Ethington data (3GED) in Appendix A that involves a Race (3 levels)-by-Grade (3 levels) design with nine outcome variables. To order the variables, then, nine eight-variable analyses would be conducted. (These normal-based analyses are conducted for illustrative purposes.) Results of the nine analyses are summarized in Table 8.7. It is found that there is no "real" Race-by-Grade interaction effect [$\Lambda \doteq .899$, $F(36, 972.4) \doteq 0.743$, $P \doteq .866$, $\tau^2_{\text{adj}} \doteq .000$]. Therefore, the relative contribution of the nine outcome variables to the interaction effect is not considered. It is judged that the Race effect [$\Lambda \doteq .799$, $F(18, 494) \doteq 3.261$, $P \doteq .000$, $\tau^2_{\text{adj}} \doteq .074$] and the Grade effect [$\Lambda \doteq .886$, $F(18, 494) \doteq 1.707$, $P \doteq .000$, $\tau^2_{\text{adj}} \doteq .026$] are both "real." From the two sets of "eye ball" variable ranks, it might be concluded that Y_5 (Student–Faculty Effort) does not contribute much to the Race effect, but does contribute some to the Grade effect. It might also be concluded

TABLE 8.7 $\Lambda_{(i)}$ Values for the Two-Factor (3 x 3) Ethington Data

	Race × Grade		Race		Grade	
Deleted	$\Lambda_{(i)}$	Rank	$\Lambda_{(i)}$	Rank	$\Lambda_{(i)}$	Rank
Y_1	.909	4.5	.808	9.5	.905	3.5
Y_2	.902	4.5	.816	4.5	.904	3.5
Y_3	.911	4.5	.851	1.5	.905	3.5
Y_4	.912	4.5	.802	9.0	.887	8
Y_5	.903	4.5	.808	7.5	.907	3.5
Y_6	.904	4.5	.813	4.5	.888	8
Y_7	.907	4.5	.814	4.5	.896	3.5
Y_8	.907	4.5	.811	4.5	.890	8
Y_9	.922	1	.848	1.5	.896	3.5
(None)	.899		.799		.886	

that Y_3 (Instruction Received) and Y_9 (Science Effort) contribute to both Race and Grade effects.

It is recognized that the two sets of outcome variable ranks for the real Race and Grade effects are not very "clear-cut." However, this approach to variable ordering in a two-factor design may very well reveal some interesting substantive information. (If the interaction effect is judged to be real, then variable ordering with respect to main effects of the two factors would be inappropriate—see Section 8.4.)

A McHenry analysis (as described in Section 6.2.2) *may* also be of interest in a two-factor ($A \times B$) design context to delete some outcome variables. One could (if the equal covariance matrix condition is met) examine the $\Lambda_{(i)}$ values or the $F_{(i)}$ values for the p outcome variables with respect to the AB effects, the A effect, and the B effect to determine if one or more outcome variables might be deleted. (Group contrast effects may also be considered.) Of course, one would delete only the variable(s) that have "low" ranks across *all* effects of interest.

8.9 SUMMARY

In this chapter we generalized the multivariate procedures discussed in the single grouping variable case (Chapters 3 to 5) to the case where analysis units are grouped on the basis of two grouping variables. Additional grouping variables could be added with little change in the data analysis steps. The additional grouping variables would just add greater complexity in the calculations and interpretation. Because our example involved an orthogonal design, the generalization was straightforward. If the design is nonorthogonal, the calculations are more complex, and additional care must be given with respect to the interpretation of the results. The use of computer software (e.g., SPSS) eliminates the computational difficulty but the interpretation cautions remain.

Technical Notes

1. *Simple Effects* One data-analytic approach that is often taken in a two-grouping-variable design, when a statistically significant interaction is identified, is to reconceptualize the MANOVA model and recalculate the sums-of-squares. Consider a study having grouping variables G and L. In the new model, the sum-of-squares associated with the interaction effect ($G \times L$) are combined with the sum-of-squares for a main effect, G, and a new set of sums-of-squares is computed. The sum-of-squares for the second main effect, L, is unaffected. The new sum of squares is used to examine differences between levels of variable G at each level of variable L. Such analyses are referred to as *simple effects*. Similar model reconceptualizations can be realized with higher order factorial designs.

In this note we only present the syntax for conducting a simple effects analysis using the SPSS MANOVA program and the output using the school stress data that were analyzed earlier. The interpretation of the results would be similar to our earlier analysis.

SPSS SYNTAX FOR SIMPLE EFFECTS

```
manova
Y1 Y2 Y3 Y4 by L(1,3) G(1,2)
/print=signif(multiv eigen dimenr efsize)
/discrim=raw cor
/design G, L w G(1), L w G(2)
/design L G w L(1) G w L(2) G w L(3).
```

/design G, L w G(1), L w G(2) this design statement requests an analysis of the differences among levels of variable *L* within level 1 of variable *G* and within level 2 of variable *G*. The second design statement requests an analysis of differences between levels of variable *G* within levels 1, 2, and 3 of variable *L*.

The design statement used here is an alternative conceptualization of the MANOVA model, which examines differences between levels of variable *G* for each level of variable *L*. Typically only one model reconceptualization is of interest. It might also be noted that the simple effects model may be of greater interest than the complete factorial model. If so, the full factorial model need not be examined. We do not include the output for the second design statement.

OUTPUT

EFFECT .. L W G(2)

Multivariate Tests of Significance (S = 2, M = 1/2, N = 24 1/2)

Test Name	Value	Approx. F	Hypoth. DF	Error DF	Sig. of F
Pillais	1.36856	28.17567	8.00	104.00	.000
Hotellings	4.35794	27.23710	8.00	100.00	.000
Wilks	.09932	27.70777	8.00	102.00	.000
Roys	.70336				

Note.. F statistic for WILKS' Lambda is exact.

- -

Multivariate Effect Size

TEST NAME	Effect Size
Pillais	.684
Hotellings	.685
Wilks	.685

- -

Eigenvalues and Canonical Correlations

Root No.	Eigenvalue	Pct.	Cum. Pct.	Canon Cor.
1	2.371	54.408	54.408	.839
2	1.987	45.592	100.000	.816

Dimension Reduction Analysis

Roots	Wilks L.	F	Hypoth. DF	Error DF	Sig. of F
1 TO 2	.09932	27.70777	8.00	102.00	.000
2 TO 2	.33480	34.43897	3.00	52.00	.000

Raw discriminant function coefficients

Variable	Function No. 1	2
Y1	.294	−.842
Y2	−.511	−.202
Y3	−.716	.579
Y4	−.396	−.315

Correlations between DEPENDENT and canonical variables

Variable	Canonical Variable 1	2
Y1	−.199	−.911
Y2	−.273	−.809
Y3	−.904	.091
Y4	−.825	.113

EFFECT .. L W G(1)

Multivariate Tests of Significance (S = 2, M = 1/2, N = 24 1/2)

Test Name	Value	Approx. F	Hypoth. DF	Error DF	Sig. of F
Pillais	1.29164	23.70454	8.00	104.00	.000
Hotellings	3.93322	24.58261	8.00	100.00	.000
Wilks	.11939	24.15018	8.00	102.00	.000
Roys	.72363				

Note.. F statistic for WILKS' Lambda is exact.

Multivariate Effect Size

TEST NAME	Effect Size
Pillais	.646
Hotellings	.663
Wilks	.654

Eigenvalues and Canonical Correlations

Root No.	Eigenvalue	Pct.	Cum. Pct.	Canon Cor.
1	2.618	66.570	66.570	.851
2	1.315	33.430	100.000	.754

Dimension Reduction Analysis

Roots	Wilks L.	F	Hypoth. DF	Error DF	Sig. of F
1 TO 2	.11939	24.15018	8.00	102.00	.000
2 TO 2	.43199	22.79105	3.00	52.00	.000

- -

Raw discriminant function coefficients

	Function No.	
Variable	1	2
Y1	.479	−1.000
Y2	−.608	.009
Y3	−.722	.456
Y4	−.395	−.363

- -

Correlations between DEPENDENT and canonical variables

	Canonical Variable	
Variable	1	2
Y1	−.071	−.956
Y2	−.215	−.742
Y3	−.868	−.080
Y4	−.828	−.023

- -

2. *Cell Contrasts* In this Note we provide the syntax and output for contrasting cell centroids. The MANOVA program in SPSS is incapable of testing contrasts among cell means, but the GLM program can contrast cell means. A disadvantage of the GLM program, however, is that it does not provide LDFs or structure r's for the contrasts.

SPSS SYNTAX FOR CELL CONTRASTS USING THE GENERAL LINEAR MODEL (GLM) PROGRAM

```
GLM
Y1 Y2 Y3 Y4  BY G L
/METHOD=SSTYPE(3)
/lmatrix "test that j1k1=j1k2" L 1 -1 0 G*L 1 -1 0 0 0 0
/lmatrix "test that j1k1=j1k3" L 1 0 -1 G*L 1 0 -1 0 0 0
/lmatrix "test that j1k2=j1k3" L 0 1 -1 G*L 0 1 -1 0 0 0
/lmatrix "test that j1k1=j2k1" G 1 -1  G*L 1 0 0 -1 0 0
/INTERCEPT=INCLUDE
/CRITERIA=ALPHA(.05)
/DESIGN=G L G*L .
```

BY G L the GLM program does not require levels of the variables to be specified. variable *G* is entered first and variable *L* is entered second. The order of variable entry is very important when requesting cell contrasts. The current study has two levels of variable *G* and three levels of variable *L*. With the current variable entry order the six cells of the design are ordered as follows: j1k1 j1k2 j1k3 j2k1 j2k2 j2k3. The levels of the second variable are changing quicker than the levels of the variable entered first.
method = SSTYPE(3) is the default SPSS system command to use the unweighted marginal means or regression approach for computing the sum-of-squares.
/lmatrix is GLM program command to request a test of a contrast.
"test that j1k1 = j1k2" the statement made within quotation marks is a label to help identify which comparison is being made.
L 1 −1 0 G*L 1 −1 0 0 0 0 are the weights identifying which cells are to be compared. This contrast compares level 1 with level 2 of variable *L* both of which are within level 1 of variable *G* (j1k1 − j1k2). Because the interest is comparing levels 1 and 2 of variable *L*, the first part of the contrast statements identifies the two levels of variable *L* that are to be contrasted L 1 −1 0. The second part of the contrast statement identifies the two cells to be compared; i.e., j1k1 vs. j1k2. The weights for these cell means are 1 and −1 and weights for the remaining cells are 0, G*L 1 −1 0 0 0 0.
G 1 −1 G*L 1 0 0 −1 0 0 these weights request the contrast between levels 1 and 2 of variable *G* at level 1 of variable *L*. The cells that are being compared are j1k1 vs. j2k1.

OUTPUT

Multivariate Tests(c)

Effect		Value	F	Hypothesis df	Error df	Sig.
Intercept						
	Pillai's	.998	5144.493(a)	4.000	51.000	.000
	Wilks'	.002	5144.493(a)	4.000	51.000	.000
	Hotelling	403.490	5144.493(a)	4.000	51.000	.000
	Roy's	403.490	5144.493(a)	4.000	51.000	.000
G	Pillai's	.928	163.942(a)	4.000	51.000	.000
	Wilks'	.072	163.942(a)	4.000	51.000	.000
	Hotelling	12.858	163.942(a)	4.000	51.000	.000
	Roy's	12.858	163.942(a)	4.000	51.000	.000
L	Pillai's	1.596	51.369	8.000	104.000	.000
	Wilks'	.040	51.313(a)	8.000	102.000	.000
	Hotelling	8.197	51.233	8.000	100.000	.000
	Roy's	4.965	64.545(b)	4.000	52.000	.000

G*L	Pillai's	.089	.602	8.000	104.000	.774
	Wilks'	.913	.594(a)	8.000	102.000	.781
	Hotelling	.094	.586	8.000	100.000	.787
	Roy's	.072	.935(b)	4.000	52.000	.451

a Exact statistic b The statistic is an upper bound on F that yields a lower bound on the significance level. c Design: Intercept+G+L+G * L

Cell contrast of male Elementary teachers (g111) versus male Middle School teachers (g112)

Contrast Results (K Matrix)(a)

Contrast			Dependent Variable			
			y1	y2	y3	y4
L1	Contrast Estimate		−3.800	−2.500	.300	.500
	Hypothesized Value		0	0	0	0
	(Estimate - Hypothesized)		−3.800	−2.500	.300	.500
	Std. Error		.483	.427	.419	.443
	Sig.		.000	.000	.478	.264
	95\%	Lower Bound	−4.768	−3.357	−.541	−.387
		Upper Bound	−2.832	−1.643	1.141	1.387

a Based on the user-specified contrast coefficients (L') matrix: test that j1k1=j1k2

Multivariate Test Results

	Value	F	Hypothesis df	Error df	Sig.
Pillai's	.572	17.064(a)	4.000	51.000	.000
Wilks'	.428	17.064(a)	4.000	51.000	.000
Hotelling	1.338	17.064(a)	4.000	51.000	.000
Roy's	1.338	17.064(a)	4.000	51.000	.000

a Exact statistic

Cell contrast of male Elementary School teachers (j1k1) versus male High School teachers (j1k3)

Contrast Results (K Matrix)(a)

Contrast			Dependent Variable			
			y1	y2	y3	y4
L1	Contrast Estimate		−2.800	−2.500	3.600	−3.500
	Hypothesized Value		0	0	0	0
	(Estimate - Hypothesized)		−2.800	−2.500	3.600	−3.500
	Std. Error		.483	.427	.419	.443
	Sig.		.000	.000	.478	.264
	95\%	Lower Bound	−3.668	−3.357	−4.441	−4.387
		Upper Bound	−1.732	−1.643	−2.759	−2.613

a Based on the user-specified contrast coefficients (L') matrix:
test that j1k1=j1k3

Multivariate Test Results

	Value	F	Hypothesis df	Error df	Sig.
Pillai's	.681	27.165(a)	4.000	51.000	.000
Wilks'	.319	27.165(a)	4.000	51.000	.000
Hotelling	2.131	27.165(a)	4.000	51.000	.000
Roy's	2.131	27.165(a)	4.000	51.000	.000

a Exact statistic

Cell contrast of male Middle School teachers (j1k2) versus male High School teachers (j1k3)

Contrast Results (K Matrix)(a)

Contrast		Dependent Variable			
		y1	y2	y3	y4
L1	Contrast Estimate	1.100	.000	−3.900	−4.000
	Hypothesized Value	0	0	0	0
	(Estimate - Hypothesized)	1.100	.000	−3.900	−4.000
	Std. Error	.483	.427	.419	.443
	Sig.	.027	1.000	.000	.000
	95\% Lower Bound	.132	−.857	−4.741	−4.887
	Upper Bound	2.068	.857	−3.059	−3.113

a Based on the user-specified contrast coefficients (L') matrix:
test that j1k2=j1k3

Multivariate Test Results

	Value	F	Hypothesis df	Error df	Sig.
Pillai's	.709	30.994(a)	4.000	51.000	.000
Wilks'	.291	30.994(a)	4.000	51.000	.000
Hotelling	2.431	30.994(a)	4.000	51.000	.000
Roy's	2.431	30.994(a)	4.000	51.000	.000

a Exact statistic

Cell contrast of male Elementary School teachers (j1k1) versus female Elementary School teachers (j2k1)

Contrast Results (K Matrix)(a)

Contrast		Dependent Variable			
		y1	y2	y3	y4
L1	Contrast Estimate	−3.300	−.300	2.500	4.500
	Hypothesized Value	0	0	0	0
	(Estimate - Hypothesized)	−3.300	−.300	2.500	4.500

Std. Error	.483	.427	.419	.443
Sig.	.000	.486	.000	.000
95\% Lower Bound	−4.268	−1.157	1.659	3.613
Upper Bound	−2.332	.557	3.341	5.387

a Based on the user-specified contrast coefficients (L') matrix: test that j1k1=j2k1

Multivariate Test Results

	Value	F	Hypothesis df	Error df	Sig.
Pillai's	.793	48.990(a)	4.000	51.000	.000
Wilks'	.207	48.990(a)	4.000	51.000	.000
Hotelling	3.842	48.990(a)	4.000	51.000	.000
Roy's	3.842	48.990(a)	4.000	51.000	.000

a Exact statistic

3. The test strategy for a two-factor design suggested here is the one often suggested in the univariate context (e.g., Maxwell and Delaney, 2000, p. 263). The strategy of examining interaction effects first, followed by either main effects (if interaction effects are not generalizable) or simple effects (if interaction effects are generalizable) is advocated for interpretation reasons, not for statistical reasons. Tukey (1991) has suggested, in a univariate context, a different strategy. He suggests the investigation of conditional simple effects. For example, simple G effects at, say, level j_1 are conditioned on the main G effects; that is, the simple effects are adjusted for the main effects, or the simple effects are investigated in the presence of the main effects. This may be opposed to the strategy above, which implies investigating simple G effects at level k_1 ignoring the main G effects. Discussions of details of investigating conditional simple effects in a multivariate context are not known to be available.

Definition *Latin square:* Roman in conservative garb.

EXERCISES

1. Consider an eight-variable, 3×4 MANOVA situation.

(a) How many *sets* of LDFs may be obtained?

(b) How many LDFs in each set?

2. Using the data in Table 8.2, provide the following vectors that would be used to compute the main effect and interaction hypothesis **SSCP** matrices:

(a) $\mathbf{y}_{.11}$

(b) $\mathbf{y}_{..1}$

(c) $\mathbf{y}_{.1.}$

(d) $\mathbf{y}_{...}$

3. Using the data in Table 8.2 compute $\mathbf{H}_G$.

Exercises 4 to 10 refer to a 2×3 $(A \times B)$ completely randomized factorial design with $n = 15$ and four outcome measures.

4. Suppose the eigenvalues for the product matrix $\mathbf{E}^{-1}\mathbf{H}_{A \times B}$ are .080 and .005. Compute:
   (a) Wilks Λ
   (b) F
   (c) ν_1 and ν_2

5. For the A main effect Λ equals .983. What is the canonical correlation between the outcome measures and variable A?

6. The eigenvalues, canonical correlations, and dimension reduction analysis is reported below for the B main effect. How many dimensions would you consider in interpreting the structure underlying the resultant group differences? On what numerical results do you base your answer?

Eigenvalues and Canonical Correlations

Root No.	Eigenvalue	Pct.	Cum. Pct.	Canon Cor.
1	1.219	97.579	97.579	.741
2	.030	2.421	100.000	.171

Dimension Reduction Analysis

Roots	Wilks L.	F	Hypoth. DF	Error DF	Sig. of F
1 TO 2	.43738	10.36923	8.00	162.00	.000
2 TO 2	.97064	.82691	3.00	82.00	.483

7. Use the eigenvalues reported in Exercise 6 to compute the Serlin adjusted ξ^2 effect size index.

8. The eigenvalue for a pairwise contrast between levels 1 and 3 of variable B equals .576. What does the squared canonical correlation equal?

9. Given the eigenvalue in Exercise 8, using any of the four multivariate test criteria, what would the F statistic for testing the hypothesis of no difference among centroids equal? For this hypothesis test, what is the number of degrees of freedom?

10. If the researcher was interested in comparing levels 1 and 2 of variable B within level 2 of variable A (cell means contrast), state the SPSS/lmatrix command to provide this analysis. Assume variable A is entered first.

Computer Applications

Exercises 11 to 20 require the analysis of the 5-group Ethington data set (5GED) described in Appendix A. For these exercises the researchers were interested in the analysis of a factorial design with Race (3 levels) and Grade (5 levels) as the grouping variables. Use the SPSS (or SAS) computer software package to compare group centroids based on the 9 outcome variables (Counselor Interaction, Writing and Speaking Skills, Self-Understanding, Instruction Received, Library Effort, Student–Faculty Effort, Interstudent Effort, Art/Music/Theater Effort, Writing Effort, and Science Effort).

11. Analyze the data set to test the assumption of homogeneous covariance matrices.

(a) What is the numerical value of the Box M statistic?

(b) What are the numerical values of χ^2, degrees of freedom, P value?

(c) Is there a "relationship" between the sample size and the log determinants of the cell covariance matrices?

(d) Based on these results, is the statistical validity of the results questionable?

12. Is there strong evidence of an interaction between Race and Grade with respect to the group centroids?

(a) Provide Λ, F, degrees of freedom, and P value.

(b) State the value of the adjusted and unadjusted ξ^2.

13. Is there strong evidence of a Race main effect?

(a) Provide statistical support for your answer including Λ, F, degrees of freedom, and P value.

(b) Provide an adjusted and unadjusted effect size index value.

14. How many dimensions are needed to describe Race group separation?

(a) What proportion of variation in the 9-variable system is explained by each of the constructs?

(b) What proportion of variation in each identified construct is explained by the Race grouping variable?

(c) What conclusion regarding the number of constructs is needed to describe group separation based on the Dimension Reduction Analysis?

15. Is there evidence of a Grade main effect?

(a) Provide statistical support for your answer including Λ, F, degrees of freedom, and P value.

(b) Provide an unadjusted and adjusted effect size index.

16. How many dimensions are needed to describe Grade group separation?

- **(a)** What proportion of variation in the 9-variable system is explained by each meaningful construct?
- **(b)** What proportion of variation in each identified construct is explained by the Grade grouping variable?
- **(c)** What conclusion is reached regarding the number of constructs needed to describe Grade group separation based on the Dimension Reduction Analysis?

17. For the Grade variable, what outcome variables seem to define each of the meaningful constructs based on the structure r's.

18. Is there evidence to indicate a difference between group mean centroids of levels 1 and 5 of the Grade variable? That is, is the contrast of group centroids between students who generally earn A grades and students who generally earn C or C− grades significant?

- **(a)** Provide statistical support for you answer including Λ, F, degrees of freedom, and P value.
- **(b)** Provide an unadjusted and adjusted effect-size index value.

19. Compare the mean centroids for the levels of Grade within the sample of Black students. Do the centroids differ?

- **(a)** Provide statistical support for your answer: Λ, F, degrees of freedom, and P value.
- **(b)** Provide an unadjusted and adjusted effect size index value.
- **(c)** What proportion of variation in the 9-variable system is explained by each of the constructs?
- **(d)** What proportion of variation in each identified construct is explained by the Race grouping variable?
- **(e)** What conclusion is reached regarding the number of constructs needed to describe group separation based on the Dimension Reduction Analysis?
- **(f)** What variables define the construct(s) that separate the Grade groups?

20. Contrast the centroids of cell means comparing Black students who generally earn A grades with Black students who generally earn C or C− grades. Is there evidence of a difference between the centroids?

- **(a)** Provide statistical support for you answer: Λ, F, degrees of freedom, and P value.
- **(b)** Provide an adjusted and unadjusted effect size index value.

CHAPTER 9

Analysis of Covariance

9.1 INTRODUCTION

Up to this point we have been presenting analysis strategies that are appropriate for studies based on research designs that may be referred to as post-test-only designs. That is, the only quantitative data available are collected following the assignment of the analysis units to a level of the grouping variable. While this design is very common, it can have two major limitations. *One* limitation of the post-test-only design is that, if analysis units are not randomly assigned to levels of the grouping variable, multiple interpretations of the results may be possible. Without random assignment, the populations being compared are likely to lack initial equivalency. The lack of initial equivalence, which is sometimes referred to as *selection bias*, introduces a number of confounding variables, any one of which can offer an alternative explanation for observed differences in the response variables, in addition to the grouping variable. The *second* limitation with the post-test-only design is that it is very inefficient. That is, to have good statistical power, large sample sizes are needed unless the grouping variable effect is very large.

One popular solution for both limitations is to obtain additional quantitative information or pretest data on the units before the assignment to the levels of the grouping variable. [The pretest data need not be the same variable as the response variable, only that the pretest data be correlated with the response variable(s).] The research design for such studies is referred to as the *multiple group pretest–posttest design*. If the units are not randomly assigned to the levels of the grouping variable, the pretest data can be used to take into account or "adjust" for initial differences that may exist among the units on the pretest variable. This "adjustment," however, may not be sufficient to compensate for a lack of initial equivalency. A lack of reliability in the pretest data may result in an underadjustment. In addition, the populations being compared may differ on other variables that have not been considered. As a result, while the pretest data may reduce the uncertainty of the interpretation to some degree, the interpretation of

Applied MANOVA and Discriminant Analysis, Second Edition, by Carl J. Huberty and Stephen Olejnik
Copyright © 2006 John Wiley & Sons, Inc.

group differences must be interpreted cautiously. [See Porter and Raudenbush (1987) for further discussion.]

When the units are randomly assigned to the levels of the grouping variable, the pretest data can improve the efficiency of the posttest-only design by reducing the error variance. How much the error variance is reduced will depend on the correlation between the pretest and posttest data. Reducing the error variance will increase the statistical power of the analysis.

The pretest data are referred to as measures on a *covariate*, and the statistical procedure for analyzing such data is called *analysis of covariance*. In this chapter, after briefly reviewing the univariate analysis of covariance, ANCOVA, we present the multivariate approach to analysis of covariance, MANCOVA. Our presentation follows the organization of Chapter 8. We introduce a new research context and a new data set. Following the univariate review, the multivariate approach is presented. The multivariate approach begins with a comment on data conditions, followed by the omnibus test, dimension reduction, and construct identification. We conclude with contrast analyses. We limit our presentation to a single grouping variable and a single covariate. The generalization of MANCOVA to the factorial design and multiple covariates is straightforward. We do, however, provide the SPSS syntax for conducting the multiple covariate analysis in the Technical Note at the end of this chapter.

9.2 RESEARCH CONTEXT

A quasi-experimental study was conducted by Baumann et al. (2003) that compared three approaches for improving student vocabulary skills. Four fifth-grade classes agreed to participate in the study. One class, $n_1 = 24$, was taught Morphemic analysis (prefixes), MO; a second class, $n_2 = 22$, was taught to use Context clues, CO; a third class, $n_3 = 21$, was taught both Morphemic analysis and Context clues, MC; and a fourth class, $n_4 = 21$, was a traditionally instructed classroom, and served as a Control group, C. Although classrooms were randomly assigned to the instructional strategies, students within the classrooms were not randomly assigned. Before beginning the instructional interventions, all students completed a 40-item vocabulary test, X, which assessed student prior knowledge of the words they would learn during the study. In addition, a standardized vocabulary test, Degrees of Word Meaning, was also completed by the students. (We consider this last variable in the Technical Note at the end of this chapter.)

Following the instruction period, several posttests were given. Data from the 88 participating students are presented in Table 9.1. (An SPSS data file containing these data, labeled BAUMANN2, is available at the Wiley website.)

Our discussion focuses on four outcome variables. The first outcome variable, Y_1, was based on a 20-item test, requiring students to provide the definition of words (a Production task). The words provided were the words taught in the Morphemic lessons. The second outcome variable, Y_2, was based on a 10-item multiple-choice test requiring students to recognize word meaning (a Recognition task). Words on this test were words taught in the Morphemic lessons. The third outcome variable,

TABLE 9.1 Vocabulary Scores for Three Treatment Interventions and a Control Group[a]

Morphemic Only					Context Only					Morphemic and Context					Control				
X	Y_1	Y_2	Y_3	Y_4	X	Y_1	Y_2	Y_3	Y_4	X	Y_1	Y_2	Y_3	Y_4	X	Y_1	Y_2	Y_3	Y_4
26	18	8	9	7	26	8	7	15	7	25	17	10	16	7	15	5	7	5	7
23	18	10	14	9	25	12	8	19	10	13	10	8	8	6	15	3	6	3	2
7	8	0	3	1	22	12	8	16	9	11	6	1	7	3	15	5	3	7	5
18	19	7	7	6	22	9	7	15	10	25	14	9	19	10	13	4	4	8	3
18	18	5	5	2	14	4	7	14	7	11	8	5	7	4	29	13	9	15	10
26	20	10	13	10	24	2	5	12	8	30	16	10	19	10	25	10	8	16	9
13	16	10	4	6	19	7	5	16	9	22	15	8	12	8	12	6	4	6	3
14	19	10	7	4	26	13	8	18	10	28	19	10	19	10	23	11	4	8	8
16	10	7	0	2	4	1	1	0	4	23	16	9	9	8	28	14	9	8	6
26	20	10	16	9	15	9	6	18	9	26	17	10	18	8	16	4	7	4	7
15	16	7	6	6	19	1	4	3	1	11	5	6	8	4	13	3	2	5	3
20	18	10	10	5	24	15	10	20	10	26	16	9	16	10	27	14	9	15	8
12	10	8	0	3	19	11	6	14	10	24	17	10	18	9	21	11	8	12	10
24	17	10	12	8	6	0	0	0	4	7	15	7	12	6	22	10	6	10	7
36	20	10	19	9	21	11	7	11	9	27	18	8	20	10	10	4	3	4	4
20	13	8	0	3	34	14	9	19	10	25	16	7	11	6	27	13	8	15	10
26	19	10	12	10	33	13	9	18	10	12	5	4	9	3	23	12	6	13	9
10	2	6	0	2	27	14	7	16	10	20	20	10	16	9	12	7	5	5	4
14	15	7	7	5	27	15	8	18	10	12	8	6	4	5	20	3	4	8	5
14	18	9	8	5	25	10	5	17	8	9	6	6	9	4	26	14	5	15	10
11	15	7	5	3	14	0	3	14	8	37	19	10	19	10	8	0	3	1	3
20	19	10	13	9	22	10	8	16	9										
24	16	10	10	10															
33	18	10	19	10															

[a] X = Pretest using a subset of lesson and transfer words; Y_1 = Morphemic Lesson Words—Production; Y_2 = Morphemic lesson words—Recognition; Y_3 = Context Lesson Words—Production; Y_4 = Context Lesson Words—Recognition.

TABLE 9.2 Means and Variances for Morphemic Only (MO), Context Only (CO), Morphemic and Context (MC), and Control (C) Groups on Five Vocabulary Tests[a]

	MO $n_1 = 24$		CO $n_2 = 22$		MC $n_3 = 21$		C $n_4 = 21$	
Vaiable	Mean	s^2	Mean	s^2	Mean	s^2	Mean	s^2
X	19.42	53.123	21.27	54.589	20.19	69.362	19.05	43.448
Y_1	15.92	19.645	8.68	26.132	13.48	25.962	7.90	20.490
Y_2	8.29	5.607	6.27	6.398	7.76	5.890	5.71	5.014
Y_3	8.29	32.737	14.05	33.474	13.14	26.329	8.71	21.614
Y_4	6.00	9.217	8.27	5.827	7.14	6.529	6.33	7.633

[a] X = Pretest using a subset of lesson and transfer words, Y_1 = Morphemic Lesson Words—Production; Y_2 = Morphemic Lesson Words—Recognition; Y_3 = Context Lesson Words—Production; Y_4 = Context Lesson Words—Recognition.

Y_3, was based on a 20-item test requiring students to provide the definition of words provided. The words on this test were taken from the Context lessons. The fourth outcome variable, Y_4, was based on a 10-item multiple choice test, similar to Y_2, requiring students to recognize the definition of the selected words. Words on this test were taken from the Context lessons.

The group means and variances on the pretest (X) and the four posttests are presented in Table 9.2. Although two pretests were administered, our discussion focuses only on the first pretest, the targeted vocabulary words, X. The data from the second pretest (Degrees of Word Meaning) are available in a data file labeled BAUMANN2 found at the Wiley website.

9.3 UNIVARIATE ANCOVA

9.3.1 Testing for Equality of Regression Slopes

The analysis of covariance model is based on the assumption that the regression slope of the outcome variable on the covariate (or pretest) is the same for all populations being compared. In general, an estimate of the within-group regression slope for Group j, $b_{jY_1|X}$, is provided by the ratio of the cross products, (CP_j), for the posttest (Y_1) and covariate (X) variables, and the sum-of-squares (SS_j) for the X variable, $[b_{jY|X} = \text{CP}_{jYX}/\text{SS}_{jX}]$. Table 9.3 provides the separate $\mathbf{SSCP}_j$ matrices for each group as well as the **E** matrix, using only one outcome variable, Y_1.

Using these data, the four regression slopes are computed as:

Morphemic Only	$b_{1Y_1\|X} = 467.833/1221.833 = .383,$
Context Only	$b_{2Y_1\|X} = 597.909/1146.364 = .522,$
Morphemic and Context	$b_{3Y_1\|X} = 680.095/1387.238 = .490,$
Control	$b_{4Y_1\|X} = 536.095/868.952 = .617.$

TABLE 9.3 *SSCP* and *E* Matrices[a]

Morphemic Only

	Y_1	X
Y_1	451.833	
X	467.833	1221.833

Context Only

	Y_1	X
Y_1	548.773	
X	597.909	1146.364

Morphemic and Context

	Y_1	X
Y_1	519.238	
X	680.095	1387.238

Control

	Y_1	X
Y_1	409.810	
X	536.095	868.952

E

	Y_1	X
Y_1	1929.654	
X	2281.932	4624.387

[a] Y_1 = Morphemic Lesson Words—Production; X = Pretest using Lesson Words.

To determine whether the observed differences in the sample slope estimates are generalizable to the population slopes, a statistical test for the equality of the population slopes can be carried out. Specifically, the null hypothesis can be written as:

$$H_0\colon \beta_{\mathrm{MO}_{Y_1|X}} = \beta_{\mathrm{CO}_{Y_1|X}} = \beta_{\mathrm{MC}_{Y_1|X}} = \beta_{\mathrm{C}_{Y_1|X}}.$$

To test this hypothesis, an F statistic can be computed as:

$$F = \frac{\dfrac{\displaystyle\sum_{j=1}^{J} \frac{(\mathrm{CP}_{j(Y_1X)})^2}{\mathrm{SS}_{j(X)}} - \frac{(\mathrm{CP}_{(Y_1X)})^2}{\mathrm{SS}_{(X)}}}{J-1}}{\dfrac{\mathrm{SS}_Y - \displaystyle\sum_{j=1}^{J} \frac{(\mathrm{CP}_{j(Y_1X)})^2}{\mathrm{SS}_{j(X)}}}{(N-J)-J}}. \tag{9.1}$$

For our data we have

$$\sum_{j=1}^{J} \frac{(\mathrm{CP}_{j(Y_1X)})^2}{\mathrm{SS}_{jX}} \doteq \frac{(467.8)^2}{1221.8} + \frac{(597.9)^2}{1146.9} + \frac{(680.1)^2}{1387.2} + \frac{(536.1)^2}{869.0} \doteq 1155.102,$$

$$\frac{(\mathrm{CP}_{(Y_1X)})^2}{\mathrm{SS}_X} \doteq \frac{(2281.932)^2}{4624.387} \doteq 1126.033,$$

$$\mathrm{SS}_{Y_1} \doteq 1929.654.$$

$$F \doteq \frac{(1155.102 - 1126.033)/(4-1)}{(1929.654 - 1155.102)/(88-4)-4} \doteq \frac{9.690}{9.682} \doteq 1.001.$$

The P value associated with $F(3, 80) \doteq 1.001$ is .396. Based on these results there is little evidence to conclude that the population slopes are unequal. This basic assumption for analysis of covariance appears to be met.

9.3.2 Omnibus Test of Adjusted Means

To test the hypothesis of the equality of postintervention means, after considering the pretest variable data, the sum-of-squares for the posttest variable is adjusted as a function of the sum-of-squares for the pretest and the square of the cross-products of X and Y. Specifically, the adjusted total sum-of-squares, $\mathrm{SS}^*_{Y_{\text{Total}}}$, is

$$\mathrm{SS}^*_{Y_{1\text{Total}}} = \mathrm{SS}_{Y_{1\text{Total}}} - \left[\frac{(\mathrm{CP}_{(Y_1X)_{\text{Total}}})^2}{\mathrm{SS}_{X_{\text{Total}}}}\right],$$

where the total **SSCP** matrix is obtained by ignoring group membership and using the grand mean of each variable to compute deviation scores (grand mean centered). For the data in Table 9.1 the grand mean for X equals 19.98 and the grand mean for Y_1 equals 11.61. The $\mathbf{SSCP}_{\text{Total}}$ for the Y_1, X matrix is

$$\mathbf{SSCP}_{\text{Total}} \doteq \begin{bmatrix} & Y_1 & X \\ Y_1 & 2924.864 & \\ X & 2221.227 & 4687.955 \end{bmatrix}$$

$$\mathrm{SS}^*_{Y_{1\text{Total}}} \doteq 2924.864 - \frac{(2221.227)^2}{4687.955} \doteq 1872.412.$$

The adjusted error sum-of-squares is computed as:

$$\mathrm{SS}^*_{Y_{1_E}} = \mathrm{SS}_{Y_{1_E}} - \left[\frac{(\mathrm{CP}_{(Y_1X)_E})^2}{\mathrm{SS}_{X_E}}\right].$$

The error **SSCP**, **E**, was computed earlier (see Table 9.3) as:

$$\mathbf{E} \doteq \begin{bmatrix} & Y_1 & X \\ Y_1 & 1929.654 & \\ X & 2281.932 & 4624.387 \end{bmatrix},$$

and

$$\mathrm{SS}^*_{Y_{1_E}} \doteq 1929.654 - \frac{(2281.932)^2}{4624.387} \doteq 803.621.$$

The adjusted hypothesis sum-of-squares, SS^*_H, is the difference between the adjusted total sum-of-squares and the adjusted error sum-of-squares:

$$\mathrm{SS}^*_H = \mathrm{SS}^*_{Y_{1\text{Total}}} - \mathrm{SS}^*_{Y_{1_E}}.$$

TABLE 9.4 Analysis of Covariance Summary Table

Source of Variance	Sum-of-Squares	Degrees of Freedom	Mean Square	F	P
Group	1068.791	3	356.264	36.80	.000
Error	803.621	83	9.682		
Total	1872.412	87			

Using the adjusted sum-of-squares computed above, the adjusted hypothesis sum-of-squares is

$$SS_H^* \doteq 1872.412 - 803.621 \doteq 1068.791.$$

Table 9.4 provides an ANCOVA summary including the mean squares, F ratio, and P value. The results of this analysis indicates that there is sufficient evidence to conclude that the observed covariate X adjusted differences among the four groups on Y_1 are generalizable to the four populations. To quantify the magnitude of the effect, η^2 may be computed as the ratio of the SS_H^*/SS_{Y_1}. For our data, generalized η^2, η_G^2, is $1068.791/2924.864 \doteq .364$. Note the denominator for η^2 is the unadjusted total sum-of-squares for the outcome variable (see Olejnik and Algina, 2003). The observed differences appear to be both statistically significant and of meaningful magnitude.

Contrasts among the adjusted means can be tested to identify specific differences between instructional approaches. An adjusted mean is computed as:

$$\overline{Y}^*_{.j} = \overline{Y}_{.j} - b(\overline{X}_{.j} - \overline{X}_{..}), \tag{9.2}$$

where $\bar{Y}_{.j}$ = posttest mean of Group j

b = pooled or common regression slope of the posttest on the covariate (i.e., pretest)

$\overline{X}_{.j}$ = covariate mean for Group j

$\overline{X}_{..}$ = grand mean on the covariate

The pooled regression slope is the ratio of the error cross-product to the sum-of-squares for the pretest variable, $b = CP_{Y_1X_{(E)}}/SS_{X_{(E)}} \doteq 2281.932/4624.387 \doteq .493$. Table 9.5 provides the adjusted means for the four participating groups.

TABLE 9.5 Adjusted Means for the Four Participating Groups

Group	Adjusted Mean
Morphemic Only (MO)	$15.92 - .493(19.42 - 19.98) \doteq 16.20$
Context Only (CO)	$8.68 - .493(21.27 - 19.98) \doteq 8.04$
Morphemic and Context (MC)	$13.48 - .493(20.19 - 19.98) \doteq 13.38$
Control (C)	$7.91 - .493(19.05 - 19.98) \doteq 8.37$

If three contrasts comparing each of the intervention classrooms with the instructed control group were of interest, the values of the three contrasts would equal

$$\hat{\psi}_1 = \overline{Y}^*_{.1} - \overline{Y}^*_{.4} \doteq 6.20 - 8.37 \doteq 7.83,$$

$$\hat{\psi}_2 = \overline{Y}^*_{.2} - \overline{Y}^*_{.4} \doteq 8.04 - 8.37 \doteq -.33,$$

$$\hat{\psi}_3 = \overline{Y}^*_{.3} - \overline{Y}^*_{.4} \doteq 13.38 - 8.37 \doteq 5.01.$$

To test whether these observed differences are generalizable to the populations, an F test can be computed:

$$F = \frac{\hat{\psi}_c^2}{\text{MSE}^* \left[\sum_{j=1}^{J} \frac{a^2}{n_j} + \frac{(\overline{X}_{.j} - \overline{X}_{.j'})^2}{\text{SS}_X} \right]} \tag{9.3}$$

For the first contrast ($c = 1$), $\hat{\psi}_1$, the F ratio equals

$$F \doteq \frac{7.83^2}{9.682[(1)^2/24 + (-1)^2/21] + (19.42 - 19.05)^2/4624.387}$$

$$\doteq 70.857$$

These results indicate that the observed difference between the Morphemic Only group and the Control group is generalizable to the corresponding populations [$F(1, 83) \doteq 70.857$, $P \doteq .000$, $d \doteq 1.633$]. For the contrast between CO and C, $F(1, 83) \doteq .113$, $P \doteq .738$, $d \doteq -.069$, and for the contrast between the MC and C groups, $F(1, 83) \doteq 27.135$, $P \doteq .000$, $d \doteq 1.045$. Using the Bonferroni adjustment for the multiple tests, the results provide sufficient evidence to conclude that the difference between MO and C and between MC and C are generalizable to the populations, while there is insufficient evidence to conclude that the observed difference between the CO and C groups is generalizable to these populations.

9.4 MULTIVARIATE ANCOVA (MANCOVA)

9.4.1 Matrix Calculations

We begin our presentation by summarizing the data in Table 9.1 in terms of the total and error sum-of-squares and cross-products (**SSCP**) matrices. Both of these matrices have the same general form, as follows.

$$\left[\begin{array}{cccc|c}
SS_{Y_1} & CP_{Y_1Y_2} & CP_{Y_1Y_3} & CP_{Y_1Y_4} & CP_{Y_1X} \\
CP_{Y_2Y_1} & SS_{Y_2} & CP_{Y_2Y_3} & CP_{Y_2Y_4} & CP_{Y_2X} \\
CP_{Y_3Y_1} & CP_{Y_3Y_2} & SS_{Y_3} & CP_{Y_3Y_4} & CP_{Y_3X} \\
CP_{Y_4Y_1} & CP_{Y_4Y_2} & CP_{Y_4Y_3} & SS_{Y_4} & CP_{Y_4X} \\
\hline
CP_{XY_1} & CP_{XY_2} & CP_{XY_3} & CP_{XY_4} & SS_X
\end{array}\right]$$

The upper left 4×4 quadrant contains the sum-of-squares and cross products for the outcome variables. The 4×1 and 1×4 vectors contain identical elements representing the cross products of the covariate X and each of the outcome variables. Finally, the lower right element is the sum-of-squares for the covariate. The Total **SSCP** matrix, **T**, is determined using the deviation scores of each observation from the grand mean of each variable. This (5×5) matrix can be summarized as:

$$\mathbf{T} = \begin{array}{c|c} \mathbf{T}_{YY} & \mathbf{T}_{YX} \\ \hline \mathbf{T}_{XY} & \mathbf{T}_{XX} \end{array} \tag{9.4}$$

Similarly, within each of the J groups, an error sum-of-squares and cross-product matrix, $\mathbf{E}_j$, is computed using deviation scores from each of the respective group means (group mean centered). These error matrices can be represented as:

$$\mathbf{E}_j = \begin{array}{c|c} \mathbf{E}_{j(YY)} & \mathbf{E}_{j(YX)} \\ \hline \mathbf{E}_{j(XY)} & \mathbf{E}_{j(XX)} \end{array}, \tag{9.5}$$

where the subscript j indicates group membership. The (5×5) error matrix, **E**, is obtained by summing the separate $\mathbf{E}_j$ matrices across the J groups:

$$\mathbf{E} = \sum_{j=1}^{J} \mathbf{E}_j = \left[\begin{array}{c|c} \mathbf{E}_{YY} & \mathbf{E}_{YX} \\ \hline \mathbf{E}_{XY} & \mathbf{E}_{XX} \end{array} \right]. \tag{9.6}$$

Table 9.6 presents these **SSCP** matrices for the total, each group error matrix, and the error matrix.

9.4.2 Testing for Equal Slopes

As presented in the univariate analysis, the analysis of covariance model assumes that the relationship between the pretest variable and each of the outcome variables is the same for all levels of the grouping variable. In our case we have four outcome variables, so a statement to test this assumption may be written as:

$$H_0 : \begin{bmatrix} \beta_{1_{Y_1|X}} \\ \beta_{1_{Y_2|X}} \\ \beta_{1_{Y_3|X}} \\ \beta_{1_{Y_4|X}} \end{bmatrix} = \begin{bmatrix} \beta_{2_{Y_1|X}} \\ \beta_{2_{Y_2|X}} \\ \beta_{2_{Y_3|X}} \\ \beta_{2_{Y_4|X}} \end{bmatrix} = \begin{bmatrix} \beta_{3_{Y_1|X}} \\ \beta_{3_{Y_2|X}} \\ \beta_{3_{Y_3|X}} \\ \beta_{3_{Y_4|X}} \end{bmatrix} = \begin{bmatrix} \beta_{4_{Y_1|X}} \\ \beta_{4_{Y_2|X}} \\ \beta_{4_{Y_3|X}} \\ \beta_{4_{Y_4|X}} \end{bmatrix}.$$

To test this hypothesis, Wilks Λ can be computed as

$$\Lambda = \frac{|\mathbf{E}'|}{|\mathbf{E}' + \mathbf{H}|}, \tag{9.7}$$

TABLE 9.6 Sum-of-Squares and Cross-Products for Grand-Mean Centered (Total), Each of the Group-Mean Centered (MO, CO, MC, and C), and the Error Matrices

$\mathbf{T} =$

	Y_1	Y_2	$Y3$	Y_4	X
Y_1	2924.864				
Y_2	1040.545	579.818			
Y_3	1475.614	720.045	3000.989		
Y_4	722.295	373.318	1233.920	686.443	
X	2221.227	1073.091	2863.977	1366.841	4687.955

$\mathbf{E}_{MO} \doteq$

	Y_1	Y_2	$Y3$	Y_4	X
Y_1	451.833				
Y_2	149.583	128.958			
Y_3	431.583	183.958	752.958		
Y_4	212.000	121.000	346.000	212.000	
X	467.833	244.083	799.083	409.000	1221.833

$\mathbf{E}_{CO} \doteq$

	Y_1	Y_2	$Y3$	Y_4	X
Y_1	548.773				
Y_2	232.909	134.364			
Y_3	482.318	258.727	702.955		
Y_4	205.909	94.364	256.727	122.364	
X	597.909	319.364	664.727	232.364	1146.364

$\mathbf{E}_{MC} \doteq$

	Y_1	Y_2	$Y3$	Y_4	X
Y_1	519.238				
Y_2	208.381	117.810			
Y_3	421.571	188.714	526.571		
Y_4	229.571	106.714	228.571	130.571	
X	680.095	300.952	696.429	363.429	1387.238

$\mathbf{E}_{C} \doteq$

	Y_1	Y_2	$Y3$	Y_4	X
Y_1	409.810				
Y_2	139.429	100.286			
Y_3	354.429	125.286	432.286		
Y_4	202.667	82.000	214.000	152.667	
X	536.095	218.286	526.286	298.667	868.952

$\mathbf{E} \doteq$

	Y_1	Y_2	$Y3$	Y_4	X
Y_1	1929.654				
Y_2	730.302	481.417			
Y_3	1689.902	756.686	2414.770		
Y_4	850.147	404.078	1045.299	617.602	
X	2281.933	1082.685	2686.525	1303.459	4624.387

where

$$\mathbf{H} = \sum_{j=1}^{J} \mathbf{E}_{j(YX)} \mathbf{E}_{j(XX)}^{-1} \mathbf{E}_{j(XY)} - \mathbf{E}_{YX} \mathbf{E}_{XX}^{-1} \mathbf{E}_{XY}$$

$$\mathbf{E}' = \mathbf{E}_{YY} - \sum_{j=1}^{J} \mathbf{E}_{j(YX)} \mathbf{E}_{j(XX)}^{-1} \mathbf{E}_{j(XY)}.$$

The Wilks Λ can be transformed to an F statistic using Eqs. (3.14) to (3.18) repeated here for convenience:

$$F = \frac{1 - \Lambda^{1/s}}{\Lambda^{1/s}} \frac{m(\mathrm{s}) - \dfrac{p(\mathrm{df}_h)}{2} + 1}{p\,(\mathrm{df}_h)},$$

where

$$m = \mathrm{df}_e - \frac{p - \mathrm{df}_h + 1}{2} \quad \text{and} \quad \mathrm{df}_h = (J-1)C,$$

with C equalling the number of covariates, and

$$s = \sqrt{\frac{p^2(\mathrm{df}_h^2) - 4}{p^2 + \mathrm{df}_h^2 - 5}}.$$

The statistic has a central F distribution with $\nu_1 = p(\mathrm{df}_h)$ and $\nu_2 = m(s) - p(\mathrm{df}_h)/2 + 1$ degrees of freedom. Because the calculation of Λ is tedious, it is best left to the computer.

9.5 COMPUTER APPLICATION I

In this section we present the SPSS syntax for the test of equal regression slope vectors and the calculation of the test statistic. Because we are interested in only testing the assumption of equal regression slope vectors, basic descriptive statistics are not requested. The MANOVA command listed here is similar to the one provided in Section 3.6 for a single-group ANOVA. The covariate is preceded with the SPSS command "with." The **/design** command identifies the sources of variation to be considered in the linear model.

SPSS SYNTAX FOR REGRESSION SLOPE VECTOR EQUALITY

```
manova Y1 Y2 Y3 Y4 by group(1, 4) with X
/analysis = Y1 Y2 Y3 Y4
/design = X, group, X by group.
```

with X SPSS separates the grouping variables from the covariate(s) with the key word "with."
analysis = Y1 Y2 Y3 Y4 specifies the outcome variables to be analyzed using the model in the design statement.

OUTPUT

Analysis: Homogeneity of Regression Slopes

EFFECT .. X BY GROUP
Multivariate Tests of Significance (S = 3, M = 0, N = 37 1/2)

Test Name	Value	Approx. F	Hypoth. DF	Error DF	Sig. of F
Pillais	.19322	1.35958	12.00	237.00	.186
Hotellings	.21527	1.35740	12.00	227.00	.188
Wilks	.81564	1.36122	12.00	204.01	.187
Roys	.12862				

- -

EFFECT .. X BY GROUP
Univariate F-tests with (3,80) D. F.

Variable	Hypoth. SS	Error SS	Hypoth. MS	Error MS	F Sig.	of F
Y1	29.10680	774.51282	9.70227	9.68141	1.00215	.396
Y2	4.37163	223.56185	1.45721	2.79452	.52145	.669
Y3	15.69371	838.34723	5.23124	10.47934	.49920	.684
Y4	14.47337	235.72726	4.82446	2.94659	1.63730	.187

- -

Interpretation: Homogeneity of Regression Slopes

The multivariate test for the equality of the regression slope vectors provides little evidence to indicate that vectors differ, $\Lambda \doteq .816$, $F(12{,}204.01) \doteq 1.361$, $P \doteq .187$. Although not necessary here, the univariate tests for regression slope homogeneity are also provided. For each of the outcome variables there is little evidence to indicate an interaction between the grouping variable and the pretest variable. It might be noted that the result for the test of equal regression slopes for Y_1 on X, $F(3, 80) \doteq 1.002$, $P \doteq .396$ is the same as that reported in Section 9.3.1.

9.6 COMPARING ADJUSTED MEANS—OMNIBUS TEST

In Sections 9.4.2 and 9.5, the assumption of equal regression slope vectors was tested. The results indicated that there was little evidence to indicate that this basic assumption of the analysis of covariance model was violated. We now proceed to test the hypothesis that the vectors of adjusted means, or the adjusted centroids, for the four populations are equal. This hypothesis may be written as:

$$H_0\colon \boldsymbol{\mu}_{1_{Y|X}} = \boldsymbol{\mu}_{2_{Y|X}} = \boldsymbol{\mu}_{3_{Y|X}} = \boldsymbol{\mu}_{4_{Y|X}},$$

where $\boldsymbol{\mu}_{j_{Y|X}}$ represents the vector of means on the Y outcome variables adjusted for the pretest variable, X, in Group j, $j = 1, 2, \ldots, J$.

Any of the four multivariate criteria can be used to test the hypothesis on the adjusted centroids. The Wilks Λ is computed as the ratio of the determinant of the adjusted error matrix, $\mathbf{E}^*$, to the determinant of the adjusted total matrix $\mathbf{T}^*$ ($\mathbf{T}^* = \mathbf{E}^* + \mathbf{H}^*$),

$$\Lambda = \frac{|\mathbf{E}^*|}{|\mathbf{T}^*|}. \tag{9.8}$$

The adjusted error matrix $\mathbf{E}^*$ is computed as:

$$\mathbf{E}^* = \mathbf{E}_{YY} - \mathbf{E}_{YX}(\mathbf{E}_{XX})^{-1}\mathbf{E}_{XY}, \tag{9.9}$$

and the adjusted total matrix $\mathbf{T}^*$ is computed as:

$$\mathbf{T}^* = \mathbf{T}_{YY} - \mathbf{T}_{YX}(\mathbf{T}_{XX})^{-1}\mathbf{T}_{XY}, \tag{9.10}$$

The Wilks Λ can then be transformed to an F statistic using Eq. (3.14) and compared to the central F distribution with degrees of freedom equaling $\nu_1 = p\,(\mathrm{df}_h)$ and $\nu_2 = m(s) - (p\,(\mathrm{df}_h)/2) + 1$. The sum-of-squares and cross-products for the Total and Error matrices presented in Table 9.6 could be substituted into Eqs. (9.9) and (9.10) and the Wilks Λ computed with Eq. (9.8).

Alternatively, an eigenanalysis of the product matrix $\mathbf{E}^{*-1}\mathbf{H}^*$ may be conducted. Using the resulting eigenvalues, the four multivariate test criteria, discussed in Section 3.5.2, may be computed (see Table 3.3). The calculations are tedious, so we will rely on SPSS to provide the appropriate statistics. In the next section we provide the SPSS syntax and the results for the omnibus test.

9.7 COMPUTER APPLICATION II

In this section we present the SPSS syntax for conducting a single grouping variable multivariate analysis of covariance. The program is essentially the same program as a single grouping variable multivariate analysis of variance (see Sections 3.6, 5.3, and 5.6). One difference, however, between analysis of variance and analysis of covariance is that the "adjusted" means are of interest in analysis of covariance. These adjusted means are obtained using the SPSS command **/pmeans**.

SPSS SYNTAX FOR MANCOVA OMNIBUS TEST AND CONSTRUCT DEFINITION

```
manova Y1 Y2 Y3 Y4 BY group(1, 4) with X
/print = signif(mulitv eigen dimenr efsize)
/discrim = raw cor
/pmeans.
```

Note: typically a researcher would be interested in additional cell information and include the following on the /**print** command: **cellinfo(all) error(sscp cov cor)**.

/**pmeans** requests that the adjusted means be reported.

OUTPUT

Analysis: Test of the Relationship Between the Covariate and the Outcome Variables

```
EFFECT . . WITHIN CELLS Regression
Multivariate Tests of Significance (S = 1, M = 1 , N = 39)
Test Name          Value      Exact F   Hypoth. DF    Error DF   Sig. of F
Pillais           .71643     50.52808         4.00       80.00        .000
Hotellings       2.52640     50.52808         4.00       80.00        .000
Wilks             .28357     50.52808         4.00       80.00        .000
Roys              .71643
Note. . F statistics are exact.
- - - - - - - - - - - - - - - - - - - - - - - - - - - - - - - - - - - - - -
Multivariate Effect Size
TEST NAME          Effect Size
 (All)                    .716
- - - - - - - - - - - - - - - - - - - - - - - - - - - - - - - - - - - - - -
Eigenvalues and Canonical Correlations
Root No.     Eigenvalue        Pct.     Cum. Pct.    Canon Cor.    Sq. Cor
       1          2.526     100.000       100.000          .846       .716
- - - - - - - - - - - - - - - - - - - - - - - - - - - - - - - - - - - - - -
```

Interpretation: Test of the Relationship Between the Covariate and the Outcome Variables

Although this test is typically of little interest to a researcher, SPSS provides a test for the vector of regression slopes that relates the outcome variables to the covariate. This test examines whether the vector of regression slopes, relating each posttest variable to the covariate, is equal to zero, H_0: $\boldsymbol{\beta} = \mathbf{0}$. In an experiment where units are randomly assigned to the levels of the grouping variable, the covariate is chosen by the researcher because it is believed that it is highly related to the outcome variables and, therefore, would reduce the error variance to increase statistical power. Alternatively, in a nonexperimental study where the populations are believed to differ on the covariate, the covariate is included because that difference is believed to offer an alternative reason for posttest differences, and it must be taken into consideration or controlled. In either case, a strong relationship between the covariate and each of the outcome variables is expected. The result of this analysis should only confirm the researcher's belief regarding the anticipated relationship.

In the present analysis, there is evidence to support the belief of a relationship between the covariate and each of the outcome variables, $\Lambda \doteq .284$, $F(4, 80) \doteq 50.528$, $P \doteq .000$. Because there is only one covariate, all of the effect size indices provide the same estimate of the strength of the relationship between the vector of outcomes and the pretest variable. Here, $\eta^2 \doteq .716$. The adjusted effect size measure equals .702 $(1 - [(88-1)/(88-4-1)][1-.716])$.

Analysis: Group Intervention Effect

EFFECT . . GROUP

Multivariate Tests of Significance (S = 3, M = 0, N = 39)

Test Name	Value	Approx. F	Hypoth. DF	Error DF	Sig. of F
Pillais	.99905	10.23537	12.00	246.00	.000
Hotellings	2.99973	19.66493	12.00	236.00	.000
Wilks	.20070	14.74665	12.00	211.95	.000
Roys	.72474				

- -

Multivariate Effect Size

TEST NAME	Effect Size
Pillais	.333
Hotellings	.500
Wilks	.415

- -

Eigenvalues and Canonical Correlations

Root No.	Eigenvalue	Pct.	Cum. Pct.	Canon Cor.
1	2.633	87.772	87.772	.851
2	.354	11.785	99.557	.511
3	.013	.443	100.000	.115

- -

Dimension Reduction Analysis

Roots	Wilks L.	F	Hypoth. DF	Error DF	Sig. of F
1 TO 3	.20070	14.74665	12.00	211.95	.000
2 TO 3	.72912	4.62017	6.00	162.00	.000
3 TO 3	.98687	.54541	2.00	82.00	.582

- -

Raw discriminant function coefficients

	Function No.	
Variable	1	2
Y1	.316	−.048
Y2	.169	−.118
Y3	−.147	−.349
Y4	−.273	.301

- -

Correlations between DEPENDENT and canonical variables

Variable	Canonical Variable 1	2
Y1	.687	−.488
Y2	.398	−.371
Y3	−.252	−.915
Y4	−.197	−.324

Adjusted and Estimated Means

Variable . . Y1

CELL	Obs. Mean	Adj. Mean	Est. Mean	Raw Resid.	Std. Resid.
1	15.917	16.196	15.917	.000	.000
2	8.682	8.045	8.682	.000	.000
3	13.476	13.373	13.476	.000	.000
4	7.905	8.366	7.905	.000	.000

Variable . . Y2

CELL	Obs. Mean	Adj. Mean	Est. Mean	Raw Resid.	Std. Resid.
1	8.292	8.424	8.292	.000	.000
2	6.273	5.971	6.273	.000	.000
3	7.762	7.713	7.762	.000	.000
4	5.714	5.933	5.714	.000	.000

Variable . . Y3

CELL	Obs. Mean	Adj. Mean	Est. Mean	Raw Resid.	Std. Resid.
1	8.292	8.620	8.292	.000	.000
2	14.045	13.296	14.045	.000	.000
3	13.143	13.022	13.143	.000	.000
4	8.714	9.257	8.714	.000	.000

Variable . . Y4

CELL	Obs. Mean	Adj. Mean	Est. Mean	Raw Resid.	Std. Resid.
1	6.000	6.159	6.000	.000	.000
2	8.273	7.909	8.273	.000	.000
3	7.143	7.084	7.143	.000	.000
4	6.333	6.597	6.333	.000	.000

Interpretation: Group Intervention Effect

The group-adjusted mean centroids are

	MO	CO	MC	C
Y_1	16.2	8.0	13.4	8.4
Y_2	8.4	6.0	7.7	5.9.
Y_3	8.6	13.3	13.0	9.3
Y_4	6.2	7.9	7.1	6.6

The SPSS MANOVA program reports the statistical results for only three of the four multivariate criteria. All three of the multivariate criteria provide evidence to indicate that the observed differences among the four group centroids are generalizable to the populations they represent, $\Lambda \doteq .201$, $F(12, 211.95) \doteq 14.747$, $P \doteq .000$, $\tau^2_{\text{adj}} \doteq .387$, $(\tau^2_{\text{adj}} \doteq 1 - [(88-1)/(88-4-1)][1-.415])$.

An examination of the dimensionality of our system of variables indicates that the first two dimensions or constructs explain 99.6 percent of the variation in the adjusted outcome variable system, while the third dimension explains less than .5 percent of the variance. The squared canonical correlation for the first dimension equals .724 ($.851^2$), which can be interpreted as meaning that 72.4 percent of the variation in the first construct is associated with the grouping variable. Similarly, the squared canonical correlation for the second dimension equals .261, indicating that 26.1 percent of the variation in the second construct is associated with the grouping variable. Less than 1 percent of the variation in the third construct is associated with the grouping variable. The dimension reduction analysis indicates that in our system of variables, at most two dimensions are needed to describe the separation of our four groups, $\Lambda \doteq .729$, $F(6, 162) \doteq 4.62$, $P \doteq .000$.

To define the two constructs that maximize the separation among the four groups, the structure r's can be examined. The outcome variable with the highest correlation (.687) with the first construct is Y_1 while the outcome variable with the highest correlation (−.915) with the second construct is Y_3. The first construct is defined by the production task (providing definitions) of the morphemic lesson words, while the second construct is defined by the production task of the context lesson words. To further understand the group separation on these two constructs, a plot of the LDF mean vectors is helpful. The LDF mean vectors are obtained by multiplying the raw discriminant function weights by the adjusted group means on the four outcome variables. The resulting LDF mean vectors for our data are

	MO	CO	MC	C
Dimension 1	3.3	−.4	1.7	.5
Dimension 2	−2.9	−3.4	−4.0	−2.4

A plot of the LDF mean vectors is presented in Figure 9.1.

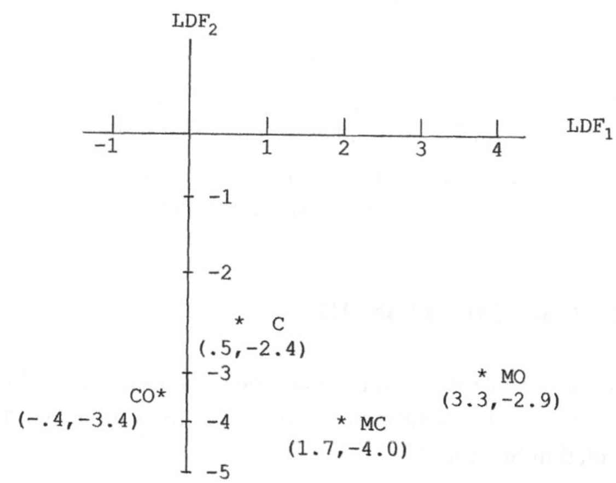

Figure 9.1 Two dimensional plot of adjusted group centroids.

Figure 9.1 shows that on the first dimension (Morphemic lesson words), the MO and MC groups differ noticeably from the CO and C groups, and the MO group differs in some degree from the MC group. For the second dimension, group separation is not quite as clear. However, the MC and CO groups appear different than the MO and C groups. Further, the MC group appears to differ to some degree from the CO group. Specific comparisons between groups can be further examined by a contrast analysis, which we present next.

9.8 CONTRAST ANALYSIS

We remind the reader that when specific comparisons, either pairwise or complex, are of interest, the omnibus test is not a necessary preliminary test for examining those contrasts. On the other hand, testing a limited number of contrasts after examining the data for "interesting" comparisons with or without the use of a Bonferroni-type adjustment for the number of hypotheses tested, is not appropriate. Testing a limited number of contrasts specified a priori is often an excellent analysis strategy, but picking comparisons because they look "significant" invalidates the probability value reported.

Hypothesis tests on specified contrasts can be conducted using a procedure similar to that presented for contrasts in the MANOVA context. That is, the adjusted hypothesis ($\boldsymbol{H}_{\hat{\psi}}{}^*$) sum-of-squares and cross-products matrix can be computed using Eq. (4.13), with the estimated contrast, $\hat{\boldsymbol{\psi}}$, based on differences among the adjusted means:

$$\boldsymbol{H}_{\hat{\psi}}{}^* = \frac{\hat{\boldsymbol{\psi}}^* \hat{\boldsymbol{\psi}}^{*\prime}}{\sum_{j=1}^{J} \frac{a_j^2}{n_j}}. \tag{9.11}$$

Here $\hat{\boldsymbol{\psi}}^* = \sum_{j=1}^{J} a_j \mathbf{y}_j^*$, and $\mathbf{y}_j^*$ is a vector of adjusted means for Group j [$\mathbf{y}_j^* = \mathbf{y}_{.j} - \mathbf{b}(\mathbf{x}_{.j} - \mathbf{x}_{..})$].

The hypothesis, H_0: $\boldsymbol{\psi}^* = 0$, can be tested using any of the four multivariate test criteria, Wilks, Bartlett–Pillai, Hotelling–Lawley, or Roy. The results of the eigenanalyis of the product matrix $\mathbf{E}^{*-1}\mathbf{H}_{\hat{\psi}}^*$ provides the eigenvalues needed to compute the test statistics. Because $\mathrm{df}_h = 1$ for all contrasts, all four tests result in the same computed F and P values. We rely on SPSS for the computations of the test statistics.

9.9 COMPUTER APPLICATION III

The SPSS syntax for MANCOVA contrasts are the same as those used for MANOVA contrasts in Section 4.5. The **/contrast** command is used along with MANCOVA commands presented in Section 9.7.

SPSS SYNTAX FOR MANCOVA CONTRAST ANALYSES

```
manova Y1 Y2 Y3 Y4 by group(1, 4)  with X
/print = signif(mulitv eigen dimenr efsize)
/discrim = raw cor
/pmeans
/contrast(group) = special(1 1 1 1, 1 0 0 −1, 0 1 0 −1, 0 0 1 −1)
/design  group(1) group(2) group(3).
```

OUTPUT

Analysis: Pairwise Contrast MC *vs.* C

EFFECT . . GROUP(3)

Multivariate Tests of Significance (S = 1, M = 1 , N = 39)

Test Name	Value	Exact F	Hypoth. DF	Error DF	Sig. of F
Pillais	.33795	10.20932	4.00	80.00	.000
Hotellings	.51047	10.20932	4.00	80.00	.000
Wilks	.66205	10.20932	4.00	80.00	.000
Roys	.33795				

Note.. F statistics are exact.

- -

Multivariate Effect Size

TEST NAME	Effect Size
(All)	.338

- -

Eigenvalues and Canonical Correlations

Root No.	Eigenvalue	Pct.	Cum. Pct.	Canon Cor.
1	.510	100.000	100.000	.581

- -

Raw discriminant function coefficients

	Function No.
Variable	1
Y1	−.227
Y2	−.196
Y3	−.193
Y4	.407

- -

Correlations between DEPENDENT and canonical variables

	Canonical Variable
Variable	1
Y1	−.800
Y2	−.534
Y3	−.583
Y4	−.140

- -

Interpretation: Pairwise Contrast MC *vs.* C

The vector of differences between the mean of the Morphemic and Context (MC) treatment group and the instructed Control (C) group is

$$\hat{\psi}_3 = \begin{bmatrix} 5.0 \\ 1.8 \\ 3.8 \\ 0.5 \end{bmatrix}.$$

The results of the hypothesis test, H_0: $\hat{\psi}_3 = 0$, indicates that there is sufficient evidence to conclude that the observed differences in the adjusted means are generalizable to the populations they represent: $\Lambda \doteq .662$, $F(4, 80) \doteq 10.209$, $P' \doteq .000$, $\eta^2_{\text{adj}} \doteq .306$. To define the single construct that maximizes the separation between the groups, the structure r's are examined. The variable with the highest correlation, $-.800$, with the construct is Y_1. These results indicate that the production of the morphemic lesson words is the primary variable defining the construct.

Analysis: Pairwise Contrasts CO *vs.* C *and* MO *vs.* C

EFFECT .. GROUP(2)

Multivariate Tests of Significance (S = 1, M = 1 , N = 39)

Test Name	Value	Exact F	Hypoth. DF	Error DF	Sig. of F
Pillais	.21672	5.53367	4.00	80.00	.001
Hotellings	.27668	5.53367	4.00	80.00	.001
Wilks	.78328	5.53367	4.00	80.00	.001
Roys	.21672				

Note.. F statistics are exact.

- -

Multivariate Effect Size

TEST NAME	Effect Size
(All)	.217

- -

Eigenvalues and Canonical Correlations

Root No.	Eigenvalue	Pct.	Cum. Pct.	Canon Cor.
1	.277	100.000	100.000	.466

- -

Raw discriminant function coefficients

	Function No.
Variable	1
Y1	.178
Y2	.059
Y3	−.318
Y4	−.101

- -

Correlations between DEPENDENT and canonical variables

Variable	Canonical Variable 1
Y1	.070
Y2	−.015
Y3	−.856
Y4	−.514

EFFECT .. GROUP(1)

Multivariate Tests of Significance (S = 1, M = 1 , N = 39)

Test Name	Value	Exact F	Hypoth. DF	Error DF	Sig. of F
Pillais	.57539	27.10214	4.00	80.00	.000
Hotellings	1.35511	27.10214	4.00	80.00	.000
Wilks	.42461	27.10214	4.00	80.00	.000
Roys	.57539				

Note.. F statistics are exact.

Multivariate Effect Size

TEST NAME	Effect Size
(All)	.575

Eigenvalues and Canonical Correlations

Root No.	Eigenvalue	Pct.	Cum. Pct.	Canon Cor.
1	1.355	100.000	100.000	.759

Raw discriminant function coefficients

Variable	Function No. 1
Y1	.325
Y2	.180
Y3	−.090
Y4	−.282

Correlations between DEPENDENT and canonical variables

Variable	Canonical Variable 1
Y1	.794
Y2	.474
Y3	−.063
Y4	−.079

Interpretation: Pairwise Contrasts MO *vs.* C *and* CO *vs.* C

We will not go into detail in interpreting the two pairwise contrasts. We will just say that the results indicate that there is sufficient evidence to conclude that the observed differences between the CO and C groups, and between the MO and C groups, are generalizable to their respective populations. Further, the contrast that separates the CO and C groups is defined primarily by the production of the Context lesson words. The construct that separates the MO and C groups is primarily defined by the production of the Morphemic lesson words.

9.10 SUMMARY

In this chapter we presented the analysis procedures for conducting the analysis of covariance via both a univariate and a multivariate approach. A close examination of the computational procedures reveals that the multivariate approach is in some ways very similar to the univariate approach. The multivariate approach, however, considers the relationships among the outcome variables. The analysis of covariance model can have considerable statistical power and, therefore, is more likely to detect group separation than posttest-only designs using analysis of variance. We have limited our discussion to a single grouping variable, but the generalization to a factorial design is straightforward. We also limited our presentation to a single covariate. We present the syntax and the computer output for analysis of covariance when two covariates are available in the Technical Note below. Because covariates are often highly correlated, the advantage of using more than one covariate in the model is generally limited. A second covariate typically does not substantially reduce the error variance beyond that gained by the first covariate. Although the results of statistical tests with a single covariate or multiple covariates typically do not differ much, the use of multiple covariates may help clarify the nature of group separation when the constructs are defined. Finally, while in nonexperimental studies the single covariate is not likely to control all sources of selection bias, the single well-chosen covariate often captures the primary sources of confounding. The use of multiple covariates does not guarantee that all sources of selection bias have been controlled.

Technical Note

Here we present the syntax and selected output for conducting a multicovariate analysis of covariance. We leave the interpretation of the results to the reader. It should be noted that the results reported here are only slightly different from those reported earlier in the chapter. For this data set, the addition of the second covariate adds little to the interpretation of the study's findings.

SPSS SYNTAX FOR TESTING THE EQUALITY OF REGRESSION PLANES WITH TWO COVARIATES

```
manova Y1 Y2 Y3 Y4 by group(1,4)  with X X2
/analysis = Y1 Y2 Y3 Y4
/design = X+X2, group, X by group + X2 by group.
```

OUTPUT

Analysis: Test for the Equality of Regression Planes with Two Covariates

EFFECT .. X BY GROUP + X2 BY GROUP
Multivariate Tests of Significance (S = 4, M = 1/2, N = 35 1/2)

Test Name	Value	Approx. F	Hypoth. DF	Error DF	Sig. of F
Pillais	.41228	1.45558	24.00	304.00	.080
Hotellings	.49589	1.47733	24.00	286.00	.073
Wilks	.63693	1.47169	24.00	255.88	.076
Roys	.21999				

SPSS SYNTAX FOR TESTING THE EQUALITY OF ADJUSTED CENTROIDS WITH TWO COVARIATES

```
manova Y1 Y2 Y3 Y4 by group(1, 4)  with X X2
/print=signif(eigen dimenr efsize)
/discrim raw cor
/pmeans
/design.
```

OUTPUT

Analysis: Test for the Equality of Adjusted Centroids with Two Covariates

EFFECT .. WITHIN CELLS Regression
Multivariate Tests of Significance (S = 2, M = 1/2, N = 38 1/2)

Test Name	Value	Approx. F	Hypoth. DF	Error DF	Sig. of F
Pillais	.81515	13.75949	8.00	160.00	.000
Hotellings	3.06941	29.92672	8.00	156.00	.000
Wilks	.23373	21.10203	8.00	158.00	.000
Roys	.74998				

Note.. F statistic for WILKS' Lambda is exact.

- -

Multivariate Effect Size

TEST NAME	Effect Size
Pillais	.408
Hotellings	.605
Wilks	.517

- -

Eigenvalues and Canonical Correlations

Root No.	Eigenvalue	Pct.	Cum. Pct.	Canon Cor.	Sq. Cor
1	3.000	97.729	97.729	.866	.750
2	.070	2.271	100.000	.255	.065

- -

Dimension Reduction Analysis

Roots	Wilks L.	F	Hypoth. DF	Error DF	Sig. of F
1 TO 2	.23373	21.10203	8.00	158.00	.000
2 TO 2	.93483	1.85893	3.00	80.00	.143

- -

EFFECT .. GROUP

Multivariate Tests of Significance (S = 3, M = 0, N = 38 1/2)

Test Name	Value	Approx. F	Hypoth. DF	Error DF	Sig. of F
Pillais	1.04903	10.88834	12.00	243.00	.000
Hotellings	3.18394	20.60719	12.00	233.00	.000
Wilks	.18471	15.58251	12.00	209.31	.000
Roys	.73370				

- -

Multivariate	Effect Size
TEST NAME	Effect Size
Pillais	.350
Hotellings	.515
Wilks	.430

- -

Eigenvalues and Canonical Correlations

Root No.	Eigenvalue	Pct.	Cum. Pct.	Canon Cor.
1	2.755	86.532	86.532	.857
2	.396	12.444	98.977	.533
3	.033	1.023	100.000	.178

- -

Dimension Reduction Analysis

Roots	Wilks L.	F	Hypoth. DF	Error DF	Sig. of F
1 TO 3	.18471	15.58251	12.00	209.31	.000
2 TO 3	.69362	5.35229	6.00	160.00	.000
3 TO 3	.96845	1.31931	2.00	81.00	.273

- -

Raw discriminant function coefficients

	Function No.	
Variable	1	2
Y1	.307	−.073
Y2	.171	−.111
Y3	−.162	−.329
Y4	−.318	.224

- -

Correlations between DEPENDENT and canonical variables

	Canonical Variable	
Variable	1	2
Y1	.647	−.564
Y2	.370	−.409
Y3	−.307	−.936
Y4	−.255	−.377

- -

Adjusted and Estimated Means

Variable .. Y1

CELL	Obs. Mean	Adj. Mean	Est. Mean	Raw Resid.	Std. Resid.
1	15.917	16.134	15.917	.000	.000
2	8.682	8.275	8.682	.000	.000
3	13.476	13.297	13.476	.000	.000
4	7.905	8.273	7.905	.000	.000

Variable .. Y2

CELL	Obs. Mean	Adj. Mean	Est. Mean	Raw Resid.	Std. Resid.
1	8.292	8.402	8.292	.000	.000
2	6.273	6.054	6.273	.000	.000
3	7.762	7.685	7.762	.000	.000
4	5.714	5.900	5.714	.000	.000

- -

Variable .. Y3

CELL	Obs. Mean	Adj. Mean	Est. Mean	Raw Resid.	Std. Resid.
1	8.292	8.479	8.292	.000	.000
2	14.045	13.820	14.045	.000	.000
3	13.143	12.848	13.143	.000	.000
4	8.714	9.047	8.714	.000	.000

- -

Variable .. Y4

CELL	Obs. Mean	Adj. Mean	Est. Mean	Raw Resid.	Std. Resid.
1	6.000	6.073	6.000	.000	.000
2	8.273	8.230	8.273	.000	.000
3	7.143	6.978	7.143	.000	.000
4	6.333	6.468	6.333	.000	.000

- -

SPSS SYNTAX FOR TESTING PAIRWISE CONTRASTS OF ADJUSTED CENTROIDS

```
manova Y1 Y2 Y3 Y4 by group(1, 4) with X X2
/print=signif(eigen dimenr efsize)
/discrim raw cor
/contrast(group) = special(1 1 1 1, 1 0 0 –1, 0 1 0 –1, 0 0 1 –1)
/design group(1) group(2) group(3).
```

OUTPUT

Analysis: Pairwise Contrasts Between Adjusted Centroids

EFFECT .. GROUP(3)

Multivariate Tests of Significance (S = 1, M = 1, N = 38 1/2)

Test Name	Value	Exact F	Hypoth. DF	Error DF	Sig. of F
Pillais	.34056	10.19956	4.00	79.00	.000

Hotellings	.51643	10.19956	4.00	79.00	.000
Wilks	.65944	10.19956	4.00	79.00	.000
Roys	.34056				

Note.. F statistics are exact.

- -

Multivariate Effect Size

TEST NAME	Effect Size
(All)	.341

- -

Eigenvalues and Canonical Correlations

Root No.	Eigenvalue	Pct.	Cum. Pct.	Canon Cor.
1	.516	100.000	100.000	.584

- -

Raw discriminant function coefficients

	Function No.
Variable	1
Y1	−.222
Y2	−.189
Y3	−.198
Y4	.380

- -

Correlations between DEPENDENT and canonical variables

	Canonical Variable
Variable	1
Y1	−.808
Y2	−.536
Y3	−.624
Y4	−.158

- -

EFFECT .. GROUP(2)

Multivariate Tests of Significance (S = 1, M = 1 , N = 38 1/2)

Test Name	Value	Exact F	Hypoth. DF	Error DF	Sig. of F
Pillais	.28541	7.88827	4.00	79.00	.000
Hotellings	.39941	7.88827	4.00	79.00	.000
Wilks	.71459	7.88827	4.00	79.00	.000
Roys	.28541				

Note.. F statistics are exact.

- -

Multivariate Effect Size

TEST NAME	Effect Size
(All)	.285

- -

Eigenvalues and Canonical Correlations

Root No.	Eigenvalue	Pct.	Cum. Pct.	Canon Cor.
1	.399	100.000	100.000	.534

- -

Raw discriminant function coefficients

	Function No.
Variable	1

Y1	.154
Y2	.075
Y3	−.299
Y4	−.222

Correlations between DEPENDENT and canonical variables

Variable	Canonical Variable 1
Y1	.000
Y2	−.052
Y3	−.874
Y4	−.607

EFFECT .. GROUP(1)

Multivariate Tests of Significance (S = 1, M = 1 , N = 38 1/2)

Test Name	Value	Exact F	Hypoth. DF	Error DF	Sig. of F
Pillais	.57710	26.95187	4.00	79.00	.000
Hotellings	1.36465	26.95187	4.00	79.00	.000
Wilks	.42290	26.95187	4.00	79.00	.000
Roys	.57710				

Note.. F statistics are exact.

Multivariate Effect Size

TEST NAME	Effect Size
(All)	.577

Eigenvalues and Canonical Correlations

Root No.	Eigenvalue	Pct.	Cum. Pct.	Canon Cor.
1	1.365	100.000	100.000	.760

Raw discriminant function coefficients

Variable	Function No. 1
Y1	.323
Y2	.181
Y3	−.094
Y4	−.297

Correlations between DEPENDENT and canonical variables

Variable	Canonical Variable 1
Y1	.804
Y2	.477
Y3	−.059
Y4	−.078

Fact A statistician baptized one of her babies and kept the other as a control.

EXERCISES

Exercises 1 to 10 refer to a randomized two-group pretest–posttest design with three outcome variables. The design is balanced with $n = 20$.

1. The following are the **SSCP** matrices for each group. For each group determine the separate regression slopes relating each of the outcome variables to X.

Group 1				
	Y_1	Y_2	Y_3	X
Y_1	43.6			
Y_2	13.9	11.8		
Y_3	38.9	14.8	59.3	
X	261.7	140.1	424.1	4424.5

Group 2				
	Y_1	Y_2	Y_3	X
Y_1	46.9			
Y_2	20.2	12.1		
Y_3	48.0	25.5	69.9	
X	409.5	209.7	553.6	7729.5

2. Using the two matrices in Exercise 1, determine the **E** matrix.

3. Determine the vector of pooled or common regession weights across the two groups.

4. How many eigenvalues will be used in the hypothesis test on the equality of the two vectors of regression coefficients?

5. If Wilks Λ for testing the equality of the between-group vectors of regression coefficients equals .959, what is the value of the F statistic? How many degrees of freedom are associated with the computed F statistic?

6. Using your results from Exercise 2 compute the adjusted error matrix, $\mathbf{E}^*$.

7. The canonical correlation between the adjusted outcome variables and the grouping variable equals .905. What is the numerical value of Wilks Λ?

8. Transform Λ to an F statistic and report its degrees of freedom.

9. The structure r's for variables Y_1, Y_2, and Y_3 are $-.432$, $-.226$, and .415, respectively. How would you define the construct separating the two groups.

10. The raw discriminant function weights are $-.284$, $-.097$, and .321 for variables Y_1, Y_2, and Y_3, respectively. Unadjusted and adjusted group means are reported below. Use this information to determine the group centroids in LDF space.

Adjusted and Estimated Means
Variable .. y1

CELL	Obs. Mean	Adj. Mean	Est. Mean	Raw Resid.	Std. Resid.
1	15.700	15.728	15.700	.000	.000
2	9.050	9.022	9.050	.000	.000

Adjusted and Estimated Means (Cont.)
Variable .. y2

CELL	Obs. Mean	Adj. Mean	Est. Mean	Raw Resid.	Std. Resid.
1	8.100	8.114	8.100	.000	.000
2	6.350	6.336	6.350	.000	.000

Adjusted and Estimated Means (Cont.)
Variable .. y3

CELL	Obs. Mean	Adj. Mean	Est. Mean	Raw Resid.	Std. Resid.
1	7.600	7.640	7.600	.000	.000
2	13.950	13.910	13.950	.000	.000

Computer Applications

Exercises 11 to 15 require the analysis of the Ethington 5-group data set (5GED) described in Appendix A. One of the questions students were asked was the number of hours per week they were employed. Because students from the three racial groups differed in the number of hours they worked, it might be useful to reanalyze the data previously analyzed in Chapter 3 to control for those differences. Exercises 11 to 15 ask specific questions regarding the results of such an analysis.

11. Do the vectors of regression slopes for each of the nine outcome variables on the Time variable differ among the three populations? Support your answer by providing:

(a) Λ

(b) F

(c) Degrees of freedom

(d) P value

12. Does adding the Time variable as a covariate change the results of the comparison of centroids for the three groups? Compare the results (e.g., Λ, F, df, P) obtained with and without the Time covariate (see Exercises 14 and 15 in Chapter 3).

13. Do the variables that define the constructs that separate the groups differ when the Time variable is added as a covariate compared with the results that do not include Time? (See Exercise 16 in Chapter 5.) Compare the structure r's for the two models.

14. What are the adjusted LDF mean centroids for the three groups?

15. Using Time as a covariate, compare the mean centroid of minority students with the mean centroid of the White students. Specifically report:

   (a) Λ

   (b) F

   (c) Degrees of freedom

   (d) P value

   (e) Variables that define the construct that separates the groups

CHAPTER 10

Repeated-Measures Analysis

10.1 INTRODUCTION

In Chapter 9 we discussed the analysis of data obtained from a research design (multiple group pretest–posttest) that can be used to address two problems often found with posttest-only research designs. The *first* problem identified was that of selection bias or nonequivalency, when analysis units are not randomly assigned to the levels of the grouping variable. The *second* problem identified was that of inefficiency; large sample sizes are needed to have good statistical power unless the effect size is "large." The one feature that both the multiple group pretest–posttest and the posttest-only design share is that both designs involve making comparisons among independent groups. Consequently, all of the designs we have discussed so far can be classified as *between-group* designs. In this chapter, we discuss an alternative research design that in many situations can also address the problems of equivalency and inefficiency. These designs are referred to as *repeated-measures* or *within-subject* designs. In this chapter we focus on the analysis of data collected on a single group of subjects who represent a single population of interest. (In this chapter "subject" is used in place of "unit"; the subscript u is used to denote a subject.) In Chapter 11 the design is extended to include multiple groups representing multiple populations and is referred to as a *mixed-model* design because there are both between-group variables and within-subject variables.

The repeated-measures design arises in three contexts. In the first context, subjects are measured at several points in time. So, for example, in a study investigating the effect of exercise on weight loss, participants might be measured weekly on their body weight while they participate in a daily exercise routine. Or, a group of individuals receiving an intervention might be measured before the start of the intervention, immediately following termination of the intervention, and some later time following the completion of the intervention. A design matrix for a study that observes subjects weekly might be presented as:

Applied MANOVA and Discriminant Analysis, Second Edition, by Carl J. Huberty and Stephen Olejnik
Copyright © 2006 John Wiley & Sons, Inc.

$$\begin{array}{c} \\ S_1 \\ S_2 \\ \vdots \\ S_N \end{array} \begin{array}{c} \begin{array}{cccccc} W_0 & W_1 & W_2 & W_3 & W_4 & W_5 \end{array} \\ \begin{bmatrix} Y_{10} & Y_{11} & Y_{12} & Y_{13} & Y_{14} & Y_{15} \\ Y_{20} & Y_{21} & Y_{22} & Y_{23} & Y_{24} & Y_{25} \\ \vdots & \vdots & \vdots & \vdots & \vdots & \vdots \\ Y_{N0} & Y_{N1} & Y_{N2} & Y_{N3} & Y_{N4} & Y_{N5} \end{bmatrix} \end{array}.$$

Here, W_t $(t = 0, 1, 2, \ldots, T)$ represents the week (time) when the measurement (score) was obtained. The 0 subscript would indicate a baseline measurement taken before an intervention (e.g., exercise program) began; S_u indicates the subject (unit) $(u = 1, 2, \ldots, N)$, and Y_{ut} indicates the measurement of subject u at time t.

A *second* repeated-measures design context occurs when each subject is exposed to each one of the J treatment conditions. For example, a researcher might be interested in studying the effect of room color on test anxiety. Each subject might be asked to complete an anxiety instrument before completing a learning task in four different rooms that vary only in wall colors. For example, one room might be a deep red, a second room a bright yellow, a third room a light green, and a fourth room painted beige. The order of the learning task and room color might be randomized. The data matrix for such an example study might be presented as:

$$\begin{array}{c} \\ S_1 \\ S_2 \\ \vdots \\ S_N \end{array} \begin{array}{c} \begin{array}{cccc} T_1 & T_2 & T_3 & T_4 \end{array} \\ \begin{bmatrix} Y_{11} & Y_{12} & Y_{13} & Y_{14} \\ Y_{21} & Y_{22} & Y_{23} & Y_{24} \\ \vdots & \vdots & \vdots & \vdots \\ Y_{N1} & Y_{N2} & Y_{N3} & Y_{N4} \end{bmatrix} \end{array}.$$

Here T_j represents treatment condition j $(j = 1, 2, \ldots, J)$, S_u indicates the subject $(u = 1, 2, \ldots, N)$, and Y_{uj} indicates the measurement of subject u under Treatment j.

A *third* repeated-measures design context in which repeated measures are used would occur if a test battery was administered to a sample of subjects, and there was an interest in comparing the mean performance on the tests. For example, the Torrence Test for Creativity consists of a series of activities that provide six scales of creativity (fluency, originality, elaboration, abstract titles, resistance to premature closure, and creative strength). A comparison of these six scales might address the question whether levels of creativity are the same for all six aspects of creativity. This type of analysis is sometimes referred to as *profile analysis*. For this type of analysis the number of items, range of possible scores, and test variances must be comparable. A data matrix for this context might be presented as

$$\begin{array}{c} \\ S_1 \\ S_2 \\ \vdots \\ S_N \end{array} \begin{array}{c} \begin{array}{cccccc} M_1 & M_2 & M_3 & M_4 & M_5 & M_6 \end{array} \\ \begin{bmatrix} Y_{11} & Y_{12} & Y_{13} & Y_{14} & Y_{15} & Y_{16} \\ Y_{21} & Y_{22} & Y_{23} & Y_{24} & Y_{25} & Y_{26} \\ \vdots & \vdots & \vdots & \vdots & \vdots & \vdots \\ Y_{N1} & Y_{N2} & Y_{N3} & Y_{N4} & Y_{N5} & Y_{N6} \end{bmatrix} \end{array}.$$

Here M_k represents measure k ($k = 1, 2, \ldots, K$), S_u indicates the subject ($u = 1, 2, \ldots, N$), and Y_{uk} indicates the measurement of subject u for Measurement k.

With the repeated-measures design, the problem of equivalence is solved because each subject is measured under all of the conditions and, therefore, serves as its own control. The problem of inefficiency is addressed by reducing the error variance as a function of the correlations among the repeated measures. That is, the error variance for a repeated-measures design ($\sigma^2_{\text{RM}_{\text{error}}}$) equals the variance of the completely randomized between-group design ($\sigma^2_{\text{BG}_{\text{error}}}$) times 1 minus the correlation (ρ) between the observations: $\sigma^2_{\text{RM}_{\text{error}}} = \sigma^2_{\text{BG}_{\text{error}}}(1 - \rho)$. Because measurements on a subject will generally be highly correlated, a substantial reduction in error variance is possible.

But the repeated-measures design is not appropriate for all research studies. There are three potential problems with this design that can provide alternative explanations for the observed differences in addition to the variable of primary interest to the researcher (e.g., time, treatment, measure). *One* potential problem is that, when subjects are asked to complete the same task repeatedly under different conditions, their performance might improve considerably over time, not because of the conditions but because of memory or practice effects. Scores may also decline, however, because of fatigue. A *second* potential problem that might occur is that, after being exposed to one condition or treatment, the effect of that treatment carries over to the next treatment or condition. Sometimes such carryover effects occur only when a certain order of treatments is provided. Such *differential-order* effects are not possible to control. A *third* potential problem with the repeated-measures design is the equivalency of repeated measurements. For example, a comparison of interventions designed to improve spelling skills would require different spelling tests. The nonequivalency of such tests could be a serious problem. Good discussions on the limitations of repeated measures are provided by Keppel and Wickens (2004, pp. 369–372) and Maxwell and Delaney (2000, pp. 481–483). In spite of these potential problems, in some disciplines the within-subject design is extremely popular.

In the next section we introduce a new research context and a new data set. Following a brief application of the univariate analysis with these data, we present the multivariate approach. We conclude the chapter with a discussion comparing the two analysis approaches to testing the omnibus hypothesis and contrast hypotheses. Our discussion focuses on a single within-subject variable. The design does generalize to more than a single within-subject variable, and we provide the SPSS syntax and the output for a study having two within-subject variables in three Technical Notes, but we do not discuss the interpretation. The interpretation should be a clear generalization from the discussion presented here.

10.2 RESEARCH CONTEXT

As part of her dissertation study, M. S. Poudevigne (2004) collected Self-Esteem data from a sample of 12 Pregnant women at the beginning of each month during the second and third trimesters (Months 4, 5, 6, 7, 8, and 9). The participants were asked to complete the Global Self-Esteem subscale from the Rosenberg Self-Esteem

TABLE 10.1 Scores from the Rosenberg Self-Esteem Inventory

	Time					
Subject	Month 4	Month 5	Month 6	Month 7	Month 8	Month 9
1	12	11	10	10	10	10
2	34	30	34	12	16	10
3	16	18	15	13	15	15
4	19	16	14	11	12	13
5	23	22	21	20	23	21
6	20	20	20	19	20	20
7	19	13	11	11	11	11
8	13	11	10	10	10	10
9	14	14	14	10	10	10
10	15	12	10	11	11	10
11	16	14	10	11	10	10
12	14	11	13	11	11	11
Mean	17.92	16.00	15.17	12.42	13.25	12.58
Standard Deviation	6.007	5.721	7.030	3.423	4.372	4.010

Inventory. Scores on this subscale can range between 10 and 40 with high scores indicating low self-esteem. The researcher was interested in determining whether these data would provide sufficient evidence to support the belief that feelings of self-esteem change over the final 6 months of pregnancy. If there was a change, the researcher was interested in determining how the relationship between Self-Esteem and Time might be characterized. Table 10.1 provides the sample data along with means and standard deviations. (An SPSS data file containing these data, labeled SELFESTEEM, is available at the Wiley website. Data were modified slightly for pedagogical purposes.)

10.3 UNIVARIATE ANALYSES

10.3.1 Omnibus Test

The data presented in Table 10.1 can be viewed as a two-way factorial design with Time, when the measurement was collected, being the first variable (column), and the second variable (row) being the Subject. With this conceptualization, there is only one measurement per cell (Subject-by-Time) and the within-cell or error variance cannot be directly estimated. To test the hypothesis that there is no difference in Self-Esteem over the 6-month period (H_0: $\mu_4 = \mu_5 = \mu_6 = \mu_7 = \mu_8 = \mu_9$), it must be assumed that there is no interaction between Subject and Time. If there is no interaction between Subject and Time, the computed mean-square interaction from the factorial design provides an estimate of the error variance. This estimate can be used as the denominator for the calculation of the F ratio for the hypothesis of interest. If there

TABLE 10.2 ANOVA Summary Table for Changes in Self-Esteem Over the Second and Third Trimesters

Source of Variation	Sum-of-Squares	Degrees of Freedom	Mean Square	F	P	η_G^2
Time (T)	287.111	$T - 1 = 5$	57.422	5.723	.000	.136
Error ($S \times T$)	551.889	$(S - 1)(T - 1) = 55$	10.034			

is an interaction between Subject and Time, the repeated-measures design would be inappropriate, and we should look for one or more subject characteristics that might help us explain why some subjects change differently than others. Perhaps, for some subjects, this is their first child, while for others this is a pregnancy for a second or third child. If this is the case, then a mixed-model design, which will be discussed in Chapter 11, would be more appropriate than the repeated-measures design.

The formulas for sums-of-squares in a two-way analysis of variance were presented in Table 8.3. Applying the formulas for a main effect and the interaction, the sum of squares for Time and the Subject-by-Time interaction, respectively, can be computed. Using the data in Table 10.1, results of the ANOVA summary table are presented in Table 10.2. The results of this analysis support the researcher's belief that feelings of Self-Esteem change over the final 6 months of pregnancy, $F(5, 55) \doteq 5.723$, $P \doteq .000$. To quantify the strength of the relationship, generalized η^2 (Olejnik and Algina, 2003) may be computed as $\eta_G^2 = \text{SS}_{\text{Time}}/[(\text{SS}_{\text{Time}} + (N - 1)\sum_{t=1}^{T} s_t^2)]$ where s_t^2 is the variance of the Self-Esteem scores at Month t. For the data in Table 10.1, the sum of the variances equals 165.146 and generalized η^2 equals .136 (287.111/2103.777). The test and effect size results indicate that the relationship between time and self-esteem is not trivial. To further characterize the nature of the relationship, focused tests for trend might be conducted.

10.3.2 Contrast Analysis

In Chapters 4 and 8, our presentation on contrasts examined both pairwise and complex contrasts. We can apply the same procedures here and compare mean Self-Esteem scores for each month with the last month of pregnancy (five pairwise contrasts), or the mean Self-Esteem score during the second trimester $[(\overline{Y}_{.4} + \overline{Y}_{.5} + \overline{Y}_{.6})/3]$ with the mean Self-Esteem score during the third trimester $[(\overline{Y}_{.7} + \overline{Y}_{.8} + \overline{Y}_{.9})/3]$, a complex contrast. But, because our data were collected monthly for 6 months with an equal amount of time passing between each measurement, it might be more interesting to construct contrasts that examine a variety of *trends* in the data. For example, the researcher might be interested in determining whether the means of Self-Esteem are changing at a consistent rate (linear), or perhaps whether means of Self-Esteem are changing at an accelerating rate (quadratic). By choosing the appropriate coefficients for the contrasts, the relationship between Self-Esteem and Time can be examined. With data collected at 6 equally spaced time points, it is possible to characterize the relationship between the Self-Esteem and Time with up to a quintic (fifth degree) polynomial. (The degree of the polynomial that can be examined is determined by the number of degrees of freedom for the repeated-measure variable.) Such a complex relationship would rarely be of interest to researchers, but linear and quadratic relationships are frequently of interest. In Table 10.3 we provide the appropriate coefficients for a first- to fifth-degree polynomial. [The coefficients for trend analysis for varying numbers of repeated measures can be found in many intermediate statistical methods textbooks, e.g., Keppel and Wickens, (2004, p. 577) and Maxwell and Delaney (2000, pp. 749–750).] Although we will only use the linear and quadratic coefficients here to demonstrate the approach, we present all of the coefficients because they will be useful in our discussion on the multivariate approach to a repeated-measures design.

A t statistic is computed as in Eq. (4.9), the ratio of the contrast to the estimated standard error of the contrast:

$$t = \frac{\hat{\psi}}{s_{\hat{\psi}}},$$

where

$$\hat{\psi} = \sum_{t=1}^{T} a_t \overline{Y}_{.t},$$

and

$$s_{\hat{\psi}} = \sqrt{\mathrm{MS}_{S \times T} \left(\sum_{t=1}^{T} \frac{a_t^2}{N} \right)}.$$

Using the data in Tables 10.1 and 10.2,

$$\begin{aligned} \hat{\psi}_{\text{Linear}} &\doteq (-5)(17.92) + (-3)(16.00) + (-1)(15.17) \\ &\quad + (1)(12.42) + (3)(13.25) + (5)(12.58) \\ &\doteq -37.7, \end{aligned}$$

TABLE 10.3 Coefficients for Linear to Quintic Polynomial Trend Analysis for Six Measurements

	Coefficient					
Trend	a_1	a_2	a_3	a_4	a_5	a_6
Linear	−5	−3	−1	1	3	5
Quadratic	5	−1	−4	−4	−1	5
Cubic	−5	7	4	−4	−7	5
Quartic	1	−3	2	2	−3	1
Quintic	−1	5	−10	10	−5	1

and

$$s_{\hat{\psi}(\text{Linear})} \doteq \sqrt{10.034\left[\frac{(-5)^2+(-3)^2+(-1)^2+(1)^2+(3)^2+(5)^2}{12}\right]}$$
$$\doteq 7.651.$$

The t-statistic value for testing the null hypothesis: H_0: $\psi_{\text{Linear}} = 0$ is

$$t \doteq \frac{-37.7}{7.651} \doteq -4.928.$$

The results of this analysis indicate support to conclude that there is a negative linear component in the relationship between Self-Esteem and Time, $t(55) \doteq -4.928$, $P \doteq .000$.

Using the coefficients for a quadratic polynomial, $\hat{\psi}_{\text{Quadratic}} \doteq 12.89$, $s_{\hat{\psi}(\text{Quadratic})} \doteq 8.381$, and $t(55) \doteq 1.539$, $P \doteq .129$. These results provide no evidence of a quadratic component in the relationship between Self-Esteem and Time. None of the tests for the remaining polynomials provided evidence to support a higher order relationship between Self-Esteem and Time [cubic: $t(55) \doteq .289$, $P \doteq .774$; quartic: $t(55) \doteq -.428$, $P \doteq .670$; quintic: $t(55) \doteq -1.453$, $P \doteq .152$]. Based on these results, we would conclude that a negative linear relationship best characterizes the change in Self-Esteem over Time. A plot of the mean Self-Esteem scores would help clarify this relationship.

10.4 MULTIVARIATE ANALYSIS

To introduce and explain the multivariate approach to the analysis of repeated measures, we begin by focusing only on the data obtained during the second trimester (Months 4, 5, and 6) of our research context. Following this presentation, we use SPSS to analyze the complete data set, Month 4 through Month 9, in Table 10.1. The multivariate approach for testing the equality of $T = 3$ population means obtained from repeated measurements (i.e., H_0: $\mu_4 = \mu_5 = \mu_6$), is to state this hypothesis as a vector of $T - 1 = 2$ contrasts. The number of "variables," p, considered in our multivariate analysis is the number of contrasts formed. For example, if H_0: $\mu_4 = \mu_5 = \mu_6$ is true, then (1) both $\mu_4 - \mu_6 = 0$ and $\mu_5 - \mu_6 = 0$; or alternatively (2) both $\mu_4 - \mu_5 = 0$ and $\mu_5 - \mu_6 = 0$. The omnibus hypothesis can be written as:

$$(1)\quad H_0\text{: } \begin{bmatrix} \mu_4 - \mu_6 \\ \mu_5 - \mu_6 \end{bmatrix} = \begin{bmatrix} 0 \\ 0 \end{bmatrix}, \quad \text{or as} \quad (2)\quad H_0\text{: } \begin{bmatrix} \mu_4 - \mu_5 \\ \mu_4 - \mu_6 \end{bmatrix} = \begin{bmatrix} 0 \\ 0 \end{bmatrix}.$$

More generally, the null hypothesis can be written as H_0: $\mathbf{A}\boldsymbol{\mu} = \mathbf{0}$, where $\mathbf{A}$ is a matrix of $(p \times T)$ contrast coefficients and $\boldsymbol{\mu}$ is a column vector of T means. For (1), $\mathbf{A}$ can be written as:

$$\mathbf{A} = \begin{bmatrix} 1 & 0 & -1 \\ 0 & 1 & -1 \end{bmatrix}.$$

For the omnibus hypothesis test, any set of $T - 1$ linearly independent contrasts may be used, under the condition that for any contrast, $\sum_{t=1}^{T} a_t = 0$. A set of contrasts is linearly independent if the coefficients of any one contrast cannot be determined from any linearly weighted combination of the coefficients of the other contrasts. For example, consider a situation where there are four repeated measures, $T = 4$. Then the coefficient matrix $\mathbf{A}$ might be

$$\mathbf{A} = \begin{bmatrix} 1 & -1 & 0 & 0 \\ 1 & 0 & -1 & 0 \\ 0 & 1 & -1 & 0 \end{bmatrix}.$$

But this set of coefficients is not linearly independent because coefficients of contrast 2 $(1 \quad 0 \quad -1 \quad 0)$ minus the coefficients of contrast 1 $(1 \quad -1 \quad 0 \quad 0)$ equal the coefficients of contrast 3: $(1 \quad 0 \quad -1 \quad 0) - (1 \quad -1 \quad 0 \quad 0) = (0 \quad 1 \quad -1 \quad 0)$. $\mathbf{A}$, therefore, would not be a valid set of coefficients. A third contrast that would be linearly independent of the first two contrasts would be $(1 \quad 0 \quad 0 \quad -1)$.

Given a set of linearly independent contrast coefficients, the omnibus hypothesis for the equality of repeated-measure means can be tested using Wilks Λ:

$$\Lambda = \frac{|\mathbf{AEA}'|}{|\mathbf{A}(\mathbf{E}+\mathbf{H})\mathbf{A}'|} = \frac{|\mathbf{AEA}'|}{|\mathbf{ATA}'|}, \tag{10.1}$$

where $\mathbf{E}$ is the error **SSCP** matrix, and $\mathbf{H}$ is the hypothesis **SSCP** matrix for the repeated-measures data.

Consider the data from the second trimester months 4, 5, and 6 in Table 10.1 and assume the contrasts are formed by subtracting means for adjacent months:

$$\mathbf{A} = \begin{bmatrix} 1 & -1 & 0 \\ 0 & 1 & -1 \end{bmatrix}.$$

The error **SSCP** matrix, $\mathbf{E}$, is computed as: $(\mathbf{Y}_{ut} - \mathbf{Y}_{.t})'(\mathbf{Y}_{ut} - \mathbf{Y}_{.t})$, where $\mathbf{Y}_{ut}$ is the $N \times T$ data matrix (Table 8.1, $N = 12$, $T = 3$) and $\mathbf{Y}_{.t}$ is a $N \times T$ matrix with each column being a vector of the mean observation for each of the T months. For the first three columns in Table 10.1,

$$\mathbf{E} \doteq \begin{bmatrix} 396.917 & 354.000 & 431.167 \\ 354.000 & 360.000 & 426.000 \\ 431.167 & 426.000 & 543.667 \end{bmatrix}.$$

Note, if the elements of $\mathbf{E}$ are divided by df_e, which equals $N - 1$, the result would be the error covariance matrix, $\mathbf{S}_e$, with the elements on the diagonal being the variance of the observations for each month.

The triple product matrix is

$$\mathbf{AEA}' \doteq \begin{bmatrix} 1 & -1 & 0 \\ 0 & 1 & -1 \end{bmatrix} \begin{bmatrix} 396.917 & 354.000 & 431.167 \\ 354.000 & 360.000 & 426.000 \\ 431.167 & 426.000 & 543.667 \end{bmatrix} \begin{bmatrix} 1 & 0 \\ -1 & 1 \\ 0 & -1 \end{bmatrix}$$

$$\doteq \begin{bmatrix} 48.917 & -11.167 \\ -11.167 & 51.667 \end{bmatrix}.$$

Therefore,

$$|\mathbf{AEA}'| \doteq (48.917)(51.667) - (-11.167)^2 \doteq 2402.693.$$

The hypothesis sum-of-squares matrix, $\mathbf{H}$, is computed as $N(\mathbf{y}_{.t} - \mathbf{y}_{..})(\mathbf{y}_{.t} - \mathbf{y}_{..})'$, where $\mathbf{y}_{.t}$ is a column vector of T means and $\mathbf{y}_{..}$ is a column vector of T grand means. Using the data in Table 10.1,

$$\mathbf{y}_{.t} - \mathbf{y}_{..} \doteq \begin{bmatrix} 17.92 - 16.36 \\ 16.00 - 16.36 \\ 15.17 - 16.36 \end{bmatrix} \doteq \begin{bmatrix} 1.56 \\ -.36 \\ -1.19 \end{bmatrix}.$$

The hypothesis **SSCP** matrix is computed as:

$$\mathbf{H} \doteq 12 \begin{bmatrix} 1.56 \\ -.36 \\ -1.19 \end{bmatrix} \begin{bmatrix} 1.56 & -.36 & -1.19 \end{bmatrix},$$

$$\doteq \begin{bmatrix} 29.037 & -6.741 & -22.296 \\ -6.741 & 1.565 & 5.176 \\ -22.296 & 5.176 & 17.120 \end{bmatrix}.$$

The denominator of (10.1) is computed as $\mathbf{ATA}'$ where $\mathbf{T} = \mathbf{E} + \mathbf{H}$. For our data

$$\mathbf{T} \doteq \begin{bmatrix} 425.954 & 347.259 & 408.871 \\ 347.259 & 358.435 & 431.176 \\ 408.871 & 431.176 & 530.787 \end{bmatrix},$$

and

$$\mathbf{ATA}' \doteq \begin{bmatrix} 1 & -1 & 0 \\ 0 & 1 & -1 \end{bmatrix} \begin{bmatrix} 425.954 & 347.259 & 408.871 \\ 347.259 & 358.435 & 431.176 \\ 408.871 & 431.176 & 530.787 \end{bmatrix} \begin{bmatrix} -1 & 0 \\ -1 & 1 \\ 0 & -1 \end{bmatrix},$$

$$\doteq \begin{bmatrix} 93 & 8 \\ 8 & 60 \end{bmatrix}.$$

The determinant of this product matrix is $|\mathbf{ATA}'| \doteq 5516$. To test the hypothesis of no change in self-reported depression during the second trimester, Wilks Λ is computed as:

$$\Lambda = \frac{|\mathbf{AEA}'|}{|\mathbf{ATA}'|} \doteq \frac{2402.693}{5516.000} \doteq .436.$$

Because the null hypothesis states that the vector of contrasts equals $\mathbf{0}$, this is equivalent to a single sample t test with $\mathrm{df}_h = 1$. Λ can be transformed to an F statistic using

$$F = \frac{1 - \Lambda}{\Lambda} \frac{\mathrm{df}_e - p + 1}{p}.$$

When the null hypothesis is true, this statistic has a central F distribution with $\nu_1 = p$ and $\nu_2 = \mathrm{df}_e - p + 1$ degrees of freedom with $\mathrm{df}_e = N - 1$. For our sample data:

$$F \doteq \frac{1 - .436}{.436} \frac{11 - 2 + 1}{2} \doteq 6.467.$$

The degrees of freedom are $\nu_1 = 2$ and $\nu_2 = \mathrm{df}_e - p + 1 = 11 - 2 + 1 = 10$. These results indicate that there is some evidence to reject the null hypothesis of no change in mean Self-Esteem, $\Lambda \doteq .436$, $F(2, 10) \doteq 6.468$, $P \doteq .016$, $\eta^2 = .564$. That is, the observed differences in Self-Esteem are generalizable to the population of women who are represented by this sample.

If all of the data in Table 10.1 are used to test the null hypothesis that there is no change in Self-Esteem over the second and third trimesters, H_0: $\mu_4 = \mu_5 = \mu_6 = \mu_7 = \mu_8 = \mu_9$, the multivariate approach would use Eq. (10.1) with the matrix of contrast coefficients being any set of five linearly independent contrasts. The $\mathbf{A}$ matrix, to test the equality of six population means, would be of order 5×6. One acceptable set of linearly independent coefficients is provided in Table 10.3. We used the first two contrasts with the univariate analysis to test for a linear and a quadratic trend in the relationship between Self-Esteem and Time. We can use this matrix along with an error **SSCP** matrix and an hypothesis **SSCP** matrix, based on all of the data collected over the 6-month period in Table 10.1 (Month 4 to Month 9), to compute Wilks Λ. Because the order of both $\mathbf{E}$ and $\mathbf{H}$ is 6×6, which makes the computations tedious, we rely on SPSS for their calculations in the next section.

10.5 COMPUTER APPLICATION I

In this section we present the SPSS syntax for conducting a repeated-measures analysis of variance with a single repeated-measures variable. Both the univariate and the multivariate test statistics are reported.

SPSS SYNTAX FOR THE OMNIBUS TEST FOR REPEATED MEASURES

```
manova month4 month5 month6 month7 month8 month9
/wsfactor = month(6)
/print = signif(efsize)
/design.
```

/wsfactor is the SPSS command that defines the repeated measures or within-subjects factor.
month is the researcher defined label for the within-subjects factor.
(6) defines the number of levels for the within-subjects factor. This number must equal the number of outcome variables listed on the MANOVA line.

OUTPUT

Analysis: Repeated Measures Omnibus Test

EFFECT .. MONTH
Multivariate Tests of Significance (S = 1, M = 1 1/2, N = 2 1/2)

Test Name	Value	Exact F	Hypoth. DF	Error DF	Sig. of F
Pillais	.75908	4.41096	5.00	7.00	.039
Hotellings	3.15069	4.41096	5.00	7.00	.039
Wilks	.24092	4.41096	5.00	7.00	.039
Roys	.75908				

Note.. F statistics are exact.

- -

Multivariate Effect Size

TEST NAME	Effect Size
(All)	.759

- -

Tests involving 'MONTH' Within-Subject Effect.
AVERAGED Tests of Significance for MONTH using UNIQUE sums of squares

Source of Variation	SS	DF	MS	F	Sig of F
WITHIN CELLS	551.89	55	10.03		
MONTH	287.11	5	57.42	5.72	.000

- -

Effect Size Measures

Source of Variation	Partial ETA Sqd
MONTH	.342

- -

Interpretation: Omnibus Test

The results of the multivariate approach for the test of no change in Self-Esteem during the second and third trimesters, H_0: $\mathbf{A}\boldsymbol{\mu} = \mathbf{0}$ (which is equivalent to the univariate hypothesis, H_0: $\mu_4 = \mu_5 = \mu_6 = \mu_7 = \mu_8 = \mu_9$) indicates that there is evidence to reject the null hypothesis, $\Lambda \doteq .241$, $F(5, 7) \doteq 4.411$, $P \doteq .039$. All of the unadjusted effect-size values are the same, .759, and the Serlin adjusted effect size is $.741[1 - [(72 - 1)/(72 - 5 - 1)](1 - .759)]$. [Note that with the repeated measures design, N is the total number of measurements in the data set, not the number of subjects (Olejnik and Algina, 2003).]

These multivariate results are consistent with those reported for the univariate approach presented in Section 10.2, and repeated here as part of the SPSS output, which also indicates that there is sufficient support for the conclusion that Self-Esteem changes over the final 6 months of pregnancy, $F(5, 55) \doteq 5.72$, $P \doteq .000$, $\eta^2 \doteq .342$, $\eta_G^2 \doteq .149$. There is quite a bit of a difference, however, in the reported P values and effect size values. To understand the reason for the difference in these results and the difference between the multivariate and univariate approaches to the analysis of a repeated-measures design, it is necessary to examine a little closer how the test statistics are computed for the two approaches to which we will turn to in the next section.

10.6 UNIVARIATE AND MULTIVARIATE ANALYSES

As we discussed earlier, the multivariate approach requires that contrasts be formed and any set of linearly independent coefficients will be acceptable with the condition that the sum of the coefficients for a contrast is 0 ($\sum_{t=1}^{T} a_t = 0$). The contrasts presented in Table 8.3 are not only linearly independent but they are also orthogonal (independent) of each other. Two contrasts, ψ_A and ψ_B, are said to be orthogonal if the sum of the products obtained by multiplying the coefficients of one contrast by the coefficients of the second contrast equal zero, $\sum_{t=1}^{T} a_t b_t = 0$. For example, the coefficients for the linear and quadratic polynomials in Table 10.3 are

Linear (a)	−5	−3	−1	1	3	5
Quadratic (b)	5	−1	−4	−4	−1	5

The product of the coefficients for the linear and quadratic polynomials are

$$
\begin{array}{lrrr}
 & (a_t) & (b_t) & \\
M_4 & (-5) & (5) = & -25 \\
M_5 & (-3) & (-1) = & 3 \\
M_6 & (-1) & (-4) = & 4 \\
M_7 & (1) & (-4) = & -4 \\
M_8 & (3) & (-1) = & -3 \\
M_9 & (5) & (5) = & 25,
\end{array}
$$

and the sum of the products equals 0 $\left(\sum_{t=1}^{T} a_t b_t = -25 + 3 + 4 - 4 - 3 + 25\right)$. Any pair of contrasts in Table 10.3 will give the same result.

In addition to being orthogonal, contrasts are sometimes normalized, which means that the sum of the squared coefficients equals 1; $\sum_{t=1}^{T} a_t^2 = 1$. To normalize a set of coefficients, each coefficient is multiplied by the positive square root of the reciprocal of the sum of squared coefficients:

$$w = \sqrt{\frac{1}{\sum_{t=1}^{T} a_t^2}}. \tag{10.2}$$

For a linear trend, the sum of the squared coefficients is

$$\sum_{t=1}^{T} a_t^2 = (-5)^2 + (-3)^2 + \cdots + (5)^2 = 70,$$

and $w = \sqrt{1/70} \doteq .1195$. Multiplying each of the coefficients for the linear polynomial by w gives the following results:

$$-.598 \quad -.359 \quad -.120 \quad .120 \quad .359 \quad .598.$$

As a check $\sum_{t=1}^{T} a_t^2 = (-.598)^2 + (-.356)^2 + \cdots + (.598)^2 \doteq 1.001$. The orthogonal contrasts that have normalized coefficients are referred to as *orthonormal contrasts*. The complete set of orthonormalized coefficients for the polynomial contrast presented in Table 10.3 is presented in Table 10.4. We will refer to this matrix of orthonormal coefficients as $\mathcal{A}$.

The univariate approach to the analysis of repeated-measures designs assumes that the variances of the orthonormal contrasts are equal and the covariances of the contrasts are 0. This assumption can be written as:

$$\mathcal{A}\boldsymbol{\Sigma}\mathcal{A}' = \sigma^2\mathbf{I}, \tag{10.3}$$

where $\mathcal{A}$ is a set of $p \times T$ orthonormal coefficients such as those presented in Table 10.4, $\boldsymbol{\Sigma}$ is the population covariance matrix for the repeated measures, σ^2

TABLE 10.4 Orthonormal Polynomial Coefficients

	Coefficient					
Trend	a_1	a_2	a_3	a_4	a_5	a_6
Linear	−.598	.359	−.120	.120	.359	.598
Quadratic	.549	−.109	−.436	−.436	−.109	.546
Cubic	−.373	.522	.298	−.298	−.522	.373
Quartic	.189	−.567	.378	.378	−.567	.189
Quintic	−.063	.315	−.630	.630	−.315	.063

is the common population variance for the orthonormal contrasts, and **I** is a $p \times p$ identity matrix. A product matrix that meets this condition is said to be *spherical.* Equivalently, the assumption of sphericity is met if the variances of difference scores for all pairs of measures are equal. For the data in Table 10.1 the error **SSCP** matrix, **E**, and error covariance matrix, $\mathbf{S}_e$, are

$$\mathbf{E} \doteq \begin{bmatrix} 396.917 & 354.000 & 431.167 & 89.417 & 170.250 & 67.583 \\ 354.000 & 360.000 & 426.000 & 116.000 & 200.000 & 108.000 \\ 431.167 & 426.000 & 543.667 & 121.167 & 222.500 & 101.833 \\ 89.417 & 116.000 & 121.167 & 128.917 & 157.750 & 145.083 \\ 170.250 & 200.000 & 222.500 & 157.750 & 215.583 & 175.250 \\ 67.583 & 108.000 & 101.833 & 145.083 & 175.250 & 176.917 \end{bmatrix},$$

and

$$\mathbf{S}_e \doteq \begin{bmatrix} 36.083 & 32.182 & 39.197 & 8.129 & 15.477 & 6.144 \\ 32.182 & 32.727 & 38.272 & 10.546 & 18.182 & 9.818 \\ 39.197 & 38.727 & 49.424 & 11.015 & 20.227 & 9.258 \\ 8.129 & 10.545 & 11.015 & 11.720 & 14.341 & 13.189 \\ 15.477 & 18.182 & 20.227 & 14.341 & 19.598 & 15.932 \\ 6.144 & 9.818 & 9.258 & 13.189 & 15.932 & 16.083 \end{bmatrix}.$$

Using the orthonormal coefficients in Table 10.4, $\mathcal{A}\mathbf{S}_e\mathcal{A}'$ is computed as:

$$\mathcal{A}\mathbf{S}_e\mathcal{A}' \doteq \begin{bmatrix} 33.560 & 2.783 & -7.924 & -0.406 & 16.577 \\ 2.834 & 2.086 & -1.829 & 0.079 & 2.842 \\ -7.924 & -1.829 & 3.509 & -0.348 & -4.964 \\ -0.406 & 0.079 & -0.348 & 0.841 & -0.139 \\ 16.577 & 2.842 & -4.964 & -0.139 & 10.242 \end{bmatrix}.$$

The elements on the diagonal of $\mathcal{A}\mathbf{S}_e\mathcal{A}'$ (contrast variances) are very different, ranging from 0.841 to 33.560. The sample covariances also vary a great deal ranging from -7.924 to 16.577. (The variance of the difference scores for months 4 and 5 equals 4.47; while the variance of the difference scores for months 4 and 9 equals 39.879.) Based on these sample data, there is some reason to doubt that the assumption of sphericity is met. That is, the sample value of $\mathcal{A}\mathbf{S}_e\mathcal{A}'$ does not appear to conform to Eq. (10.3). If the assumption of sphericity is violated, the F value from the univariate approach computed as the ratio of the mean square for Time to the mean-square interaction ($F = \mathrm{MS}_{\mathrm{Time}}/\mathrm{MS}_{S\times T}$) does *not* have a central F distribution with $(T-1)$ and $(N-1)(T-1)$ degrees of freedom. The actual distribution of F is shifted somewhat to the right of this distribution, which means using the central F distribution with $(T-1)$ and $(N-1)(T-1)$ degrees of freedom will yield a P value that is too small. That is, the test of the null hypothesis will be liberal.

It should be pointed out that the mean of the diagonal elements in the $\mathcal{A}\mathbf{S}_e\mathcal{A}'$ product matrix equals the $\mathrm{MS}_{S\times T}$ used in the calculation of the F ratio for the univariate hypothesis test $[(33.560 + 2.086 + 3.509 + 0.841 + 10.242)/5 \doteq 10.048]$.

The numerator of the univariate F ratio can also be obtained from the diagonal of the triple product matrix, $\mathcal{A}\mathbf{H}\mathcal{A}'$.

The $\mathbf{H}$ matrix is computed by $N(\mathbf{y}_{.t} - \mathbf{y}_{..})(\mathbf{y}_{.t} - \mathbf{y}_{..})'$. Using the means in Table 10.1, the hypothesis matrix $\mathbf{H}$ is then computed as:

$$\mathbf{H} \doteq (12) \begin{bmatrix} 3.363 \\ 1.443 \\ 0.613 \\ -2.137 \\ -1.307 \\ -1.977 \end{bmatrix} \begin{bmatrix} 3.363 & 1.443 & 0.613 & -2.137 & -1.307 & -1.977 \end{bmatrix},$$

and

$$\mathcal{A}\mathbf{H}\mathcal{A}' \doteq \begin{bmatrix} 243.711 & -77.788 & -14.367 & 21.294 & 65.016 \\ -77.788 & 24.829 & 4.586 & -6.797 & -20.752 \\ -14.367 & 4.586 & 0.847 & -1.255 & -3.833 \\ 21.294 & -6.797 & -1.255 & 1.860 & 5.681 \\ 65.016 & -20.752 & -3.833 & 5.681 & 17.345 \end{bmatrix}$$

The sum of the diagonal elements $\mathcal{A}\mathbf{H}\mathcal{A}'$ equals the univariate sum of squares for the Time main effect $(243.711 + 24.829 + 0.847 + 1.860 + 17.345 \doteq 288.592)$. Divide this sum by the degrees of freedom $(T - 1)$ for the repeated measures factor (Time) to obtain the univariate mean square for the Time factor $288.592/5 \doteq 57.718$.

The MS_{Time} and $\text{MS}_{S\times T}$ used for the calculation of the univariate F ratio can, therefore, be obtained directly from the sum of the elements on the main diagonal (i.e., the trace) of the matrices $\mathcal{A}\mathbf{H}\mathcal{A}'$ and $\mathcal{A}\mathbf{E}\mathcal{A}'$, respectively, divided by the respective degrees of freedom. The multivariate test, on the other hand, involves the determinants of these two matrices, which consider not only the trace but the off-diagonal elements (CP elements) as well. If the assumption of sphericity is met, the CP elements provide no additional information, and the univariate approach would provide a more powerful statistical hypothesis test.

10.7 TESTING FOR SPHERICITY

A statistical test for the sphericity assumption was developed by Mauchly (1940) and is reported in SPSS. Mauchly's test statistic is

$$W = \frac{|\mathcal{A}\mathbf{S}_{\mathbf{e}}\mathcal{A}'|}{\left[\frac{\text{trace}(\mathcal{A}\mathbf{S}_{\mathbf{e}}\mathcal{A}')}{p}\right]^{p}}, \tag{10.4}$$

where p is the number of contrasts or $T - 1$.

The statistic W can be transformed to a statistic having a χ^2 distribution with degrees of freedom, $\text{df} = T(T - 1)/2 - 1$. The transformation of W is

$$\chi^2 = -(N - J)d[\ln(W)],$$

where d is computed as:

$$d = 1 - \left[\frac{2T^2 - 3T + 3}{6(N - J)(T - 1)}\right].$$

For our data:

$$\begin{aligned}
J &= 1, \\
|\mathcal{A}\mathbf{S}_e\mathcal{A}'| &= 32.640, \\
\mathrm{tr}(\mathcal{A}\mathbf{S}_e\mathcal{A}') &\doteq 102396.17, \\
p &= 6 - 1 = 5, \\
W &\doteq \frac{32.640}{(102396.17/5)^5} \doteq .0003, \\
d &= 1 - \left[\frac{6(12 - 1)(6 - 1)}{2(6)^2 - 3(6) + 3}\right] \doteq .827, \\
J &= 1, \\
\chi^2 &\doteq -(12 - 1)(.827)\ln(.0003) \doteq 73.792 \text{ and} \\
\mathrm{df} &= \frac{6(6 - 1)}{2} - 1 = 14.
\end{aligned}$$

These results provide strong evidence that the sphericity assumption has been violated, $W \doteq .0003$, $\chi^2(14) \doteq 73.792$, $P \doteq .000$. Consequently, the reported P value from the univariate analysis is very likely to be too small.

There are two limitations associated with the Mauchly test. *First*, it assumes multivariate normality. When this assumption is violated, it can lead to a rejection of the sphericity assumption even though the assumption has been met. *Second*, simulation studies have shown that the Mauchly test is not very sensitive to important violations of the sphericity assumption and, consequently, cannot be relied on to provide adequate guidance as to whether the univariate or multivariate approaches should be adopted. A more powerful test to detect a violation of the sphericity assumption has been developed (Cornell et al., 1992), but it has not been incorporated into computer software packages like SPSS or SAS.

An approximate solution, for the potentially liberal univariate hypothesis test for repeated measures, is to adjust the degrees of freedom for the computed F statistic that are used to determine the actual P value. The adjustment is achieved by multiplying the numerator and denominator degrees of freedom by ε, which measures the degree to which condition (10.3) is violated. Three estimators of ε have been suggested. *One* estimator of ε is the reciprocal of the degrees of freedom for the repeated measures factor $[1/(T - 1)]$ and is often represented by ε^* (Geisser and Greenhouse, 1958). For our problem, $\varepsilon^* = 1/(6 - 1) = .20$. This estimator is referred to as the *lower bound adjustment* in SPSS. The problem with this solution is that it overadjusts for the assumption violation, and therefore the resulting P value will be too large. In other words, using the lower bound adjustment provides a conservative hypothesis test for differences in means for the repeated-measures variable. A *second* estimator for ε is

$$\varepsilon' = \frac{\left[\text{trace}(\mathcal{A}\mathbf{S}_e\mathcal{A}')\right]^2}{(T-1)\left(\sum_{r=1}^{R}\sum_{c=1}^{C} x_{cr}^2\right)}, \tag{10.5}$$

where $[\text{trace}(\mathcal{A}\mathbf{S}_e\mathcal{A}')]^2$ is the squared sum of the orthonormal contrast variances, and $\sum_{r=1}^{R}\sum_{c=1}^{C} x_{cr}^2$ is the sum of squared c column by r row elements of the $\mathcal{A}\mathbf{S}_e\mathcal{A}'$ triple product matrix. SPSS refers to this estimator of ε as the *Greenhouse–Geisser adjustment*. Using our data and the $\mathcal{A}\mathbf{S}_e\mathcal{A}'$ triple-product matrix reported above, ε is estimated as:

$$\begin{aligned}
[\text{trace}(\mathcal{A}\mathbf{S}_e\mathcal{A}')]^2 &\doteq [33.560 + 2.086 + 3.509 + .841 + 10.242]2 \\
&= [50.238]2 \doteq 2523.857, \\
\sum_{r=1}^{R}\sum_{c=1}^{C} x_{cr}^2 &\doteq 33.560^2 + 2.783^2 + \cdots + 16.577^2 \\
&\quad + 2.834^2 + \cdots + 2.842^2 + \cdots + 10.242^2 \\
&\doteq 2012.53,
\end{aligned}$$

and

$$\varepsilon' \doteq \frac{2523.857}{(6-1)(2012.530)} = .251.$$

A problem with ε' is that it also underestimates ε and provides a slightly conservative test for the equality of means for the repeated-measures variable.

A *third* estimator of ε, proposed by Huynh and Feldt (1976), is

$$\tilde{\varepsilon} = \frac{N(T-1)\varepsilon' - 2}{(T-1)[(N-J)-(T-1)\varepsilon']}. \tag{10.6}$$

For our data and ε' computed above, $\tilde{\varepsilon}$ is computed as:

$$\tilde{\varepsilon} \doteq \frac{12(6-1).251 - 2}{(6-1)[(12-1)-(6-1).251]} \doteq \frac{13.06}{48.725} \doteq .268.$$

A problem with the Huynh and Feldt estimator is that it overestimates ε and can yield a P value smaller than the true P value (a liberal test). When condition (10.3) holds, $\varepsilon = 1$, the Huynh–Feldt $\tilde{\varepsilon}$ can be greater than 1. In those situations, the Huynh–Feldt $\tilde{\varepsilon}$ would be set equal to 1.

Using these estimates to adjust the numerator and denominator degrees of freedom by multiplying these estimates of ε by the degrees of freedom for our problem gives the following numerator and denominator degrees of freedom:

$$\text{Lower bound} = .2(6-1), .2(6-1)(12-1) = 1, 11,$$
$$\text{Greenhouse–Geisser} \doteq .251(6-1),\ .251(6-1)(12-1) \doteq 1.255, 13.805,$$
$$\text{Huynh–Feldt} \doteq .268(6-1),\ .268(6-1)(12-1) \doteq 1.340, 14.740.$$

Using our computed F statistic (5.723) obtained previously in Section 10.3, the corresponding P values are:

$$\text{Lower bound}: F(1, 11) \doteq 5.723,\ P \doteq .036,$$
$$\text{Greenhouse–Geisser}: F(1.255, 13.805) \doteq 5.723,\ P \doteq .026,$$
$$\text{Huynh–Feldt}: F(1.34, 14.740) \doteq 5.723,\ P \doteq .023.$$

With the adjusted degrees of freedom to evaluate the observed univariate F ratio, the results are now similar to those reported for the multivariate analysis. In this case all of the analyses lead to the same conclusion: There is evidence to indicate that Self-Esteem changes over the second and third trimesters of pregnancy. This many not always be the case. In this example, while the conclusion is the same, the reported P values are somewhat different.

Some data analysts always prefer the multivariate approach for the analysis of repeated measures; others prefer the univariate approach using the adjusted degrees of freedom test. The multivariate approach provides an exact hypothesis test while the adjusted univariate approach provides an approximate test that is conservative. The multivariate test, however, can be less powerful than the univariate test with adjusted degrees of freedom. With small sample sizes, the multivariate hypothesis test can have substantially fewer degrees of freedom than the adjusted univariate criterion. Maxwell and Delaney (2000, p. 603), who prefer the multivariate approach, have suggested that the minimum sample size for the multivariate approach be $T + 10$. That is, there should be at least 10 more individuals participating in the study than the number of observations per individual. In the current study, with 6 observations per individual a minimum sample size would be 16. We only had a total of 12 individuals participating. In our example, the slightly larger P value reported for the multivariate approach may reflect its slightly lower statistical power.

10.8 COMPUTER APPLICATION II

In this section we present the SPSS syntax command to obtain the Mauchly test for sphericity and the three estimates of ε. While the Mauchly test is reported along with the estimates of ε, it does not report the P values associated with the adjusted degrees of freedom tests. The SPSS General Linear Model program, however, does report the results of the Mauchly test, the estimates of ε, and the P values for the F statistic associated with the adjusted degrees of freedom test.

SPSS SYNTAX FOR OBTAINING THREE ESTIMATES OF ε

```
manova month4 month5 month6 month7 month8 month9
/wsfactor=month(6)
/print=homogeneity
/design.
```

/print = homogeneity in the repeated measures context, **homogeneity** requests the calculation of Mauchly test for sphericity and the three estimates of ε.

OUTPUT

Analysis: Mauchly Test for Sphericity and Estimates of ε

```
Tests involving 'MONTH' Within-Subject Effect.
Mauchly sphericity test, W =          .00032
 Chi-square approx. =               73.25517 with 14 D. F.
 Significance =                         .000

 Greenhouse-Geisser Epsilon =         .25099
 Huynh-Feldt Epsilon =                .26802
 Lower-bound Epsilon =                .20000

AVERAGED Tests of Significance that follow multivariate tests are
equivalent to univariate or split-plot or mixed-model approach to
repeated measures. Epsilons may be used to adjust d.f. for the
AVERAGED results.
- - - - - - - - - - - - - - - - - - - - - - - - - - - - - - - - - - -
 EFFECT .. MONTH
 Multivariate Tests of Significance ( S = 1,  M = 1 1/2,  N = 2 1/2)

 Test Name      Value    Exact F   Hypoth. DF   Error DF  Sig. of F

 Pillais        .75908   4.41096      5.00         7.00      .039
 Hotellings    3.15069   4.41096      5.00         7.00      .039
 Wilks          .24092   4.41096      5.00         7.00      .039
 Roys           .75908
 Note.. F statistics are exact.

- - - - - - - - - - - - - - - - - - - - - - - - - - - - - - - - - - -
Tests involving 'MONTH' Within-Subject Effect.

 AVERAGED Tests of Significance for month using UNIQUE sums of squares
 Source of Variation          SS         DF       MS        F   Sig of F
```

WITHIN CELLS	551.89	55	10.03		
MONTH	287.11	5	57.42	5.72	.000

Interpretation: Mauchly Test for Sphericity

The results of the Mauchly test for sphericity provide evidence to indicate that the sphericity assumption is not met with these data, $W \doteq .0003$, $\chi^2(14) \doteq 73.255$, $P \doteq .000$. These results are consistent, within rounding, of the results we provided in Section 10.7. Because SPSS does not adjust the degrees of freedom for the F statistic from the univariate approach, the reported P value is too small. It would be necessary to multiply the reported degrees of freedom by the selected estimate of ε and estimate the P value by using tabled values of F distributions.

10.9 CONTRAST ANALYSIS

In Section 10.3 we examined both linear and quadratic trend contrasts for the univariate approach using a t test. That is, we tested the null hypothesis: H_0: $\psi = 0$ by computing the ratio of the contrast of interest to its estimated standard error, $t = \hat{\psi}/s_{\hat{\psi}}$, where

$$\hat{\psi} = \sum_{t=1}^{T} a_t \overline{Y}_{.t},$$

and

$$s_{\hat{\psi}} = \sqrt{\text{MS}_{S\times T}\left(\frac{\sum_{t=1}^{T} a_t^2}{N}\right)}.$$

For the multivariate approach, contrasts can be tested in a manner similar to the univariate approach presented above. The value of the contrast, $\hat{\psi}$, is the same for both approaches, but the estimated standard errors for the contrasts are different. As presented above, for the univariate approach the estimated standard error is computed using $\text{MS}_{S\times T}$. In Section 10.5 we demonstrated that $\text{MS}_{S\times T}$ equals the mean of the values on the main diagonal of the $\mathcal{A}\text{S}_e\mathcal{A}'$ product matrix. The values on the main diagonal of the $\mathcal{A}\text{S}_e\mathcal{A}'$ product matrix are the variances for the contrasts specified by the coefficients provided in the $\mathcal{A}$ matrix. That is, $\text{MS}_{S\times T}$ is the mean othonormal contrast variance. For a set of orthonormal pairwise contrasts, then, all pairwise univariate contrasts would have the same estimated standard error. The multivariate approach provides a different estimated standard error for each contrast because each contrast uses its own variance. For convenience, the $\mathcal{A}\mathbf{S}_e\mathcal{A}'$ product matrix is restated here:

$$\mathcal{A}\mathbf{S}_e\mathcal{A}' \doteq \begin{bmatrix} 33.560 & 2.783 & -7.924 & -0.406 & 16.577 \\ 2.834 & 2.086 & -1.829 & 0.079 & 2.842 \\ -7.924 & -1.829 & 3.509 & -0.348 & -4.964 \\ -0.406 & 0.079 & -0.348 & 0.841 & -0.139 \\ 16.577 & 2.842 & -4.964 & -0.139 & 10.242 \end{bmatrix}.$$

The variance of the linear contrast, MS_{Linear}, is 33.560 and the variance for the quadratic contrast, $MS_{\text{Quadratic}}$, is 2.086. The multivariate approach computes the t statistic for the linear trend as:

$$\begin{aligned} \hat{\psi}_{\text{Linear}} &\doteq (-.598)(17.92) + (-.359)(16.00) + (-.120)(15.17) \\ &\quad + (.120)(12.42) + (.359)(13.25) + (.598)(12.58) \\ &\doteq -4.51 \end{aligned}$$

$$\begin{aligned} s_{\hat{\psi}(\text{Linear})} &\doteq \sqrt{33.56\frac{(-.598)^2 + (-.359)^2 + (-.120)^2 + (.120)^2 + (.359)^2 + (.598)^2}{12}} \\ &\doteq 1.672 \end{aligned}$$

$$t \doteq \frac{-4.51}{1.672} \doteq -2.697.$$

For the quadratic trend, $\hat{\psi}_{\text{Quadratic}} \doteq 1.436$, $s_{\hat{\psi}_{\text{Quadratic}}} \doteq .417$, and $t \doteq 3.444$.

These results indicate that there is sufficient evidence to indicate a statistically significant negative linear component, $t(11) \doteq -2.697$, $P \doteq .021$ and a quadratic relationship between Self-Esteem and Time, $t(11) \doteq 3.444$, $P \doteq .006$. Note that with the multivariate approach, the degrees of freedom are 11 (i.e., $N - 1$). Additional tests using the coefficients in Table 10.4 provide little support for a higher order polynomial trend between Time and Self-Esteem. Based on these results we would conclude that a quadratic trend best characterizes the relationship between Self-Esteem and Time.

It might come as a surprise to some researchers to find that the univariate test of quadratic trend provided little evidence of a quadratic trend, while the multivariate approach indicated a quadratic trend. An explanation for these contradictory results might be provided by examining the variance estimates of the two contrasts. The variance of the univariate quadratic contrast uses the mean variance across all orthonormal contrasts, $MS_{S \times T}$. As was pointed out earlier, the elements on the main diagonal of the $\mathcal{A}\mathbf{S}_e\mathcal{A}'$ matrix varied considerably leaving the mean variance quite large (10.034). On the other hand, the multivariate approach uses the estimated variance of the quadratic trend alone, and it was only 2.096. If the sphericity assumption (equal contrast variances) had been met, then the use of the mean variance across the contrasts would provide greater statistical power because the degrees of freedom would have been greater. In our example, the variances of the contrasts are not equal so the univariate approach is inappropriate.

10.10 COMPUTER APPLICATION III

In this section we present the SPSS syntax for polynomial trend analysis. We present the appropriate coefficients here, but the same results could be obtained by **/contrast(month)=polynomial**. We provide the specific coefficients to demonstrate how a researcher may request any set of orthogonal linearly independent contrasts. Readers should take notice of the warning that appears at the beginning of the analysis output.

SPSS SYNTAX FOR POLYNOMIAL TREND CONTRASTS

```
manova month4 month5 month6 month7 month8 month9
/wsfactor=month(6)
/contrast(month)=special
(1 1 1 1 1 1,
−5 −3 −1 1 3 5,
5 −1 −4 −4 −1 5,
−5 7 4 −4 −7 5,
1 −3 2 2 −3 1,
−1 5 −10 10 −5 1)
/print=transform signif(averf)
/rename=con linear quadratic cubic quartic quintic
/design.
```

/contrast(month) is an SPSS command code requesting contrasts among levels of the month variable is computed.
Special (111111, −5 −3 −1135,...) specifies the coefficients for the contrasts of interest. Because polynomial contrasts are of interest, the same results could have been obtained without stating the coefficients **if = polynomial** replaces **=special** and the contrast coefficients.
/print = transform requests the printing of the orthonormalizing coefficients.
/rename provides labels for the transformed contrasts.

Analysis: Contrast Tests

```
>Warning # 12252 in column 18.  Text: SPECIAL
>Special contrasts were requested for a WSFACTOR.  MANOVA automatically
>orthonormalizes contrast matrices for WSFACTORS.  If the special contrasts
>that were requested are nonorthogonal, the contrasts actually fitted are
>not the contrasts requested.  See the transformation matrix for the actual
>contrasts fitted.  Use TRANSFORM instead of WSFACTORS to produce
>nonorthogonal contrasts for within subjects factors.  Multivariate and
>averaged tests remain valid.
```

Orthonormalized Transformation Matrix (Transposed)

	CON	LINEAR	QUADRATI	CUBIC	QUARTIC	QUINTIC
MONTH4	.408	−.598	.546	−.373	.189	−.063
MONTH5	.408	−.359	−.109	.522	−.567	.315
MONTH6	.408	−.120	−.436	.298	.378	−.630
MONTH7	.408	.120	−.436	−.298	.378	.630
MONTH8	.408	.359	−.109	−.522	−.567	−.315
MONTH9	.408	.598	.546	.373	.189	
.063						

Estimates for LINEAR MONTH

Parameter	Coeff.	Std. Err.	t-Value	Sig. t	Lower −95%	CL- Upper
1	−4.502	1.670	−2.695	.021	−8.178	−.826

Estimates for QUADRATI MONTH

Parameter	Coeff.	Std. Err.	t-Value	Sig. t	Lower −95%	CL- Upper
1	1.409	.41795	3.372	.006	.489	2.329

Estimates for CUBIC MONTH

Parameter	Coeff.	Std. Err.	t-Value	Sig. t	Lower −95%	CL- Upper
1	.267	.541	.494	.631	−.924	1.458

Estimates for QUARTIC MONTH

Parameter	Coeff.	Std. Err.	t-Value	Sig. t	Lower −95%	CL- Upper
1	−.394	.265	−1.488	.165	−.976	.189

Estimates for QUINTIC MONTH

Parameter	Coeff.	Std. Err.	t-Value	Sig. t	Lower −95%	CL- Upper
1	−1.202	.924	−1.301	.220	−3.235	.831

Interpretation: Contrast Tests

The computer output begins by reporting the orthonormalizing transformation matrix. These results correspond to the transformation matrix reported in Table 10.4 except for the first column. The first column in the output labeled CON is the orthonormal coefficients for a vector of 1's that represents the intercept or the constant in a regression model. It provides no useful information for the repeated-measures design. It is always a good idea for researchers to print the transformation matrix to be sure that the intended analysis is actually computed by the program. For example, if a researcher requests a set of special contrasts comparing each level of the repeated measures variable with the last level (e.g. $1\ \ 1\ \ 1\ \ 1\ \ 1\ \ 1, 1\ \ 0\ \ 0\ \ 0\ \ 0\ \ -1, 0\ \ 1\ \ 0\ \ 0\ \ 0\ \ -1, \ldots, 0\ \ 0\ \ 0\ \ 0\ \ 1\ \ -1$), the program actually computes *Helmert* contrasts. That is, the contrasts actually compare the mean of each level of the repeated-measures variable with the mean of the remaining levels. So, rather than comparing the mean of Month 4 with the mean of Month 9, the first contrast would compare Month 4 with the mean of Months 5, 6, 7, 8, and 9. Although a warning is provided, without examining the transformation matrix, a researcher could easily interpret the results for the contrasts but for the wrong hypotheses. The pairwise contrasts we requested above are not orthogonal.

The results for the linear through the quintic trend analyses are reported with the value of the contrast, $\hat{\psi}$, reported as the *Coeff.* Because multiple tests are being examined, it is generally advisable to "adjust" the reported P values by considering the number of hypotheses tested (Bonferroni adjustment). If the researcher had no a priori hypothesis regarding the nature of the relationship between Self-Esteem and Time, and intended to examine all five trends, each of the obtained P values should be multiplied by 5. But if the researcher had decided a priori that only linear and quadratic trends would be interpretable, then the obtained P values would be multiplied by 2. If we assume the latter *a priori* hypotheses, the results would be interpreted as evidence of a quadratic trend in the means for Self-Esteem, $t(11) \doteq 3.372$, $P' \doteq .012$. A plot of the observed monthly self-esteem means would further help interpret the findings.

10.11 SUMMARY

In this chapter we examined the analysis of data from a repeated-measures or within-subject research design. This design occurs when the subjects are measured over several time points, measured under different treatment conditions, or measured with a test battery. Such data can be analyzed using either a univariate or a multivariate approach. If a reasonable sample size is available, the multivariate approach should be preferred because it provides an exact test of the hypothesis of the equality of the repeated-measures means. The univariate approach requires a certain structure (sphericity) for the relationship between the repeated measurements. When that structure is not appropriate, the univariate hypothesis test will be liberal. That is, the reported P value will be too small. This structure is generally not met when the measurements obtained are time linked. Although adjustments to the degrees of freedom have been suggested, these adjustments provide only an approximate P value. In addition to the accuracy of the P value, Maxwell and Delany (2000, p. 601) point out that the multivariate approach for contrast analysis is consistent with the multivariate approach to testing the omnibus hypothesis. When the sphericity assumption is violated, contrasts must use separate variance estimates for each contrast tested. The univariate hypothesis test uses the mean variance across all orthonormal contrasts and compensates for the assumption violation by adjusting the degrees of freedom. This inconsistency between the omnibus test and the contrast tests is undesirable. As a result, the multivariate approach to the analysis of repeated measures should be a researcher's first choice for data analysis.

In this chapter we considered a context where there was a single repeated-measures variable. It is possible to have more than one repeated-measures variable. For example, in our context we considered the monthly measures of Self-Esteem as a single variable with six levels. Alternatively, the researcher could have considered the six measures as representing two variables, trimester and month. That is, the researcher might have been interested in comparing the mean Self-Esteem in the second trimester with the mean Self-Esteem in the third trimester. In addition, the Month variable could be thought of as representing the beginning, middle, and end of a trimester. Thinking about the data this way would create a 2×3 factorial structure for the repeated measures. An interaction between Trimester and Month could be tested as well as the Trimester main effect. In the Technical Notes we provide the SPSS syntax to conduct a factorial analysis of the repeated measures and we provide the output of the analysis. We believe the interpretation is self-explanatory. It might be noted that the same hypotheses could be tested with the procedures presented in this chapter by the appropriate choice of coefficients for contrast analyses.

Technical Notes

1. SPSS syntax for a repeated-measures design having two within-subject factors. *One* factor is *Trimester* having two levels (months 4, 5, and 6 to represent one trimester and months 7, 8, and 9 as a second trimester). The *second* factor is *Month* having three levels representing early, mid, and late points in a trimester. The design is described as a 2×3 repeated-measures design.

SPSS SYNTAX FOR A FACTORIAL DESIGN WITH REPEATED MEASURES

```
manova  month4 month5 month6 month7 month8 month9
/wsfactor=trimester(2) month(3)
/print= signif(efsize)
/design.
```

/wsfactor=trimester(2) month(3) splits the six repeated measures into two variables. The order in which the measures are listed on the MANOVA line is very important. Months 4, 5, and 6 are the first, second, and third months of the second Trimester; while Months 7, 8, and 9 are the first, second, and third months of the third Trimester. To indicate this ordering of data, Month is listed last so levels of this variable are changing quicker than the levels of Trimester. That is, Month 4 is the first month of the second Trimester, Month 5 is the second month of the second Trimester, while Month 7 is the first month of the third Trimester.

Note: there are 2 levels for the TRIMESTE effect. Average tests are identical to the univariate tests of significance.

Tests involving 'TRIMESTE' Within-Subject Effect.

Tests of Significance for T2 using UNIQUE sums of squares

Source of Variation	SS	DF	MS	F	Sig of F
WITHIN CELLS	466.94	11	42.45		

TRIMESTE	234.72	1	234.72	5.53	.038

- -

Effect Size Measures

Source of Variation	Partial ETA Sqd
TRIMESTE	.335

- -

Tests involving 'MONTH' Within-Subject Effect.

Mauchly sphericity test, W = .66483
Chi-square approx. = 4.08229 with 2 D. F.
Significance = .130
Greenhouse-Geisser Epsilon = .74897
Huynh-Feldt Epsilon = .84062
Lower-bound Epsilon = .50000

AVERAGED Tests of Significance that follow multivariate tests are equivalent to univariate or split-plot or mixed-model approach to repeated measures. Epsilons may be used to adjust d.f. for the AVERAGED results.

- -

EFFECT .. MONTH

Multivariate Tests of Significance (S = 1, M = 0, N = 4)

Test Name	Value	Exact F	Hypoth. DF	Error DF	Sig. of F
Pillais	.53517	5.75671	2.00	10.00	.022
Hotellings	1.15134	5.75671	2.00	10.00	.022
Wilks	.46483	5.75671	2.00	10.00	.022
Roys	.53517				

Note. . F statistics are exact.

- -

Multivariate Effect Size

TEST NAME	Effect Size
(All)	.535

- -

Tests involving 'MONTH' Within-Subject Effect.

AVERAGED Tests of Significance for MO using UNIQUE sums of squares

Source of Variation	SS	DF	MS	F	Sig of F
WITHIN CELLS	34.81	22	1.58		
MONTH	20.19	2	10.10	6.38	.007

- -

Effect Size Measures

Source of Variation	Partial ETA Sqd

MONTH .367

- -

Tests involving 'TRIMESTE BY MONTH' Within-Subject Effect.

Mauchly sphericity test, W =	.77809
Chi-square approx. =	2.50909 with 2 D. F.
Significance =	.285
Greenhouse-Geisser Epsilon =	.81839
Huynh-Feldt Epsilon =	.94206
Lower-bound Epsilon =	.50000

AVERAGED Tests of Significance that follow multivariate tests are equivalent to univariate or split-plot or mixed-model approach to repeated measures. Epsilons may be used to adjust d.f. for the AVERAGED results.

- -

EFFECT .. TRIMESTE BY MONTH

Multivariate Tests of Significance (S = 1, M = 0, N = 4)

Test Name	Value	Exact F	Hypoth. DF	Error DF	Sig. of F
Pillais	.70342	11.85876	2.00	10.00	.002
Hotellings	2.37175	11.85876	2.00	10.00	.002
Wilks	.29658	11.85876	2.00	10.00	.002
Roys	.70342				

Note.. F statistics are exact.

- -

Multivariate Effect Size

TEST NAME	Effect Size
(All)	.703

- -

Tests involving 'TRIMESTE BY MONTH' Within-Subject Effect.

AVERAGED Tests of Significance for MO using UNIQUE sums of squares

Source of Variation	SS	DF	MS	F	Sig of F
WITHIN CELLS	50.14	22	2.28		
TRIMESTE BY MONTH	32.19	2	16.10	7.06	.004

- -

Effect Size Measures

Source of Variation	Partial ETA Sqd
TRIMESTE BY MONTH	.391

- -

2. The following SPSS syntax would be used to test the simple effect of differences among Months during the second Trimester.

SPSS SYNTAX FOR REPEATED MEASURES SIMPLE EFFECT

```
manova month4 month5 month6
/wsfactor= month(3)
/contrast(month)=polynomial
/print= transform signif( efsize)
/rename con linear quadratic
/design .
```

Orthonormalized Transformation Matrix (Transposed)

	CON	LINEAR	QUADRATI
MONTH4	.577	−.707	.408
MONTH5	.577	.000	−.816
MONTH6	.577	.707	.408

- - - Tests involving 'MONTH' Within-Subject Effect.

Mauchly sphericity test, W =	.90153
Chi-square approx. =	1.03667 with 2 D. F.
Significance =	.596
Greenhouse-Geisser Epsilon =	.91035
Huynh-Feldt Epsilon =	1.00000
Lower-bound Epsilon =	.50000

AVERAGED Tests of Significance that follow multivariate tests are equivalent to univariate or split-plot or mixed-model approach to repeated measures. Epsilons may be used to adjust d.f. for the AVERAGED results.

- -

EFFECT .. MONTH

Multivariate Tests of Significance (S = 1, M = 0, N = 4)

Test Name	Value	Exact F	Hypoth. DF	Error DF	Sig. of F
Pillais	.56442	6.47891	2.00	10.00	.016
Hotellings	1.29578	6.47891	2.00	10.00	.016
Wilks	.43558	6.47891	2.00	10.00	.016
Roys	.56442				

Note.. F statistics are exact.

- -

Multivariate Effect Size

TEST NAME	Effect Size
(All)	.564

Tests involving 'MONTH' Within-Subject Effect.

AVERAGED Tests of Significance for MO using UNIQUE sums of squares

Source of Variation	SS	DF	MS	F	Sig of F
WITHIN CELLS	59.61	22	2.71		

MONTH	47.72	2	23.86	8.81	.002

- -

Effect Size Measures

Source of Variation	Partial ETA Sqd
MONTH	.445

- -

Estimates for LINEAR
MONTH

Parameter	Coeff.	Std. Err.	t-Value	Sig. t	Lower −95%	CL- Upper
1	−1.945	.544	−3.572	.004	−3.143	−.746

- -

Estimates for QUADRATI
MONTH

Parameter	Coeff.	Std. Err.	t-Value	Sig. t	Lower −95%	CL- Upper
1	.442	.394	1.123	.285	−.425	1.309

- -

3. The following SPSS syntax would be used to test the simple effect of differences among Months during the Third Trimester.

SPSS SYNTAX FOR A THIRD TRIMESTER SIMPLE EFFECT

```
manova month7 month8 month9
/wsfactor = month(3)
/contrast(month)=polynomial
/print= transform signif( efsize)
/rename con linear quadratic
/design.
```

Orthonormalized Transformation
Matrix (Transposed)

	CON	LINEAR	QUADRATI
MONTH7	.577	−.707	.408
MONTH8	.577	.000	−.816
MONTH9	.577	.707	.408

- -

Tests involving 'MONTH' Within-Subject Effect.

Mauchly sphericity test, W =	.76662
Chi-square approx. =	2.65763 with 2 D. F.
Significance =	.265

Greenhouse-Geisser Epsilon =	.81078
Huynh-Feldt Epsilon =	.93079
Lower-bound Epsilon =	.50000

AVERAGED Tests of Significance that follow multivariate tests are equivalent to univariate or split-plot or mixed-model approach to repeated measures. Epsilons may be used to adjust d.f. for the AVERAGED results.

- - - - - - - - - - - - - EFFECT .. MONTH

Multivariate Tests of Significance (S = 1, M = 0, N = 4)

Test Name	Value	Exact F	Hypoth. DF	Error DF	Sig. of F
Pillais	.26640	1.81572	2.00	10.00	.212
Hotellings	.36314	1.81572	2.00	10.00	.212
Wilks	.73360	1.81572	2.00	10.00	.212
Roys	.26640				

Note.. F statistics are exact.

- - - - - - - - - - - - -

Multivariate Effect Size

TEST NAME	Effect Size
(All)	.266

- -

Tests involving 'MONTH' Within-Subject Effect.

AVERAGED Tests of Significance for MO using UNIQUE sums of squares

Source of Variation	SS	DF	MS	F	Sig of F
WITHIN CELLS	25.33	22	1.15		
MONTH	4.67	2	2.33	2.03	.156

- -

Effect Size Measures

Source of Variation	Partial ETA Sqd
MONTH	.156

- -

Estimates for LINEAR

MONTH

Parameter	Coeff.	Std. Err.	t-Value	Sig. t	Lower −95%	CL- Upper
1	.118	.244	.484	.638	−.418	.654

- -

Estimates for QUADRATI

MONTH

Parameter	Coeff.	Std. Err.	t-Value	Sig. t	Lower −95%	CL- Upper
1	-.612	.364	−1.682	.121	−1.414	.189

- -

Definition *Gamma function*: A sorority dance.

EXERCISES

Exercises 1 to 10 refer to a single-group repeated-measures study of weight loss. Twenty obese men participated in a 5-month diet–exercise program designed to reduce weight.

1. If the researcher's interest was in examining monthly differences from the baseline (Month 1):

(a) What would be the order of the **A** transformation matrix?

(b) What would the elements of the transformation matrix **A** equal?

(c) Are the contrasts identified in b orthogonal? Independent?

2. What does the **H** matrix equal if the following are the average monthly weights of the participants?

$$\begin{bmatrix} 252.2 & 253.9 & 254.5 & 249.8 & 243.2 \end{bmatrix}$$

3. For this data set, Wilks Λ equals .326. Transform Λ to an F statistic and determine the degrees of freedom.

4. Suppose the researcher was interested in testing hypotheses for linear and quadratic trends in weight loss. The following are the coefficients for fourth-degree polynomials.

$$\begin{bmatrix} 2 & -1 & 0 & 1 & 2 \\ 2 & -1 & -2 & -1 & 2 \\ -1 & 2 & 0 & -2 & 1 \\ 1 & -4 & 6 & -4 & 1 \end{bmatrix}$$

(a) Verify that the first and second contrasts are orthogonal.

(b) Normalize this set of coefficients.

5. Do you think the researcher should be concerned regarding a violation of the sphericity assumption if the following is the $\mathcal{A}\mathbf{S}_e\mathcal{A}'$ matrix? Why?

$$\mathcal{A}\mathbf{S}_e\mathcal{A}' \doteq \begin{bmatrix} 121.2 & -22.1 & -35.0 & 3.5 \\ -22.1 & 45.8 & 8.2 & 4.9 \\ -35.0 & 8.2 & 38.6 & 4.1 \\ 3.5 & 4.9 & 4.1 & 31.4 \end{bmatrix}$$

6. The determinant of the $\mathcal{A}\mathbf{S}_e\mathcal{A}'$ is 4.27×10^6. Using the Mauchly test for sphericity, compute
   (a) W
   (b) χ^2
   (c) df
   Do these results indicate the sphericity assumption has been violated? Are the results consistent with your assessment in Exercise 5?

7. Using the $\mathcal{A}\mathbf{S}_e\mathcal{A}'$ matrix reported in Exercise 5, compute
   (a) ε^*
   (b) ε'
   (c) $\tilde{\varepsilon}$

8. If the univariate approach to the analysis of the repeated-measures design was adopted, what would $\text{MS}_{S\times T}$ equal?

9. Using the monthly weight means from Exercise 2, orthonormalized coefficients from Exercise 4, and the $\mathcal{A}\mathbf{S}_e\mathcal{A}'$ matrix in Exercise 5, test for linear and quadratic trends in changes in weight over the 5-month period. Compute the following:
   (a) $\hat{\psi}_{\text{linear}}$
   (b) $s_{\hat{\psi}_{\text{linear}}}$
   (c) t
   (d) df
   (e) $\hat{\psi}_{\text{quadratic}}$
   (f) $s_{\hat{\psi}_{\text{quadratic}}}$
   (g) t
   (h) df

10. If the sphericity assumption had been met and the univariate approach was taken, how would the calculation of the estimated standard errors for the linear trend differ from the estimated standard error computed in Exercise 9?

Computer Applications

Exercises 11 to 14 require the analysis of a hypothetical data set A3, labeled FLEX, and is described in Appendix A. Use these data and a computer software package (e.g., SPSS or SAS) to answer the questions in these exercises.

11. Is there evidence to indicate that the sphericity assumption is violated in this data set?
    (a) State the numerical value of
        i. Mauchly's W

ii. χ^2

iii. Degrees of freedom

iv. P value

(b) Huynh–Feldt $\tilde{\varepsilon}$ value

12. Is there evidence to indicate Behavior changed over the 5-week period? Support your answer with appropriate statistical data:

(a) Wilks Λ

(b) F

(c) Degrees of freedom

(d) P value

13. Provide an adjusted and an unadjusted effect size index for the relationship between Behavior and the Week variable.

14. How would you characterize the change in Behavior over the 5-week period? Provide the statistical evidence for a:

(a) Fourth-degree polynomial

(b) Cubic trend

(c) Quadratic trend

(d) Linear trend

CHAPTER 11

Mixed-Model Analysis

11.1 INTRODUCTION

Our discussion in the previous chapter on the analysis of repeated measures was limited to data obtained from a single sample. In the present chapter we extend the analysis of repeated measures to include multiple samples. The multiple samples represent multiple populations. Because this design includes both the within-subjects data and at least one grouping variable or between-groups data, the design is referred to as a *mixed-model design* and is also sometimes referred to as a *split-plot design* reflecting its origination in agricultural research.

The grouping variable in a mixed-model design may be either a manipulated factor such an intervention (e.g., Reading Program) or it may be a nonmanipulated factor such as a characteristic of the participants (e.g., Gender) in the study. In Chapter 10 we stated that the repeated measures may be obtained from three different contexts. *One*, analysis units are observed at several time points, and changes in behavior over time are of interest (e.g., growth modeling). Our research context in Chapter 10 involving pregnant women during the second and third trimesters is an example of such a repeated-measures design. *Two*, units might be exposed to different treatments in a random order. In the *third* context, units may have completed subscales of a test battery, and comparisons among the subscales are of interest (i.e., *profile analysis*). In all three of these contexts the inclusion of multiple samples would allow a researcher to investigate an interaction effect involving the grouping variable and the repeated-measures variable. This would be particularly important for the first repeated-measures context where the meaning of changes over time may not be clear. Observed changes over time may have multiple explanations such as maturation, fatigue, or memory. The addition of a sample from a second population can help address this issue by serving as a comparison or control group.

Applied MANOVA and Discriminant Analysis, Second Edition, by Carl J. Huberty and Stephen Olejnik
Copyright © 2006 John Wiley & Sons, Inc.

The inclusion of a grouping variable may also be added to the repeated-measures design because the researcher is interested in differences among the populations across all levels of the repeated-measures variable. For example, individuals may be randomly assigned to each of two weight reduction programs, and weight loss may be recorded each month over a 6-month intervention period. The pattern of weight loss may be similar for the two programs (i.e., no interaction) but one program may result in a greater average or total weight loss than the other program (i.e., grouping variable main effect).

The inclusion of a grouping variable allows the researcher to make comparisons between levels of the grouping variable similar to those that were made in a one-factor between-groups design. If more than one grouping variable is included in the study, then comparisons similar to those addressed in the factorial design (Chapter 8) can be addressed. Furthermore, one or more covariates could also be added to the analysis to address questions similar to those discussed in Chapter 9. In the present chapter we limit our discussion to a single between-groups variable and a single repeated-measures variable. We begin with a brief review of the univariate approach to the mixed-model design and then present in some detail the multivariate analysis. Because comparisons between levels of the grouping variable are made across levels of the repeated-measures factor, the between-groups variable is a univariate data analysis problem. The analysis of the repeated-measures variable and the interaction between the grouping variable and the repeated-measures variable, however, can be addressed using multivariate procedures.

The generalization to multiple grouping variables and multiple repeated-measures variables is, however, straightforward. Covariates that are obtained prior to the beginning of the study or time-varying covariates (obtained concurrently with each level of the repeated-measures variable) can also be included in a mixed-model design. We will not discuss these added complexities, but interested readers may find useful discussions in Keppel and Wickens (2004, pp. 510–529), Kirk (1995, pp. 512–586), and Winer (1971, pp. 796–809).

11.2 RESEARCH CONTEXT

Continuing the research context introduced in Section 10.2, M. S. Poudevigne (2004) also collected Self-Esteem data from a sample of Nonpregnant women matched with respect to age, height, weight, and race to the sample of Pregnant women as described in Chapter 10. The researcher was primarily interested in comparing changes in Self-Esteem over the 6-month period for Pregnant women with changes in Self-Esteem for the Nonpregnant women. Table 11.1 presents the data for both samples of women, and Table 11.2 provides the descriptive statistics (means and standard deviations) within each group and across both groups. (An SPSS data file containing these scores, labeled SELFESTEEM2 is available at the Wiley website. The data have been modified slightly for pedagogical purposes.)

TABLE 11.1 Self-Esteem Scores for Pregnant and Nonpregnant Women

Group	Subject	Month 4	Month 5	Month 6	Month 7	Month 8	Month 9	Mean
Pregnant	1	12	11	10	10	10	10	10.50
	2	34	30	34	12	16	10	22.67
	3	16	18	15	13	15	15	15.33
	4	19	16	14	11	12	13	14.17
	5	23	22	21	20	23	21	21.67
	6	20	20	20	19	20	20	19.83
	7	19	13	11	11	11	11	12.67
	8	13	11	10	10	10	10	10.67
	9	14	14	14	10	10	10	12.00
	10	15	12	10	11	11	10	11.50
	11	16	14	10	11	10	10	11.83
	12	14	11	13	11	11	11	11.83
Nonpregnant	13	17	20	20	20	20	19	19.33
	14	20	20	22	20	20	20	20.33
	15	13	10	10	11	15	10	11.50
	16	14	14	15	12	13	14	13.67
	17	10	10	10	11	12	11	10.67
	18	21	20	21	22	22	22	21.33
	19	16	17	16	20	20	18	17.83
	20	12	10	11	10	10	10	10.50
	21	18	15	18	15	17	15	16.33
	22	16	17	17	13	17	17	16.17
	23	11	16	16	15	19	20	16.17
	24	13	10	11	10	10	10	10.67

TABLE 11.2 Self-Esteem Means and (Standard Deviations) by Group and Month

Group	Month 4	Month 5	Month 6	Month 7	Month 8	Month 9	Mean
Pregnant	17.92	16.00	15.17	12.42	13.25	12.58	14.557
	(6.007)	(5.721)	(7.030)	(3.423)	(4.372)	(4.010)	
Nonpregnant	15.08	14.92	15.58	14.92	16.25	15.50	15.375
	(3.502)	(4.100)	(4.295)	(4.461)	(4.181)	(4.442)	
Mean	16.500	15.458	15.375	13.667	14.750	14.042	14.967

11.3 UNIVARIATE ANALYSIS

The data in Tables 11.1 and 11.2 can be used to test three omnibus hypotheses: (1) changes in Self-Esteem over time is the same among Pregnant and Nonpregnant women; (2) for all women, there is no change over time in Self-Esteem; and (3) the

TABLE 11.3 Formulas for the Univariate Sum-of-Squares for the Mixed-Model Analysis of Variance[a]

Source of Variation	Sum-of-Squares
Within subjects	
Time (T)	$nJ\sum_{t=1}^{T}(\overline{Y}_{..t}-\overline{Y}_{...})^2$
Group × time ($G \times T$)	$n\sum_{j=1}^{J}\sum_{t=1}^{T}(\overline{Y}_{.jt}-\overline{Y}_{..t}-\overline{Y}_{.j.}+\overline{Y}_{...})^2$
Time × Subjects|Group ($T \times S\|G$)	$\sum_{j=1}^{J}\sum_{t=1}^{T}\sum_{u=1}^{n_j}(Y_{ujt}-\overline{Y}_{.jt}-\overline{Y}_{uj.}+\overline{Y}_{.j.})^2$
Between groups	
Group (G)	$nT\sum_{j-1}^{J}(\overline{Y}_{.j.}-\overline{Y}_{...})^2$
Subjects|Group ($S\|G$)	$T\sum_{j=1}^{J}\sum_{u=1}^{n_j}(\overline{Y}_{uj.}-\overline{Y}_{.j.})^2$

[a]where n = number of individuals in a group (we assume an equal number of individuals in each group)
J = number of group
T = number of time points
Y_{ujt} = observation on units, u, at Time t, in Group j
$\overline{Y}_{..t}$ = mean observation across all units at time t
$\overline{Y}_{.j.}$ = mean of observations of all units in Group
$\overline{Y}_{.jt}$ = mean observation across all units in Group j at time t
$\overline{Y}_{uj.}$ = mean observation across Time for unit u in Group j
$\overline{Y}_{...}$ = mean observation across Time and across all units (grand mean of all observations)

mean reported Self-Esteem is the same for Pregnant and Nonpregnant women. The first two hypotheses involve the repeated-measures variable (Time), and the third hypothesis involves only the between-groups variable. The formulas and the computations of the sum-of-squares for the sources of variation needed to test the three hypotheses are reported in Table 11.3. The ANOVA summary presenting the degrees of freedom, sum-of-squares, mean squares, F ratios, and P values for this analysis are presented in Table 11.4. The results of these analyses provide some evidence to indicate that changes in Self-Esteem over Time may not have been the same

TABLE 11.4 Univariate Analysis of Variance Summary Table for the Mixed Model

Source of Variation	Sum-of-Squares	Degrees of Freedom	Mean Square	F	P	η_G^2
Within subjects						
Time	128.357	5	25.671	4.142	.002	.041
Group × time	174.955	5	34.991	5.646	.000	.098
Time × subjects|group	681.667	110	6.197			
Between groups						
Group	24.088	1	24.088	.232	.634	.008
Subjects|groups	2284.376	22	103.835			

for Pregnant and Nonpregnant women [$F(5, 110) \doteq 5.646$, $P \doteq .000$, $\eta_G^2 \doteq .098$]. [The effect size is estimated using generalized η^2 (Olejnik and Algina, 2003).] Because the results indicate that the difference in Self-Esteem between Pregnant and Nonpregnant women is not consistent across the 6-month period, the main effect for Time or Group would be of little interest.

Because there is considerable evidence that Group and Time interact, we might consider two alternative approaches to further analyzing the data. One popular approach is to test for simple effects. That is, test for differences between levels of one variable within each level of the second variable. Because the primary interest in this research context is a comparison between Pregnant and Nonpregnant women, these simple effects would be a series of t tests between the two populations for each level of Time. In other research contexts the simple effects would examine differences between levels of the repeated-measures variable within each level of the grouping variable. This latter analysis would be equivalent to conducting a series of repeated-measures analyses (see Chapter 10). We will not pursue either of these analyses here. Kirk (1995, pp. 535–539) provides an excellent discussion and application of the simple effects analysis for a univariate mixed-model analysis of variance.

An alternative to simple effects is to conduct interaction contrasts. In the current research context there is an interest in characterizing the nature of the change in Self-Esteem. In Chapter 10 the results of a contrast analysis within the Pregnant sample provided some evidence to indicate that for this population the nature of the change in Self-Esteem could be characterized as a quadratic trend. Given the interaction effect, an examination of group differences in trend across the 6-month period might be of interest. We will address this question of differences in trend following our discussion of the multivariate approach to the mixed-model design.

11.4 MULTIVARIATE ANALYSIS

A multivariate approach may be taken for testing the hypotheses involving the repeated-measures variable. Because the effect of the grouping variable is estimated across all levels of the repeated-measures variable, the test for differences among levels of the grouping variable is a univariate analysis and will not be discussed further here. If the assumption of sphericity is met (see Section 10.7), the univariate approach would be more powerful than the multivariate approach. With a mixed-model design we assume that the separate-group covariance matrices are equal (i.e., $\boldsymbol{\Sigma}_1 = \boldsymbol{\Sigma}_2 = \cdots = \boldsymbol{\Sigma}_J$, where J is the number of levels of the grouping variable). The Box M test (see Section 3.3) may be used to test this assumption. Assuming equal covariance matrices, the assumption of sphericity can be tested with the Mauchly statistic using the error covariance matrix. It should be recalled, however, that the Mauchly test, Eq. (10.7), is sensitive to multivariate nonnormality and can be insensitive to some violations of the sphericity assumption that can affect the statistical validity of the univariate tests involving the repeated-measures variable(s). We do not demonstrate the calculation of the Mauchly test here, but we will request this test in the computer application section. At this point we simply state that for the Mauchly test, $W \doteq .007$, $\chi^2(14) \doteq 98.761$, $P \doteq .000$. The results of the Mauchly test indicates that the sphericity assumption may be violated, and the reported P value for the univariate hypothesis tests for the interaction between Time and Group and the Time main effect may be underestimated.

Additional evidence of a violation of the sphericity assumption is provided by the Greenhouse–Geisser ε' [see Eq. (10.5)] and the Huynh–Feldt $\tilde{\varepsilon}$ [see Eq. (10.6)] statistics. As noted in Chapter 8, these statistics overadjust for a violation of the sphericity assumption. We do not demonstrate the calculation of these statistics here but simply state that the Greenhouse–Geisser ε' is .308 and the Huynh-Feldt $\tilde{\varepsilon}$ is .342. These results further indicate that the sphericity assumption may be violated.

The multivariate approach to the analysis of repeated measures considers the relationships among the repeated measures when computing the multivariate test criterion (e.g., Wilks Λ), so it does not assume sphericity. For our context, two hypotheses involve the repeated-measures variable: (1) the interaction between the grouping variable (Pregnant vs. Nonpregnant women) and the repeated-measures variable (Time), and (2) the main effect for Time. We will begin our analysis by examining the interaction effect and then present the test for the repeated-measures main effect.

11.4.1 Group-by-Time Interaction

It should be recalled from Chapter 10 that the multivariate approach to the analysis of repeated-measures data is to form a set of linearly independent contrasts. The omnibus multivariate hypothesis test examines whether the vectors of the contrasts are equal to $\mathbf{0}$, H_0: $\mathbf{A}\boldsymbol{\mu} = \mathbf{0}$. The multivariate interaction hypothesis test examines whether the vector of contrasts within each population are equal: H_0: $\mathbf{A}\boldsymbol{\mu}_1 = \mathbf{A}\boldsymbol{\mu}_2 = \cdots = \mathbf{A}\boldsymbol{\mu}_j (j = 1, 2, \ldots, J)$. This hypothesis can be tested using

$$\Lambda = \frac{|\mathbf{AEA}'|}{|\mathbf{A}(\mathbf{E} + \mathbf{H}_{T \times G})\mathbf{A}'|}, \qquad (11.1)$$

where $\mathbf{A}$ = desired $(p \times T)$ matrix of contrast coefficients, with $p = T - 1$, with T being the number of levels of the repeated-measures variable
$\mathbf{E}$ = error **SSCP**
$\mathbf{H}_{T \times G}$ = **SSCP** interaction hypothesis matrix.

Within each Group j the error **SSCP** matrix, $\mathbf{E}_j$ is computed by deviating unit scores at Time t, $\mathbf{Y}_{ujt}$, from the group mean scores at Time t, $\mathbf{Y}_{.jt}$:

$$\mathbf{E}_j = (\mathbf{Y}_{ujt} - \mathbf{Y}_{.jt})'(\mathbf{Y}_{ujt} - \mathbf{Y}_{.jt}),$$

where $\mathbf{Y}_{ujt} = n_j \times T$ matrix of observations in Group j
$\mathbf{Y}_{.jt} = n_j \times T$ matrix of means in Group j

This is similar to the calculation of $\mathbf{E}$ for the repeated-measures design (see Section 10.4), but in the mixed-model design there are J separate error matrices. The separate

TABLE 11.5 Separate Group and Summed Error SSCP Matrices

$$\mathbf{E}_1 \doteq \begin{bmatrix} 396.917 & 354.000 & 431.167 & 89.417 & 170.250 & 67.583 \\ 354.000 & 360.000 & 426.000 & 116.000 & 200.000 & 108.000 \\ 431.167 & 426.000 & 543.667 & 121.167 & 222.500 & 101.833 \\ 89.417 & 116.000 & 121.167 & 128.917 & 157.750 & 145.083 \\ 170.250 & 200.000 & 222.500 & 157.750 & 210.250 & 175.250 \\ 67.583 & 108.000 & 101.833 & 145.083 & 175.250 & 176.917 \end{bmatrix}$$

$$\mathbf{E}_2 \doteq \begin{bmatrix} 134.917 & 126.083 & 143.417 & 136.083 & 114.750 & 115.500 \\ 126.083 & 184.917 & 187.563 & 183.917 & 171.250 & 191.500 \\ 143.417 & 187.583 & 202.917 & 183.583 & 167.250 & 190.500 \\ 136.083 & 183.912 & 183.583 & 218.917 & 189.250 & 195.500 \\ 114.750 & 171.250 & 167.250 & 189.250 & 192.250 & 189.500 \\ 115.500 & 191.500 & 190.500 & 194.500 & 189.500 & 217.000 \end{bmatrix}$$

$$\mathbf{E} \doteq \begin{bmatrix} 531.833 & 480.083 & 574.583 & 225.500 & 285.000 & 183.083 \\ 480.083 & 544.917 & 613.583 & 299.917 & 371.250 & 299.500 \\ 574.583 & 613.583 & 746.583 & 304.750 & 389.750 & 292.333 \\ 225.500 & 299.917 & 304.750 & 347.833 & 347.000 & 339.583 \\ 285.000 & 371.250 & 389.750 & 347.000 & 402.500 & 364.750 \\ 183.083 & 299.500 & 292.333 & 339.583 & 364.750 & 393.917 \end{bmatrix}$$

error matrices are then summed across the J groups ($\mathbf{E} = \sum_{j=1}^{J} \mathbf{E}_j$). Table 11.5 presents the separate error matrices along with the summed error matrix.

The **SSCP** matrix for the interaction hypothesis is obtained by:

$$\mathbf{H}_{T\times G} = \sum_{j=1}^{J} \mathbf{H}_{jT\times G}$$

$$= \sum_{j=1}^{J} n_j(\mathbf{y}_{.jt} - \mathbf{y}_{..t})(\mathbf{y}_{.jt} - \mathbf{y}_{..t})',$$

where n_j is the number of units in Group j, $\mathbf{y}_{.jt}$ is a vector of T means for Group j, and $\mathbf{y}_{..t}$ is vector of T means across the J groups. The interaction **SSCP** matrix is obtained by summing the separate **SSCP** matrices across the J groups. The computations for interaction **SSCP** for the two groups and the pooled $\mathbf{H}_{T\times G}$ matries are

$$\mathbf{y}_{.1t} - \mathbf{y}_{..t} \doteq \begin{bmatrix} 17.92 - 16.500 \\ 16.00 - 15.458 \\ 15.17 - 15.375 \\ 12.42 - 13.667 \\ 13.25 - 14.750 \\ 12.58 - 14.042 \end{bmatrix} \doteq \begin{bmatrix} 1.420 \\ .542 \\ -.205 \\ -1.247 \\ -1.500 \\ -1.462 \end{bmatrix},$$

$$\mathbf{H}_{1_{T\times G}} \doteq 12 \begin{bmatrix} 1.420 \\ .542 \\ -.205 \\ -1.247 \\ -1.500 \\ -1.462 \end{bmatrix} \begin{bmatrix} 1.420 & .542 & -.205 & -1.247 & -1.50 & -1.462 \end{bmatrix},$$

$$\doteq \begin{bmatrix} 24.197 & 9.236 & -3.493 & -21.249 & -25.560 & -24.912 \\ 9.236 & 3.525 & -1.333 & -8.110 & -9.756 & -9.509 \\ -3.493 & -1.333 & 0.504 & 3.068 & 3.690 & 3.597 \\ -21.249 & -8.110 & 3.068 & 18.660 & 22.446 & 21.877 \\ -25.560 & -9.756 & 3.690 & 22.446 & 27.000 & 26.316 \\ -24.912 & -9.509 & 3.597 & 21.877 & 26.316 & 25.649 \end{bmatrix},$$

$$\mathbf{H}_{2_{T\times G}} \doteq \begin{bmatrix} 24.197 & 9.236 & -3.493 & -21.249 & -25.560 & -24.912 \\ 9.236 & 3.525 & -1.333 & -8.110 & -9.756 & -9.509 \\ -3.493 & -1.333 & 0.504 & 3.068 & 3.690 & 3.597 \\ -21.249 & -8.110 & 3.068 & 18.660 & 22.446 & 21.877 \\ -25.560 & -9.756 & 3.690 & 22.446 & 27.000 & 26.316 \\ -24.912 & -9.509 & 3.597 & 21.877 & 26.316 & 25.649 \end{bmatrix},$$

$$\mathbf{H}_{T\times G} \doteq \begin{bmatrix} 48.394 & 18.471 & -6.986 & -42.498 & -51.120 & -49.825 \\ 18.471 & 7.050 & -2.667 & -16.221 & -19.512 & -19.018 \\ -6.986 & -2.667 & 1.009 & 6.135 & 7.380 & 7.193 \\ -42.498 & -16.221 & 6.135 & 37.320 & 44.892 & 43.755 \\ -51.120 & -19.512 & 7.380 & 44.892 & 54.000 & 52.632 \\ -49.825 & -19.018 & 7.193 & 43.755 & 52.632 & 51.299 \end{bmatrix}.$$

To test the interaction hypothesis, Wilks Λ can be computed using Eq. (11.1). For the data in Table 11.1 the error **SSCP** matrix, **E**, is reported in Table 11.5, the hypothesis **SSCP** matrix, $\mathbf{H}_{T\times G}$, is presented above, and a set of orthonomalized contrast coefficients is provided in Table 10.4. Using these results, $\Lambda_{T\times G}$ is computed as:

$$\begin{aligned} \Lambda_{T\times G} &= \frac{|\mathcal{A}\mathbf{E}\mathcal{A}'|}{|\mathcal{A}(\mathbf{E}+\mathbf{H}_{T\times G})\mathcal{A}'|} \\ &\doteq \frac{3.48\times 10^{8}}{8.82\times 10^{8}} \\ &\doteq .3945. \end{aligned}$$

(It should be noted that any set of linearly independent contrast coefficients would provide the same value of Λ.)

Because the degrees of freedom for the grouping variable equals 1, $\text{df}_G = 1$, the Wilks Λ can then be transformed using

$$
\begin{aligned}
F_{T\times G} &= \frac{1 - \Lambda_{T\times G}}{\Lambda_{T\times G}} \frac{\text{df}_e - p + 1}{p} \\
&\doteq \frac{1 - .3945}{.3945} \frac{22 - 5 + 1}{5} \\
&\doteq 5.525
\end{aligned}
$$

Assuming data assumptions have been met, this statistic has a central F distribution with $\nu_1 = p$, $\nu_2 = \text{df}_e - p + 1$. If $\text{df}_G = 2$, Eq. (3.17) would be used with degrees of freedom $\nu_1 = 2p$, $\nu_2 = 2(\text{df}_e - p + 1)$. More generally, when $r \geq 2$ where $r = \min(p, \text{df}_G)$, Eqs. (3.18) to (3.20) would be used. For our computed F ratio with degrees of freedom equaling 5 and 18, the P value is .003. The data do provide evidence to indicate that changes in Self-Esteem over the 6-month period is not the same for Pregnant and Nonpregnant women. The Huynh–Feldt $\tilde{\varepsilon}$ adjusted degrees-of-freedom univariate test for the Time $\times$ Group interaction resulted in $F(1.710, 37.616) \doteq 5.627$, and $P \doteq .010$. Both the basic multivariate test and the test using the Huynh–Feldt $\tilde{\varepsilon}$ adjusted degrees-of-freedom lead to the same conclusion, but here the multivariate test provides a smaller P value.

Because there is evidence of an interaction between Group and Time, an examination of changes in Self-Esteem across all women would not be considered, we will continue with the repeated-measures variable (Time) to demonstrate the application of the multivariate approach to the mixed-model design.

11.4.2 Repeated-Measures Variable Main Effect

Assuming that the Time main effect was interpretable and of interest to the researcher, the null hypothesis of no change in Self-Esteem over time is essentially equivalent to the hypothesis tested for Time in the repeated-measures analysis presented in Chapter 10. If there is no interaction between Time and Group, and if the separate covariance matrices are equal, the F ratios would be approximately equal (if the two conditions are exactly met, the computed F ratios would be identical), and the two tests would differ only in the error degrees of freedom, ν_2.

To test the repeated-measures main effect, Eq. (11.1) can be used again with only a slight modification. The **SSCP** for **H** must be changed to reflect the hypothesis of interest. For the main effect, we are interested in the extent to which the mean monthly Self-Esteem scores differ from the Self-Esteem grand mean ($\overline{Y}_{...} \doteq 14.966$). The **SSCP** for the repeated-measures Time main effect, $\mathbf{H}_{\text{Time}}$, is computed identically to that presented in Chapter 10 for the repeated-measures effect: $\mathbf{H}_{\text{Time}} = N(\mathbf{y}_{..t} - \mathbf{y}_{...})(\mathbf{y}_{..t} - \mathbf{y}_{...})'$, where $\mathbf{y}_{..t}$ is a column vector of the monthly means averaged over the J groups and $\mathbf{y}_{...}$ is a $p \times 1$ grand mean Self-Esteem vector across all units. Using the monthly means averaged across the two samples and the grand mean from Table 11.2, the **SSCP** for the Time main effect, $\mathbf{H}_{\text{Time}}$, can be computed as:

$$\mathbf{y}_{..t} - \mathbf{y}_{...} \doteq \begin{bmatrix} 16.50 - 14.966 \\ 15.46 - 14.966 \\ 15.38 - 14.966 \\ 13.67 - 14.966 \\ 14.75 - 14.966 \\ 14.04 - 14.966 \end{bmatrix} \doteq \begin{bmatrix} 1.534 \\ .494 \\ .414 \\ -1.296 \\ -.216 \\ -.926 \end{bmatrix}$$

$$\mathbf{H}_{\text{Time}} \doteq 24 \begin{bmatrix} 1.534 \\ .494 \\ .414 \\ -1.296 \\ -.216 \\ -.926 \end{bmatrix} \begin{bmatrix} 1.534 & .494 & .414 & -1.296 & -.216 & -.926 \end{bmatrix}$$

$$\doteq \begin{bmatrix} 56.476 & 18.187 & 15.242 & -47.714 & -7.952 & -34.092 \\ 18.187 & 5.857 & 4.908 & -15.365 & -2.561 & -10.979 \\ 15.242 & 4.908 & 4.114 & -12.877 & -2.146 & -9.201 \\ -47.714 & -15.366 & -12.877 & 40.311 & 6.718 & 28.802 \\ -7.952 & -2.561 & -2.146 & 6.718 & 1.120 & 4.800 \\ -34.092 & -10.979 & -9.201 & 28.802 & 4.800 & 20.579 \end{bmatrix}.$$

Wilks Λ can be computed using Eq. (11.2) by substituting the **SSCP** for **E** presented in Table 11.5, $\mathbf{H}_{\text{Time}}$ computed above, and using the orthonormal contrast coefficients from Table 10.4:

$$\Lambda_{\text{Time}} = \frac{|\mathcal{A}\mathbf{E}\mathcal{A}'|}{|\mathcal{A}(\mathbf{E} + \mathbf{H}_{\text{Time}})\mathcal{A}'|} \tag{11.2}$$

$$\doteq \frac{3.48 \times 10^8}{7.39 \times 10^8} \doteq .4730.$$

Wilks Λ can be transformed to a statistic that has a central F distribution, assuming that the data assumptions have been met, with $\nu_1 = p$ and $\nu_2 = \text{df}_e - p + 1$ degrees of freedom with $\text{df}_e = N - J$ using:

$$F_{\text{Time}} = \frac{1 - \Lambda_{\text{Time}}}{\Lambda_{\text{Time}}} \frac{\text{df}_e - p + 1}{p}$$

$$\doteq \frac{1 - .4730}{.4730} \frac{22 - 5 + 1}{5}$$

$$\doteq 4.011.$$

The results of this analysis provides some evidence to support the belief that Self-Esteem changes over the 6-month period, $F_{\text{Time}}(5, 18) \doteq 4.011$, $P \doteq .013$. In the next section, we present the SPSS program to carry out the analyses presented in this section, along with the output.

11.5 COMPUTER APPLICATION I

In this section we present the SPSS syntax for the multiple group repeated-measures design. The program is essentially the same as the program for the single-group repeated-measures design presented in Section 10.5. The difference between the programs is the addition of **by group(1,2)** on the **manova** line to identify the multiple group structure. Because we now have multiple groups, the equality of the covariance matrices should be tested with the Box test.

SPSS SYNTAX FOR A MIXED-MODEL DESIGN

```
manova
month4 month5 month6 month7 month8 month9 by group(1,2)
/wsfactor = month(6)
/print = signif(efsize) homogeneity (box)
/design.
```

by group(1,2) identifies the between-subjects factor for the mixed-model analysis. **/print = signif(averf)** requests that the univariate hypothesis tests on the repeated measures variable be reported.

OUTPUT

Analysis: Test of the Equality of Covariance Matrices

```
- - - - - - - - - - - - - - - - - - - - - - - - - - - - - - - - - - - - - - - -
Cell Number .. 1
Determinant of Covariance matrix of dependent variables = 724.84417
LOG(Determinant) = 6.58596
- - - - - - - - - - - - - -
Cell Number .. 2
Determinant of Covariance matrix of dependent variables =  207.06094
LOG(Determinant) = 5.33301
- - - - - - - - -
Determinant of pooled Covariance matrix of dependent vars.=4583.94552
LOG(Determinant) = 8.43032
- - - - - - - - - - - - - - - - - - - - - - - - - - - - -
Multivariate test for Homogeneity of Dispersion matrices
Boxs M =                     54.35827
F WITH (21,1780) DF =         1.80998, P =  .014 (Approx.)
Chi-Square with 21 DF =      38.65085, P =  .011 (Approx.)
- - - - - - - - - - - - - - - - - - - - - - - - - - - - - - - - - - - - - - - -
```

Interpretation: Test of the Equality of Covariance Matrices

The Box M test for the equality of covariance matrices is reported for both the χ^2 approximation (see Section 3.3) and an F approximation. The F approximation requires additional computations beyond those for the χ^2 approximation. The results presented here provide some evidence that the covariance matrices differ. In the present context, however, it is noted that both samples included 12 participants. The balanced design minimizes the threat to statistical validity for the hypothesis tests. In addition, as discussed earlier, the Box test is very sensitive to even small departures of covariance equality. Also, the Box test is sensitive to departures from multivariate normality. As a result we will proceed with our analysis assuming that the violation of covariance equality assumption does not invalidate statistical tests.

Analysis: Test for Sphericity

```
Tests involving 'MONTH' Within-Subject Effect.
  Mauchly sphericity test, W =        .00735
  Chi-square approx. =              98.76100 with 14 D. F.
  Significance =                        .000

  Greenhouse-Geisser Epsilon =        .30818
  Huynh-Feldt Epsilon =               .34196
  Lower-bound Epsilon =               .20000

AVERAGED Tests of Significance that follow multivariate tests are
equivalent to univariate or split-plot or mixed-model approach to
repeated measures. Epsilons may be used to adjust d.f. for the
AVERAGED results.
```

Interpretation: Test for Sphericity

The Mauchly test for sphericity provides some evidence to indicate that the assumption necessary for testing hypotheses involving the repeated-measures variable is violated, $W \doteq .007$, $\chi^2(14) \doteq 98.761$, $P \doteq .000$. Based on these results we would expect that the P values reported for the hypothesis tests involving the repeated-measures variable would be too small. In addition, the Greenhouse–Geisser $\varepsilon' \doteq .308$ and the Huynh–Feldt $\tilde{\varepsilon} \doteq .342$ provide a further indication of the assumption of sphericity that is violated. If the univariate hypothesis tests for the repeated-measures variable are to be interpreted, the degrees of freedom for the test statistics should be adjusted by one of the epsilon values.

Analysis: Test for the Within-Subjects Factors

```
EFFECT . . GROUP BY MONTH
Multivariate Tests of Significance (S = 1, M = 1 1/2, N = 8 )
Test Name       Value    Exact F   Hypoth. DF   Error DF   Sig. of F
Pillais        .60532   5.52139         5.00      18.00        .003
```

Hotellings	1.53372	5.52139	5.00	18.00	.003
Wilks	.39468	5.52139	5.00	18.00	.003
Roys	.60532				

Note.. F statistics are exact.

- -

Multivariate Effect Size

TEST NAME	Effect Size
(All)	.605

- -

EFFECT .. MONTH

Multivariate Tests of Significance (S = 1, M = 1 1/2, N = 8)

Test Name	Value	Exact F	Hypoth. DF	Error DF	Sig. of F
Pillais	.52871	4.03854	5.00	18.00	.012
Hotellings	1.12182	4.03854	5.00	18.00	.012
Wilks	.47129	4.03854	5.00	18.00	.012
Roys	.52871				

Note.. F statistics are exact.

- -

Multivariate Effect Size

TEST NAME	Effect Size
(All)	.529

- -

Tests involving 'MONTH' Within-Subject Effect.

AVERAGED Tests of Significance for MO using UNIQUE sums of squares

Source of Variation	SS	DF	MS	F	Sig of F
WITHIN CELLS	682.76	110	6.21		
MONTH	128.45	5	25.69	4.14	.002
GROUP BY MONTH	174.62	5	34.92	5.63	.000

- -

Effect Size Measures

Source of Variation	Partial ETA Sqd
MONTH	.158
GROUP BY MONTH	.204

- -

Interpretation: Test for the Within-Subjects Factors

The multivariate test for the interaction between the grouping variable and the repeated-measures variable provides some evidence to indicate that the monthly changes in Self-Esteem are not the same for the Pregnant and Nonpregnant women, Wilks $\Lambda \doteq .395$, $F(5, 18) \doteq 5.521$, $P \doteq .003$. The unadjusted multivariate effect-size index equals .605 and the Serlin-adjusted effect size equals .591 $(1 - [(144 - 1)/(144 - 5 - 1)]\ (1 - .605))$. The univariate test using the Huynh–Feldt $\tilde{\varepsilon}$ adjusted degrees of freedom also provides support for the conclusion that changes in Self-Esteem is not the same for Pregnant and Nonpregnant women; $F(1.71, 37.616) \doteq 5.627$, $P \doteq .010$. If the main effect for Time was interpretable, the multivariate test provides support for the conclusion that the Self-Esteem of women did change over the 6-month period; $F(5, 18) \doteq 4.039$, $P \doteq .012$. The Huynh–Feldt $\tilde{\varepsilon}$ adjusted degrees of freedom test also provides support for the same conclusion, $F(1.71, 37.616) \doteq 4.139$, $P \doteq .029$.

Analysis: Test for the Between-Group Factor

Tests of Between-Subjects Effects.
Tests of Significance for T1 using UNIQUE sums of squares

Source of Variation	SS	DF	MS	F	Sig of F
WITHIN CELLS	2284.82	22	103.86		
GROUP	24.17	1	24.17	.23	.634

- -

Effect Size Measures

Source of Variation	Partial ETA Sqd
GROUP	.010

- -

Interpretation: Test for the Between-Group Factor

The between-group effect is a univariate hypothesis test comparing the two populations across the 6-month observational period. This test is equivalent to conducting a single-factor analysis of variance test, where the unit of analysis is the individual's mean Self-Esteem across the time points. The results indicate that there is insufficient evidence to conclude that the two populations differ in the mean Self-Esteem [$F(1, 22) \doteq .233$, $P \doteq .634$]. The Pregnant women's mean Self-Esteem score was 14.56 while the Nonpregnant women had a mean Self-Esteem score of 15.38. Because the Group-by-Time interaction indicated that the difference between the two populations was not the same for all time points, the interpretation of this main effect may be misleading.

11.6 CONTRAST ANALYSIS

To identify specific differences between levels of the between-group variable (Group) we would proceed with specific contrasts as outlined in Section 4.4. For our data set, there was insufficient evidence to indicate group differences (Pregnant vs. Nonpregnant women) on Self-Esteem, so an examination of contrasts is not demonstrated here. In the computer application section below, however, we do request the pairwise contrast and interpret the results to demonstrate an application. Although we only present the pairwise contrast, if more groups were available, complex contrasts could also be requested.

For the repeated-measures variable, two sets of contrasts can be examined. One set can be formed to compare levels of the repeated-measures variable across all levels of the grouping variable. These are contrasts for the main effect for the repeated-measures variable and would be computed as demonstrated in Section 10.9. A second set of contrasts can be examined involving the interaction of the between-group variable and the repeated-measures variable. For example, in our context, it might be of interest to compare the nature of change in Self-Esteem among Pregnant women with changes among Nonpregnant women. We may have hypothesized that a quadratic trend may describe changes in Self-Esteem among Pregnant women, but no change in behavior among the Nonpregnant women. Or, perhaps, we may hypothesize some trend other than quadratic for the Nonpregnant women.

The interaction contrasts can be formed by computing the desired contrasts within each group as was demonstrated in Section 10.9. For tests of trend within each group, the product of the $p \times T$ orthonormalized contrast coefficients (Table 10.4) and the $T \times 1$ vector of monthly means would be computed: $\hat{\boldsymbol{\psi}}_j = \mathcal{A}\mathbf{y}_{.jt}$. For the Pregnant women ($j = 1$) the resulting vector is

$$\hat{\boldsymbol{\psi}}_1' \doteq \begin{bmatrix} -4.506 & 1.438 & .267 & -.394 & -1.202 \end{bmatrix}.$$

The first value (−4.506) represents the linear trend, the second value (1.438) represents the quadratic trend, and so forth. For the Nonpregnant women ($j = 2$), a similar set of contrasts can be formed:

$$\hat{\boldsymbol{\psi}}_2' \doteq \begin{bmatrix} .648 & .003 & -.342 & -.362 & -.814 \end{bmatrix}.$$

The vector of interaction contrasts is the difference between the two sets of contrasts:

$$\hat{\boldsymbol{\psi}}_1' - \hat{\boldsymbol{\psi}}_2' = \hat{\boldsymbol{\psi}}' \doteq \begin{bmatrix} -5.154 & 1.435 & .609 & -.032 & -.388 \end{bmatrix}.$$

The first value in this vector is the difference in the linear trend between the two groups. The second value is the difference in the quadratic trend, and so forth.

To test the hypotheses that the trends differ, the estimated standard error for each trend is needed. The error **SSCP** matrix, **E**, is given in Table 11.5. Multiplying this matrix by the reciprocal of the error degrees of freedom, $1/(N - J)$, gives the error covariance matrix $\mathbf{S}_e$. In our example the error degrees of freedom is 22 (= 24 − 2). Multiplying **E** by the scalar .0455 (1/22) gives the following results:

$$\mathbf{S}_e \doteq \begin{bmatrix} 24.198 & 21.844 & 26.144 & 10.260 & 12.968 & 8.330 \\ 21.844 & 24.793 & 27.918 & 13.633 & 16.892 & 13.627 \\ 26.144 & 27.918 & 33.970 & 13.866 & 17.734 & 13.301 \\ 10.260 & 13.633 & 13.866 & 15.826 & 15.789 & 15.451 \\ 12.968 & 16.892 & 17.734 & 15.789 & 18.314 & 16.596 \\ 8.330 & 13.627 & 13.301 & 15.451 & 16.596 & 17.923 \end{bmatrix}.$$

Pre- and postmultiplying $\mathbf{S}_e$ by $\mathcal{A}$, $\mathcal{A}\mathbf{S}_e\mathcal{A}'$ gives the variance of each contrast (MS_q, $q = 1, 2, \ldots, p$) on the main diagonal:

$$\mathcal{A}\mathbf{S}_e\mathcal{A}' \doteq \begin{bmatrix} 19.362 & .859 & -3.468 & -1.085 & 8.865 \\ .859 & 1.548 & -1.009 & .112 & 1.091 \\ -3.468 & -1.009 & 3.260 & -.226 & -2.793 \\ -1.085 & .112 & -.226 & 1.087 & -.101 \\ 8.865 & 1.091 & -2.793 & -.101 & 5.847 \end{bmatrix}.$$

The estimated standard error for contrast q is computed as:

$$s_{\hat{\psi}_q} = \sqrt{\text{MS}_q \sum_{j=1}^{J}\sum_{t=1}^{T} \frac{a_{jt}^2}{n_j}},$$

where a_{jt}^2 is the contrast coefficient for the tth time point for Group j, and n_j is the number of units in Group j. For our context, the estimated standard error for the linear interaction contrast ($q = 1$) is computed as:

$$\begin{aligned} s_{\hat{\psi}_{\text{Linear}\times G}} &\doteq \sqrt{19.362\frac{[(-.598)^2 + \cdots + (.598)^2] + [(-.598)^2 + \cdots + (.598)^2]}{12}} \\ &\doteq \sqrt{19.362\frac{2}{12}} \\ &\doteq 1.796. \end{aligned}$$

The t test is computed as the ratio of the contrast to its estimated standard error:

$$\begin{aligned} t &\doteq \frac{-5.154}{1.796} \\ &\doteq -2.870. \end{aligned}$$

Similarly, the estimated standard error for the quadratic interaction contrast is

$$\begin{aligned} s_{\hat{\psi}_{\text{Quad}\times G}} &\doteq \sqrt{1.548\frac{(.546)^2 + (-.109)^2 + \cdots + (.546)^2 + (.546)^2 + \cdots + (.546)^2}{12}} \\ &\doteq \sqrt{1.548\frac{2}{12}} \\ &\doteq .508. \end{aligned}$$

The computed t statistic is

$$\begin{aligned} t &\doteq \frac{1.435}{.508} \\ &\doteq 2.825. \end{aligned}$$

Interaction hypotheses comparing other patterns of means could be computed in a similar manner. However, patterns in means other than linear or quadratic are frequently of little interest to researchers.

Under the null hypothesis of no group-by-trend interaction, each computed t statistic has a Student t distribution with $N - J$ degrees of freedom. For the linear interaction hypothesis, $t(22) \doteq -2.870$, $P \doteq .009$, and for the quadratic interaction hypothesis, $t(22) \doteq 2.825$, $P \doteq .010$. If these two contrasts constitute a family of contrasts of interest to the researcher, then the researcher might conclude, considering a Bonferroni adjustment (i.e., $P' \doteq .018$ and $P' \doteq .020$), that there is sufficient evidence to indicate that the quadratic trend is not the same for the Pregnant and Nonpregnant women. A plot of the monthly Self-Esteem means for each group would be helpful in further interpreting these results.

11.7 COMPUTER APPLICATION II

In this section we have added the **/contrast** command for both the repeated-measures variable and the between-subjects variable. Here, rather than specifying the coefficients first presented in Table 10.3, we just specify **polynomial**. For the between-subjects variable only two groups are considered, so the only contrast possible is pairwise.

SPSS SYNTAX FOR BETWEEN-GROUPS AND WITHIN-SUBJECTS CONTRASTS

```
manova
month4 month5 month6 month7 month8 month9 by group(1,2)
/wsfactor=month(6)
/contrast(month)=polynomial
/contrast(group)=special (1 1, 1 -1)
/print=transform
/rename=ave linear quadratic cubic quartic quintic
/design.
```

/contrast(time)=polynomial requests contrasts be computed on sources of variation involving the repeated-measures factor, Time.
/contrast(group)=special(1 1, 1 −1) requests a specific contrast between levels of between-subjects factor Group. Alternatively, = **simple** could replace special and the coefficients would not be needed.

OUTPUT

Analysis: Contrast Tests

Orthonormalized Transformation Matrix (Transposed)

	AVE	LINEAR	QUADRATI	CUBIC	QUARTIC	QUINTIC
MO4	.408	−.598	.546	−.373	.189	−.063
MO5	.408	−.359	−.109	.522	−.567	.315
MO6	.408	−.120	−.436	.298	.378	−.630
MO7	.408	.120	−.436	−.298	.378	.630
MO8	.408	.359	−.109	−.522	−.567	−.315
MO9	.408	.598	.546	.373	.189	.063

- -

Tests of Between-Subjects Effects.

Tests of Significance for AVE using UNIQUE sums of squares

Source of Variation	SS	DF	MS	F	Sig of F
WITHIN CELLS	2284.82	22	103.86		
GROUP	24.17	1	24.17	.23	.634

- -

Estimates for AVE

GROUP

Parameter	Coeff.	Std. Err.	t-Value	Sig. t	Lower −95%	CL- Upper
2	−2.0072208	4.16044	−.48245	.63425	−10.63544	6.62100

- -

Estimates for LINEAR

MONTH

Parameter	Coeff.	Std. Err.	t-Value	Sig. t	Lower −95%	CL- Upper
1	−1.9273061	.89669	−2.14936	.04286	−3.78692	−.06769

Group BY MONTH

Parameter	Coeff.	Std. Err.	t-Value	Sig. t	Lower −95%	CL- Upper
2	−5.1494433	1.79337	−2.87137	.00887	−8.86867	−1.43021

- -

Estimates for QUADRATIC

MONTH

Parameter	Coeff.	Std. Err.	t-Value	Sig. t	Lower −95%	CL- Upper
1	.6910233	.25479	2.71216	.01273	.16263	1.21942

Group BY MONTH

Parameter	Coeff.	Std. Err.	t-Value	Sig. t	Lower −95%	CL- Upper
2	1.43660111	.50957	2.81922	.00999	.37981	2.49339

- -

Estimates for CUBIC

MONTH

Parameter	Coeff.	Std. Err.	t-Value	Sig. t	Lower −95%	CL- Upper
1	−.0372678	.36819	−.10122	.92029	−.80084	.72631

Group BY MONTH

Parameter	Coeff.	Std. Err.	t-Value	Sig. t	Lower −95%	CL- Upper
2	.6087074	.73638	.82663	.41732	−.91844	2.13586

- -

Estimates for QUARTIC

MONTH

Parameter	Coeff.	Std. Err.	t-Value	Sig. t	Lower −95%	CL- Upper
1	−.37796447	.21222	−1.78100	.08873	−.81808	.06215

Group BY MONTH

Parameter	Coeff.	Std. Err.	t-Value	Sig. t	Lower -95%	CL- Upper
2	−.03149704	.42444	−.07421	.94152	−.91173	.84874

- -

Estimates for QUINTIC

MONTH

Parameter	Coeff.	Std. Err.	t-Value	Sig. t	Lower −95%	CL- Upper
1	−1.0079053	.49351	−2.04233	.05329	−2.03138	.01557

Group BY MONTH

Parameter	Coeff.	Std. Err.	t-Value	Sig. t	Lower −95%	CL- Upper
2	−.38846349	.98702	−.39357	.69768	−2.43541	1.65848

- -

Interpretation: Contrast Tests

The contrast between Pregnant and Nonpregnant women (Group), which follows the orthonormalizing transformation matrix and the analysis-of-variance table, is presented first. While in this example the test of statistical significance is redundant with the ANOVA F test (because there are only two populations being compared), the contrast analysis also reports the analysis in terms of a .95 confidence interval. The use of a confidence interval (CI) is an excellent way of presenting the findings. Although not stated on the output, the degrees of freedom value for the t test is the same as that for the denominator of the F test, $N - J$. The results presented here indicate that, on the average, across the 6-month reporting period, Nonpregnant women reported a mean of two points higher on the Self-Esteem scale than the Pregnant women with a .95CI of $(-10.635, 6.621)$. This observed difference is no greater than what might be expected due to sampling error, and the observed difference cannot be generalized to the populations they represent, $t(22) \doteq -.482$, $P \doteq .634$. It should be recalled that on the Rosenberg Self-Esteem Inventory, lower scores indicate higher Self-Esteem.

For the within-subject variable the results are reported for both the main effect (Time) and the interaction effect (Group × Time). The t tests and confidence intervals are reported for the $T - 1$ trends. Examining the interaction contrasts first and using the Bonferroni adjustment, assuming no specific interaction trend had been hypothesized ($P' = 5P$), the interaction t test provides some evidence to indicate that the quadratic trend is different for Pregnant and Nonpregnant women, $t(22) \doteq 2.819$, $P' \doteq .049$. It should be noted that the confidence interval reported on the output does not adjust for the number of confidence intervals estimated. That is, each contrast represents a unique family. A Bonferroni-adjusted confidence interval can be obtained, but it would require the researcher to recompute the intervals using the appropriate critical t value, which considers the number of confidence intervals considered.

If the interaction contrast had not been judged to be generalizable, and only linear or quadratic trends involving the means averaged over the groups were of interest, then using the Bonferroni adjustment, $P' = 2P$, the results would provide sufficient evidence to conclude a quadratic trend is generalizable to all women during this period, $t(22) \doteq -2.712$, $P' \doteq .025$.

11.8 SUMMARY

In this chapter we generalized the analysis of a single-sample repeated-measures analysis (Chapter 10) to multiple samples. Because the multiple samples permit comparisons between populations, and multiple measurements on each unit permit comparisons on the repeated measures, we have a mixed model. Hypotheses involving only between-group variables are tested using univariate procedures. Hypotheses involving the interaction(s) of the repeated-measures variable(s) with between-group variable(s) and the main effect(s) for the repeated-measures variable(s) can be tested using either the univariate or the multivariate procedures. The multivariate procedures are generally preferred because these tests are typically more powerful, make fewer assumptions, and reported P values are exact, not approximate, as is the case with the univariate procedures.

As demonstrated here, the tests of omnibus hypotheses involving a repeated-measures variable (interactions and main effects) use the same multivariate criterion (i.e., Wilks Λ) as those used in the between-group multivariate analyses. With the repeated-measures analysis, however, the data are transformed to construct contrasts, and the vector of contrast outcomes become the multiple outcomes analyzed. Because the multivariate approach considers not only the variance of the multiple outcomes as well as the covariances, a specified structure in the relationships among the outcomes is unnecessary.

When specific contrasts, pairwise or complex, involving the repeated-measures variable are of interest, the multivariate approach estimates the standard error for a contrast by using the separate variance for each contrast. The univariate approach uses the interaction Mean Square, $\text{MS}_{G \times T}$, to estimate the standard error of a contrast. If the sphericity assumption is violated, these estimated standard errors are incorrect. Using separate contrast variances for focused tests is inconsistent with the univariate omnibus tests. Maxwell and Delaney (2000, p. 601) point out there is no relationship between such focused tests and the omnibus tests. Consequently, it is possible to obtain contradictory results, a very undesirable consequence.

In this chapter we considered a single repeated-measures variable and a single between-group variable. Generalization to multiple repeated-measures variables (see Technical Notes in Chapter 10) and multiple between-group variables is straightforward. Covariates may also be added to adjust for between-group differences. In the Technical Note below we provide the SPSS syntax for a covariate analysis in a mixed-model design using the same data set discussed in this chapter but with a new variable, Age, as a covariate. We do not provide an interpretation of these results, but they should be clear from the interpretation presented in this chapter.

Technical Note

In many research situations it is desirable to include covariates to increase statistical power or to "adjust" for *nuisance* variables (e.g., confounding variables). With a mixed-model design the covariates may be either fixed or varied with the repeated measures variable. A *fixed covariate* is one in which the covariate is obtained at a single point in time. A *varying covariate* is one in which the covariate measure is observed at each level of the repeated-measures variable. As an example of a fixed covariate, suppose in the study described in this chapter the Nonpregnant women were older than the Pregnant women. In such a situation it might be recommended that the researcher "control" for the age differences between groups. As an example of a varying covariate, suppose the researcher obtained each woman's weight along with her Self-Esteem at the beginning of each month. The researcher might want to "control" for variation Weight when examining changes in Self-Esteem over the 6-month period. The analysis below uses a fixed covariate. It is useful to compare the results presented here with those presented earlier in the chapter. (The input data file labeled SELFESTEEM3 is available at the Wiley website.)

SPSS SYNTAX FOR MANCOVA WITH A FIXED COVARIATE IN A MIXED-MODEL DESIGN

```
manova
month4 month5 month6 month7 month8 month9 by group (1,2)
with age4 age5 age6 age7 age8 age9
/wsfactor=month(6)
/print= signif(averf efsize) homogeneity(box)
/design
```

with age4 age5 age6 age7 age8 age9 identifies the covariate, Age. With a fixed covariate in a mixed-model design, the single covariate score is repeated as frequently as there are levels of the repeated-measures variable. Although the values are the same, the variable name must change.

OUTPUT

Note: The only values that change are those associated with the between-group variable.

```
Multivariate test for Homogeneity of Dispersion matrices
Boxs M =                          54.35827
F WITH (21,1780) DF =              1.80998, P =   .014 (Approx.)
Chi-Square with 21 DF =           38.65085, P =   .011 (Approx.)
- - - - - - - - - - - - - - - - - - - - - - - - - - - - - - - - - - -
```

Tests of Between-Subjects Effects.
Tests of Significance for T1 using UNIQUE sums of squares

Source of Variation	SS	DF	MS	F	Sig of F
WITHIN CELLS	1048.09	21	49.91		
REGRESSION	1236.73	1	1236.73	24.78	.000
GROUP	252.08	1	252.08	5.05	.035

- -

Effect Size Measures

Source of Variation	Partial ETA Sqd
Regression	.541
GROUP	.194

- -

Regression analysis for WITHIN CELLS error term
- - - Individual Univariate .9500 confidence intervals
Dependent variable .. T1

COVARIATE	B	Beta	Std. Err.	t-Value	Sig. of t
T7	.94684	.77219	.190	4.978	.000

COVARIATE	Lower −95%	CL- Upper	ETA Sq.
T7	.551	1.342	.541

- -

Tests involving 'MONTH' Within-Subject Effect.
Mauchly sphericity test, W = .00735
Chi-square approx. = 98.76100 with 14 D. F.
Significance = .000

Greenhouse-Geisser Epsilon = .30818
Huynh-Feldt Epsilon = .34196
Lower-bound Epsilon = .20000

AVERAGED Tests of Significance that follow multivariate tests are equivalent to univariate or split-plot or mixed-model approach to repeated measures. Epsilons may be used to adjust d.f. for the AVERAGED results.

- -

EFFECT .. GROUP BY MONTH
Multivariate Tests of Significance (S = 1, M = 1 1/2, N = 8)

Test Name	Value	Exact F	Hypoth. DF	Error DF	Sig. of F
Pillais	.60532	5.52139	5.00	18.00	.003
Hotellings	1.53372	5.52139	5.00	18.00	.003
Wilks	.39468	5.52139	5.00	18.00	.003
Roys	.60532				

Note.. F statistics are exact.

- -

Multivariate Effect Size

TEST NAME	Effect Size
(All)	.605

- -

EFFECT .. MONTH

Multivariate Tests of Significance (S = 1, M = 1 1/2, N = 8)

Test Name	Value	Exact F	Hypoth. DF	Error DF	Sig. of F
Pillais	.52871	4.03854	5.00	18.00	.012
Hotellings	1.12182	4.03854	5.00	18.00	.012
Wilks	.47129	4.03854	5.00	18.00	.012
Roys	.52871				

Note.. F statistics are exact.

- -

Multivariate Effect Size

TEST NAME	Effect Size
(All)	.529

- -

Tests involving 'MONTH' Within-Subject Effect.

AVERAGED Tests of Significance for MO using UNIQUE sums of squares

Source of Variation	SS	DF	MS	F	Sig of F
WITHIN CELLS	682.76	110	6.21		
MONTH	128.45	5	25.69	4.14	.002
GROUP BY MONTH	174.62	5	34.92	5.63	.000

- -

Effect Size Measures

Source of Variation	Partial ETA Sqd
MONTH	.158
GROUP BY MONTH	.204

- -

Fact: Julius Ceasar was a Latin square.

EXERCISES

Exercises 1 to 9 are based on the following context. An exercise physiologist was interested in determining the optimal amount of time needed for an adequate warm-up routine before beginning a vigorous exercise program in order to minimize postexercise muscle soreness. The researcher believed, however, that the amount of time needed for warm-up would vary depending on an exerciser's weight. The researcher designed a study in which participants, who where classified as normal weight, overweight, or obese, varied the number of minutes in warm-up activities: 5 minutes, 10 minutes, 15 minutes, or 20 minutes before beginning an exercise program. Over four consecutive weeks the warm-up period was varied randomly across 15 male participants from each of the weight groups. The morning following each warm-up and exercise program the participants were contacted and asked several questions regarding the muscle soreness and discomfort. Responses were scored on a 20-point scale.

1. The multivariate approach to the mixed-model design assumes that covariance matrices are equal. How many covariance matrices are there in this context?

2. For this context what is the numerical value for d when computing the Mauchly test for sphericity? (See Section 10.7.)

3. If the value of the Mauchly W statistic is .728, what is the numerical value of the χ^2 statistic and the degrees of freedom? (See Section 10.7.)

4. For these data, the Greenhouse–Geisser ε' is .840. What is the numerical value of the Huynh–Feldt $\tilde{\varepsilon}$? (See Section 10.7.)

5. To test the interaction between the Grouping variable and the Minutes variable, Wilks Λ was computed to equal .818. Transform the Λ value to an F statistic value and give the appropriate degrees of freedom.

6. For the Minutes main effect, across levels of the grouping variable, the means were computed as 14.8, 13.5, 12.7, and 12.6 for warm-up periods 5, 10, 15, and 20 minutes, respectively. What does the hypothesis **SSCP** for the Minutes main effect, $\mathbf{H}_{\text{Minutes}}$, equal?

7. For the Minutes main effect, Λ was computed to equal .452. Compute the a ξ^2_{adj} statistic using the Serlin procedure.

For Exercises 8 and 9 use the following orthonormal transformation matrix for polynomial trend, $\mathcal{A}$, and the $\mathcal{A}\mathbf{S}_e\mathcal{A}$ covariance matrix:

$$\mathcal{A} \doteq \begin{bmatrix} -.671 & -.224 & .224 & .671 \\ .500 & -.500 & -.500 & .500 \\ -.224 & .671 & -.671 & .224 \end{bmatrix} \quad \text{and} \quad \mathcal{A}\mathbf{S}_e\mathcal{A} \doteq \begin{bmatrix} 2.70 & -.32 & .92 \\ -.32 & 1.62 & .47 \\ .92 & 0.47 & 2.99 \end{bmatrix}.$$

8. What is the estimated standard error for the interaction contrast comparing the linear trends of normal weight males with obese males?

9. Use the $\mathcal{A}$ and $\mathcal{A}\mathbf{S}_e\mathcal{A}$ matrices provided above, as well as the means provided in Exercise 6, to
   - **(a)** Compute the t statistic to test for a linear trend.
   - **(b)** Compute the t statistic to test for a quadratic trend.
   - **(c)** How would you interpret these results in light of the researcher's interest?

Computer Applications

Exercises 10 to 15 require the analysis of a hypothetical data set A4, labeled FLEX2 described in Appendix A. Use these data and a computer software package (e.g., SPSS) to answer the questions in these exercises.

10. Is there sufficient evidence to indicate that the covariance matrices of the three groups differ? In defending your response, provide the following:

(a) Box M
(b) χ^2
(c) Degrees of freedom
(d) P value

11. Is there evidence to indicate that the sphericity assumption is violated in this data set? In defending your response, provide the following:
   (a) State the numerical value
      i. Mauchly W
      ii. χ^2
      iii. Degrees of freedom
      iv. P value
   (b) Huynh–Feldt epsilon

12. Is there evidence of an interaction between the grouping variable and the Time variable? In defending your response, provide the following:
   (a) Wilks Λ
   (b) F
   (c) Degrees of freedom
   (d) P value
   (e) ξ^2_{adj}

13. Contrast the changes in behavior between the Group 1 and Group 3 (interaction contrast). What degree polynomial best describes the differences between these groups? Support your responses with appropriate statistics including:
   (a) F
   (b) Degrees of freedom
   (c) P value

14. Assuming differences between groups averaged over the 5-week period was of interest, is there evidence to indicate that the groups differ? Support your answer with appropriate information:
   (a) F
   (b) Degrees of freedom
   (c) P value
   (d) η^2

15. Do each of the treatment groups (Group 1 and Group 2) differ from the control group (Group 3)? Support your answer with appropriate information:
   (a) t
   (b) Degrees of freedom
   (c) P value

PART IV

Group Membership Prediction

The variable sets in this part of the book consist of several predictor (X) variables on the one hand and a single criterion variable on the other hand. The collection of techniques presented constitutes what we label predictive discriminant analysis (PDA). The criterion variable is a grouping variable with at least two levels. Measures on the X variables for analysis units are used to develop a rule to predict unit group membership; that is, to classify units into groups. Methods for assessing the goodness of the rule are presented in Chapter 16. The problems of predictor and predictor ordering are discussed in Chapter 17. Some nonnormal rules are reviewed in Chapter 19. This part concludes with suggestions for reporting results of a PDA (Chapter 20) and with definitions of some PDA-related analyses (Chapter 21).

The goals of the reader for this part are to be able to (1) critically evaluate an application of PDA and (2) write up a report of a study in which a PDA is applied. While studying Part IV, the reader may refer to an overview given in the form of a flowchart in Figure 20.1.

Applied MANOVA and Discriminant Analysis, Second Edition, by Carl J. Huberty and Stephen Olejnik
Copyright © 2006 John Wiley & Sons, Inc.

CHAPTER 12

Classification Basics

12.1 INTRODUCTION

The notions of explanation and prediction are in close alignment. In many scientific contexts some (philosophers?) have argued that explanation—the identification of "patterns" or of "structure"—is a necessary antecedent of prediction; others have argued that the converse is the case. The position taken here is noncommittal; prediction was originally viewed as a means of enhancing an explanation and is currently viewed as a solution to a *practical* problem. In many contexts, explanations and predictions are made more plausible by conclusions drawn from data. Our interest now is on predictions based on data; the identification of structure was discussed in Part II.

An approach commonly used in making empirical, or statistical, predictions is *multiple regression*. Multiple regression techniques are appropriate in a situation involving, on the one hand, a set of p predictor (random or fixed) variables, X_1, $X_2, \ldots, X_p$, and on the other hand, a single criterion (random) variable, Y. (Note that here we are dealing with a *single* group of N analysis units, for each of which we have $p + 1$ response measures.) One goal of a multiple regression analysis is to set up a rule, based on an $N \times (p + 1)$ data matrix, to be used in predicting (or estimating) a criterion variable measure, given measures on the p predictors. It turns out that this amounts to determining a set of (regression) weights, $b_1, b_2, \ldots, b_p$, corresponding to a given set of p predictor variable measures to yield a linear composite value that is essentially a predicted value of the criterion variable. The predicted criterion measure for analysis unit u may be represented as:

$$\hat{Y}_u = b_0 + b_1 X_{1u} + b_2 X_{2u} + \cdots + b_p X_{pu},$$

where b_0 is the *regression constant*. The composite may also be expressed as:

$$\hat{Y}_u = b_0 + \sum_{i=1}^{p} b_i X_{iu},$$

Applied MANOVA and Discriminant Analysis, Second Edition, by Carl J. Huberty and Stephen Olejnik
Copyright © 2006 John Wiley & Sons, Inc.

or as

$$\hat{Y}_u = b_0 + \mathbf{b}'\mathbf{x}_u,$$

where $\mathbf{b}'$ is the $1 \times p$ row vector of regression weights, and $\mathbf{x}_u$ is the $p \times 1$ column vector of predictor variable measures for unit u.

When the X measures are based on different metrics, it is sometimes desirable to remove the effect of the varying metrics on the regression weights. This may be accomplished by using *standardized weights*:

$$b_i^* = b_i \frac{s_i}{s_Y},$$

where s_i and s_Y are the estimated standard deviations of X_i and Y, respectively. These weights are used to predict a standardized Y measure,

$$\hat{Z}_Y = \mathbf{b}^{*\prime}\mathbf{z}_u,$$

where $\mathbf{z}_u$ is a $p \times 1$ vector of standardized X measures. (Note that $b_0^* = 0$. Also, the b^* values may be obtained by standardizing the X and Y measures and finding the least-squares solutions.)

Another approach used in making empirical predictions involves an aspect of discriminant analysis called *predictive discriminant analysis* (PDA). Techniques of PDA are appropriate in a multiple-group setting in which we have p X measures for each unit belonging to one of J groups. It is assumed that the J groups of n_j units represent J meaningful populations. In such a setting the criterion variable is a dichotomous or polytomous grouping variable. One goal of PDA is to set up a rule, based on J $n_j \times p$ data matrices that would predict population membership for a unit. Unless otherwise noted, it is generally assumed in this book that any unit to be classified does in fact belong to one of the J criterion populations. (The problem of initially misclassified units is discussed in Section 23.4.)

A rule in PDA, termed a *classification rule*, can, as we shall see, take three different forms. *One* form is that of a composite of the predictor measures; a *second* form is that of an estimated probability of population membership; a *third* form is that of a distance between two points. Before discussing the three forms of a classification rule, some preliminaries are needed.

12.2 NOTION OF DISTANCE

Implicit in many, and explicit in some, of the rules discussed here is the notion of *distance*. Consider the distance between two points from a simple geometric view. In a bivariate (X_1, X_2) space we might represent the distance $\tilde{d}_{AB}$ between points A: (X_{1A}, X_{2A}) and B: (X_{1B}, X_{2B}) as shown in Figure 12.1.

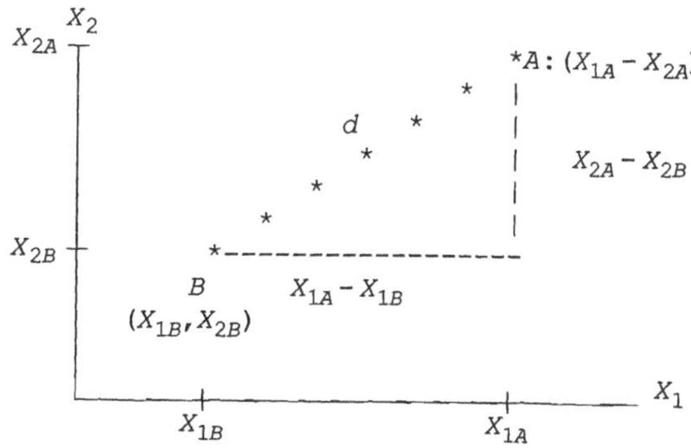

Figure 12.1 Distance in a plane.

By the Pythagorean theorem, we get the usual geometric (i.e., Euclidean) index of distance:

$$\tilde{d}_{AB}^2 = (X_{1A} - X_{1B})^2 + (X_{2A} - X_{2B})^2$$
$$= \sum_{i=1}^{2} (X_{iA} - X_{iB})^2.$$

For example, for A: (6, 5) and B: (2, 3), we get

$$\tilde{d}_{AB}^2 = (6-2)^2 + (5-3)^2$$
$$= 20$$

or

$$\tilde{d}_{AB} = \sqrt{20} \doteq 4.5.$$

Note that $\tilde{d}_{AB}^2$ may be expressed as:

$$[\mathbf{x}_A - \mathbf{x}_B]'[\mathbf{x}_A - \mathbf{x}_B],$$

where $\mathbf{x}_A$ and $\mathbf{x}_B$ are 2×1 column vectors of scores, and $\mathbf{x}_A - \mathbf{x}_B$ is a 2×1 column vector of differences. [The order of the $(1 \times 2)(2 \times 1)$ vector product is 1×1.] That is,

$$\tilde{d}_{AB}^2 = [X_{1A} - X_{1B} \quad X_{2A} - X_{2B}]\begin{bmatrix} X_{1A} - X_{1B} \\ X_{2A} - X_{2B} \end{bmatrix}$$
$$= (X_{1A} - X_{1B})^2 + (X_{2A} - X_{2B})^2.$$

For the example above,

$$\tilde{d}_{AB}^2 = [6-2 \quad 5-3]\begin{bmatrix} 6-2 \\ 5-3 \end{bmatrix} = [4 \quad 2]\begin{bmatrix} 4 \\ 2 \end{bmatrix} = 20.$$

This index is appropriate if two conditions are assumed: (1) Measures on X_1 and X_2 are uncorrelated (i.e., $\rho_{12} = .00$)[1]; and (2) measures on X_1 and X_2 have unit variances (i.e., $\sigma_1^2 = \sigma_2^2 = 1.0$).[2]

Extending the Euclidean distance idea to a general p-variate space, we may write

$$\tilde{d}_{AB}^2 = \sum_{i=1}^{p}(X_{iA} - X_{iB})^2$$

or

$$\tilde{d}_{AB}^2 = [\mathbf{x}_A - \mathbf{x}_B]'[\mathbf{x}_A - \mathbf{x}_B],$$

where $\mathbf{x}_A$ and $\mathbf{x}_B$ are $p \times 1$ vectors. (The latter expression for $\tilde{d}^2$ is a row vector of p differences multiplied by a column vector of the same p differences.) Similar to the bivariate case, this index is based on the assumption of uncorrelated variables, all with unit variances. That is, the $p \times p$ covariance matrix, $\mathbf{\Sigma}$, for the p variables is an identity matrix:

$$\mathbf{\Sigma} = \begin{bmatrix} \sigma_1^2 & \rho_{12}\sigma_1\sigma_2 & \cdots & \rho_{1p}\sigma_1\sigma_p \\ \rho_{21}\sigma_2\sigma_1 & \sigma_2^2 & \cdots & \rho_{2p}\sigma_2\sigma_p \\ \vdots & & & \vdots \\ \rho_{p1}\sigma_p\sigma_1 & & \cdots & \sigma_p^2 \end{bmatrix} = \begin{bmatrix} 1 & 0 & \cdots & 0 \\ 0 & 1 & \cdots & 0 \\ \vdots & \vdots & & \vdots \\ 0 & & \cdots & 1 \end{bmatrix}.$$

(Recall that the covariance of X_i and $X_{i'}$ is $\sigma_{ii'} = \rho_{ii'}\sigma_i\sigma_{i'}$.) So much for the special case of uncorrelated variables with variances of 1.0. Because this context is unrealistic, we do not discuss it further.

The basic requirement in comparing distances involving measures on two (or more) variables is that the same metric is used in computing the distances. One way this is assured is, of course, if all standard deviations, or variances, are equal. The variance-equal-one condition is, indeed, a special case. If this is not the case, the unequal variances must be "taken into consideration." This is accomplished by dividing the measures by the corresponding standard deviation (see Exercise 1 at the end of this chapter).

Because empirical scientists usually deal with intercorrelated variables (albeit the intercorrelations are often "modest"), these intercorrelations, too, must be "taken into consideration" in assessing distances. This is accomplished in a more complicated manner. Consideration of both unequal variances and nonzero intercorrelations is accomplished by using

$$\Delta_{AB}^2 = [\mathbf{x}_A - \mathbf{x}_B]'\mathbf{\Sigma}^{-1}[\mathbf{x}_A - \mathbf{x}_B] \tag{12.1}$$

[1]Uncorrelated variables may be depicted geometrically using axes that are perpendicular (i.e., orthogonal) to each other.

[2]When using the $\tilde{d}^2$ index above, we assume that the measurement metrics (reflected by standard deviations) on the two axes are not only identical but that both standard deviations are 1.0.

as an index of the (squared) distance between point A (defined by the column vector $\mathbf{x}_A$) and point B (defined by the column vector $\mathbf{x}_B$). In expression (12.1), Σ is the population covariance matrix, and Δ_{AB} is a *generalized distance* index that is often attributed to an Indian statistician, P. C. Mahalanobis (Huberty, 2005).

Given the special case where the variables are uncorrelated with variances of 1.0, and knowing that the inverse of an identity matrix is an identity matrix, it should be clear that

$$\tilde{d}^2_{AB} = [\mathbf{x}_A - \mathbf{x}_B]'[\mathbf{x}_A - \mathbf{x}_B]$$

is a special case of (12.1).

12.3 DISTANCE AND CLASSIFICATION

As discussed in Section 12.2, Δ is used as an index of the distance between two points in a p-dimensional space. In that discussion, the two points represented two vectors of p observations each, that is, "profiles" of two analysis units.[3] There are two other Mahalanobis-type distance indices that are of importance in discriminant analysis. *One* is an index of distance between two points where each point represents a vector of means on the p variables. Having two populations, with centroids $\boldsymbol{\mu}_1$ and $\boldsymbol{\mu}_2$, the "distance between the populations," that is, the distance between the two centroids, may be represented by:

$$\Delta_{12} = [(\boldsymbol{\mu}_1 - \boldsymbol{\mu}_2)'\boldsymbol{\Sigma}^{-1}(\boldsymbol{\mu}_1 - \boldsymbol{\mu}_2)]^{1/2}. \qquad (12.2)$$

Here, Σ is the covariance matrix common to the two populations; that is, the covariance matrices in the two populations are assumed to be equal.[4]

A *second* Mahalanobis-type index is appropriate when one point represents a vector of p observations on an analysis unit and the other point represents a centroid for a population. Suppose that there are J populations of interest; the distance between $\mathbf{x}_u$, the observation vector for unit u, and $\boldsymbol{\mu}_j$, the centroid for Population j, may be represented by:

$$\Delta_{uj} = [(\mathbf{x}_u - \boldsymbol{\mu}_j)'\boldsymbol{\Sigma}_j^{-1}(\mathbf{x}_u - \boldsymbol{\mu}_j)]^{1/2}, \qquad (12.3)$$

where Σ_j is the covariance matrix for Population j. This distance index is of particular interest in classification analyses because a goal in classification is to classify a unit into that population to which the unit is nearest. That is, unit u is classified into Population j if Δ_{uj} is smaller than $\Delta_{uj'}$ for all $j' \neq j$ and $j, j' = 1, 2, \ldots, J$.

[3]The index of distance between two observation vectors is of particular interest in *cluster analysis*, where clusters of analysis units are determined by values of Δ^2—units u_1 and u_2 are placed in a cluster if the value of $\Delta^2_{u_1,u_2}$ is "small." In this context, Δ^2 is used as an index of proximity or similarity (or, more appropriately, *dis*similarity).

[4]This distance index is useful in a descriptive discriminant analysis, where interest is on group separation or group comparisons (see Section 3.2.2).

In the univariate case, Δ_{uj} in (12.3) may be expressed as:

$$\Delta_{uj} = \left[(X_u - \mu_j)' \frac{1}{\sigma_j^2}(X_u - \mu_j)\right]^{1/2} = \left[\frac{(X_u - \mu_j)^2}{\sigma_j^2}\right]^{1/2}.$$

It will be shown in Chapter 13 that by using Δ_{uj} values for classifying, distance is measured in a probabilistic sense. That is, by stating that unit u is a distance of Δ_{uj} from the centroid of Population j, we are claiming that it is more probable that the unit (when randomly selected from Population j) is from Population j than if the distance index value is larger. Further, we shall see that the use of Δ_{uj} in classification is appropriate, theoretically, when we are dealing with continuous, normally distributed predictor variables.

In sum, then, there are three types of distances: (1) unit to unit [see (12.1)], (2) centroid to centroid [see (12.2)], and (3) unit to centroid [see (12.3)]. It is the third type on which emphasis is given in PDA. In Chapter 13 we return to the notion of distance and see how it ties in specifically with the classification problem. But first, let us discuss the classification problem in more general terms.

12.4 CLASSIFICATION RULES IN GENERAL

The basic purpose of a PDA may be described as follows: Suppose that we have samples from J populations of size n_j, $j = 1, 2, \ldots, J$, with p measures on each of the $N(= \Sigma n_j)$ units. Using this $N \times p$ data matrix, we want to determine from which of the J populations an $(N + 1)$st unit is *most likely* to have been randomly sampled. To accomplish this task we use the information in the given $N \times p$ data matrix to set up a rule for making the assignment.[5]

12.4.1 Maximum Likelihood

A decision, classification, assignment, or identification rule that is commonly used is based on the *maximum-likelihood principle*: Assign a unit to the population in which its observation vector has the greatest likelihood of occurrence. This may be viewed in terms of likelihood functions, density functions, or probability functions.

Let us sidetrack a bit and discuss the notion of the likelihood of an observation. To simplify the discussion, consider a single (continuous) variable X and two populations. Suppose that (theoretical) models for the two populations of X scores are represented as in Figure 12.2. In this figure we see graphical representations of two

[5]This situation is similar to that in *multiple regression studies*, where one is typically predicting a score on a continuous variable instead of predicting group membership. Note that in both situations a rule based on a given data matrix is derived and may be used with "new" units.

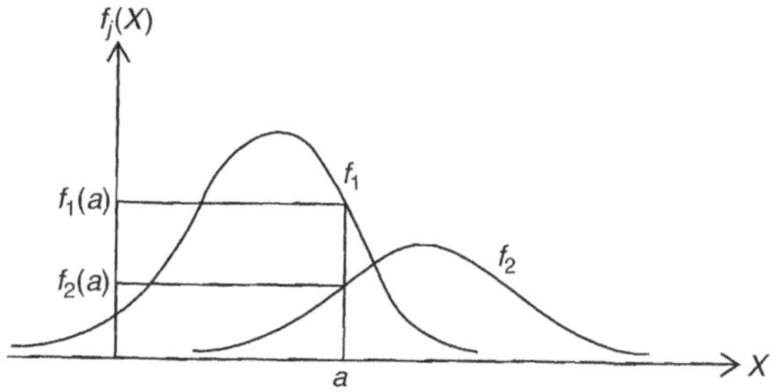

Figure 12.2 Graphical representations of two density functions.

(probability) *density functions*, f_1 and f_2. These may be thought of as "smoothed-out" relative frequency polygons for two distributions of X scores. They are graphs of "density" functions because the area above the X axis and under each curve is 1.0, and the mass in any interval on the X axis under the curve is numerically equal to the probability that X will assume a value in that interval. Given that the unit is randomly selected from Population 1, the *likelihood* of an observation $X = a$ is denoted by $f_1(a)$.

Thus, with a univariate two-group classification problem as in the preceding paragraph, applying the maximum-likelihood principle we arrive at the following rule: Assign a unit with $X = a$ to Population 1 if $f_1(a) > f_2(a)$, that is, if the likelihood of an observation $X = a$ is greater for Population 1 than for Population 2; otherwise, assign the unit to Population 2.

Let us now return to the general multivariate J-group classification problem. We will assume that the "form" of the density function is the same for all J populations: for example, that they are all multivariate normal (discussed in Chapter 13). Let f denote this common density function. Then the *maximum-likelihood rule* is: Assign analysis unit u to Population j if the likelihood of the observation vector, $\mathbf{x}_u$, is greater for Group j than for any other group. This rule may be stated as follows:

Assign unit u to Population j if

$$f(\mathbf{x}_u | j) > f(\mathbf{x}_u | j') \qquad (12.4)$$

for $j' \neq j$.

12.4.2 Typicality Probability

The rule may also be stated in terms of something called *inverse probabilities*, rather than in terms of likelihoods. Such a probability is denoted as $P(\mathbf{x}|j)$ and may be viewed as the proportion of units in Population j that have score vectors "near" $\mathbf{x}$. That is, $P(\mathbf{x}|j)$ denotes the probability that a randomly selected unit has a profile close to $\mathbf{x}$, given that the unit is a member of Population j; $P(\mathbf{x}|j)$ values are termed *typicality probabilities* in this book. [A more "down-to-earth" interpretation of $P(\mathbf{x}|j)$ is given in Section 14.4.] It turns out that $P(\mathbf{x}|j)$ is, in the limit, proportional to $f(\mathbf{x}|j)$. Therefore, a second statement of the maximum-likelihood rule may be given in terms of these typicality probabilities:

> Assign unit u to Population j if
>
> $$P(\mathbf{x}_u|j) > P(\mathbf{x}_u|j') \tag{12.5}$$
>
> for $j' \neq j$.

The notion of $P(\mathbf{x}_u|j)$ is discussed in greater detail in the Technical Note in Chapter 14.

12.4.3 Posterior Probability

Another view of the rule is taken by considering the probability of unit u belonging to Group j, given that the unit has a particular observation vector, $\mathbf{x}_u$. This probability, denoted $P(j|\mathbf{x}_u)$, is called the *posterior probability* of membership in Population j; "posterior" in the sense that this is a probability of population membership conditioned on knowing $\mathbf{x}_u$, that is, *after* the p X values are obtained. [According to David (1995), "posterior probability" was first used by Wrinch and Jefferys (1921).] With this view we see the necessity of the assumption stated in Section 12.1, namely, that the unit to be classified does in fact belong to one of the J criterion populations. It seems reasonable that a unit be assigned to that population for which $P(j|\mathbf{x})$, the posterior probability of membership, is greatest. Now, the probability that a unit belongs to Population j (given an observed score vector) is equal to the ratio of the probability of its score vector in Population j to the sum of the probabilities associated with its score vector in all J groups. [This follows from the multiplication rule for probabilities; see the Technical Note at the end of this chapter wherein the $\pi_j \equiv P(j)$ values are common.] That is,

$$P(j|\mathbf{x}_u) = \frac{P(\mathbf{x}_u|j)}{\sum_{j'=1}^{J} P(\mathbf{x}_u|j')}. \tag{12.6}$$

Consequently, a third statement of the maximum-likelihood rule [following (12.4) and (12.5)] is

> Assign unit u to Population j if
>
> $$P(j|\mathbf{x}_u) > P(j'|\mathbf{x}_u) \tag{12.7}$$
>
> for $j' \neq j$, where $P(\mathrm{j}|\mathbf{x}_u)$ is defined as in (12.6).

No matter which view of the maximum-likelihood rule is taken, J values need to be estimated for each unit; J values of $f(\mathbf{x}_u|j)$ for (12.4), or J values of $P(\mathbf{x}_u|j)$ for (12.5), or J values of $P(j|\mathbf{x}_u)$ for (12.7). Furthermore, for a given unit, the denominator of (12.6) is constant across all of the J $P(j|\mathbf{x}_u)$ values. Thus, (12.7) is equivalent to (12.5) because the J $P(j|\mathbf{x}_u)$ values are proportional to the J $P(\mathbf{x}_u|j)$ values.

12.4.4 Prior Probability

By examining the two latter statements of the rule, (12.7) and (12.5), it is clear that the adequacy of the rule depends on the goodness of the estimates of the typicality probabilities $P(\mathbf{x}_u|j)$, $j = 1, 2, \ldots, J$, and that goodness in turn depends on the size (and representativeness) of the J original (or training) samples on which the estimates are based. Thus, it may be well to take the relative sizes of the populations into consideration. Let π_j denote the proportion of units in the total universe (i.e., the aggregate of the J populations) that is in Population j. That is, if a unit is randomly selected from the universe, the probability that it would be from Population j is π_j. The symbol π_j is used to denote the *prior probability* of membership in Population j, "prior" in the sense that this is a probability of population membership before $\mathbf{x}_u$ is known. [The first use of the term "prior probability" is also attributed to Wrinch and Jeffreys (1921).] The π_j values have also been termed *base rates*; also, too, sometimes $P(j)$ is used instead of π_j.

It is reasonable that these prior probabilities be taken into consideration when arriving at values of $P(j|\mathbf{x}_u)$ in (12.6). The product, $\pi_j \cdot P(\mathbf{x}_u|j)$, denotes the joint probability that a randomly selected unit belongs to Population j and at the same time has a score vector "close" to $\mathbf{x}_u$. These products may be used to arrive at values of $P(j|\mathbf{x}_u)$ by employing a rule in probability due to Reverend T. Bayes (1701–1761). Incorporating prior probabilities, the posterior probability of unit u belonging to Population j, given a score vector $\mathbf{x}_u$, is

$$P(j|\mathbf{x}_u) = \frac{\pi_j \cdot P(\mathbf{x}_u|j)}{\sum_{j'=1}^{J} \pi_{j'} \cdot P(\mathbf{x}_u|j')}. \tag{12.8}$$

(See the Technical Note at the end of this chapter.) Note that (12.6) is a special case of (12.8), where the J π_j values are identical. The *maximum (Bayesian) probability rule* is thus stated as:

> Assign unit u to Population j if
>
> $$P(j|\mathbf{x}_u) > P(j'|\mathbf{x}_u) \tag{12.9}$$
>
> for $j \neq j'$, where $P(j|\mathbf{x}_u)$ is defined as in (12.8).

Again, J values of $P(j|\mathbf{x}_u)$ in (12.8) need to be determined for each unit. Because the denominator in (12.8) is constant for all populations, the rule could more simply be based on the J values of $\pi_j \cdot P(\mathbf{x}_u | j)$. Also, because $P(\mathbf{x}_u | j)$ values are proportional to $f(\mathbf{x}_u | j)$ values, we could focus on the J values of $\pi_j \cdot f(\mathbf{x}_u | j)$. The latter is what the Indian statistician C. R. Rao, in 1973, calls *discriminant scores*. In fact, (12.8) may be equivalently stated as:

$$P(j | \mathbf{x}_u) = \frac{\pi_j \cdot f(\mathbf{x}_u | j)}{\sum_{j'=1}^{J} \pi_{j'} \cdot f(\mathbf{x}_u | j')}. \tag{12.10}$$

By using the conditional Bayesian posterior probabilities [as in (12.8)], the total *number* of misclassification errors is minimized. By using the maximum-likelihood rule (12.7), on the other hand, the total *proportion* of misclassification errors is minimized.

A third criterion, in addition to number of errors and proportion of errors, may be considered by further refinement of the Bayes procedure. This refinement yields a decision-theory-based rule that minimizes the total *cost* of misclassification errors. Misclassification costs are difficult to assess in many areas of study. The incorporation of misclassification costs is discussed in Section 13.7.

12.5 COMMENTS

The discussion of likelihoods, typicality probabilities, posterior probabilities, and prior probabilities provides some rationale for a single classification rule in general form, namely, that stated in (12.9). To use this rule, the two sets of probabilities expressed on the right-hand side of (12.8), namely, π_j and $P(\mathbf{x}_u | j)$, need to be estimated. Estimates of the J prior probabilities, π_j, are sometimes based on the sample sizes: $\hat{\pi}_j = q_j = n_j / N$. These estimates are appropriate, of course, only if the sample sizes are in proportion to the population sizes. This may be the case when a stratified or quota sampling procedure is employed, and the relative population sizes remain fairly stable over time. Even though the n_j values are identical, however, the q_j values need not be. Furthermore, if the populations are equally numerous, equal priors would be the choice regardless of the n_j values.

To estimate the $P(\mathbf{x}_u | j)$ values, a *model* must be specified for the distribution of $\mathbf{x}$ in each of the J populations. An example of a model specification was given earlier in this section for the univariate case (see Fig. 12.2). The popular multivariate normal model is discussed in Chapter 13.

It should be noted that there are three types of *generalized distance* indices to consider: (1) from one observed vector to another observed vector [see (12.1)], (2) from one centroid to another [see (12.2)], and (3) from an observed vector to a centroid [see (12.3)]. The first type of distance is basic for cluster analysis. The second type of distance plays an important role in descriptive discriminant analysis (DDA), while the third type is involved in predictive discriminant analysis (PDA).

There are two purposes for a PDA. *One* purpose is to estimate the prediction/classification accuracy (i.e., hit rates) for the data on hand; that is, to assess the predictive "power" of a set of predictor variables. The *second* purpose of conducting a PDA is to develop as good a rule as possible using the data on hand to be used with "new" units. These ideas will be discussed in subsequent chapters.

Technical Note

The relationship between the two *conditional probabilities*, $P(j|\mathbf{x}_u)$ and $P(\mathbf{x}_u|j)$, in (12.8) may be derived by using the multiplication rule for probabilities. Namely, for independent events A and B, $P(A \cap B) = P(B) \cdot P(A|B) = P(A) \cdot P(B|A)$. Regarding j and $\mathbf{x}_u$ as the events of interest in the classification problem, we have that

$$P(j \cap \mathbf{x}_u) = P(\mathbf{x}_u) \cdot P(j|\mathbf{x}_u) = \pi_j \cdot P(\mathbf{x}_u|j),$$

and solving for $P(j|\mathbf{x}_u)$ yields

$$P(j|\mathbf{x}_u) = \frac{\pi_j \cdot P(\mathbf{x}_u|j)}{P(\mathbf{x}_u)}. \tag{12.11}$$

But, because

$$\sum_{j'=1}^{J} P(j'|\mathbf{x}_u) = 1 = \sum_{j'=1}^{J} \frac{\pi_{j'} \cdot P(\mathbf{x}_u|j')}{P(\mathbf{x}_u)},$$

we have that

$$P(\mathbf{x}_u) = \sum_{j'=1}^{J} \pi_{j'} \cdot P(\mathbf{x}_u|j').$$

Substituting this expression for $P(\mathbf{x}_u)$ into (12.11), we arrive at (12.8):

$$P(j|\mathbf{x}_u) = \frac{\pi_j \cdot P(\mathbf{x}_u|j)}{\sum_{j'=1}^{J} \pi_{j'} \cdot P(\mathbf{x}_u|j')}.$$

Further Reading

Anderson (2003, Chapter 6) gives a theoretical discussion of PDA—his first edition appeared in 1958(!).

David and Edwards (2001, pp. 223,228) mention that the expressions "prior probability" and "posterior probability" date back to 1921, if not to 1830.

McLachlan (1992, pp. 22–26) reviews a number of indices of distance between two groups other than the Mahalanobis distance.

***Definition Goodness-of-fit*:** The variance between your uniform and your distribution.

EXERCISES

1. In the "standardized" (X_1, X_2) space, find the distance between the two points representing $(-2, 1)$ and $(2, 7)$. [*Note*: Here, expression (12.1) may be used with $\mathbf{\Sigma} = \mathbf{I}$, or the expression for d_{AB}^2 may be used.]

2. Consider the (X_1, X_2) space having a metric defined by the covariance matrix

$$\mathbf{\Sigma} = \begin{bmatrix} 5 & 3 \\ 6 & 4 \end{bmatrix}.$$

Find the distance between the two points representing $(-2, 1)$ and $(2, 7)$.

$$\left(\text{Note that } \mathbf{\Sigma}^{-1} = \begin{bmatrix} 2 & -\frac{3}{2} \\ -3 & \frac{5}{2} \end{bmatrix}. \right)$$

[*Note*: Expression (12.1) may be used.]

3. Consider, again, a bivariate situation, with a set of measures on X_1 and X_2. Suppose that there are two populations of (X_1, X_2) pairs of scores, with $\mu_{11} = 10$, $\mu_{21} = 12$, $\mu_{12} = 2$, and $\mu_{22} = 3$, and

$$\mathbf{\Sigma}_1 = \mathbf{\Sigma}_2 = \begin{bmatrix} 5 & 3 \\ 6 & 4 \end{bmatrix}.$$

   (a) Find the distance from an observed (X_1, X_2) score vector, (6, 7), to the centroid of the first population, (10, 12). To the centroid of the second population. [*Note*: Here, expression (12.3) is used.]

   (b) Find the distance between the two centroids. [*Note*: Here, expression (12.2) is used.]

4. Let Y_1 denote arithmetic achievement and Y_2 denote reading comprehension. Suppose that $\sigma_1^2 = 25$, $\sigma_2^2 = 9$, and $\rho_{12} = .60$. Then

$$\mathbf{\Sigma} = \begin{bmatrix} 25 & 9 \\ 9 & 9 \end{bmatrix} \quad \text{and} \quad \mathbf{\Sigma}^{-1} = \frac{1}{144} \begin{bmatrix} 9 & -9 \\ -9 & 25 \end{bmatrix}.$$

[Verify that $\mathbf{\Sigma}\mathbf{\Sigma}^{-1} = \mathbf{\Sigma}^{-1}\mathbf{\Sigma} = \mathbf{I}$.] Find the three pairwise squared distances (Δ^2 values) involving Leah: (30, 20), Joe: (30, 30), and John: (20, 25).

5. Let Y_1 denote arithmetic achievement and Y_2 denote grip strength. If age is held somewhat constant, it is reasonable to assume that $\rho_{Y_1 Y_2}$ is near zero. Suppose that $\sigma_{Y_1}^2 = 25$, $\sigma_{Y_2}^2 = 9$. Let $X_1 = Y_1/5$ and $X_2 = Y_2/3$; then $\sigma_{X_1}^2 = \sigma_{X_2}^2 = 1.0$. Suppose that Sherrie's scores on Y_1 and Y_2 are 20 and 6, respectively, that Kama's scores are 30 and 3, and that Sandy's scores are 25 and 9. Find the three pairwise

"distances" (d^2 values) between Sherrie, Kama, and Sandy in the (Y_1, Y_2) space and again in the (X_1, X_2) space. What is your conclusion regarding the proximity of Sherrie, Kama, and Sandy?

6. Consider a univariate situation with measures on, say, X. Suppose that all X values of interest are in the interval (10, 60). Very simply, how would you graphically represent the points that are a distance of 5 from the "point" 23?

7. Consider now a bivariate situation with pairs of measures on, say, X_1 and X_2. Restrict X_1 values to (10, 60) and X_2 values to (20, 60), with $\overline{X}_1 = 35$ and $\overline{X}_2 = 40$.

(a) Assume that the correlation between the X_1 values and the X_2 values is zero, and that $s_1^2 = s_2^2 = 1.0$. How would you graphically represent the points that are a distance of 15 from the "point" (35, 40)?

(b) Note that in Exercise 6, the "locus" of points a constant distance from a point is a pair of points. In Exercise 7(a), the locus of points of interest is a circle. It can be shown using analytic geometry that the locus of points a constant distance from a fixed point is an *ellipse* when the correlation is not zero and when the standard deviations are not unity. That is, the graphical representation of Δ^2 [in (12.2)] is an ellipse if $p = 2$ and is an ellipsoid if $p > 2$.

8. Consider, again, the bivariate situation described in Exercise 3. Let an observed score vector be $\mathbf{x}' = (6, 7)$.

(a) Assume that $P(\mathbf{x}|1) = .30$. What is the value of $P(\mathbf{x}|2)$?

(b) Assuming equal priors (i.e., $\pi_1 = \pi_2 = .5$), to which group would the unit having the observed score vector be assigned? [*Note*: Here, rule (12.5) is used.]

(c) Find $P(1|\mathbf{x})$ and $P(2|\mathbf{x})$. [*Note*: Here, expression (12.6) is used.]

(d) What is the value of $P(1|\mathbf{x}) + P(2|\mathbf{x})$? Will this always be the case?

(e) Assume, now, that $\pi_1 = .70$. What is the value of π_2?

(f) With the priors of part (e), and the typicality probabilities of part (a), what is the value of $P(1|\mathbf{x})$? Of $P(2|\mathbf{x})$? [*Note*: Here, expression (12.8) is used.]

(g) Based on your answers for part (f), to which group would the unit having the observed score vector be assigned? [*Note*: Here, rule (12.9) is used.]

9. Consider the research situation you described in Exercise 2 of Chapter 1 that involves prediction of group membership. What do you think are the relative sizes of the populations? (That is, specify π_j values.) Do you think the relative sizes will change in the near or distant future? If you had resources to collect predictor measures on 200 units, how many units would you sample from each population? That is, would you use equal sample sizes, sample sizes proportional to population sizes, or some other sampling plan?

CHAPTER 13

Multivariate Normal Rules

13.1 INTRODUCTION

As stated in Section 12.4, the *maximum probability rule* involving posterior probabilities of group membership—see (12.6) or (12.8)—will minimize the total number of misclassification errors. This optimal rule can be applied only if the probability density functions, the ranges of which are denoted by $f(\mathbf{x}|j)$, are known; that is, if all distribution parameters are known. [Recall that the $P(\mathbf{x}|j)$ values are proportional to the $f(\mathbf{x}|j)$ values.] Of course, the distribution parameters, $\boldsymbol{\Sigma}$'s and $\boldsymbol{\mu}$'s, are usually not known; one then must use estimates and be satisfied with a less than optimal rule. Three approaches may be taken to construct a classification rule that uses estimates of the density values, $\hat{f}(\mathbf{x}_u|j)$. The *first* approach discussed is to specify a theoretical probability distribution model, assume that the data on hand fit the model, estimate the model parameters using the data, and construct a rule using these estimates. The *second* approach is to estimate the density values directly from the data with no prior model specification, and construct a rule using these estimates. The *third* approach is sort of a combination of the other two. With this approach, a Bayesian framework is used to obtain density estimates for a particular model given the available data—see Section 12.4.4.

The *first* approach is most commonly used. It, along with the third approach, is discussed in the present chapter, where the underlying model is that of multivariate normality. The first approach is also discussed in Section 19.3 in the context of a multinomial model. The *second* approach is discussed briefly in Section 19.2.

13.2 NORMAL DENSITY FUNCTIONS

The family of univariate normal probability density functions is defined by:

$$f(X|j) = \frac{1}{\sqrt{2\pi}\sqrt{\sigma_{jX}^2}} \exp\left[-\frac{1}{2}\frac{(X-\mu_{jX})^2}{\sigma_{jX}^2}\right], \tag{13.1}$$

Applied MANOVA and Discriminant Analysis, Second Edition, by Carl J. Huberty and Stephen Olejnik
Copyright © 2006 John Wiley & Sons, Inc.

where μ_{jX} and σ^2_{jX} are the mean and variance of variable X, respectively, for population j. [The expression exp(·) is an alternative to $e^{(\cdot)}$, where e is the irrational number 2.71828..., an endless nonrepeating decimal. This number is the base for natural logarithms.] Note that if μ_{jX} and σ^2_{jX} are specified, a particular member of the family of density functions is completely determined. The graph of a particular density function is the familiar two-dimensional bell-shaped curve. (See Fig. 12.2 for two such graphs.) For a particular density function, the right-hand side of (13.1) yields the ordinate of the normal curve for the abscissa X. For example, with $\mu_{1X} = 50$ and $\sigma^2_{1X} = 100$, for $X = 60$, we get, using a hand calculator, $f(60|1) \doteq .024$; that is, .024 is the height of the normal (50, 100) curve for a Population 1 score value of $X = 60$ (and also at $X = 40$).

The reader will recall from the study of elementary statistics that the normal model was often referred to in univariate analyses to arrive at tail-area probability estimates (i.e., P values). In essence, what was done was to insert estimates of μ_{jX} and σ^2_{jX}, based on sample data, to arrive at an estimate of $f(X|j)$:

$$\hat{f}(X_u|j) = \frac{1}{\sqrt{2\pi}\sqrt{s^2_{jX}}} \exp\left[-\frac{1}{2}\frac{(X_{uj} - \overline{X}_{.j})^2}{s^2_{jX}}\right],$$

where

$$\overline{X}_{.j} = \frac{1}{n_j}\sum_{u=1}^{n_j} X_{uj},$$

$$s^2_{jX} = \frac{1}{n_j - 1}\sum_{u=1}^{n_j}(X_{uj} - \overline{X}_{.j})^2, \tag{13.2}$$

and π is the irrational number 3.14159....

The generalization of (13.1) to the multivariate case may be made by analogy to arrive at the family of p-variate normal probability density functions, which are defined by:

$$f(\mathbf{x}|j) = \frac{1}{\sqrt{(2\pi)^p}\sqrt{|\mathbf{\Sigma}_j|}} \exp\left[-\tfrac{1}{2}(\mathbf{x} - \boldsymbol{\mu}_j)'\mathbf{\Sigma}_j^{-1}(\mathbf{x} - \boldsymbol{\mu}_j)\right]. \tag{13.3}$$

The $p \times 1$ vector of predictor scores, $\mathbf{x}$, is a generalization of X in the univariate case. The $p \times p$ population covariance matrix, $\mathbf{\Sigma}_j$, is the generalization of σ^2_j in the univariate case; the $p \times 1$ mean vector, $\boldsymbol{\mu}_j$, is the generalization of μ_j. The determinant of $\mathbf{\Sigma}_j$, $|\mathbf{\Sigma}_j|$, is called the *generalized variance* of the set of p variables see Technical Note 1 at the end of this chapter. Analogous to the univariate case, specification of $\boldsymbol{\mu}_j$ and $\mathbf{\Sigma}_j$ completely determine the multivariate normal density function for Population j. The graph of a particular function is a $(p + 1)$-dimensional "bell-shaped" surface. The right-hand side of (13.3) yields the "height" of the surface corresponding to the observation vector $\mathbf{x}$ in Population j.

It is obvious from examining (13.3) that calculating $f(\mathbf{x}|j)$ values with a hand calculator is no simple task when $p > 2$. The difficulty is in calculating the value of the determinant, $|\boldsymbol{\Sigma}_j|$, and the value of the *quadratic form*,

$$(\mathbf{x} - \boldsymbol{\mu}_j)'\boldsymbol{\Sigma}_j^{-1}(\mathbf{x} - \boldsymbol{\mu}_j). \tag{13.4}$$

It is of interest to note that the expression (13.4) is a distance index discussed in Section 12.3 [see (12.3)]. If we consider an observation vector for an analysis unit u, from Population j, then

$$\Delta_{uj}^2 = (\mathbf{x}_u - \boldsymbol{\mu}_j)'\boldsymbol{\Sigma}_j^{-1}(\mathbf{x}_u - \boldsymbol{\mu}_j),$$

denotes the square of the distance from the point representing $\mathbf{x}_u$ to the point representing the centroid of Population j, $\boldsymbol{\mu}_j$.

Because in data analysis situations we seldom know parameter values, we need to determine likelihood estimates, $\hat{f}(\mathbf{x}|j)$. These are obtained in the usual manner of inserting estimates of $\boldsymbol{\mu}_j$ and of $\boldsymbol{\Sigma}_j$ into expression (13.3):

$$\hat{f}(\mathbf{x}|j) = \frac{1}{\sqrt{(2\pi)^p}\sqrt{|\mathbf{S}_j|}} \exp\left[-\tfrac{1}{2}(\mathbf{x} - \bar{\mathbf{x}}_j)'\mathbf{S}_j^{-1}(\mathbf{x} - \bar{\mathbf{x}}_j)\right], \tag{13.5}$$

where $\bar{\mathbf{x}}_j$ is the $p \times 1$ vector of means for Group j, and $\mathbf{S}_j$ is the $p \times p$ covariance matrix for Group j. The main diagonal elements of $\mathbf{S}_j$ are the p variances [see (13.2)], and the off-diagonal elements are the $p(p-1)/2$ covariances (see Section 2.4).

A sample Mahalanobis index for the squared distance between an observation vector for unit u and the centroid for Group j may be written, then, as:

$$D_{uj}^2 = (\mathbf{x}_u - \bar{\mathbf{x}}_j)'\mathbf{S}_j^{-1}(\mathbf{x}_u - \bar{\mathbf{x}}_j). \tag{13.6}$$

Expression (13.5) may then be stated for unit u as:

$$\hat{f}(\mathbf{x}_u|j) = (2\pi)^{-p/2} \cdot |\mathbf{S}_j|^{-1/2} \exp\left(-\tfrac{1}{2}D_{uj}^2\right). \tag{13.7}$$

13.3 CLASSIFICATION RULES BASED ON NORMALITY

The form of the posterior probabilities used in the maximum probability rule on which we will base our introduction of normal rules is that given in (12.10) and repeated here:

$$P(j|\mathbf{x}_u) = \frac{\pi_j \cdot f(\mathbf{x}_u|j)}{\sum_{j'=1}^{J} \pi_{j'} \cdot f(\mathbf{x}_u|j')}.$$

Working with parameter estimators as is typical, we have

$$\hat{P}(j|\mathbf{x}_u) = \frac{q_j \cdot \hat{f}(\mathbf{x}_u|j)}{\sum_{j'=1}^{J} q_{j'} \cdot \hat{f}(\mathbf{x}_u|j')}, \tag{13.8}$$

where $q_j = \hat{\pi}_j$. Substituting from (13.7) for the p-variate case, and noting that $(2\pi)^{-p/2}$ is a factor common to both numerator and denominator, we may write

$$\hat{P}(j|\mathbf{x}_u) = \frac{q_j \cdot |\mathbf{S}_j|^{-1/2} \exp\left(-\frac{1}{2}D_{uj}^2\right)}{\sum_{j'=1}^{J} q_{j'} \cdot |\mathbf{S}_{j'}|^{-1/2} \exp\left(-\frac{1}{2}D_{uj'}^2\right)}. \tag{13.9}$$

Therefore, the maximum-probability rule for the p-variate normal case may be expressed as:

> Assign unit u to Population j if
>
> $$\hat{P}(j|\mathbf{x}_u) > \hat{P}(j'|\mathbf{x}_u) \tag{13.10}$$
>
> for $j \neq j'$, where $\hat{P}(j|\mathbf{x}_u)$ is defined as in (13.9).

In other words, assign unit u to that population whose sample yields the largest value of $\hat{P}(j|\mathbf{x}_u)$.

Consider the special case where we assume that the J population covariance matrices are equal; that is,

$$\boldsymbol{\Sigma}_1 = \boldsymbol{\Sigma}_2 = \cdots = \boldsymbol{\Sigma}_J = \boldsymbol{\Sigma}. \tag{13.11}$$

In this case an estimator for $\boldsymbol{\Sigma}$ is the error (or pooled) sample $(p \times p)$ covariance matrix, $\mathbf{S}_e$. The main-diagonal elements of $\mathbf{S}_e$ are the p error sample variances as found in univariate analysis of variance. That is, the ith main-diagonal element in $\mathbf{S}_e$ is the error sum-of-squares for variable i divided by $N - J$. The (i, i') off-diagonal element is the error sum-of-products for variables i and i' divided by $N - J$. The squared distance of unit u from the centroid of Group j may then be expressed as a special case of (13.6):

$$D_{uj}^{*2} = (\mathbf{x}_u - \bar{\mathbf{x}}_j)'\mathbf{S}_e^{-1}(\mathbf{x}_u - \bar{\mathbf{x}}_j). \tag{13.12}$$

Thus, the estimated likelihood in (13.7) becomes

$$\hat{f}(\mathbf{x}_u|j) = (2\pi)^{-p/2} \cdot |\mathbf{S}_e|^{-1/2} \exp\left(-\frac{1}{2}D_{uj}^{*2}\right). \tag{13.13}$$

Because the product $(2\pi)^{-p/2} \cdot |\mathbf{S}_e|^{-1/2}$ would be common to the numerator and the denominator of (13.8), we have

$$\hat{P}(j|\mathbf{x}_u) = \frac{q_j \cdot \exp\left(-\frac{1}{2}D_{uj}^{*2}\right)}{\sum_{j'=1}^{k} q_{j'} \cdot \exp\left(-\frac{1}{2}D_{uj'}^{*2}\right)}. \tag{13.14}$$

Thus, the maximum-probability rule for the p-variate normal case, under condition (13.11), is

> Assign unit u to Population j if
>
> $$\hat{P}(j|\mathbf{x}_u) > \hat{P}(j'|\mathbf{x}_u) \tag{13.15}$$
>
> for $j \neq j'$, where $\hat{P}(j|\mathbf{x}_u)$ is defined as in (13.14).

The estimated probabilities given by (13.14) are calculated and reported by the two computer program packages, SAS and SPSS; values of (13.9) for *all* groups may be obtained via the SAS package. This type of computer output is discussed later in this chapter.

In Section 12.4 it was mentioned that density values, $\hat{f}(\mathbf{x}|j)$, are proportional to the probability values, $\hat{P}(\mathbf{x}|j)$. Thus the posterior probability estimates of (13.8) may be written as:

$$\hat{P}(j|\mathbf{x}_u) = \frac{q_j \cdot \hat{P}(\mathbf{x}_u|j)}{\sum_{j'=1}^{J} q_{j'} \cdot \hat{P}(\mathbf{x}_u|j')}, \tag{13.16}$$

and the rule in (13.15) may be thought of as a rule that incorporates typicality probability estimates. Furthermore, for a given unit, the value of the denominator in (13.16) is the same for all groups. Therefore, for classification purposes, the denominator may be ignored, and rules (13.10) and (13.15) may be stated as follows:

> Assign unit u to Population j if
>
> $$q_j \hat{P}(\mathbf{x}_u|j) > q_{j'} \hat{P}(\mathbf{x}_u|j') \tag{13.17}$$
>
> for $j \neq j'$.

13.4 CLASSIFICATION FUNCTIONS

13.4.1 Quadratic Functions

As just stated, the forms of the rule stated in (13.10) and (13.15) may equivalently be stated in terms of only the numerators in (13.9) and (13.14). That is, (13.10) may equivalently be stated in terms of maximizing

$$q_j \cdot |\mathbf{S}_j|^{-1/2} \exp\left(-\tfrac{1}{2} D_{uj}^2\right).$$

Now, maximizing $q_j \cdot |\mathbf{S}_j|^{-1/2} \exp(-\frac{1}{2} D_{uj}^2)$ is equivalent to maximizing the natural logarithm of this product:

$$Q_{uj} = \ln q_j - \tfrac{1}{2} \ln |\mathbf{S}_j| - \tfrac{1}{2} D_{uj}^2. \tag{13.18}$$

Thus, the maximum-probability rule for the p-variate normal case may be expressed as:

> Assign unit u to Population j if
>
> $$Q_{uj} > Q_{uj'} \qquad (13.19)$$
>
> for $j \neq j'$, where Q_{uj} is defined as in (13.18).

Some fairly extensive matrix manipulation would lead one to conclude that the expression Q_{uj} is quadratic in $\mathbf{x}_u$ and hence is called a *quadratic classification function* (QCF)—for $J = 2$, Q_{uj} may also be expressed as a difference of two quadratic forms (see Technical Note 2). So, for each unit u, J QCF values are found, and unit u is assigned to that population whose sample yields the largest QCF value. This is termed a *quadratic classification rule.*

13.4.2 Linear Functions

Consider, again, the special case of equal population covariance matrices, with $\mathbf{S}_e$ being the estimator for the common covariance matrix. Maximizing $\hat{P}(j|\mathbf{x}_u)$ in (13.14) is equivalent to maximizing $q_j \cdot \exp(-\frac{1}{2}D^{*2}_{uj})$. This, in turn, is equivalent to maximizing the natural logarithm of the product:

$$\ln q_j - \tfrac{1}{2}D^{*2}_{uj} = \ln q_j - \tfrac{1}{2}(\mathbf{x}_u - \mathbf{x}_j)'\mathbf{S}_e^{-1}(\mathbf{x}_u - \overline{\mathbf{x}}_j). \qquad (13.20)$$

Matrix algebra yields a term, $-\frac{1}{2}\mathbf{x}_u'\mathbf{S}_e^{-1}\mathbf{x}_u$, that would, for a given unit u, be common for all j, and hence may be ignored for classification purposes. Thus, maximizing (13.20) is equivalent to maximizing

$$\begin{aligned} L_{uj} &= \left[\overline{\mathbf{x}}_j'\mathbf{S}_e^{-1}\right]\mathbf{x}_u - \tfrac{1}{2}\overline{\mathbf{x}}_j'\mathbf{S}_e^{-1}\overline{\mathbf{x}}_j + \ln q_j \\ &= \left[\overline{\mathbf{x}}_j'\mathbf{S}_e^{-1}\right]\mathbf{x}_u + \left[-\tfrac{1}{2}\overline{\mathbf{x}}_j'\mathbf{S}_e^{-1}\overline{\mathbf{x}}_j + \ln q_j\right]. \end{aligned} \qquad (13.21)$$

Thus, the maximum-probability rule for the p-variate normal, equal covariance matrices case may be expressed as:

> Assign unit u to Population j if
>
> $$L_{uj} > L_{uj'} \qquad (13.22)$$
>
> for $j \neq j'$, where L_{uj} is as defined in (13.21).

The expression L_{uj} is linear in $\mathbf{x}_u$ and hence is called a *linear classification function* (LCF), and the rule in (13.22) is a *linear classification rule.* From (13.21) it may be

TABLE 13.1 LCFs for the 3-Group Ethington Data

Variable	LCF_1	LCF_2	LCF_3
counsum	−.50	−.32	−.24
gainsum	.04	.06	−.01
learnsum	.44	.49	.57
qelib	.06	.07	.06
qefac	.35	.24	.32
qestacq	.11	.12	.09
qeamt	.53	.55	.44
qewrite	.35	.37	.35
qesci	.22	.16	.12
(constant)	−16.12	−15.78	−15.80

concluded that L_{uj} can be written as a linear composite of the X scores with the row vector of weights

$$\mathbf{b}'_j = \overline{\mathbf{x}}'_j \mathbf{S}_e^{-1}, \tag{13.23}$$

and constant

$$c_j = -\tfrac{1}{2}\overline{\mathbf{x}}'_j \mathbf{S}_e^{-1} \overline{\mathbf{x}}_j + \ln q_j. \tag{13.24}$$

That is,[1]

$$L_{uj} = \mathbf{b}'_j \cdot \mathbf{x}_u + c_j.$$

The weights given by (13.23) are typically reported in computer package output, similarly for the constant values of (13.24). These weights and constants are applicable to raw data. For example, using the 3-group Ethington data (3GED), the three sets of LCF weights (and constants) are given in Table 13.1. (How to calculate these weights will be illustrated in Chapter 14.)

Thus, the maximum-probability rule boils down to: Assign unit u to that population whose sample yields the largest QCF score [with unequal covariance matrices; see (13.19)], or the largest LCF score [with equal covariance matrices; see (13.22)].

13.4.3 Distance-Based Classification

In terms of observed vector-to-centroid distance (see Section 13.3), rule (13.22) may be viewed as: Assign unit u to that population to which it is "closest." Closeness is used in terms of distances of observation vectors from sample centroids. Referring

[1]Another form of (13.21) is

$$L_{uj} = b_{1j}X_{1u} + b_{2j}X_{2u} + \cdots + b_{pj}X_{pu} + c_j.$$

to the quadratic classification function in (13.18), we see that maximizing Q_{uj} is equivalent to minimizing

$$\begin{aligned} d_{uj} &= -2Q_{uj} \\ &= \ln|S_j| + D^2_{uj} - 2\ln q_j. \end{aligned} \quad (13.25)$$

Thus, the maximum-probability (or minimum-distance) rule may be expressed as:

> Assign unit u to population j if
>
> $$d_{uj} < d_{uj'} \quad (13.26)$$
>
> for $j \neq j'$, where d_{uj} is defined as in (13.25).

This rule is equivalent to rule (13.19).

When the equal-covariance-matrices condition is reasonably met, minimizing (13.25) is minimizing $\ln|\mathbf{S}_e| + D^{*2}_{uj} - 2\ln q_j$, which is equivalent (because $\ln|\mathbf{S}|$ is a constant for all u and all j) to minimizing

$$d^*_{uj} = D^{*2}_{uj} - 2\ln q_j, \quad (13.27)$$

where D^{*2}_{uj} is the squared distance based on the error covariance matrix, $\mathbf{S}_e$ [see (13.12)]. Thus the maximum-probability (or minimum-distance) rule for the equal covariance matrices case may be expressed as:

> Assign unit u to Population j if
>
> $$d^*_{uj} < d^*_{uj'} \quad (13.28)$$
>
> for $j \neq j'$, where d^*_{uj} is defined as in (13.27).

The SAS DISCRIM program terms the expressions in (13.25) and (13.27) the *generalized squared distance function*.

Of course, if the equal prior probability condition is also imposed, rule (13.26) is simplified to minimizing

$$\ln|\mathbf{S}_j| + D^2_{uj}, \quad (13.29)$$

and rule (13.28) is simplified to minimizing

$$D^{*2}_{uj}. \quad (13.30)$$

These are precisely the statistics that Tatsuoka (1988, pp. 351–358) uses with his "minimum chi-squared" rules.

13.5 SUMMARY OF CLASSIFICATION STATISTICS

The discussion in Sections 13.3 and 13.4 and a cursory reading of the many textbooks and articles dealing with classification methods and rules might lead one to conclude that there are several different rules. This is not so when restricting the discussion to multivariate normal-based rules. The general statistic is that involving the estimated posterior probability $\hat{P}(j|\mathbf{x})$ in (13.9) or, equivalently, that involving the quadratic classification function (QCF) in (13.18) or, still equivalently, that involving the generalized distance in (13.25).

Special cases of the general statistic may be obtained when two conditions are considered. *One* condition pertains to the J population covariance matrices. If these matrices are not assumed, or cannot tenably be concluded, to be equal, we have a *quadratic* rule, as stated in (13.10) in terms of $\hat{P}(j|\mathbf{x})$ values, in (13.19) in terms of QCF values, and in (13.26) in terms of distance values; otherwise, the

TABLE 13.2 Classification Statistics

Quadratic

$$\hat{P}(j|\mathbf{x}_u) = \frac{q_j \cdot |\mathbf{S}_j|^{-1/2} \cdot \exp\left(-\frac{1}{2}D_{uj}^2\right)}{\sum_{j'=1}^{J} q_{j'} \cdot |\mathbf{S}_{j'}|^{-1/2} \cdot \exp\left(-\frac{1}{2}D_{uj'}^2\right)} \qquad (13.9)$$

$$Q_{uj} = \ln q_j - \tfrac{1}{2}\ln|\mathbf{S}_j| - \tfrac{1}{2}D_{uj}^2 \qquad (13.18)$$

$$d_{uj} = \ln|\mathbf{S}_j| + D_{uj}^2 - 2\ln q_j \qquad (13.25)$$

Linear

$$\hat{P}(j|\mathbf{x}_u) = \frac{q_j \cdot \exp\left(-\frac{1}{2}D_{uj}^{*2}\right)}{\sum_{j'=1}^{J} q_{j'} \cdot \exp\left(-\frac{1}{2}D_{uj'}^{*2}\right)} \qquad (13.14)$$

$$L_{uj} = [\bar{\mathbf{x}}_j'\mathbf{S}_e^{-1}]\mathbf{x}_u - \tfrac{1}{2}\bar{\mathbf{x}}_j'\mathbf{S}_e^{-1}\bar{\mathbf{x}}_j + \ln q_j \qquad (13.21)$$

$$d_{uj}^* = D_{uj}^{*2} - 2\ln q_j \qquad (13.27)$$

Note: $D_{uj}^2 = (\mathbf{x}_u - \bar{\mathbf{x}}_j)'\mathbf{S}_j^{-1}(\mathbf{x}_u - \bar{\mathbf{x}}_j)$ (13.6)

$D_{uj}^{*2} = (\mathbf{x}_u - \bar{\mathbf{x}}_j)'\mathbf{S}_e^{-1}(\mathbf{x}_u - \bar{\mathbf{x}}_j)$ (13.12)

TABLE 13.3 Alternative Forms of Classification Statistics

	Covariance Matrices	
Prior Probabilities	Unequal (Quadratic Rule)	Equal (Linear Rule)
Unequal	$\dfrac{q_j \cdot \lvert \mathbf{S}_j \rvert^{-1/2} \cdot \exp\left(-\frac{1}{2}D_{uj}^2\right)}{\sum_{j'=1}^{J} q_{j'} \cdot \lvert \mathbf{S}_{j'} \rvert^{-1/2} \cdot \exp\left(-\frac{1}{2}D_{uj'}^2\right)}$	$\dfrac{q_j \cdot \exp\left(-\frac{1}{2}D_{ugj}^{*2}\right)}{\sum_{j'=1}^{J} q_{j'} \cdot \exp\left(-\frac{1}{2}D_{uj'}^{*2}\right)}$
	$\ln q_j - \frac{1}{2}\ln \lvert \mathbf{S}_j \rvert - \frac{1}{2}D_{uj}^2$	$\ln q_j - \frac{1}{2}D_{uj}^{*2}$
	$\ln \lvert \mathbf{S}_j \rvert + D_{uj}^2 - 2\ln q_j$ [a]	$D_{uj}^{*2} - 2\ln q_j$ [a]
Equal	$\dfrac{\lvert \mathbf{S}_j \rvert^{-1/2} \cdot \exp\left(-\frac{1}{2}D_{uj}^2\right)}{\sum_{j'=1}^{J} \lvert \mathbf{S}_{j'} \rvert^{-1/2} \exp\left(-\frac{1}{2}D_{uj'}^2\right)}$	$\dfrac{\exp\left(-\frac{1}{2}D_{uj}^{*2}\right)}{\sum_{j'=1}^{J} \exp\left(-\frac{1}{2}D_{uj'}^{*2}\right)}$
	$-\frac{1}{2}\ln \lvert \mathbf{S}_j \rvert - \frac{1}{2}D_{uj}^2$	$-\frac{1}{2}D_{uj}^{*2}$
	$\ln \lvert \mathbf{S}_j \rvert + D_{uj}^2$ [a]	D_{uj}^{*2} [a]

[a] For classification purposes, the maximum of the J values for all statistics is considered except for these four, where the minimum is considered.

linear rule is employed. The *second* condition pertains to the J prior probabilities. If the priors are taken to be equal, the statistics in (13.29) and (13.30) may be employed.

For ease of reference, the classification statistics are summarized in Table 13.2. The three quadratic statistics yield identical classification results, similarly for the three linear statistics. Another summarization of classification statistics is given in Table 13.3. The three statistics in each "quadrant" yield identical classification results. Note that if all the q terms in the upper part of Table 13.3 are constant (i.e., equal priors), the constant prior may be formally ignored in any form of the classification rule, as found in the lower part of Table 13.3.

13.6 CHOICE OF RULE FORM

13.6.1 Normal-Based Rule

When dealing with continuous predictor variables there are two data conditions that may be of some concern. *One* condition is that of multivariate normality of the observation vectors in each group. Does this condition need to be met to use any of the statistics in Table 13.2 or 13.3? In the sense of optimality, the answer is yes. If yes, we have to rely on a way of "testing" the condition of joint normality. Empirical and graphical techniques for checking multivariate normality are discussed by Fan (1996, p. 169) and Rencher (2002, pp. 92–99).

One thing that may be done is to check the *necessity* of multivariate normality, that is, check for the normality of each variable distribution. (It should be recognized that marginal univariate normality is not *sufficient* for joint normality.) This may be done graphically using stem-and-leaf or normal probability plots that may be obtained via, for example, the SPSS MANOVA program. If one is satisfied with multiple univariate normality, then one may proceed as though the multivariate normality condition is met. Or, one can proceed "blindly," *assuming*, without any checks, that the condition is met. If it can be assumed that the sample observations used are typical of those in the populations, logically there is no problem with "working in the dark," so to speak. If a normal rule is built on nonnormal data and then used on similar nonnormal data, the rule used may not be the most efficient one. However, it may be claimed that "this is what I will get *if* I assume normality."

If one is bothered by working in the dark, there are other classification rules available. Some light will be shed on these rules in Chapter 19.

13.6.2 Covariance Matrix Equality

The *second* condition, often referred to as an *assumption*, is really of no great concern in classification analyses. This condition pertains to the equality of the J population covariance matrices. It might be recalled that when group centroids are compared using multivariate analysis of variance (see Chapter 3), it is assumed that J population covariance matrices are equal. If sample sizes are unequal and covariance matrices are unequal, the reported P values may underestimate or overestimate the actual P value. In the context of predictive discriminant analysis the inequality of the covariance matrices is of no concern simply because we have statistics available to us that take this condition "into consideration" (see Tables 13.2 and 13.3). There is the question, however, of how to decide whether or not the condition is met.

The approach typically recommended is to test the multivariate hypothesis

$$\mathbf{\Sigma}_1 = \mathbf{\Sigma}_2 = \cdots = \mathbf{\Sigma}_J,$$

a generalization of the univariate hypothesis

$$\sigma_1^2 = \sigma_2^2 = \cdots = \sigma_J^2,$$

using an approximate χ^2 (Bartlett) or an F (Box) statistic (see Section 3.3). As was noted earlier, there are two serious problems with either test, however. *First*, as Olson (1974) found in a Monte Carlo study, the basic statistic is quite sensitive to lack of normality; that is, the hypothesis may be rejected because of nonnormality rather than because of unequal covariance matrices. *Second*, either test is quite powerful in that rejection is very often concluded, especially if N/p is "large." So what does one do? If normality is tenable, and a test of covariance homogeneity is desired (or if one is compelled to use a statistical test), use the Bartlett or Box test with $\alpha < .01$. (This α value should be decreased further with a larger N.) Something that is often recommended when the covariance homogeneity condition is in question is to use equal sample sizes and rely on (yet unconvincingly concluded) robustness. This is fine if, in the design of the study, use of equal sample sizes makes sense.

The SPSS DISCRIMINANT and MANOVA programs both use the (Box) F statistic. In addition, MANOVA uses the (Bartlett) chi-squared statistic. [The SAS DISCRIM procedure also uses a chi-squared statistic; however, the latter statistic is slightly different from the chi-squared statistic typically given in textbooks (e.g., Tatsuoka, 1988, p. 99).]

Similar to the univariate situation, the tests for covariance matrix homogeneity are sensitive to the lack of multivariate normality. A simultaneous test for the two conditions has been proposed by Hawkins (1981). This test is reviewed by McLachlan (1992, pp. 169–172), who also discusses separate tests for the two conditions.

There is another data condition that should be considered when assessing covariance matrix heterogeneity. As is known in the univariate context, an outlying observation may drastically affect a variance estimate. Similarly, multivariate outliers may affect estimates of $\mathbf{\Sigma}_j$, $j = 1, 2, \ldots, J$. Hence data sets should be examined for possible outliers *prior* to the homogeneity test. Detection of outliers is discussed in Sections 14.4, 15.6, and 23.3.

Another assessment of the equal covariance matrix condition may be made by examining the (natural) logarithms of the $J + 1$ covariance matrices (for each of the J groups, and for the error matrix). These logarithms are outputted by the SPSS DISCRIMINANT, and SAS DISCRIM programs. What is done, then, is to make an eyeball assessment of the equality of the $J + 1$ logarithms. If they are "in the same ballpark," then one can proceed with the *linear* rule. In addition, one may examine the $J + 1$ matrix *traces*—the sum of the p variances for each of the $J + 1$ matrices. If these $J + 1$ traces are "in the same ballpark," one can proceed with the linear rule. No criteria are advanced for either decision—these are judgment calls for the researcher to make; one may consult with a methodologist or a statistician.

13.6.3 Rule Choice

The choice of linear versus quadratic classification is discussed by McLachlan (1992, pp. 132–137) and Meshbane and Morris (1995). The general conclusion is that if the $n_j : p$ ratios are small, then a linear rule is preferred even with covariance heterogeneity, whereas if the $n_j : p$ ratios are large and heterogeneity is clearly present, then a quadratic rule is preferred. A reason for the preference of a linear rule for small (or even moderate) samples is the potential for greater across-sample stability of results (with or without normality)—see Huberty and Curry (1978), Michaelis (1973), and

Wahl and Kronmal (1977). Very little guidance as to definitions of "small" and "large" is proffered. Greene and Rayens (1989) and Nakanishi and Sato (1985) reach the same conclusion from simulation studies. Based on some asymptotic expansions of linear and quadratic functions, Wakaki (1990) also reached that conclusion. It may be noted that virtually all linear–quadratic comparisons have been studied in the two-group context—see, also, Rubin (1990) and Further Reading of Chapter 22.

13.6.4 Priors

Whether or not to use equal priors (i.e., q values) is, to some extent, a judgment call to be made by the researcher. A prior probability of group membership reflects how likely it is for a case to emanate from the given group. It may be argued, therefore, that the priors be based on the relative sizes of the respective populations. It is the responsibility of the researcher to make this assessment; such an assessment may be based on the associated substantive theory, on previous research or records, on experience, or simply on a guesstimate. In any case, it is suggested that the priors should *not* correspond to the relative sample sizes, unless, of course, a proportional sampling plan was utilized. Note that it may be reasonable to use unequal priors when equal sample sizes are involved.

13.7 COMMENTS

When the multivariate normal model is imposed, the general form of a classification rule stated in Section 12.4 [i.e., (12.9)] may be expressed in any one of three forms: (1) in terms of a posterior probability [see (13.10) or (13.15)], (2) in terms of a variable composite [see (13.19) or (13.22)], or (3) in terms of a multivariate distance [see (13.26) or (13.28)]. The choice of the two alternatives for each form may depend on the tenability of the equal covariance matrix condition, with the second alternative [i.e., (13.19) or (13.22) or (13.28)] being chosen if the condition is tenable. There is complete equivalence of the three forms (13.15), (13.22), and (13.27) of the *linear* rules (i.e., when the covariance homogeneity condition is tenable); similarly, for the three forms of the *quadratic* rule [(13.10), (13.19), (13.26)]—equivalence in the sense of giving identical classification results.

Finally, in considering the covariance matrices and prior probabilities in a classification rule, one might also consider differential *costs of misclassification*. For example, if a unit actually belongs to Population 1, it may be more costly if that unit is assigned to Population 3 than if assigned to Population 2. Or, it may be more costly if a Population 1 unit is assigned to Population 2 than vice versa. The incorporation of misclassification costs into a classification rule may be a bit cumbersome in the $J > 2$ case [see Johnson and Wichern (2002, pp. 612–616)]. In the two-group case, only relative costs need be specified.

A general discussion of how misclassification costs may be incorporated into a prediction rule is as follows. In terms of typicality probabilities we have rule (13.17).

Equivalently, we have

Assign unit u to Population j if

$$\sum_{\substack{j'=1 \\ j' \neq j}}^{J} q_{j'} \cdot \hat{P}(\mathbf{x}_u | j') \tag{13.31}$$

is a minimum.

Rules (13.17) and (13.31) are used when misclassification costs are ignored (i.e., when the costs are equal). Now, let $C(j|j')$ denote the cost of assigning a unit to Group j when, in fact, it belongs to Group j'. Incorporating misclassification costs, the prediction rule becomes

Assign unit u to Population j if

$$\sum_{\substack{j'=1 \\ j' \neq j}}^{J} q_{j'} \cdot \hat{P}(\mathbf{x}_u | j') \cdot C(j|j') \tag{13.32}$$

is a minimum.

For example, for $J = 3$, assign unit u to Population 1 if $[q_2 \hat{P}(\mathbf{x}_u|2) \cdot C(1|2) + q_3 \hat{P}(\mathbf{x}_u|3) \cdot C(1|3)]$ is smaller than $[q_1 \hat{P}(\mathbf{x}_u|1) \cdot C(2|1) + q_3 \hat{P}(\mathbf{x}_u|3) \cdot C(2|3)]$ and $[q_1 \hat{P}(\mathbf{x}_u|1) \cdot C(3|1) + q_2 \hat{P}(\mathbf{x}_u|2) \cdot C(3|2)]$. Note that this rule takes into consideration $C(1|2), C(1|3), C(2|1), C(2|3), C(3|1)$, and $C(3|2)$. Neither of the two computer program packages allows for incorporating misclassification costs directly; for a two-group problem, however, the specified priors may reflect relative costs (see Section 18.2).

The rules discussed in this chapter are those most often used by researchers. The popularity of these rules is undoubtedly due to the fact that the computations for them have been readily accessible through computer program packages. The normal-based rules are applicable with continuous predictor variables and, as discussed in Section 19.3, with some discrete predictors. When there is serious doubt about the multivariate normality of the predictors, some modifications of classification statistics are needed (see Section 19.2).

The rules discussed in this chapter are sometimes referred to as *estimative* rules. This is because the method common to all of the rules emphasizes first obtaining estimates of the unknown parameters and then substituting these estimates into the classification algorithms. The rules are initially developed under the assumption that all parameters are known. An alternative approach—termed the *predictive method*—has been suggested by S. Geisser (1929–2004) and is presented in the first edition of this book (Huberty, 1994c, pp. 66–67).

Technical Notes

1. Consider that the scalar, $|\mathbf{\Sigma}|$, as a definition of *generalized variance*, appears to be universally accepted. If $p = 1$, then clearly $|\mathbf{\Sigma}| = \sigma^2$. There is another generalization, however, that makes some sense; this is the *trace* of $\mathbf{\Sigma}$, $\text{tr}(\mathbf{\Sigma})$, which is the sum of the main diagonal elements of $\mathbf{\Sigma}$. Again, if $p = 1$, $\text{tr}(\mathbf{\Sigma}) = \sigma^2$. That it makes sense to consider the trace *along with* the determinant as an index of generalized variance may be seen by examining the test statistic for H_0: $\mathbf{\Sigma} = \mathbf{I}$, which is a function of $|\mathbf{S}|$ *and* $\text{tr}(\mathbf{S})$, where $\mathbf{S} = \hat{\mathbf{\Sigma}}$ (Morrison, 1990, p. 769). For elaboration, see Huberty (1983).

2. As indicated in (13.18), there is a quadratic expression for each $j = 1, 2, \ldots, J$. Each of the J quadratic expressions involves a set of quadratic weights (i.e., weights for elements of $\mathbf{x}_u^2$), a set of variable cross-product weights, a set of linear weights, and a constant. These J complex quadratic composites are typically of limited interest to the applied researcher. They may, however, be of interest to the researcher who intends to use a quadratic rule (built on a design sample) with new units (see Section 16.9).

3. To conduct a *linear* PDA, the SPSS DISCRIMINANT program uses a "data reduction" approach. The analysis begins with determining $J - 1$ linear discriminant functions (LDFs). (This is like doing a principal component analysis prior to the analysis of interest.) Scores on these LDFs are then used as input for the linear PDA. The J linear classification function (LCF) scores are linear composites of the LDF scores. It turns out that these linear PDA results are the same as those obtained when using LCF scores based on the original predictor variable scores. The bad news is that SPSS uses LDF scores for their quadratic PDA—their quadratic results are incorrect.

Further Reading

Bar-Hen (1996) proposes a test to determine whether or not an analysis unit (whose membership of one of the J a priori populations is unknown) belongs to a new population.

Flury (1995) discusses (with references) common principal component (CPA) discrimination and proportional discrimination with $\neq$ Σ's.

Glover (1990) reviews the linear programming (LP) approach to classification and presents a demonstration of added flexibility of some LP models.

Grouven et al. (1996) develop a "user-friendly" PC program that performs linear and quadratic PDA for two or more groups allowing for the incorporation of misclassification costs.

McLachlan (1992) reviews theoretical aspects of quadratic rules plus a number of other classification rules (in Chapter 3), compares linear and quadratic rules (pp. 132–137), and presents an extensive review of studies that assess the robustness of linear and quadratic classification to violations of normality (pp. 152–161).

Rencher (2002, Chapter 7) provides a detailed discussion of the test of equal Σ's.

Schott (1993) proposes the use of composite scores as input for a quadratic classification analysis, and further suggests a "dimension reduction" in the sense of using fewer than $r = \min(p, J - 1)$ composite scores as input for the QCFs.

Smith (1947) is one of the first to apply the normal-based quadratic rule; he also first proposed the *resubstitution* error rate estimator.

Fact There are three kinds of people in this world—Those who can count and those who can't.

EXERCISES

1. What is the height of a *univariate* standard normal curve corresponding to a z value of 1.20? That is, what is the value of $f(1.20)$ where f is the univariate standard normal density function (with $\mu = 0$ and $\sigma^2 = 1$)? [See (13.1).] What is another z value with an $f(z)$ value the same as $f(1.20)$?

2. Find the ordinate of a $N(2, 16)$ curve for an abscissa of 2; for an abscissa of 6. [Refer to (13.1).]

3. How is a posterior probability (estimate) used in a PDA?

4. Consider the question: Assuming that profiles of high school students remain fairly stable, how well can we "fit" a given student profile with profiles of typical students who opt for one of, say, four postsecondary experiences? Now the priors to be used in this situation would depend to some extent on the locale. Construct a (real?) classification situation (i.e., specify predictors and criterion groups) and defend a set of priors that would be utilized.

5. Specify priors that seem reasonable to you for your personal data set—Exercise 2 in Chapter 1.

CHAPTER 14

Classification Results

14.1 INTRODUCTION

The statistic that is predominantly used in data analysis packages to decide group assignment is the estimated posterior probability of group membership. (An exception is when the single-nearest-neighbor rule is used; see Section 19.2.2.) When multivariate normality of the distribution of predictor vectors is assumed, one can equivalently use the value of a classification function (linear or quadratic) for assignment purposes. Whether a probability value or function value is used, a value for each analysis unit is found for each population category. The J function values for a unit are of little use other than for the purpose of deciding into which population the unit will be classified. Posterior probability estimates, however, yield information about each unit in addition to that of indicating predicted population membership. This added information pertains to characterizations of the individual analysis units and will be discussed in Section 14.4. Information pertaining to numbers or proportions in *groups* of units correctly classified into the population categories is presented in Section 14.5. Prior to these two discussions, listings of SPSS and SAS commands used in analyzing a real data set are given.

14.2 RESEARCH CONTEXT

Illustrations of PDA results given in this chapter (and some succeeding chapters) are based on a real data set that involves a sample of community college students. For the exemplary PDAs, we are using Grade as the grouping variable, with three levels: A ($n_1 = 66$), B ($n_2 = 122$), and C or C− ($n_3 = 76$). Thus, we have $N = 264$. [This is the 3-group Ethington data set (3GED).] We are using respective priors of .25, .50, and .25 (guesstimates!).

Applied MANOVA and Discriminant Analysis, Second Edition, by Carl J. Huberty and Stephen Olejnik
Copyright © 2006 John Wiley & Sons, Inc.

14.3 COMPUTER APPLICATION

In this section we provide the SPSS and SAS syntax for determining classification functions. The prior probabilities are based on researcher experiences with similar students.

SPSS SYNTAX FOR A LINEAR PDA USING 3 GRADE GROUPS AND 9 PREDICTOR VARIABLES

```
discriminant
/group = grade(1,3)
/variables = counsum to qesci
/priors .25 .50  .25
/statistics all
/plot = cases
/classify = pooled.
```

discriminant is the SPSS system command for discriminant analysis.
/groups = grades(1,3) identifies the classification variable with the lower and upper value limits for the classification variable.
/variables = counsum to qesci identifies the nine predictor variables. If variables are consecutive, the "to" function can eliminate the listing of individual variables.
/priors = .25 .50 .25 if base rates are known, they may be specified with this statement (see Section 12.4.4). Alternatively, **= equal** or **= size** may be requested if equal base rates, or base rates proportional to the current group sizes, are desired.
/statistics = all requests a variety of statistics including: means, standard deviations, Box test, classification function coefficients, and classification results.
/plot = cases requests individual posterior and typicality probabilities as well as Mahalanobis D^2 statistics for the two largest posterior probabilities.
/classify = pooled requests that the linear rule be used.

SAS SYNTAX FOR A LINEAR PDA USING 3 GRADE GROUPS AND 9 PREDICTOR VARIABLES

```
proc discrim pool=test crosslist posterr;
class grade;
var counsum - - qesci;
priors '1'=.25 '2'=.50 '3'=.25;
run;
```

proc discrim is the SAS system command to request the discriminant analysis procedure.
pool=test requests the χ^2 test for covariance matrix equality. By default, if the results are not statistically significant at the .10 level, the linear rule is applied. If the results are significant at the .10 level, the quadratic rule is applied. If a linear rule is known to be desired, *test* should be changed to *yes*. If a quadratic rule is known to be desired, *test* should be changed to *no*.
crosslist requests the individual classification results on a cross-validation sample obtained using the leave-one-out method (see Section 15.3.3).
posterr requests the posterior probabilities be reported for all J groups.
class grade specifies the classification variable.
consum - - qesci lists the predictor variables. When variables are consecutively listed, the - - function can be used rather than listing all individual predictors.
priors '1'=.25 '2'=.50 '3'=.25 lists the known group base rates.

The SPSS DISCRIMINANT program uses a classification based on statistics identical in form to (13.4) and (13.9), but the "score" input is not raw data on the p predictors. Rather, input for each analysis unit consists of a set of scores on linear composites (actually, LDFs) of the p predictor scores—see Huberty (1984). It should be noted that the results from SPSS quadratic classification are *not* the same as from SAS, which uses raw data input to (13.9). It turns out that the SPSS "quadratic" results are *not* correct.

It should also be noted that SAS DISCRIM does *not* output the typicality probability estimates.

14.4 INDIVIDUAL UNIT RESULTS

The SAS package outputs the J normal-based posterior probability estimates [i.e., $\hat{P}(j|\mathbf{x})$ values] for each unit; SPSS DISCRIMINANT yields only the two largest $\hat{P}(j|\mathbf{x})$ values. As discussed in Section 15.3, such calculation of $\hat{P}(j|\mathbf{x})$ values is related directly to an *internal* predictive discriminant analysis. These so-called internal $\hat{P}(j|\mathbf{x})$ values are those referred to above as part of the output of SAS DISCRIM (with some options) and SPSS DISCRIMINANT. For an *external* analysis, on the other hand, the $\bar{\mathbf{x}}_j$ and $\mathbf{S}_j$ values used in the squared distances in (13.6) would be based on only some of the available unit score vectors. This would determine $\hat{P}(j|\mathbf{x})$ values different from the internal values. Such external $\hat{P}(j|\mathbf{x})$ values are obtainable via SAS DISCRIM with options CROSSVALIDATE and CROSSLIST.

The output of SAS DISCRIM may be abbreviated by listing the posterior probabilities only for the misclassified units. This is accomplished by using LISTERR (for an internal analysis) or CROSSLISTERR (for an external analysis).

Results obtained on individual units depend not only on whether an internal or external analysis is conducted, but on whether a linear or quadratic classification rule

TABLE 14.1 Some Unit Classification Results for the 3-Group Ethington Data

Student	Actual Group	Predicted Group (j)	Typicality Probability $\hat{P}(\mathbf{x}\|j)$	Posterior Probablities $\hat{P}(1\|\mathbf{x})$	$\hat{P}(2\|\mathbf{x})$	$\hat{P}(3\|\mathbf{x})$
35	1	1	.350	.605	.360	.035
43	1	2	.005	.114	.811	.075
57	1	2	.531	.089	.595	.316
96	2	1	.336	.676	.260	.063
105	2	2	.321	.092	.710	.198
153	3	2	.277	.097	.455	.448

is employed. A small portion of linear external SPSS DISCRIMINANT output is given in Table 14.1. These are linear leave-one-out (L-O-O discussed in Chapter 15) classification results. It should be noted that the SPSS output provides only two estimated posterior probability values; but, with $J = 3$ groups, the third value is easily obtained because the sum of the J values is 1.00. That is, we assume that every student belongs to one (and only one) of the criterion populations; thus, the sum of the J $\hat{P}(j|\mathbf{x})$ values is unity.

By examining the J $\hat{P}(j|\mathbf{x})$ values, we can assess, in a probabilistic sense, the closeness of each unit to the centroid of each of the J groups. For example, referring to Table 14.1, we see that the 9-element score vector for Student 35 from Group 1 (grade of A) is closest to the centroid for Group 1; the posterior probability estimate of $\hat{P}(1|\mathbf{x}_{35}) \doteq .605$ is clearly the largest of the three. This would be considered a "hit." That is, a CCSEQ student with a score vector like that of Student 35 (in Group 1) would be predicted to earn an A grade. Another "hit" would be Student 105. On the other hand, a student with a score vector like that of Student 57, who was grouped with the A students, would be predicted to earn a B. This is clear from the $\hat{P}(2|\mathbf{x}_{57})$ value of .595. This would be considered an "error," as would Student 96. Other entries in Table 14.1 will be subsequently discussed.

It is very rare in PDA applications to see the reporting of all unit results. For an exception, see Seshia et al. (1983) who report the largest posterior probability for each of 104 analysis units.

14.4.1 In-Doubt Units

An in-doubt unit (or, a "fence rider") is one with approximately—usually two—equal $\hat{P}(j|\mathbf{x})$ values. That is, if two such values are "close," then it can be concluded that the unit's score vector is about the same distance from centroids of two groups. Such fence riders and associated score vectors could be studied for peculiarities. The profiles (i.e., score vectors) of fence riders might reveal some interesting reasons why they resemble the typical member of one group—reflected by the group centroid—about as much as the typical member of another group.

An example of an in-doubt unit is given in Table 14.1. It may be obvious that for Student 153 (a C or C− student), the posterior probability estimates, $\hat{P}(2|\mathbf{x}_{153}) \doteq .455$

and $\hat{P}(3|\mathbf{x}_{153}) \doteq .448$ are "close." Thus, the grade prediction for Student 153 is not clear, a C or C− or a B?

An examination of in-doubt units may very well reveal some interesting information—for separate groups as well as for the total group. Predictor score profiles for collections of in-doubt units may provide descriptions of units for which group identification is not clear. In a real, practical group prediction situation, such unit descriptions may be informative—in addition to the focus on hit rates typically considered in a PDA study by practicing researchers.

If a large number of in-doubt units resulted, this might suggest the possibility of the actual existence of an additional group, one in-between two of the original groups.

Based on the obtained $\hat{P}(j|\mathbf{x})$ values, it might be decided that some units should not be assigned to any of the J criterion populations being studied. The basis for such a decision may be a minimum, or *threshold*, $\hat{P}(j|\mathbf{x})$ value. The use of a threshold value in the estimation of hit rates is discussed in Section 15.6.

14.4.2 Outliers

An *outlier* is a unit that may not belong to any group; that is, a unit that is not close to the typical member of any group. The typicalness of the observed score vector for unit u, $\mathbf{x}_u$, belonging to Group j is reflected by the proportion of units in Group j that have vectors close to $\mathbf{x}_u$; this proportion is denoted as $\hat{P}(\mathbf{x}_u|j)$ and is called a *typicality probability*. Rather than dealing directly with these as "probabilities," let us view typicalness in terms of distance. Now, assuming common population covariance matrices, the squared distance between points representing $\mathbf{x}_u$ and the centroid of Group j, $\overline{\mathbf{x}}_j$, is

$$D_{uj}^{*2} = (\mathbf{x}_u - \overline{\mathbf{x}}_j)'\mathbf{S}_e^{-1}(\mathbf{x}_u - \overline{\mathbf{x}}_j), \tag{14.1}$$

a repeat of expression (13.12).

A difficulty with using such a distance index is in assessing the largeness of a D^{*2} value. It may be more attractive to use a probability associated with the distance index. This would be well and good if even the approximate distribution of D_{ug}^{*2} was known.[1] The distribution of another squared distance is, however, known and is considered by SPSS. This is the distance from the point representing a unit and a point representing a group centroid, both *in the space defined by the linear discriminant functions*; these LDFs are discussed in detail in Section 5.2. A squared distance between a unit and a centroid in the LDF space has a chi-squared distribution with df $= r =$ number of LDFs. In a one-factor layout as we have here, df $= J - 1$. The probability in which we are interested for a given unit is the tail area of a $\chi^2(J-1)$ distribution. Such a tail area may be interpreted as the probability that any member of Group j would yield a squared distance that is equal to, or greater than, the squared distance that unit

[1] Some writers refer D_{uj}^{*2} to $\chi^2(p)$, while Afifi and Azen (1979, p. 523) and Morrison (1990, p. 549) suggest two different F transformations of D_{uj}^{*2}; McLachlan (1992, pp. 181–182) discusses an F transformation of D_{uj}^{*2} and an F transformation of D_{uj}^{2} that may be used in outlier detection.

u is from the centroid of the group to which it is assigned. (See the Technical Note at the end of this chapter.)

These tail areas, which are labeled *typicality probabilities* for each unit, are given by the SPSS DISCRIMINANT procedure (but, *not* by SAS DISCRIM). As can be seen in Table 14.1, there is one $\hat{P}(\mathbf{x}_u|j)$ value for each student. This value may reveal a potential outlier. For example, check Student 43 results. The three $\hat{P}(j|\mathbf{x}_{43})$ values are .114, .811, and .075. Thus, it may be concluded that the 9-element vector of CCSEQ scores for Student 43 is closest to the CCSEQ centroid for Group 3; thus, a student with a vector of CCSEQ scores similar to those of Student 43 would be (correctly?) predicted to belong to Group 2—a "hit." It may be noted, however, that $\hat{P}(\mathbf{x}_{43}|2) \doteq .005$. Thus, it may be concluded that a student with a score vector like that of Student 43 is not very close to the centroid of Group 2. That is, Student 43 may be considered an outlier.

A great number of units similarly identified *may* indicate that there actually is a $(J+1)$st population, one that is "outside" the original groups. Reaching such a conclusion would depend on the configuration of the group centroids in a p-dimensional space, a discussion of which was given in Section 5.5. How potential outliers might be dealt with in terms of estimating classification accuracy is discussed in Section 15.6. Outliers are discussed in more detail in Section 23.3.

14.5 GROUP RESULTS

Results of a classification analysis are often reported in the form of a *classification table*, as in Table 14.2. (Such a table is sometimes labeled a *confusion matrix*.) The entry $n_{jj'}$ in the (j, j') cell is the number of units in Group j that are assigned to (or predicted to be in) Group j'. A *hit* results when a unit emanating from Group j is assigned (by means of the classification rule used) to Group j. The *hit rate* for Group j is given by n_{jj}/n_j and the total-group hit rate is $\Sigma n_{jj}/N$.

An example of an actual classification table, based on the 3-group Ethington data set (3GED), is given in Table 14.3. These results are based on the use of statistic (13.14) with $q_1 = q_3 = .25$ and $q_2 = .50$. The separate-group hit rates are indicated in parentheses on the main diagonal. The total-group hit rate is $(18 + 103 + 15)/264 \doteq .515$. More details on how such a table is determined and interpreted are given in Chapter 15.

TABLE 14.2 Classification Table for J = 3

		Predicted Group			
		1	2	3	Total
Actual Group	1	n_{11}	n_{12}	n_{13}	n_1
	2	n_{21}	n_{22}	n_{23}	n_2
	3	n_{31}	n_{32}	n_{33}	n_3
Total		$n_{.1}$	$n_{.2}$	$n_{.3}$	N

TABLE 14.3 Classification Table for the 3-Group Ethington Data

		Predicted Group			
		1	2	3	n_j
Actual Group	1	18 (.237)	53	5	76
	2	9	103 (.844)	10	122
	3	6	45	15 (.227)	66
Total		33	201	30	$N = 264$

14.6 COMMENTS

Examination of analysis unit probabilities may be informative for a reason in addition to identifying in-doubt units and outlying units. Suppose that a unit in the *training* sample is misclassified on the basis of a "large" posterior probability associated with a group other than the group in which the unit was originally located. For example, consider a Group 2 unit assigned to Group 3 with, say, $\hat{P}(3|\mathbf{x}) \doteq .938$. Such a posterior probability may suggest a questionable initial group identification for the unit. As an example of such a result based on the 3-group Ethington data group predictions (see Table 14.1), with Student 43 from Group 1 (grade of A) we get $\hat{P}(2|\mathbf{x}_{43}) \doteq .811$, a "miss." Might Student 43 have been "overevaluated"?

Finally, PDA results may be helpful in the context of a cluster analysis (CA) study. An initial decision made in CA pertains to determining the number of clusters. A PDA *may* help in making this decision. Suppose the number of clusters initially considered are 4, 5, and 6. Then what may be done is to conduct three PDAs using the set of p response variables as predictors and $J = 4$, $J = 5$, and $J = 6$. The number of clusters to retain for further analysis would be that J value that yields the "best" hit rate(s)—a researcher judgment call. These analyses may, however, be troublesome because of unknown priors. PDA may be more helpful as a posttypology analysis. Once the final cluster analysis is completed, a classification rule may be determined to use with new analysis units having the same p variable measures. (Again, special efforts need to be made in estimating the set of J priors.) This latter analysis was advocated by Huberty et al. (2005) and Huberty and Lowman (1997).

Technical Note

Suppose that a classification rule is built on a set of data, and then the task is to classify a new analysis unit. In such a situation it may be of interest to determine whether or not the new unit may be designated an outlier. Such a designation may be accomplished as follows. Let

$r = \min(p, \text{df}_h)$ denote the number of linear discriminant functions (LDFs) (i.e., the dimension of the LDF space)

$\mathbf{S}_e = p \times p$ error covariance matrix (in the p-variable space)

$\mathbf{V} = p \times r$ matrix of the r sets of LDF weights

Then $\mathbf{V}'\mathbf{S}_e\mathbf{V}$ is the $r \times r$ error covariance matrix based on the r LDF scores for each unit.

If $\mathbf{C} = \mathbf{V}'\mathbf{S}_e\mathbf{V}$, the squared distance from a point representing a unit's vector of r LDF scores to a point representing a vector of LDF means for Group j is

$$D_{uj}^2 = (\mathbf{f}_u - \bar{\mathbf{f}}_j)'\mathbf{C}^{-1}(\mathbf{f}_u - \bar{\mathbf{f}}_j),$$

where $\mathbf{f}_u$ is an $r \times 1$ vector of LDF scores for unit u, and $\bar{\mathbf{f}}_j$ is the centroid of LDF means for Group j. A D_{uj}^2 value may be referred to the $\chi^2(r)$ distribution to obtain the tail area associated with unit u. This tail area is what is referred to as the *typicality probability*, $\hat{P}(\mathbf{x}_u|j)$. The $\hat{P}(\mathbf{x}_u|j)$ value is the estimated probability of having a score vector near $\mathbf{x}_u$ or one more extreme.

Definition ***Minimax***: Dress for ambivalent women.

EXERCISES

1. Consider a multivariate *three*-group classification (i.e., PDA) context. For unit 8 suppose that $\hat{P}(1|\mathbf{x}_8) = .30$, $\hat{P}(2|\mathbf{x}_8) = .65$, $\hat{P}(\mathbf{x}_8|1) = .50$, and $\hat{P}(\mathbf{x}_8|2) = .25$.
   (a) Into which group should unit 8 be assigned?
   (b) What do the numbers .30 and .65 represent? Explain.
   (c) What do the numbers .50 and .25 represent? Explain.
   (d) What is the value of $\hat{P}(3|\mathbf{x}_8)$?
   (e) What is the value of $\hat{P}(\mathbf{x}_8|3)$?

2. Why is it recommended that PDA results for individual analysis units be examined in addition to the separate-group and/or total-group hit rates?

3. **(a)** Explain why an internal classification analysis would yield biased results.
   (b) Relate this bias to that in a multiple regression analysis.

Computer Applications

4. Conduct a linear PDA (using SAS DISCRIM and SPSS DISCRIMINANT) on the 3-group Ethington data set (3GED, see Appendix A)—use the commands given in Section 14.3.

(a) What is the reported value of $\hat{P}(3|\mathbf{x}_{24})$? Of $\hat{P}(\mathbf{x}_{24}|3)$?

(b) Identify two fence riders, using the criterion of the two $\hat{P}(j|\mathbf{x})$ values being within .02 of each other.

(c) Identify two outliers (via SPSS), using the criterion of the $\hat{P}(\mathbf{x}|j)$ values being less than .05. [Note: the .02 and .05 criteria are not to be universially used.]

(d) Locate the classification table for this analysis. What percent of the Group 2 units were correctly (re)classified? Percent of the Group 3 units? What is the percent of hits over all three groups? (How is this percent determined?)

5. Conduct a PDA on your personal data set using the statistical package of choice; delete any categorical predictors. Examine the results for (a) outliers, (b) in-doubt cases, (c) separate group hit rates, and (d) total group hit rate. Finally, express your confidence of the hit rate estimates in terms of accuracy and precision.

CHAPTER 15

Hit Rate Estimation

15.1 INTRODUCTION

The reader who has studied multiple correlation analysis (MCA) might recall that the sample multiple correlation coefficient, R, is a *biased* estimator for its population counterpart, ρ. In fact, R is positively biased; that is, on the average, it overestimates the true degree of relationship—$E(R) > \rho$.[1] The positive bias is due, in part, to the fact that the linear composite of the X variables is determined in such a manner that the sample correlation, R, between the linear composite scores and the Y-variable scores is maximized. That is, the sample data used to determine the linear composite are *also* used to assess the relationship between the composite and the Y variable.

The same estimation problem exists in multiple regression analysis (MRA). Here the multiple correlation coefficient, R, is used as an index of accuracy of prediction, that is, as an estimator for the true validity coefficient, ρ_v. The parameter ρ_v is obtained by correlating, in the population, the criterion (Y) scores with a specific sample linear composite of the predictors (X variables). Thus, there is a ρ_v value corresponding to each sample. (This is opposed to the single ρ value, which is the correlation between the population Y scores and the linear composite of the X scores determined *in the population*; in other words, ρ is the multiple correlation coefficient based on known parameter information.) The parameter ρ_v might be described as a *conditional multiple correlation coefficient*; that is, it is the true correlation conditioned on a particular sample linear composite. Similar to the multiple correlation problem, in multiple regression it is true that $E(R) > \rho_v$.

So, if for no reason other than bias, it would seem reasonable to use an estimator other than R when estimating ρ or ρ_v. Much work has involved the derivation and study of transformations of R, or of R^2, to arrive at less biased estimators for

[1] $E(R)$ denotes the "expected value of R." Suppose that all possible samples (of a fixed size) are selected and that for each sample, an R value is calculated; the mean of the collection of R values is the expected value of R. $E(R)$ is sometimes termed the *long-run average* R. If the statistic R is used as an estimator for a parameter θ and $E(R) = \theta$, it is said that "R is an unbiased estimator for θ."

Applied MANOVA and Discriminant Analysis, Second Edition, by Carl J. Huberty and Stephen Olejnik
Copyright © 2006 John Wiley & Sons, Inc.

ρ and for ρ_v. These transformations yield adjusted or "shrunken" R^2 values. See Huberty (2003) and Huberty and Hussein (2001) for discussions of some adjusted R^2 statistics.

There is a similar estimation problem in predictive discriminant analysis (PDA) when it comes to assessing degree of classification accuracy. Making this assessment amounts to estimating a true hit rate. The estimation process involves seeking answers to three questions: (1) How accurately can a classification rule based on population information be expected to perform? (2) How accurately can a rule based on a given sample be expected to classify analysis units in future samples? (3) How accurately can a rule based on any sample of a fixed size be expected to classify units in future samples?

15.2 TRUE HIT RATES

There are essentially three probabilities of correct classification (i.e., three population hit rates) that may be considered for estimation purposes. The three true hit rates to be discussed correspond directly to the three questions posed above. Each of these hit rates is viewed initially as the proportion of correct classifications across *all* subpopulations. The *first* is the *optimal* hit rate, denoted here by $P^{(o)}$. This is the hit rate obtained when a classification rule based on known parameters (i.e., the J subpopulation mean vectors and the common covariance matrix) is applied to the population. The *second* true hit rate is the *actual* hit rate (sometimes called the conditional hit rate), denoted here by $P^{(a)}$. This is the hit rate obtained by applying a rule based on a particular (i.e., training) sample to future samples (taken from the same population, of course). That is, $P^{(a)}$ may be thought of as the expected proportion of correct classifications over future samples yielded by a rule based on statistics from a particular sample. This hit rate would be of interest after the training sample has been determined. The *third* true hit rate considered is the *expected actual* hit rate (sometimes called the unconditional hit rate), denoted here by $P^{(e)}$. This is the expected proportion of correct classifications over all possible samples (of size $N = \Sigma\, n_j$, where n_j is the size of Group j). Note that $P^{(e)} = E(P^{(a)})$. This hit rate would be of interest before any sample is determined.

These three hit rates may be related to some parameters in the context of multiple correlation/regression. The true multiple correlation coefficient, ρ, is analogous to $P^{(o)}$. The true validity coefficient for a training sample (i.e., the mean result of applying the training sample prediction equation to repeated, or future, samples), ρ_v, is analogous to $P^{(a)}$. Herzberg 1969 terms ρ_v the "population cross validity," while Browne (1975) considers ρ_v^2 a "measure of predictive precision." When considering repeated training samples, ρ_v is a random variable, and one might consider the corresponding parameter to be $E(\rho_v)$. This is analogous to $P^{(e)}$ because $P^{(e)} = E(P^{(a)})$. It seems reasonable that $E(\rho_v) \doteq \rho$; similarly, Sorum (1972a) considers the estimation of $P^{(o)}$ and $P^{(e)}$ as a single problem, "or at least as two closely related problems" (p. 315). The relationship between indices of interest in PDA and MRA/MCA is summarized in Table 15.1.

TABLE 15.1 PDA and MCA/MRA Indices

Condition	PDA True Hit Rate	MCA/MRA Coefficient
All parameters known	$P^{(o)}$ (optimal)	ρ
Sample-based rule applied to population	$P^{(a)}$ (actual)	ρ_v

15.3 HIT-RATE ESTIMATORS

The literature on estimation in predictive discriminant analysis varies somewhat with respect to definitions, terms, and notation. The variety from Hills (1966) to Glick (1978) and Dillon (1979) through Toussaint (1974) is staggering. In many empirical comparisons of classification estimators, no distinction among some parameters is made, a reminder of some writings in multiple regression/correlation. There is even some disagreement on the parameter–estimator matchups (see, e.g., Fatti et al., 1982; Glick, 1978; Sorum, 1972b).

The current attempt is made to present generally acceptable definitions while keeping terms and notation as straightforward as reasonable. Many writers focus on classification error rates or misclassification, as opposed to classification hit rates or correct classification. The latter view has been adopted in this book.

The methods of hit-rate estimation discussed here are all based on the condition of equal misclassification costs; for example, the consequences of classifying a Group 1 analysis unit into Group 2 are no more or less serious than when classifying the unit into Group 3. (See Section 13.17 for a discussion of the incorporation of misclassification costs.)

15.3.1 Formula Estimators

Estimating $P^{(o)}$ For $J = 2$ with assumed multivariate normality of the predictors and known population parameters (i.e., mean vectors and a common covariance matrix), the two-population *optimal* hit rates are

$$P_1^{(o)} = 1 - \phi\left(\frac{\Gamma - \Delta^2/2}{\Delta}\right), \qquad P_2^{(o)} = 1 - \phi\left(\frac{-\Gamma - \Delta^2/2}{\Delta}\right),$$

where ϕ is the standard normal distribution function, that is, $\phi(z) = \text{Prob}(Z \leq z)$; $\Gamma = \ln(\pi_2/\pi_1)$, π_j is the prior probability of membership in Population j; and Δ^2 is the population Mahalanobis distance index (Afifi and Azen, 1979, p. 293). If $\pi_1 = \pi_2 = \frac{1}{2}$, then $\ln(\pi_2/\pi_1) = 0$ and

$$P^{(o)} = P_1^{(o)} = P_2^{(o)} = 1 - \phi(-\Delta/2) = \phi(\Delta/2).$$

Sample values for Γ and Δ may be used to obtain "plug-in" hit-rate estimates. Researcher estimates are used to replace π_j; let $q_j = \hat{\pi}_j$. If D^2, the sample counterpart of Δ^2, is used to replace Δ^2, the estimator obtained for $P^{(o)}$ is biased optimistically. This is the so-called *D method*, generally recognized as a poor method for estimating $P^{(o)}$. Various adjustments of D^2 have been proposed as plug-in values for Δ^2 (see Sorum, 1972b). A nearly unbiased estimator for Δ^2 is

$$\tilde{D}^2 = \frac{N-p-3}{N-2}D^2 - \frac{pN}{n_1 n_2}, \tag{15.1}$$

where p is the number of predictors. Thus, formula estimators for $P_j^{(o)}$, $j = 1, 2$, are

$$\hat{P}_1^{(o)} = 1 - \phi\left(\frac{K - \tilde{D}^2/2}{\tilde{D}}\right), \qquad \hat{P}_2^{(o)} = 1 - \phi\left(\frac{-K - \tilde{D}^2/2}{\tilde{D}}\right), \tag{15.2}$$

where $K = \hat{\Gamma} = \ln(q_2/q_1)$. An estimator for the total population optimal hit rate $P^{(o)}$ is

$$\hat{P}^{(o)} = q_1\hat{P}_1^{(o)} + q_2\hat{P}_2^{(o)}. \tag{15.3}$$

If $q_1 = q_2$, then

$$\hat{P}^{(o)} = \hat{P}_1^{(o)} = \hat{P}_2^{(o)} = 1 - \phi(-\tilde{D}/2) = \phi(\tilde{D}/2).$$

For more than two populations, no known formula has been proposed for estimating $P^{(o)}$.

The estimation of $P^{(o)}$ in predictive discriminant analysis is analogous to the estimation of ρ (or ρ^2) in multiple correlation analysis. One way to estimate ρ^2 is to use an adjusted R^2 estimator such as that proposed by R. J. Wherry, Sr. (1904–1981). Similarly, an adjusted D^2 estimator, $\tilde{D}^2$, is used to estimate $P^{(o)}$. Only from SAS procedures DISCRIM and CANDISC is it possible explicitly to get values of D^2 (or D) from the packages. Even with DISCRIM, one has to do some additional calculations if unequal priors are used. With unequal priors the squared distances yielded by DISCRIM are of the form

$$D_{jj'}^2 - 2\ln q_{j'}. \tag{15.4}$$

Estimating $P^{(a)}$ The use of a formula for estimating the *actual* hit rate, $P^{(a)}$, has also been restricted to the two-group multivariate normal case. McLachlan (1992, p. 368) derived a formula that allows for estimates of $P_1^{(a)}$ and $P_2^{(a)}$. The McLachlan estimator for the jth subpopulation actual hit rate is $\hat{P}_j^{(a)} = 1 - Q_j$, with

$$Q_j = \phi\left(-\frac{D}{2}\right) + f\left(-\frac{D}{2}\right)\left\{\frac{p-1}{Dn_j} + \frac{D}{32m}[4(4p-1) - D^2] + \frac{(p-1)(p-2)}{4Dn_j^2} + \frac{p-1}{64mn_j}\left[-D^3 + 8D(2p+1) + \frac{16}{D}\right] + \frac{D}{12,288m^2} \times [3D^6 - 4D^4(24p+7) + 16D^2(48p^2 - 48p - 53) + 192(-8p+15)]\right\}, \tag{15.5}$$

where f is the standard normal density function, that is, $(z, f(z))$ is a point on the standard normal curve; and $m = n_1 + n_2 - 2$. A number of analyses conducted indicated that for most practical use, the last term in the multiplier of $f(-D/2)$ may be ignored. This leaves a more workable *shrinkage formula*:

$$\hat{P}_j^{(a)} = 1 - \phi\left(-\frac{D}{2}\right) - f\left(-\frac{D}{2}\right)\left\{\frac{p-1}{Dn_j} + \frac{D}{32m}[4(4p-1) - D^2] + \frac{(p-1)(p-2)}{4Dn_j^2} + \frac{p-1}{64mn_j}\left[-D^3 + 8D(2p+1) + \frac{16}{D}\right]\right\}. \tag{15.6}$$

It may be noted that the amount of shrinkage is inversely related to values of $N(=m+2)$ and D, and directly related to the value of p, the number of predictors.

To arrive at an estimate of the actual total population hit rate $P^{(a)}$ for equal costs of misclassification, one may use $\hat{P}^{(a)} = q_1\hat{P}_1^{(a)} + q_2\hat{P}_2^{(a)}$. No known formula has been proposed for estimating $P^{(a)}$ in the multiple-group case.

The estimation of $P^{(a)}$ in PDA is analogous to the estimation of ρ_v (or ρ_v^2) in multiple regression analysis. One way to estimate ρ_v^2 is to use a "shrunken R^2" estimator such as that proposed by F. M. Lord (1912–2000) (see Huberty and Mourad, 1980). Similarly, a "shrunken-D^2" estimator may be used to estimate $P^{(a)}$ (Dorans, 1988).

15.3.2 Internal Analysis

Classification results often obtained using computer programs are calculated as follows. The user specifies either a linear or a quadratic rule and the priors to be employed (see Table 13.2). Suppose, for the moment, that a rule of the form of (13.22), involving a linear classification function (LCF), is to be used. The available sample data are used to determine the J vectors of weights, $\mathbf{b}'_j = \bar{\mathbf{x}}'_j S_e^{-1}$, and the J constants, $c_j = -\frac{1}{2}\bar{\mathbf{x}}'_j S_e^{-1}\bar{\mathbf{x}}_j + \ln q_j$. Then these *same* data are used to determine, for each analysis unit, the J LCF values and then to (re)classify the N units into the

J populations. Thus the sampled units are classified using parameter estimates based on the study samples.[2]

Such an analysis may be termed an *internal classification analysis*. Recall from our discussion in Section 13.4 that optimal classification rules were derived so as to minimize the total number, or proportion, of classification errors. An internal analysis can be expected to be negatively biased in the sense that the observed error rate can be expected to be less than the true error rate. Or, in terms of hit rates, an internal analysis can be expected to be positively biased. Results of an internal analysis (see, e.g., Tables 14.2 and 14.3) may thus be misleading—misleading in that the observed hit rates may be spuriously high. The observed rates have been called *apparent* or *resubstitution hit rates*.

The apparent hit rate has been used by many applied researchers as a general hit-rate estimate; that is, no reference is typically made to $P^{(o)}$, $P^{(a)}$, or $P^{(e)}$. No matter which true hit rate is considered, the apparent hit rate yields a positively biased estimate (Frank et al., 1965; Hand, 1997, pp. 121–122; McLachlan, 1977; Michaelis, 1973). Similarly, the sample multiple correlation, R, is a positively biased estimator for ρ_v. Reasoning behind the positive bias of an apparent hit rate is somewhat similar to that for R. As mentioned earlier, one form of the classification rule consists of a set of J linear composites called linear classification functions (LCFs). With known parameters (mean vectors and common covariance matrix), weights for these composites are determined so as to maximize proportions of correct classifications; some optimality is lost when using parameter estimates. The near-optimizing process makes the greatest possible use of any and all idiosyncrasies of the data on hand. That is, as Mosteller and Tukey (1977, p. 769) state, "optimization capitalizes on chance." Thus, from an inference viewpoint, something other than apparent hit rates need to be used as valid hit-rate estimates.[3]

15.3.3 External Analysis

Whereas in an internal analysis the units classified are the units considered in rule formulation, in an *external classification analysis* the classification rule is determined from one set of units and then used to classify another set of units. This approach exemplifies the traditional *cross-validation* idea. Two ways of carrying out an external analysis are considered here—both analyses involve sample splitting and may be accomplished using computer program packages. The results of an external analysis are also typically given in the form of a classification table.

Holdout Method One way of carrying out an external analysis is to use a single splitting of the available sample into two subsamples: (1) a *training* or design sample and (2) a *test* or holdout sample. A classification rule is determined using the training sample data and then applied to the test sample data. This is called the *holdout method*.

[2]Just as with a least-squares regression equation, the LCF "fit" to the data is one that maximizes the predictive power of the rule *for the particular sample being studied*.

[3]Exceptions to this "rule" exist when the N/p ratio is "very large," when internal hit rates are "close" to external hit rates.

A hit-rate estimate is the proportion of the test sample units that are correctly classified (using the rule developed on the training sample).

A holdout analysis may be accomplished using the SPSS DISCRIMINANT procedure or the SAS DISCRIM procedure. With SAS DISCRIM the user must externally generate the units to comprise the training and test samples. Test samples are generated internally with the SPSS program. Holdout classification results from each of the two programs appear in the form of a classification table.

Four problems with the holdout method are pointed out by Lachenbruch and Mickey (1968, pp. 2–3). A basic requirement is that of "large" samples. *One* drawback of this method is that in some applications, large samples are not available. A *second* drawback is that the classification rule that is validated is not the one that would actually be used—the rule that is validated would be based on, say, 75 percent of the total sample, but the rule that should actually be used would be based on the total sample. A *third* drawback is that there are problems connected with the size of the test sample—if it is large, a good assessment of the performance of the classification rule will be obtained, but the rule itself is likely to be poor, whereas if the test sample is small, the rule will be better but its performance will be highly variable. *Finally*, this method is quite uneconomical with real data sets. A larger sample than is necessary to obtain a good classification rule must be selected to obtain hit-rate estimates, and not all of the data available are used in the estimation. With the holdout method one needs to decide what portion of the data to include in the test sample. Statistical theory gives little guidance, and a handy rule of thumb is yet to be established. Only in the two-group case has a suggestion been given. Some asymptotic theory developed by Schaafsma and van Vark (1979, p. 776) suggests that the ratio of the test-sample size to the training-sample size is a function of p, the number of predictors: $[1 + (p - 1)^{1/2}]^{-1}$. This led Schaafsma and van Vark to "the feeling that it might be reasonable to recommend" a test-to-training sample size ratio of $3/10$, "at least if $p > 10$." This, in turn, implies that the training-to-total sample size ratio should be approximately $3/4$. Using a test sample of 25 to 35 percent of the total sample seems reasonable, at least in the $J = 2$ case.

Which true hit rate is being estimated by using the holdout method is somewhat puzzling. In early empirical studies, the holdout method was often included in hit-rate estimation comparisons. A holdout hit rate, however, is not an appropriate estimator for any of $P^{(o)}$, $P^{(a)}$, or $P^{(e)}$. A holdout hit rate is a good estimator for $P^{(a)}$ only when the classification rule is considered to be based on a sample the size of the training sample, not the total original sample. For examples of reported studies in which this cross-validation method was employed, see Chapman et al. (1977), LaRocco et al. (1977), and Ware and Williams (1977).

Leave-One-Out Method A second way of carrying out an external analysis is to use the leave-one-out (L-O-O) method popularized by P. A. Lachenbruch (initially proposed in his 1965 dissertation and then published in 1967). The method involves a two-step process. *First*, one unit is deleted and LCFs are determined on the remaining $N - 1$ units. *Second*, these LCFs are used to classify the deleted unit into one of the J criterion groups. This process is carried out N times, and the proportions of deleted units correctly classified are used as hit-rate estimates. For such classification it may be considered that a training sample of size $N - 1$ and a test sample of size 1 are being used.

Hit-rate estimates based on the L-O-O method may be obtained using the SPSS DISCRIMINANT and SAS DISCRIM programs. With the SAS DISCRIM program, L-O-O (therein termed *cross validation*) results may be obtained by specifying CROSSVALIDATE (or, CROSSLIST) as an option; with SPSS DISCRIMINANT, the command is CROSSVALID.

Just as for a hit-rate estimate obtained using the holdout method, the L-O-O estimator, strictly speaking, is not appropriate for estimating any of $P^{(o)}$, $P^{(a)}$, or $P^{(e)}$. It was originally designed to estimate $P^{(a)}$. However, because one unit is deleted, it is estimating a hit rate conditioned on a rule based on a sample of size $N - 1$ rather than N. Unless N is extremely small, however, a L-O-O estimator yields a reasonable estimate of $P^{(a)}$.

There is some evidence that suggests a possible drawback with the L-O-O estimation method. The results of two simulation studies (Glick, 1978; Hora and Wilcox, 1982) indicate that the L-O-O method may yield hit-rate estimates that have relatively high variability over repeated sampling. This relatively high variability may be due to the reuse of the original data—the N sets of J LCFs are derived from nearly identical samples. Glick (1978, p. 221) goes so far as to conclude that the L-O-O estimator "should be eschewed." Hora and Wilcox (1982) are a bit more cautious about avoiding the use of the L-O-O method. Hand (1997, p. 126) concludes that the "632 jackknife method" is favored over the L-O-O method.

It is of interest to note that the L-O-O notion of estimation has also been used in multiple regression analysis (see Allen, 1971; Gollob, 1967; Huberty and Mourad, 1980). In regression analysis, the L-O-O estimator—termed the PRESS statistic by Allen—is used to estimate ρ_v.

The methods of hit-rate estimation discussed to this point involve the determination of a distance, a classification function score, or a posterior probability value for each group for each unit. Each unit is assigned to a particular group if its distance value for that group is minimum, if the function score for that group is maximum, or if the probability value for that group is maximum relative to the scores or values for the other groups. A hit-rate estimate is obtained by counting the number of such assignments for each group. It is not surprising, then, that such estimators are called *counting estimators*. Such an estimator is the type predominately used by applied researchers. Another type of estimator is the so-called *plug-in* or *formula estimator* (for two-group situations) discussed earlier in this section. A third type of estimator is called a *posterior probability estimator*, a discussion of which is turned to next.

15.3.4 Maximum-Posterior-Probability Method

Most researchers who have considered the idea of cross validation in assessing predictive accuracy[4] have used either the holdout or L-O-O counting methods of estimating

[4]The expression *predictive accuracy* is used in a descriptive, as opposed to an estimative, sense. The same goes for the use of *classification accuracy* in this book, particularly in Chapters 15 to 19.

a true hit rate. There is another method that was proposed in the early 1970s by K. Fukunaga and D. L. Kessell in the engineering literature and discussed by Glick (1978) and Hora and Wilcox (1982) (see, also, Dillon and Goldstein, 1984, pp. 406–409). This method might be termed the *maximum-posterior-probability* (M-P-P) method. (The reason for the use of "maximum" is implied by the definition of an estimator for $P_j^{(a)}$ and is explicit in the definition of an estimator for $P^{(a)}$.) The M-P-P estimator for $P_j^{(a)}$ is simply a "mean" of the estimated posterior probabilities for units from all groups assigned to population j by the classification rule used. The sum of these estimated posteriors is divided by $N \cdot q_j$.

The M-P-P estimator for $P_j^{(a)}$ may be expressed as:

$$\hat{P}_j^{(a)} = \frac{1}{N \cdot q_j} \sum_{j'=1}^{J} \left\{ \sum_{u=1}^{n_{j'}} \left[\begin{matrix} \text{post. prob. for all } \mathbf{x}_u \text{ in} \\ \text{Group } j' \text{ assigned to Group } j \end{matrix} \right] \right\}. \tag{15.7}$$

The total group true hit rate, $P^{(a)}$, can be estimated using

$$\hat{P}^{(a)} = \sum_{j=1}^{J} q_j \hat{P}_j^{(a)} = \frac{1}{N} \sum_{u=1}^{N} \max \left[\begin{matrix} \hat{P}(1|\mathbf{x}_u), \hat{P}(2|\mathbf{x}_u), \ldots, \\ \hat{P}(j|\mathbf{x}_u), \ldots, \hat{P}(J|\mathbf{x}_u) \end{matrix} \right]. \tag{15.8}$$

That is, this estimate of $P^{(a)}$ is calculated from the mean of the maximum estimated posterior probabilities for each unit.

The estimated posterior probabilities, $\hat{P}(j|\mathbf{x}_u)$, in (15.8) may be determined via either an internal analysis or an external analysis. Internal $\hat{P}(j|\mathbf{x}_u)$ values are found by resubstitution and may be obtained from either the SAS DISCRIM procedure or the SPSS DISCRIMINANT procedure. The internal hit-rate estimator will herein be denoted M-P-P/I. External $\hat{P}(j|\mathbf{x}_u)$ values may be obtained via the SAS DISCRIM procedure using the CROSSVALIDATE and CROSSLIST options. The external hit-rate estimator will be denoted M-P-P/L-O-O.

Hora and Wilcox (1982) conclude that if the multivariate normality condition is tenable, the M-P-P/L-O-O method is preferred to the usual L-O-O method, the latter being preferred when that condition is untenable. There is some evidence that M-P-P estimators have reasonably good accuracy (i.e., they have low bias) and decent precision (i.e., they have low sampling variability). Some Monte Carlo results obtained by Glick (1978) indicate that the M-P-P/I estimator has relatively low bias and lower sampling variability than the L-O-O estimator for univariate prediction with two criterion groups. Hora and Wilcox (1982) conclude that the M-P-P/L-O-O estimator has greater accuracy than the M-P-P/I estimator.

In nearly all studies comparing hit/error rate estimators, normal-based classification rules are used. That is, estimators of posterior probabilities of group membership, $\hat{P}(j|\mathbf{x})$, are, as mentioned in Chapter 13, based on multivariate normal distributions. Thus, hit-rate estimators, such as M-P-P estimators, that involve actual $\hat{P}(j|\mathbf{x})$ numerical values may be viewed as "correct" only when multivariate normality is reasonable. The "usual" L-O-O estimators obtained from SAS DISCRIM are *counting estimators*; the actual $\hat{P}(j|\mathbf{x})$ numerical values are considered only in an ordinal sense (for a given unit).

The M-P-P/L-O-O estimator is considered by McLachlan (1992, p. 767) to be a "smoothed" estimator. The M-P-P/I and M-P-P/L-O-O estimators are considered *unstratified* (smoothed) estimators by SAS. Also defined by SAS is an "estimator *stratified* over the group" from which the units emanate:

$$\hat{P}_j^{(a)}\,\text{(stratified)}$$
$$= \frac{1}{q_j} \cdot \sum_{j'=1}^{J} \left\{ \frac{q_{j'}}{n_{j'}} \cdot \sum_{u=1}^{n_{j'}} \left[\begin{array}{c} \text{post. prob. for all } \mathbf{x}_u \text{ in} \\ \text{Group } j' \text{ assigned to Group } j \end{array} \right] \right\}. \tag{15.9}$$

The posterior-probability values would be based on an internal analysis for the M-P-P/I estimator or on a L-O-O analysis for the M-P-P/L-O-O estimator.

Whether one uses stratified or unstratified M-P-P estimates depends on the confidence one has in the prior-probability estimates used. If the priors used are based on considerable knowledge of relative population sizes, the stratified estimates are to be preferred. It may be noted that if the priors used are proportional to the group sizes, the stratified and unstratified estimates will be equal. It should also be noted that the separate group M-P-P *error* rate estimates (stratified or unstratified) may be negative; this is sometimes due to discrepancies between priors used and n_j/N ratios, particularly when hit-rate estimates are "high."

As is well known, estimates of any kind are only as "good" as the samples used. In a PDA context, an argument may be made for proportional sampling, that is, for using group sizes proportional to respective (sub)population sizes. With such sampling, the n_j/N ratios will be good estimates of prior probabilities of group membership. It is conjectured that more precise M-P-P hit-rate estimates are obtained if the priors used are approximately proportional to the actual group sizes.

It may be noted that, as Hora and Wilcox (1982, p. 58) point out, "an interesting and unique feature (of the total-group M-P-P estimator) is that it can be calculated by using units of unknown classification." Data on as yet unclassified units may be used to estimate $P^{(a)}$ just as though group membership were known. Applied researchers should, perhaps, keep this unique feature of the M-P-P estimator in mind. There may very well be research situations where group membership of some units is not clear, yet group-membership prediction is a very reasonable objective. As convenient as this characteristic of the total group M-P-P estimator is, there is a problem with the current (as of the date of this writing) version of SAS DISCRIM. To obtain a stratified estimate according to (15.9), knowledge of group membership is necessary. SAS DISCRIM output will, however, contain a stratified estimate using classification of ungrouped observations in the test data set.

15.4 COMPUTER APPLICATION

Illustrations of the use of the various hit rate estimators are based on analyses of the 3-group Ethington data (3GED). The SAS and SPSS programs listed in Section 14.3

TABLE 15.2 Leave-One-Out Results for the 3-Group Ethington Data

		Predicted Group				
		1	2	3	n_j	q_j
Actual Group	1	16 (.211)	55	5	76	.25
	2	12	99 (.811)	11	122	.50
	3	6	47	13 (.197)	66	.25
Total		34	201	29	264 = N	

yield some results that would be used to estimate hit rates. We will not deal much with internal hit-rate estimates—because of resulting expected positive bias in the obtained hit rates. We will focus on the L-O-O external analysis. (SAS syntax and SPSS syntax for conducting a linear L-O-O PDA are given in Section 14.3.)

Results of the application of the linear L-O-O method on the Ethington data obtained using SAS DISCRIM or SPSS DISCRIMINANT are given in Table 15.2. It is clear that, by using the 9 predictors described in Appendix A, only the Group 2 hit rate is "respectable." That is, based on the available (264 × 9) data set, only the prediction of B students is judged to be reasonable. As mentioned earlier, group prediction results may be obtained using not only a linear external rule, but linear internal, quadratic external, and quadratic internal as well.

A summary of hit-rate estimates for the 3-group Ethington data is given in Table 15.3. The linear L-O-O estimates may be obtained from Table 15.2. The quadratic L-O-O and M-P-P/L-O-O results are obtained using SAS DISCRIM with

TABLE 15.3 Hit-Rate Estimates for the 3-Group Ethington Data

	Internal	L-O-O	M-P-P/I[a]	M-P-P/L-O-O[a]
		Linear		
Group 1	.237	.211	.242	.251
Group 2	.844	.811	.832	.829
Group 3	.227	.197	.228	.219
Total Group	.515	.473	.533	.532
		Quadratic		
Group 1	.447	.290	.427	.499
Group 2	.787	.623	.775	.733
Group 3	.470	.288	.573	.540
Total Group	.610	.443	.638	.626

[a] Stratified estimates.

the commands CROSSVALIDATE and POOL = NO. The internal and quadratic results are reported merely for comparison purposes. The internal results would seldom, if ever, be reported. The quadratic results would be reported if the group covariance matrices are *clearly* unequal (which is not the case with the current data set).

The M-P-P/L-O-O estimates may be obtained with simple (but numerous) calculations involving probability estimates yielded by SAS DISCRIM using (15.7), (15.8), and (15.9). For an example, see Huberty (1994c, p. 92). Complements of the (unstratified) hit-rate estimates may be obtained directly via SAS DISCRIM using POOL = YES and POSTERR options. The stratified M-P-P/I estimates may also be obtained by simple (but tedious) arithmetic. Or they may be obtained using SAS DISCRIM with the same options indicated in the preceding sentence.

One might compare the results given in Table 14.3 with those in Table 15.2. The former results are based on linear *internal* classification rule. As might be expected, the three separate-group linear internal hit rates are higher. This does *not* imply that the linear internal rule is better. As discussed in Section 15.3.2, internal hit rates are *positively* biased hit-rate estimators, more so with "small" group sizes.

15.5 CHOICE OF HIT-RATE ESTIMATOR

To this point in the discussion, some 12 varieties of hit-rate estimators have been presented, six linear and six quadratic. The six are: (1) apparent (or resubstitution), (2) L-O-O, (3) M-P-P/I unstratified, (4) M-P-P/I stratified, (5) M-P-P/L-O-O unstratified, and (6) M-P-P/L-O-O stratified. The linear-versus-quadratic issue was discussed in Section 13.6. There appears to be no good reason for using the resubstitution estimator (unless the min n_j/p ratio is very large). Similarly, the M-P-P/I estimators have limited appeal (to some at least). So it may appear that the choice is among the usual L-O-O (linear or quadratic) estimator and the two (stratified or unstratified) M-P-P/L-O-O estimators. That is, choose a linear or quadratic L-O-O rule with either a stratified or unstratified M-P-P/L-O-O estimator. The choice is a difficult one, indeed. Some evidence obtained by Hora and Wilcox (1982) suggests that the linear M-P-P/L-O-O (unstratified) estimator is the one to be favored. The unstratified-versus-stratified issue depends on one's confidence in the priors used. The jury is still out; the case is reviewed briefly in Section 22.2, where a tentative judgment is made.

15.6 OUTLIERS AND IN-DOUBT UNITS

The identification of outlying units and of in-doubt cases (or fence riders) was discussed in Section 14.4. But how might such cases be taken into consideration in the process of estimating hit rates? This question has not been considered to any great extent in the literature by statisticians (see, however, Habbema et al., 1978) or by methodologists (see, however, Huberty and Wisenbaker, 1992a) and therefore not by substantive researchers. (Check some of the Further Reading in this chapter for more details on outliers.) Such a consideration of data diagnostics, quite popular in multiple regression/correlation analysis (see, e.g., Myers, 1990), is recommended in PDA.

15.6.1 Outliers

As mentioned repeatedly in this chapter and emphasized in subsequent chapters, either total-group or multiple separate-group classification accuracy may be of interest. To deal with outliers, then, one might consider doing multiple PDAs, an analysis with the suspected outliers included, and analyses with the suspected outliers discarded. Whether or not an outlier candidate is to be included in estimating the hit rate(s) of interest is a judgment call on the part of the researcher. A question then arises: Are the potential outliers deleted one at a time, two at a time, or all at once? No foolproof strategy is known. Here again, good reasoning and sound judgment will well serve researchers. Of course, what is to be determined is if any or all of the potential outliers have an *influence* on the results.

Parenthetically, it might be tempting to set a typicality probability cutoff at, say, .10. Then a unit having a typicality probability less than .10 would be considered a potential outlier. It may not be wise, however, to use a specific cutoff for all data sets. Would it be reasonable if a given cutoff identified 15 percent of the units as potential outliers? More on outliers is discussed in Section 23.3.

15.6.2 In-Doubt Units

The role of in-doubt analysis units in hit-rate estimation is, perhaps, more crucial. A classification matrix such as that in Table 15.2 has been called a *forced classification matrix* (Habbema et al., 1978). Such a matrix is based on all N units of a study; that is, each unit is assigned to a group, correctly or not. For most data sets, however, there are some units whose group membership is questionable. There are two ways in which group membership may be questionable. *One* is if the unit is a potential outlier; this was discussed above. The *second* is if the posterior probabilities for a unit do not clearly establish membership in one group. The posterior probabilities for two groups may be nearly the same, or the largest probability may not be very "large." These situations suggest that it may be reasonable to establish an "in-doubt" group of units. A way of establishing such a group is to require that the largest posterior probability for any one unit be at least as great as some *threshold* value. How to set a threshold value is, again, a judgment call. For situations involving approximately equally numerous populations, the threshold value should be greater than $1/J$, where J is the number of groups. For unequal priors, a threshold value somewhere near the maximum prior may be reasonable. It may also depend on the extent of the overlap of the groups; little overlap would imply a larger threshold value to use. In medical research (see, e.g., Anderson and Boyle, 1968), where consequences of misclassification may be quite serious, threshold values of at least .90 have been used. A consideration of in-doubt units was made in an interesting application by Heyck and Klecka (1973).

The SAS DISCRIM program is the only one that allows for the input of a THRESHOLD value. The SAS DISCRIM command is included in the PROC DISCRIM line: for example, THRESHOLD = .45.

The SAS linear L-O-O results using a threshold value of .45 with the 3-group Ethington data (3GED) are given in Table 15.4. In this example, a student is assigned to a Grade group, correctly or not, only if the largest posterior probability is greater than .45. Classification errors reflected in this table are "serious errors" in the sense that a student is predicted to an incorrect grade group if the posterior probability exceeds the threshold value. These errors are serious because group assignments are made only when the classification rule would lead one to have a great deal of confidence in any assignment; that is, the posterior probability of group membership is greater than .45 for *any* group. With our example, 214 out of 264 students were classified, while the remaining 50 were not classified.

The impact of in-doubt units on predictive accuracy may be assessed in terms of separate-group hit rates and/or in terms of the total-group hit rate. For illustrative purposes, let us focus on Group 2 of the Ethington data. From Table 15.4 it can be seen that with a threshold value of .45, the obtained hit rate for Group 2 is $86/103 \doteq .835$, the *doubt rate* is $19/122 \doteq .156$, and the *serious error rate* is $17/103 \doteq .165$. The 17 errors are serious in the sense that the 17 $\hat{P}(j|\mathbf{x})$ values for Group 1 and Group 3—the "wrong groups"—are all at least .45. These three kinds of rates may be considered in deciding on a reasonable threshold value to use. Results on two threshold values (.45 and .60) are given in Table 15.5.

Implications of such analyses are, of course, dependent on the data set on hand. For threshold values of .45 and .60 with the 3-group Ethington data, the *serious error rate* does not appear to be too high. For Group 1 and Group 3, however, the serious error rates are .873 and .821, respectively, using a threshold value of .45. When choosing a threshold value, one must try to "balance" the three kinds of rates in a way that "makes sense" for the given problem; again, a judgment call.

TABLE 15.4 SAS Linear L-O-O Results for the 3-Group Ethington Data with THRESHOLD = .45[a]

		Predicted Group			Predicted	In-Doubt	Grand
		1	2	3	Total	Students	Total
Actual Group	1	7 (.127)	45	3	55	21	76
	2	9	86 (.835)	8	103	19	122
	3	6	40	10 (.179)	56	10	66
Total		22	171	21	214	50	264

[a]Separate group hit rates are given in parentheses; the total group hit rate is $103/214 \doteq .481$.

TABLE 15.5 Threshold Classification Rates for Group 2 of the 3-Group Ethington Data

	Threshold		
	None	.45	.60
Hit rate	.811 (99/122)	.835 (86/103)	.857 (24/28)
Doubt rate	N/A	.156 (19/122)	.770 (94/122)
Serious error rate	N/A	.165 (17/103)	.143 (4/28)

15.7 SAMPLE SIZE

How large a sample is needed for valid hit-rate estimation? Of course, validity is a matter of degree; representativeness assumed, the larger the sample size, the greater the validity of estimation. That is, the larger the sample, the better the hit-rate *assessment*. Just as for the ratio of test sample size to design sample size, only limited guidance has been given in terms of a rule-of-thumb for minimum total sample size. This guidance, too, has been given only for the special two-group, multivariate-normal, assumed equal-covariance-matrix case. Based upon results of a number of two-group data simulations, Foley (1972) found that when n/p was greater than 3, the average internal (i.e., apparent) hit rate, across replications, was "reasonably close" to the true hit rate, $P^{(a)}$ (n is the number of units of each of the two groups). Hecker and Wegener (1978, p. 751) and Jain and Chandrasekaran (1982) indicate, however, that this criterion is too weak. It is herein proposed that to estimate true hit rates validly using an *internal* analysis, the minimum number of units in the smallest group should be at least five times the number of predictors: $\min(n_j) \geq 5p$.

Let us check to see if this sample-size criterion is achieved with the 3-group Ethington data (3GED). Here $p = 9$ and $\min_j(n_j) = 66$—when $j = 3$—and $5p = 45$. Obviously, this data set does meet this criterion because $66 > 45$. Therefore, with this data set we may rely on an internal analysis to estimate true hit rates.

To estimate hit rates validly using the L-O-O method, it is suggested that $\min_j(n_j) > 3p$. While considering the equal n, two-group case, Lachenbruch (1968) found that an n/p ratio of approximately 3 is necessary to obtain reasonable hit rate estimates (see Section 18.5). A minimum total sample size of $N = 3Jp$ is thus necessary to use the L-O-O method.

The minimum n/p ratio of approximately 5 might also be considered in defining a "large" N to decide whether the holdout method might be used. Here the minimum total sample size would be $N = 5Jp$. Using a *training* sample of at least $.60N$ implies that the classification rule being validated would be based on a total of $3Jp$ units. If these guides are appropriate, one might conclude that a *holdout* analysis would seldom, if ever, be used.

TABLE 15.6 Smallest Group Sizes for a PDA

		Rule Type			
		Linear Internal	Linear External	Quadratic Internal	Quadratic External
Expected Hit Rate	Large	5p	3p	7p	5p
	Small	7p	5p	9p	7p

It should be noted that the sample-size recommendations above were formulated on the basis of research done for the two-group case. They have not been empirically tested to any great extent and may very well be somewhat conservative. Another "factor" to consider in designing a study in which a PDA would be conducted is the expected hit rate. Doing so, the summary in Table 15.6 may be considered in designing a PDA study. According to Jain et al. (2000, p. 541), these suggestions may be a bit conservative—they suggest the $\min_j(n_j)$ be at least $10p$.

It is, perhaps, worth repeating a comment made near the end of Section 15.3. If feasible, it is suggested that the group sizes be proportional to the corresponding population sizes. For example, suppose that three population sizes are in the ratio 3 : 4 : 1 and resources are available to select a total sample of 240 units. Then if the group sizes are also in the ratio 3 : 4 : 1, the actual group sizes should be about 90, 120, and 30, respectively. (Of course, the number of predictors, p, needs to be considered.) An exception to this suggestion may be when one population is very small relative to the other involved population(s). For example, if one were attempting to predict membership into populations with a size ratio of, say, 20 : 1, then it may be suggested that the group sizes have a ratio of 10 : 1. The reader can undoubtedly think of some real-life research situations having a 20 : 1 population size ratio.

15.8 COMMENTS

In all data analysis circles, the method one uses to estimate parameter values depends, of course, on the definition of the parameter as well as on the definition of the population. Just as parameter definition is an important consideration in multiple regression analysis, it is equally important in predictive discriminant analysis. Of the three types of true hit rates (optimal, actual, expected actual), the first seems to be of least interest and the second of most interest. Of course, interest depends on individual problem situations, but in the (practical?) situation where one is interested in developing a group membership prediction rule to be used with nondesign units, the actual hit rate is the one on which we could focus.

If a sample size is large enough [$\min_j(n_j) \geq 5p$], we may justifiably use an internal analysis to estimate a true actual hit rate. Separate group as well as total group hit rates may be of concern. If a sample size is small, the approach to obtaining estimates of a true actual hit rate suggested here is the L-O-O method of estimation. This external classification analysis can be accomplished via the SAS DISCRIM procedure.

Further Reading

Choi (1986) edits and contributes the lead article in a special issue of *Computers and Mathematics with Applications* that deals with statistical methods of discrimination and classification.

Habbema et al. (1978, 1981), Habbema and Hilden (1981), and Hilden et al. (1978a, 1978b) develop a series of five articles in *Methods of Information in Medicine* that cover the basic classification problem through utility considerations to general recommendations, all pertaining to PDA as applied in the context of medical diagnosis.

Hand (1986) summarizes the state of the art of error-rate estimation up to the mid-1980s. Varieties of bootstrap estimators and average-conditional estimators are emphasized.

Hand (1987) proposes a total-group error rate formula estimation for the two-group situation that is based on a L-O-O classification rule.

Hand (1994; 1997, Chapter 6) discusses, in some detail, the assessment/evaluation of classification rules.

Hand (1997, Chapter 7) provides a summary of misclassification estimation that includes the L-O-O, bootstrap, and jackknife methods. He favors the 632 bootstrap method (p. 126).

Jain and Chandrasekaran (1982) give a fairly comprehensive review of the sample-size problem from a pattern recognition perspective.

James (1985, pp. 70–72) and Hand (1981, pp. 197–199) suggest the idea of a classification rule with a "reject option" when referring to the use of a threshold posterior probability value for deciding whether or not a unit is to be classified. References of study about in-doubt cases are cited in these two books.

Lachenbruch (1968) discusses the relationship between estimation in PDA and in multiple correlation/regression analysis.

Lesaffre et al. (1989) introduce an approximate L-O-O approach to estimation for a logistic classification rule.

McDonald et al. (1976) propose a test to decide if a unit originated from either of two populations. Yet another referent distribution for a D_{ug}^{*2} value is given (see footnote 2 in Chapter 14).

McLachlan (1992) presents a very thorough review of hit-rate—or error-rate—estimation (in Chapter 10), and devotes a chapter (Chapter 11) to the very interesting problem of assessing the precision of estimating posterior probabilities of group membership.

Rencher (1998, Chapter 6; 2002, Chapter 9) provides detail on hit/error-rate estimation methods.

Sadek (1992) studied the influence of a single outlier in a two-group PDA context. The effect of two kinds of outlier, local and global, on change and precision of classification results depended on group separation, group size, and outlier location.

Schiavo and Hand (2000) review fairly recent research on estimating error rates in PDA.

Williams et al. (1990) conclude that on the basis of a simulation study, the following factors interact in their effect on hit-rate accuracy and precision: (1) number of predictors, (2) covariance structure (i.e., "system variability"), (3) group separation, and (4) group sizes.

Definition ***Goodness of fit***: Used only in the finest clothing stores.

EXERCISES

1. The sample R^2 in multiple regression analysis is used as an index of predictive accuracy. What is an analogous index in predictive discriminant analysis?

2. For which of the following would the ratio of minimum group size to number of predictors in a PDA need to be the largest for "good" hit rate estimation?
   (a) Linear internal
   (b) Linear L-O-O
   (c) Quadratic internal
   (d) Quadratic L-O-O

3. Give a meaning of "typicality probability" (aside from how it is used).

4. When *might* an apparent hit rate be an acceptable estimate of a "true" hit rate?

5. Suppose that a classification rule was developed using 2004 data with particular group priors. Looking ahead, suppose that it is planned to apply the 2004 rule using predictor values in year 2008; but it is believed that the relative sizes of the populations will change from 2004 to 2008.
   (a) Which priors would you use for the year 2008 predictor values?
   (b) If you think new priors should be used, would you be in a position (in 2004) to estimate any hit rate(s) for year 2008? Why or why not?

6. Briefly discuss *internal* versus *external* PDA.

Computer Applications

7. Consider the first two groups in the 3-group Ethington data (3GED) ($p = 9$, $n_1 = 76$, $n_2 = 122$, and $N = 198$) with $q_1 = .333$ and $q_2 = .667$. Two-group analyses are considered for this exercise.

(a) Using SAS DISCRIM with POOL = YES obtain a D_{12}^2 value [see (14.4)], and a value of $\hat{D}_{12}^2$ [see (15.1)].

(b) Obtain the internal (or apparent or resubstitution) hit rates for Group 1 and Group 2, and the total group hit rate from the DISCRIM output.

(c) Obtain the two linear L-O-O hit rates from SAS DISCRIM output; these are obtained using the CROSSVALIDATE and POOL = YES options.

(d) Again, obtain SAS DISCRIM (with options CROSSVALIDATE, POOL = YES, CROSSLIST, and POSTERR) output using only the first two groups. Determine the L-O-O and unstratified M-P-P/L-O-O hit rates for Group 1, Group 2, and total group.

(e) Compare the results of parts (b) through (d).

8. Conduct a linear L-O-O analysis with the 3-group Ethington data (3GED) using SPSS DISCRIMINANT.

(a) Locate the listing of estimated posterior probabilities; you should find two listed values of (13.14). What is the value of $\hat{P}(1|\mathbf{x}_{21})$? Of $\hat{P}(2|\mathbf{x}_{21})$? Of $\hat{P}(3|\mathbf{x}_{21})$?

(b) What is the reported value of $\hat{P}(\mathbf{x}_{21}|2)$? [This is a typicality probability; see (12.5).]

(c) Give the linear classification function (LCF) for Group 2 (round all values to two decimal places). [The set of LCF weights and constant are expressed in (13.23) and (13.24).] Compare your weights and constant with those in Table 13.1.

9. Estimate predictive accuracy for the 5-group Ethington data (5GED). (It is recognized that this is a "loose" request!)

10. In Exercise 2 in Chapter 14 you identified two fence riders in the 5-group Ethington data set (5GED).

(a) Identify three more fence riders and three more outliers.

(b) Determine a new (linear) classification table based on the remaining 545 − 10 = 535 units.

(c) Compare the new table with that found in Exercise 2 in Chapter 14. About which table would you "feel better"?

11. For this exercise consider a linear L-O-O anlaysis using the 5-group Ethington data (5GED).

(a) Develop a table like Table 15.4 using a threshold value of .40.

(b) What is the serious error rate for Group 3?

12. Conduct a *quadratic* L-O-O analysis with the 3-group Ethhington data (3GED) using SAS DISCRIM.

(a) What is the value for $\hat{P}(2)$ $(=\hat{\pi}_2 = q_2)$? For $\hat{\pi}_3(=q_3)$?

(b) Locate the listing of estimated posterior probabilities of group membership. What is the value of $\hat{P}(1|\mathbf{x}_{88})$? Of $\hat{P}(3|\mathbf{x}_{88})$? Of $\hat{P}(5|\mathbf{x}_{88})$?

(c) Into which group is OBS 300 (or unit 300) classified?

13. Refer to the output for Exercise 4. What result would have had to occur for you to conclude that the $J = 3$ covariance matrices were not equal? (Assuming what other data conditions is met?)

14. For the data set you have for Exercise 2 in Chapter 1, estimate the separate group and total group hit rates in whatever manner that makes sense to you. (Disregard any categorical predictors.)

CHAPTER 16

Effectiveness of Classification Rules

16.1 INTRODUCTION

In a prediction study where the criterion variable is normally distributed, the predictors have a joint-normal distribution, and a multiple regression analysis is carried out, the researcher typically addresses the following question: Are the results of the prediction better than what could be obtained by chance? Results "obtained by chance" in a multiple regression context generally imply results are obtained in a situation of no correlation between the criterion variable and a linear composite of the predictors. *Chance predictions* would be made by assigning the mean Y value to all analysis units, regardless of each unit's set of predictor values. In essence what is being done is deciding whether knowledge of predictor values is helpful in predicting a value of the criterion variable.

Discussion in the preceding paragraph refers to the statistical significance of the multiple correlation coefficient. In this sense, chance prediction would accompany a correlation of zero (see Technical Note 1). Of course, statistical significance does not necessarily imply any particular degree of prediction effectiveness. Assessing the degree of prediction effectiveness is a matter of judgment. One *aid* for judgment is the magnitude of the square of the sample correlation coefficient, R^2, or better yet, the magnitude of an adjusted R^2—see Huberty and Hussein (2003). Another aid might be the size of interval estimates of the true squared correlation coefficient and of true predicted criterion values for various predictor combinations. After all is said and done, nonetheless, degree of effectiveness is a matter of judgment.

The story is somewhat similar in predictive discriminant analysis. Again, what is meant by "chance" must be understood; and again, a judgment must be made regarding practical or meaningful significance versus statistical significance. That is, assessing the degree of prediction effectiveness is still a matter of judgment.

Two different interpretations of chance classification are presented, one involving statistical significance, the other not. The matter of comparing the effectiveness of

Applied MANOVA and Discriminant Analysis, Second Edition, by Carl J. Huberty and Stephen Olejnik
Copyright © 2006 John Wiley & Sons, Inc.

alternate classification rules, statistical and otherwise, is also discussed. Finally, the contribution of prior probabilities to classification effectiveness is examined.

16.2 PROPORTIONAL CHANCE CRITERION

16.2.1 Definition

Consider the following situation. We have two criterion populations of approximately the same size. A sample of analysis units of a given size is taken from each population. It is reasonable to expect, then, that by using the flip of a fair coin to decide group membership, we could correctly classify about 50 percent of the analysis units? Similarly, with three populations and samples of the same size, we could expect to classify correctly about one-third of the units by chance. In general, with J populations and samples of common sizes, we could expect to classify correctly about $1/J$ of the units *by chance.*

The foregoing conclusion may be justified in the following manner. Consider a classification table as in Table 16.1, a repeat of Table 14.2. In this table, each main diagonal element, n_{jj}, in cell (j, j), $j = 1, 2, 3$, represents the number of hits for the respective group. Thus n_{jj}/n_j is the hit rate for Group j (e.g., n_{22}/n_2 is the hit rate for Group 2). If predicted group membership is independent of actual group membership, the *expected* frequency in cell (j, j) is given by:

$$e_j = q_j n_j,$$

where q_j is the estimated prior probability of membership for Group j and n_j is the size of Group j. If $n_1 = n_2 = \cdots = n_J = n$, then $N = Jn$. Let $q_j = n_j/N = n/N$. The overall chance frequency of hits would then be

$$e = \sum_{j=1}^{J} e_j = \sum_{j=1}^{J} q_j n_j = \sum_{j=1}^{J} \frac{n}{N} n = J \frac{n}{N} n = n.$$

Thus the chance hit rate, e/N, would be $n/N = 1/J$.

TABLE 16.1 Classification Table Notation

		Predicted Group			
		1	2	3	
Actual Group	1	n_{11}	n_{12}	n_{13}	$n_{1.} = n_1$
	2	n_{21}	n_{22}	n_{23}	$n_{2.} = n_2$
	3	n_{31}	n_{32}	n_{33}	$n_{3.} = n_2$
Total		$n_{.1}$	$n_{.2}$	$n_{.3}$	$n_{..} = N$

Now, consider a situation where the J populations are of varying sizes. Again, let q_j denote the estimated prior probability of membership in Group j. And the chance frequency of hits for Group j is

$$e_j = q_j n_j. \tag{16.1}$$

The overall (or total group) chance frequency of hits would be given as:

$$e = \sum_{j=1}^{J} q_j n_j, \tag{16.2}$$

and the overall (expected) chance hit rate is

$$H_e = \frac{e}{N} = \frac{1}{N}\sum_{j=1}^{J} q_j n_j. \tag{16.3}$$

Expression (16.3) is the *proportional chance criterion* for the total-group hit rate. That is, an observed total-group hit rate may be compared with e/N to determine if we have better than chance classification. For separate-group classification, $e_j/n_j = q_j$ would be the proportional chance criterion.

16.2.2 Statistical Test

So, now we have one definition of *chance classification* for individual groups and for the total group. A question then arises: Is the observed classification accuracy[1] better than what may be expected by chance? To address this question, let the total observed frequency of hits be denoted as:

$$o = \sum_{j=1}^{J} n_{jj},$$

which is the sum of the main-diagonal elements of the classification table (see Table 16.1). Then, from a statistical significance point of view, the null hypothesis is that the number of units correctly classified does not exceed the number correctly classified by chance. When the overall number of correctly classified units, o, is less than or equal to the chance number, e, it is not necessary to test for significance. Hence, the significance test involves a *directional* alternative hypothesis.

Now o is a statistic that can take on values from 0 to N. In classification problems, N is generally large enough so that a normal probability distribution may be used to approximate the distribution of o. Thus a standard normal statistic may be used to test the null hypothesis:

$$z = \frac{o - e}{\sqrt{e(N - e)/N}}. \tag{16.4}$$

[1] See footnote 4 in Chapter 15.

The lower bound of a one-sided interval estimate of the true overall frequency of hits may be determined using

$$o - z_{1-\alpha}\sqrt{e(N-e)/N}, \tag{16.5}$$

where $z_{1-\alpha}$ is the $100(1-\alpha)$ centile of the standard normal distribution.

Even though the observed total number of correct classifications, o, may be significantly greater than the chance number, e, it may be suspected that not all separate-group predictions are significantly greater than those to be expected by chance. For Group j, the standardized normal test statistic would be

$$z = \frac{n_{jj} - e_j}{\sqrt{e_j(n_j - e_j)/n_j}}. \tag{16.6}$$

It was argued in Section 15.3 that the results of an external analysis yield better (in the sense of accuracy) hit rate estimates than those of an internal analysis. Thus, to address the better-than-chance question, the results of an *external* analysis should be used.

To illustrate testing for better-than-chance classification accuracy, consider the results given in Table 15.2, which are repeated in Table 16.2. With this data set, reasonable prior probabilities are .25, .50, and .25 for the three respective groups. From Table 16.2, we get $o = 16 + 99 + 13 = 128$; $e_1 = .25(76) = 19$, $e_2 = 61$, $e_3 = 16.5$, and $e = 96.5$. Because $o > e$, we proceed.

The value of the overall statistic (16.4) is 4.16, which clearly indicates a better-than-chance result. The lower bound for a 99 percent confidence interval for the true frequency of total-group hits is [see (16.5)] $128 - 2.326(7.807) \doteq 109.84$. For individual groups we get the following results:

Group	n_{jj}	n_j	e_j	z	P	Lower Bound for 99% Interval
1	16	76	18	NA[a]	NA	NA
2	99	122	61	6.88	.000	86.65
3	13	66	16.5	NA[a]	NA	NA

[a]Not applicable because 16 < 18 and 13 < 16.5.

TABLE 16.2 Linear L-O-O Results for the 3-Group Ethington Data

		Predicted Group				
		1	2	3	n_j	q_j
Actual Group	1	16 (.211)	55	5	76	.25
	2	12	99 (.811)	11	122	.50
	3	6	47	13 (.197)	66	.25
Total		34	201	29	$264 = N$	

In this case, the total-group hit rate as well as the hit rate for Group 2 are significantly better than what may be expected by chance. This is not true for the hit rates of Groups 1 and 3. That is, information yielded by the nine predictors enable us to classify students into grade level 2 (B) statistically better than chance.

If the J separate-group better-than-chance tests are of interest, then the J/P values should be adjusted. A sensible adjustment for multiple testing in this setting may be accomplished by multiplying each of the individual P values by J (the Bonferroni adjustment). Judgment would then be made about the "smallness" of each product.

16.3 MAXIMUM-CHANCE CRITERION

Consider now a second meaning of chance classification. Suppose we have a two-group situation that yielded the (external) results given in Table 16.3. Here we have $n_1 = 22$ and $n_2 = 78$. Assume that $q_1 = .20$ and $q_2 = .80$ are reasonable priors. Using the proportional chance criterion, we would expect to correctly classify [see (16.2)]

$$e = .20(22) + .80(78) = 66.8$$

of the 100 units. We find that $o = 9 + 77 = 86$ yields a standard normal statistic value [see (16.4)] of 4.08 ($P \doteq .000$). With such results one might be tempted to conclude that acceptable classification accuracy was obtained. Consider, however, a researcher who states, "I could get a hit rate of .78 simply by assigning all of the 100 units to Group 2." Such an interpretation of "chance" leads to the *maximum chance criterion* of

$$\max(q_1, q_2).$$

For the situation above,

$$\max(.20, .80) = .80,$$

which implies that 80 units could be correctly classified by "chance." Using this as the expected hit rate and the results of Table 16.3, the standard normal statistic value is only $-.50 (P \doteq .308)$. As such, one *might* question whether the rule used (to get the results in Table 16.3) yielded classification accuracy better than "chance."

TABLE 16.3 Hypothetical Classification Table

		Predicted Group		
		1	2	Total
Actual	1	9	13	22
Group	2	1	77	78
Total		10	90	100

16.4 IMPROVEMENT OVER CHANCE

A statistical assessment of whether or not actual classification results obtained are better than those obtained by "chance" was discussed above. Suppose one concludes that the total-group results obtained are statistically better than chance, as found previously with $z \doteq 4.16$. It seems natural now to ask: How much better than chance can we predict group membership? The answer to this question may be addressed by using the index

$$I = \frac{H_o - H_e}{1 - H_e}, \tag{16.7}$$

where H_o is the observed hit rate, and H_e is the hit rate expected by chance. For the (external) results given in Table 16.2 and using the proportional chance criterion for total group classification, $H_o = 128/264 \doteq .485$, $H_e = 96.5/264 \doteq .366$, and $I \doteq .188$. That is, by using a linear classification rule, about 19 percent fewer classification errors would be made than if classification were done by chance. The index I may thus be viewed as a proportional reduction in error or *improvement over chance* descriptive statistic.

Of course, the I index may be used with any definition of chance (with H_e thus determined) and with any rule whatever to determine H_o. For example, Walters (1986) used I in conjunction with a rule that was based solely on an arbitrary cutoff score on a single predictor, with $H_e = .50$.

Many data analysis methodologists recommend reporting the value of an "effect size" index along with a test statistic value and a P value. This is the case in the two-group t-test situation and in the ANOVA F-test situation. In the current situation of testing to determine if a hit rate is statistically significantly better than chance [using (16.4) or (16.6)], it makes sense to consider the value of I in (16.7) as the value of an effect-size index, an index that reflects "meaningfulness." As with any effect-size index, a judgment has to be made regarding the magnitude of I.

16.5 COMPARISON OF RULES

There are times when a researcher may be interested in comparing the effectiveness of two (or more) classification rules using a given set of analysis units. The rules may be of a different form or of the same form but with different predictor variables. For example, it may be of interest to compare the results of a linear rule with those of a quadratic rule, or to compare the results of using 16 predictors with those using, say, 6 predictors. Another example, discussed in Section 16.6, involves a comparison of a rule that employs equal prior probabilities of group membership with a rule using unequal priors.

For the moment, discussion will be restricted to comparing total-group classification accuracy for only two rules. Data to be considered in a comparison consist of a 1 (correct classification) or a 0 (incorrect classification) for each unit in each of J

TABLE 16.4 Comparison of Rules

		Rule 2		
		Hit	Miss	Total
Rule 1	Hit	a	b	$a+b$
	Miss	c	d	$c+d$
Total		$a+c$	$b+d$	N

groups. (The "design" here may be viewed as a simple repeated-measures design, the "within" factor having two levels.) The $N \times 2$ data matrix may be summarized in a 2×2 table (Table 16.4). There are a units correctly classified by both rules, b units correctly classified by Rule 1 and incorrectly classified by Rule 2, and so on.

A comparison question, then, is whether the proportion correctly classified by Rule 1, $(a+b)/N$, is significantly different from the proportion correctly classified by Rule 2, $(a+c)/N$. From a statistical testing viewpoint, the null hypothesis is that the true proportion of correct classifications for Rule 1 is the same as the proportion of Rule 2. One approach to answering the question is to use the test statistic proposed by Q. McNemar (1900–1986) (see, e.g., Agresti, 2002, p. 411):

$$\frac{(b-c)^2}{b+c},$$

which, if $b+c$ is "large," is an approximate chi-squared statistic with 1 degree of freedom. (Alternatively, a standard normal statistic that is the positive square root of the above expression may be used.)

16.6 COMPUTER APPLICATION I

As an example, consider comparing a linear rule with a quadratic rule using the 3-group Ethington data (3GED)—internal analyses are used. The 264×2 matrix of 1's and 0's may be obtained using, for example, SAS programming as follows:

SAS SYNTAX FOR COMPARING LINEAR AND QUADRATIC INTERNAL HIT RATES

```
Proc discrim data=one pool=yes out=two noprint;
class grade;
var counsum - - qesci;
```
(*continued on next page*)

```
priors '1'=.25 '2'=.50 '3'=.25;

Proc discrim data=one pool=no out=three noprint;
class grade;
var counsum - - qesci;
priors '1'=.25 '2'=.50 '3'=.25;

data settwo;
set two;
lgp=_into_;
keep grade lgp;

data setthree;
set three;
qcp=_into_;
keep grade qcp;

data match;
merge settwo setthree;
if grades eq lgp then lg=1;
if grades le lgp then lg=0;
if grades eq qcp then qg=1;
if grades ne qcp then qg=0;

proc print data=match;
var grade  lgp qcp;

proc freq data=match;
tables grade*(lgp qcp);
tables lg*qg / all;

run;
```

OUTPUT

Analysis:* 2 × 2 *Frequency Table of Hit Rates (Table 16.5)

Interpretation:* 2 × 2 *Frequency Table of Hit Rates

A summary of the 264×2 matrix of 1's and 0's is given in Table 16.5. The value of the McNemar chi-squared statistic (with df $= 1$) is 8.33, $P \doteq .002$. Thus it may be concluded that the quadratic internal hit rate is statistically higher than the linear internal hit rate. A word of caution is in order. Note that the list of SAS commands above yields a comparison of *internal* classification results. Estimates—posterior probabilities and hit rates—based on an internal analysis are not as "good" as those based on an *external* analysis (see Section 15.3). Therefore, when comparing classification rules, external results should be used. That the use of external results may lead to rule

TABLE 16.5 Summary of Linear and Quadratic L-O-O Classification Results for the 3-Group Ethington Data

		Quadratic Rule		
		Hit	Miss	Total
Linear Rule	Hit	111	25	136
	Miss	50	78	128
Total		161	103	264

comparison conclusions different from those based on internal results was found by Morris and Meshbane (1995).

Note that if $b = c = 0$—when the two rules yield identical results—the data would obviously support the null hypothesis, and a statistical test would not be needed.

The ideas above could be utilized in the comparison of two rules with respect to classification accuracy for a particular group, say, Group j (as opposed to total-group classification accuracy). If Group j accuracy is of interest, the SAS programming would have to be modified. The McNemar statistic would be applied to an $n_j \times 2$ data matrix.

If it is of interest to compare classification results of more than two rules on the same data set, a chi-squared statistic proposed by W. G. Cochran (1909–1980) may be employed (Cochran's Q; see Agresti, 2002, p. 459). An overall comparison of multiple rules may be accomplished by using the SPSS NPAR TESTS procedure to get a Q value along with the df value. It is recommended (Myers et al., 1982) that the df value be adjusted to determine a P value. Pairwise comparisons of rules may be made in lieu of the omnibus comparison; see Fleiss et al. (2003, Chapter 9) for details.

An alternative approach to comparing more than two rules is to consider the results of K rules as coming from a simple repeated-measures design (see Myers et al., 1982; Looney, 1988). Here we have an $N \times K$ matrix of 1's and 0's. The omnibus multivariate analysis may be accomplished using SAS GLM or SPSS MANOVA (see Section 10.7).

16.7 EFFECT OF UNEQUAL PRIORS

As noted in Section 13.5 (see Table 13.2), prior probability of membership in each group plays a role in the classification rule used. It was also argued (in Section 12.4.4) that group prior probability estimates should be based on the relative sizes of the respective populations. [See Tatsuoka 1988, p. 359) for a brief discussion of the potential problem of "perpetuating the status quo."] Seldom would the case be that priors should be based on observed sample sizes—unless a proportional group sampling plan was used. Of course, complete lack of knowledge of relative population sizes would suggest the use of equal priors—at least two references may be found

TABLE 16.6 Linear L-O-O Results for the 3-Group Ethington Data Using Equal Priors

		Predicted Group			
		1	2	3	n_j
Actual Group	1	38 (.500)	24	14	76
	2	43	38 (.311)	11	122
	3	15	15	36 (.545)	66
Total		96	77	91	264

with a recommendation that caution be urged with general use of unequal priors (Lindeman et al., 1980, pp. 207–212; Meshbane and Morris, 1995).

The use of equal priors (.333, .333, .333) with the 3-group Ethington data (3GED) yields the linear L-O-O results reported in Table 16.6. The total-group hit rate is $(38 + 38 + 36)/264 \doteq .424$, which is a little lower than that with priors of .25, .50, and .25 (when .485 was obtained).

What is the possible effect of using unequal priors versus using equal priors? The comparison of linear L-O-O PDA results using the 3-group Ethington data is given in Table 16.7. The unequal group priors are, respectively, .25, .50, and .25, while the equal priors used are .333, .333, and .333. This summary was obtained counting "by hand" using the two computer outputs. The value of the McNemar statistic is $(45 - 62)^2/(45 + 62) \doteq 2.70$, which yields $P \doteq .10$. Thus, for this data set, it does not appear that there is much of a difference in total-group hit rate in the results using equal priors versus unequal priors. Comparing the hit rates for Group 2 in Tables 16.2 and 16.6 ($99/122 \doteq .811$ versus $38/122 \doteq .311$), however, indicates a "substantial" difference. Using equal priors, hit rates for the two smaller groups actually increase when going from unequal priors to equal priors. Thus the general idea: For groups of unequal sizes that tend to reflect relative population sizes (in an order sense), use of unequal priors will increase the hit rates for the larger groups and decrease the hit rates for the

TABLE 16.7 Summary of Total-Group Linear L-O-O Results for the 3-Group Ethington Data

		Unequal Priors		
		Hit	Miss	Total
Equal Priors	Hit	66	45	111
	Miss	62	91	153
Total		128	136	264

smaller groups. It is not that one set of results is "good" and the other "bad." The moral of the story may be: Use the best estimates for the priors that are available.

It should be noted that a statistical assessment of the difference between the hit rates for Group 2 with and without equal priors may be accomplished by generating a summary table of classifications (like Table 16.6). Comparative classification accuracies for Groups 1 and 3 may be assessed similarly.

16.8 PDA VALIDITY/RELIABILITY

As pointed out by Hand (1997, p. 475), many terms in the PDA context are used differently by different writers. One such term is *validity*. An alternative is *imprecision*; another is *reliability*. The view of validity/reliability we consider herein relates directly to the question: Are the PDA results (i.e., the hit/error rate(s) of interest) free from sample bias? One approach to addressing this question is to confirm the obtained results by using multiple collections of analysis units. This can be very expensive from an applied researcher standpoint. Another approach is to examine split-sample results. Even with "limited" group sizes, one can use a resampling approach (e.g., bootstrap; McLachlan, 1992, p. 767) to assess the variability of PDA results. Although not stating the use of bootstrapping, Jain and Jain (1994) considered a "series of simulated training samples" (p. 149) based on a variety of sample splitting percents of the total group size (ranging from 10 to 80). For example, 30 percent was used for the "training sample" and 70 percent for the "test sample." For each training-test split, 40 training samples were simulated. Their simulated results suggested that a 50/50 split gave "the desirable results" (p. 150).

16.9 APPLYING A CLASSIFICATION RULE TO NEW UNITS

The focus of the discussion of predictive discriminant analysis (PDA) to this point has been on estimating hit rates and assessing the classification results. There may, however, be a more basic and practical purpose behind predicting group membership. A more complete, practical view of PDA may be expressed by the following steps:

1. For the data on hand (i.e., the training sample), develop a classification rule (Chapter 13).
2. Obtain the classification results (Chapter 14).
3. Using the results, arrive at separate-group and total-group hit rate estimates (Chapter 15).
4. Assess the hit-rate estimates (Chapter 16).
5. Assuming that the assessment in step 4 indicates a reasonable rule, apply the rule to a new set of analysis units whose group membership is unknown and is to be predicted.

For example, consider a clinical setting in which units (people or animals) in the criterion groups are treated according to group membership (see Section 23.6). Suppose that a training sample is available for which group membership of the units is known. A rule could be developed and assessed. This rule, then, could be utilized to classify a new unit so that group-specific treatments could be applied to the new unit.

It should be noted that what is being presented in this section does *not* pertain to cross validation, an approach to hit-rate estimation mentioned in Section 15.3. What is being discussed is the situation where the rule has already been derived and assessed, and the rule is then to be applied to a new unit in a "real-life" setting.

A developed classification rule could be applied to new units, one at a time if needed. Benchmarks for classification pertaining to in-doubt units and outlying units would have been determined with the training sample data. Such characterizations may be useful in describing a new unit and in determining a subsequent activity or treatment or decision applicable to a new unit.

As indicated in the SAS User's Guide, with SAS DISCRIM, the application rule may be "stored" and then applied to any new unit(s) by using the TESTDATA option. Two approaches may be taken with SPSS DISCRIMINANT. One is to use the SELECT subcommand with each set of new units—each analysis with a new set of units will rebuild the original classification rule. The other approach is to use the MATRIX subcommand, which "stores" the original rule, which may then be applied to the new unit(s).

16.9.1 Computer Application II

The application of a linear classification rule to five "new" community college students may be accomplished using the following SAS syntax. Table 16.8 provides a list of five hypothetical students used to demonstrate this application. These data are entered as a second SAS data file. We label this data set NEW.

TABLE 16.8 Scores on Nine Predictor Variables for Five Hypothetical New Students

Id	coun sum	gain sum	learn sum	qelib	qefac	qestacq	qeamt	qewrite	qesci
1	3	12	24	11	12	8	8	17	9
2	4	18	17	14	16	12	9	21	18
3	6	17	22	7	13	7	6	23	9
4	1	8	9	7	14	9	6	9	9
5	2	12	10	8	12	7	6	19	20

The first SAS syntax creates the classification rule and the second SAS syntax applies the rule to the new data file.

SAS SYNTAX FOR A LINEAR PDA USING THREE GRADE GROUPS AND NINE PREDICTOR VARIABLES

```
proc discrim pool=yes crosslist posterr outstat=ethstat;
class grade;
var consum - - qesci;
priors '1'=.25 '2'=.50 '3'=.25;
run;
```

outstat=ethstat stores the classification function weights in a SAS file named ethstat. The remaining commands were defined in Section 14.3

SAS SYNTAX FOR CLASSIFYING NEW STUDENTS

```
proc discrim data=ethstat testdata=new testlist;
class grade;
testid id;
var consum - - qesci;
run;
```

data=ethstat identifies the data file containing the classification function weights.
testdata=new identifies the data file containing the students to be classified.
testlist requests the listing of the classification results and the posterior probabilities.
testid id lists the classification results by the "id" variable.

OUTPUT (Table 16.9)

16.9.2 Computer Application III

To illustrate the application of a quadratic classification rule to a new community college student, a subset of three of the nine predictors in the 3GED data set will be used. The three predictors used are X_1 (counsum), X_2 (gainsum), and X_3 (learnsum).

TABLE 16.9 Classification Results for New Students

Id	Classified into Grade	Posterior Probabilities $\hat{P}(1\|\mathbf{x})$	$\hat{P}(2\|\mathbf{x})$	$\hat{P}(3\|\mathbf{x})$
1	3	.118	.388	.494
2	2	.316	.508	.176
3	3	.078	.417	.505
4	1	.460	.304	.236
5	1	.490	.387	.123

The SAS DISCRIM program will be used—SPSS DISCRIMINANT will not yield the needed information.

SAS SYNTAX TO OBTAIN QUADRATIC CLASSIFICATION FUNCTION WEIGHTS USING THREE GRADE GROUPS AND THREE PREDICTORS

```
proc discrim outstat=wts method=normal pool=no  listerr crosslist
posterr;
class grade;
var counsum gainsum learnsum;
priors '1'=.25 '2'=.50 '3'=.25;
run; proc print data=wts; run;
```

Most of the SAS commands were defined in Section 14.3. Only the additional commands to obtain the quadratic weights are defined here.

outstat=wts requests an output file be generated labeled wts. The file name (i.e., wts) is user generated. A variety of statistics are reported.
method=normal requests that the output file (wts) include the weights for the quadratic classification function.
pool=no requests that the quadratic rule be used.
proc print data=wts requests that the ouput file be printed.

OUTPUT

The relevant SAS output for the quadratic classification functions for A students (Grade = 1) are:

Grade	Type	NAME	counsum	gainsum	learnsum
1	QUAD	counsum	−0.1328	0.0293	−0.0023
1	QUAD	gainsum	0.0293	−0.0267	0.0115
1	QUAD	learnsum	−0.0023	0.0115	−0.0245
1	QUAD	LINEAR	0.1346	0.3032	0.4458
1	QUAD	CONST	−11.9811	−11.9811	−11.9811

The quadratic classification function for Group 1 is, therefore, the following:

$$\begin{aligned} Z \doteq\ & -11.9811 + .1346X_1 + .3032X_2 + .4458X_3 \\ & - .1328X_1^2 - .0267X_2^2 - .0245X_3^2 \\ & + .0293X_1X_2 - .0023X_1X_3 + .0115X_2X_3 \end{aligned}$$

The same SAS syntax used to develop the linear rule and classify new students (see Section 16.9.1) can be used to develop a quadratic rule and classify new students with one exception. To obtain the quadratic rule **pool=yes** must be changed to **pool=no.**

Table 16.10 presents the results of classifying the five new students using the quadratic rule with the nine variables in the 3GED data set. Using these data, only the classification of the individual with id=1 changes from Grade level 3 to Grade level 2. Note, however, that with the quadratic rule id=1 might be considered an in-doubt unit. This would not be the case when the linear rule is used.

It should be noted that there are two "forms" of a "classification rule." *One* is for J sets of LCF/QCF weights. With this form, a linear/quadratic classification function score is determined for each group for each new unit. The LCF weights may be obtained by using either SPSS DISCRIMINANT or SAS DISCRM. The QCF weights may be obtained via SAS DISCRIM. The *second* form of a classification rule involves posterior probabilities, one for each group for each unit. These probabilities may be obtained via the SPSS and SAS packages for a linear rule. It may be noted the SPSS DISCRIMINANT program also outputs a typicality probability for each new unit (using a *linear* rule).

TABLE 16.10 Classification Results for New Students Using a Quadratic Rule

		Posterior Probabilities		
Id	Classified into Grade	$\hat{P}(1\|\mathbf{x})$	$\hat{P}(2\|\mathbf{x})$	$\hat{P}(3\|\mathbf{x})$
1	2	.053	.504	.443
2	2	.169	.554	.277
3	3	.084	.351	.566
4	1	.603	.277	.120
5	1	.477	.392	.132

In many practical situations it would be desirable to assess the outcome of applying a classification rule to new units. This would call for some record keeping and/or observation on someone's part.

16.10 COMMENTS

Discussion of assessments of classification rule accuracy in this chapter focused on the use of counting estimators of hit rates. This focus is, of course, not necessary. Rather, one might focus on the use of a posterior probability estimator (of separate-group or total-group hit rates); see Section 15.3 for a discussion of such estimators. The statistical test procedures for posterior probability estimators are the same as for counting estimators; the same goes for the use of the improvement over chance index.

It was stated at the end of Section 15.3 that the use of a total-group posterior probability estimator does not require knowledge of unit group membership. If group membership of some units is not known, however, a definition of chance group prediction may be problematic. If the q_j values are not known, how might "chance" be defined? Suppose that one has a good handle on prior probabilities of group membership. Then one *might* use $(q_1^2 + q_2^2 + q_3^2)$ for the overall (proportional) chance hit rate in a three-group situation—this implicitly assumes the use of a sampling plan where group sizes are proportional to population sizes. Finally, unknown group membership for some units would preclude the use of the rule comparison strategy discussed in Section 16.5.

Technical Notes

1. In multiple regression, the sample squared correlation coefficient, R^2, is often used as an index of accuracy of prediction. Now, when the population multiple correlation coefficient is zero, it is known that the long-run mean of R^2 is given by $E(R^2) = p/(N - 1)$ (Huberty and Hussein, 2001), where p is the number of predictors and N is the sample size. Thus, some methodologists might consider a chance R^2 to be equal to $p/(N - 1)$ rather than zero (see Huberty, 1994b).

2. A second approach to the better-than-chance question (see Section 16.2) is to consider it a "matching problem" (Mosteller and Bush, 1954, pp. 307–311). If a rule indicates that a unit be classified into the same group from which it emanated, the rule yielded a *match*. The question addressed here is: Is the total group proportion of matches higher than what may be attributed to chance? The test statistic proposed is

$$z = \frac{o - e}{\hat{\sigma}_e}, \tag{16.8}$$

which, for large N, is approximately a standard normal statistic. Using the notation of Table 16.1, the denominator in (16.8) is the positive square root of

$$\hat{\sigma}_e^2 = \frac{1}{N^2(N-1)}\left[\left(\sum_{j=1}^{J} n_j n_{.j}\right)^2 - N\sum_{j=1}^{J} n_j n_{.j}(n_j + n_{.j}) + N^2\sum_{j=1}^{J} n_j n_{.j}\right].$$

Using the results from Table 16.2, we get $\hat{\sigma}_e^2 \doteq 45.125$; therefore, the value of (16.8) is $(128 - 95.5)/6.7175 \doteq 4.84$. This z-value is in the same ballpark as the value of (16.4) found earlier, 4.16.

A third approach to the better-than-chance question is to employ one of the Cohen (1968,1972) kappa statistics. These statistics were applied to classification problems by Altman et al., (1976) and by Weidemann and Fenster (1978). [Formulas in the latter article are in error; see Hubert (1978).] As pointed out by Hoffmann and Overall (1976), however, kappa statistics are to be used with caution.

3. It should be noted that it is *inappropriate* to answer the better-than-chance question by applying the Pearson chi-squared statistic to a classification table. This is because only the main-diagonal entries are of interest as far as the significance question is concerned. If the observed hits yield significance, so would the Pearson statistic. However, a significant Pearson statistic does not imply a better-than-chance result. That is, a significant Pearson statistic is necessary but *not sufficient* to conclude that a rule yields better than chance classification.

Further Reading

Huberty (1984) mentions some indices that may be considered as alternatives to I in (16.7); a reference dealing with a statistical test for the I index is also given.

Park and Kshirsagar (1996) discuss "chance" classification in a (mathematical) context different from our context.

Press (1972 p. 773) presents a $\chi^2(1)$ statistic for testing the hypothesis of chance classification; this statistic is a special case of the square of (16.4) when a chance proportion of correct classifications is $1/J$.

Rudolph and Karson (1988) use real two-group data to study the effects of using unequal priors and differential misclassification costs. They report (on p. 80) that the "use of the true population prior probabilities clearly reduces the error rates." They also concluded that classification results can be distorted when the "true" costs are quite unequal.

Walters (1986) uses the I index [see (16.7)] to assess the goodness of a screening procedure in a non-PDA context.

A variety of indices similar to the I index in (16.7) have been proposed:

Costner (1965) and Suich and Turek (1989) discuss indices of proportional reduction in error (PRE) that are equivalent to the I index. Turek and Suich (1983) present a test of significance for a special PRE index.

Fleiss et al. (2003, Chapter 18) discuss statistical properties of varieties of the Cohen kappa index.

Klecka (1980, p. 760) suggests yet another form of a PRE index to be used in a PDA context that is equivalent to I.

Light (1971) presents a summary of a number of "agreement" indices, including those proposed in the 1950s and 1960s by J. Cohen (1923–1998), L. A. Goodman, W. H. Kruskal, and W. A. Scott (1926–1991).

Lykken and Rose (1963) advance an improvement-over-chance statistic to be used in the context of actuarial prediction.

Definition ***Stepwise regression*:** Thumbsucking by a street-smart kid.

EXERCISES

1. Given the classification table below (assume group sizes reflect priors):

		Predicted Group					
		1	2	3	n_j	q_j	e_j
Actual Group	1	115	20	15	150	.75	112.5
	2	2	20	3	25	.125	3.125
	3	1	4	20	25	.125	3.125
Total		118	44	38	200		118.75

(a) Is the total-group hit rate significantly better than chance (using the proportional chance criterion)?

(b) How much better than chance is the total-group hit rate?

(c) Same as (a) except use the maximum chance criterion.

2. Specify two questions in the context of a PDA that may be addressed via statistical tests.

3. What statistical concept is shared among DDA, PDA, and MCA/MRA?

Computer Applications

4. Using the 5-group Ethington data set (5GED), conduct a SAS DISCRIM quadratic L-O-O PDA with respective group priors of .15, .20, .30, .20, and .15.
   - **(a)** Find the five e_j values [see (16.1)].
   - **(b)** Find e [see (16.2)], and o.
   - **(c)** Find the value of the $N(0, 1)$ statistic in (16.4) and the lower bound of a 95 percent confidence interval as in (16.5).
   - **(d)** What do you conclude from your calculations?
   - **(e)** How much better than chance is the observed total group hit rate?
   - **(f)** What interpretation of 'chance' was used for the assessment above?

5. Consider, again, the 5-group Ethington data set (5GED). Construct a table like Table 16.7 based on a linear external (L-O-O) rule and on a quadratic external rule. (For the first analysis, use either SPSS DISCRIMINANT or SAS DISCRIM; use SAS DISCRIM for the second analysis.)
   - **(a)** What is the linear Group 1 hit rate?
   - **(b)** What is the quadratic Group 1 hit rate?
   - **(c)** Are the two hit rates statistically different?

6. Consider the quadratic *un*stratified M-P-P/L-O-O results from the SAS output in Exercise 5.
   - **(a)** Using the e_j values and e value from Exercise 4, determine if the quadratic *un*stratified M-P-P/L-O-O observed separate-group and total-group hit rates are better than chance.
   - **(b)** How much better than chance?

7. In some way (randomly?) delete five students from each group of the 3-group Ethington data set (3GED). Develop a rule using the "test sample" of 249 (= 264 − 15). Then apply the developed rule to the 15 "new" students. Are there any in-doubt students? Outliers?

8. Using your personal data (without any categorical predictors):
   - **(a)** Are your separate-group and total-group hit rate estimates (see Exercise 14 of Chapter 15) better than chance?
   - **(b)** How much better than chance?

CHAPTER 17

Deleting and Ordering Predictors

17.1 INTRODUCTION

In multiple correlation and regression analyses it is well known that the value of the sample multiple correlation coefficient, R, cannot decrease as the number, p, of "X variables" increases. That is, the apparent degree of variation in the "Y variable" attributable to the p X variables is typically enhanced as p is increased. However, R becomes an increasingly positively biased estimator with an increase in the p/N ratio, where N is the sample size (Morrison, 1990, p. 400).

In classification analyses (i.e., PDAs), an increase in the number of predictors affects the results in one way different from, and in another way similar to, that in multiple regression. *First*, unlike regression, it may very well happen that as p is increased, the hit rates (separate-group and/or total-group) will decrease. This is particularly true if the variables to be added do not contribute substantially to the intergroup differences. *Second*, similar to regression, as p increases, the positive bias of the internal hit rates increases.

Thus, one good reason to seek a subset of the p predictors (i.e., to delete some "poor" predictors) is to determine a rule that will yield a high degree of classification precision as well as predictive accuracy. If one purpose of a classification analysis is to set up a relatively precise rule to be used with subsequent analysis units, the number of predictors should be small relative to the size of the sample. And if this purpose pertains to enhancing the predictive accuracy of a set of variables, it may be desirable to delete some of the variables.

Another good reason for deleting some predictor variables is to reduce the complexity of the problem. If a purpose of a classification analysis is to supplement the interpretation of grouping variable effects (see Technical Note 1 in Chapter 4), parsimony may be a goal to keep in mind.

An analysis problem closely related to variable deletion is that of variable ordering. In a classification context, predictor variables might be ordered with respect to their contribution to the classification accuracy of interest—separate-group or total-group.

Applied MANOVA and Discriminant Analysis, Second Edition, by Carl J. Huberty and Stephen Olejnik
Copyright © 2006 John Wiley & Sons, Inc.

17.2 PREDICTOR DELETION

It should be noted at the outset that the discussion in this section pertains to deleting predictors in the context of classification (i.e., predictive discriminant analysis), *not* in the context of group separation or grouping-variable effects (where descriptive discriminant analysis is pertinent). Variable deletion in the latter context was discussed in Section 6.2.

The deletion/selection of response variables in a PDA context is sometimes referred to as *feature extraction* (see, e.g., Hand, 1997, p. 387; McLachlan, 1992, Chapter 12; Webb, 2002, Chapter 9). This expression is quite popular in a *pattern recognition* context.

17.2.1 Purposes of Deletion

Why, in a PDA context, is it desirable to consider the deletion of some predictors? *One* reason for deleting predictors may be a very practical one. If fewer than the original p predictors may be used in the classification rule (assuming that no appreciable decrease in hit rate is incurred), it would be less costly in collecting data on the predictors for the purpose of classifying new analysis units.

Another reason for deleting variables in a PDA context is based on a combination of a practical consideration and a theoretical consideration. In PDA, as in other data analysis contexts, estimation of parameters is quite involved; parameters such as interpopulation distance and population LCF weights. Data analysis models with which we associate parameter estimators with low bias and high precision are, of course, preferred. PDA models with few predictors, relative to N, yield relatively more accurate (i.e., less biased) and more precise estimators (see, e.g., Hora and Wilcox, 1982). Theory aside, deletion of some predictors (i.e., using a classification rule with fewer predictors) may very well yield a higher hit rate than if all of the original predictors were included in the rule. Simply stated, a better classification rule may be proposed if some predictors are deleted.

17.2.2 Deletion Methods

Little research dealing with variable deletion in classification problems has been reported. One article in the behavioral literature dealing in part with the issue is that by Henschke and Chen (1974). In this article the authors discuss a forward stepwise algorithm that uses an *estimated expected loss* as a stopping criterion. It turns out that this criterion is the complement of the total-group (internal) hit rate. Their algorithm is built on a linear composite of the predictors rather than on the "raw" predictors.

The SAS and SPSS packages do not contain a deletion program of any type where classification accuracy (in a descriptive sense) is an explicit criterion. Both packages do, however, contain stepwise discriminant analysis programs. The criterion for entering variables into these analyses is based on group separation. Only if a linear normal rule, when equal priors is appropriate, is the separation criterion "equivalent" to the classification criterion.

17.2.3 Package Analyses

While reading reports involving a multiple regression analysis, one often finds results of a "stepwise analysis." Sometimes the analysis is not clearly described—there is a "forward stepwise analysis" and a "backward stepwise analysis," elaborations of which are not given herein. The same may be stated with regard to stepwise discriminant analyses. Because we do not support stepwise analysis in a PDA context, descriptions and use of such analyses will not be given herein (see Rencher, 2002, pp. 311–313). A "forward selection" strategy was developed by Smith (1984); it is described in Huberty (1994c, pp. 119–122).

17.2.4 All Possible Subsets

There are two problems with a stepwise analysis and the forward selection analysis: (1) the "best subset" of a given size may not emerge, and (2) only one "good" subset of each size is suggested. There is an alternative analysis approach that deals with these two problems. An all-possible-subset approach is certainly possible, although feasibility of such an approach may be questioned in some situations. (For p predictors, a total of $2^p - 1$ predictor subsets would need to be assessed.) These are situations in which there are large numbers of predictors, and no prior knowledge of worthwhileness of any subset of predictors. In some situations, a researcher will be familiar with the relative worthwhileness of at least some of the predictors; if so, the better predictors may be considered de facto members of subsets to be assessed with respect to classification accuracy. This scheme could drastically reduce the total number of subsets to assess.

Furthermore, a researcher could very well decide that a minimum predictor subset size would be desirable. For example, suppose that one has 10 predictors, and it may be judged—judgment based on previous research, experience, measurement, and so on—that it is desirable to include three of the predictors. Further, it may be desirable to have a final subset of at least five. Thus the number of subsets to consider is reduced from $2^{10} - 1 = 1023$ to $2^7 - 8 = 120$. That is, at most 120 analyses would need to be conducted, obtaining a number of subsets of size $5, 6, \ldots, 9$.

It turns out that conducting a "large" number of multiple analyses is, at this point in time, not a concern. A computer program very useful for conducting an all-possible-subsets analysis has been written by J. D. Morris (Florida Atlantic University); the Morris program is available at the Wiley website.

17.3 COMPUTER APPLICATION

As an example of the use of the Morris program, consider the 3-gourp Ethington data set (3GED). The nine variable definitions are given in Appendix A. From these definitions, it may be reasonable that the least three predictors (X_7, X_8, X_9) should be in any determined subset. If so, then the Morris program will determine the best predictor subsets of sizes $4, 5, \ldots, 8$.

The subset analyses are based on linear L-O-O PDAs. The program has some reasonable limitations including the following four: maximum number of predictors, 14; maximum number of groups, 10; maximum number of best subsets to identify, 280; maximum number of best subsets of each size, 30. With a Morris analysis, a subset of the original predictors may be "forced in" at the start.

The Morris program provides prompts to the user to define the number of variables, the variables to be FORCED in, number of groups, group identification code, the data file name, a FORTRAN format statement, and priors if unequal. The data file can be saved using MICROSOFT NOTEPAD. We saved the 3-group Ethington data in a file labeled tda1.dat

MORRIS PROMPTS FOR ALL-POSSIBLE-SUBSET ANALYSES

```
Enter number of discriminating variables (Max set to 14)
9
Enter # (0 to 8) of variables to be "FORCED" to remain in the subsets
3
Enter 3 variable indices (separated by spaces) to be "FORCED"
7 8 9
Enter 3 group indices (separated by spaces)
1 2 3
Total number of best subsets to identify (MAX set to 280)
10
Number of best subsets of each size (MAX set to 31)
10
Enter listwise missing data code
-9
Enter the data file name
tda1.dat
Enter FORTRAN- type Format (GROUP MEMBERSHIP LAST)
(T2,F1.0,8F2.0,T1,F1.0)
Enter1 (0 OTHERWISE) for unequal Priors
1
Enter Priors for Groups separated by spaces
.25 .50 .25
Enter to submit the program
```

OUTPUT

Analysis: Morris Best-Subset Identification

The total-group linear L-O-O results of the Morris best-subset analysis of the 3-group Ethington data set are given in Table 17.1 and Figure 17.1.

TABLE 17.1 Total-Group L-O-O Hit Rates for Variable Subsets from the 3-Group Ethington Data

Subset Size	Best Subset	Hit Rate
3	X_7, X_8, X_9	.462
4	X_5, X_7, X_8, X_9	.481
5	X_3, X_5, X_7, X_8, X_9	.489
6	$X_3, X_5, X_6, X_7, X_8, X_9$	.504
7	$X_1, X_2, X_3, X_5, X_7, X_8, X_9$	.511
8	$X_1, X_2, X_3, X_4, X_5, X_7, X_8, X_9$	.492
9	$X_1, X_2, \ldots, X_9$	.485

Interpretation: Morris Best-Subset Analysis

From these results, it appears that the total-group hit rate increases as one, two, three, and four variables (from .462 to .511) are supplemented to the subset of X_7, X_8, and X_9. But, as the fifth (X_4) and sixth (X_6) variables are added, the hit rate decreases (from .511 to .485). Going "strictly by the numbers," it *may* be concluded that the subset to be retained is that excluding X_4 and X_6.

It should be noted that there may very well be two (or more) subsets of a given size that yield hit rates that are "close." The subset of choice may be based on predictor set collections—a researcher judgment call.

If one is into basing predictor deletion on the data on hand, the recommendation here is to use the all-possible-subsets approach. This approach may be used with or without forcing some predictors to be common to all subsets to be considered. With the all-possible-suspects approach, one should also think seriously about considering multiple subsets of a given size when determining the final subset of predictors. It is

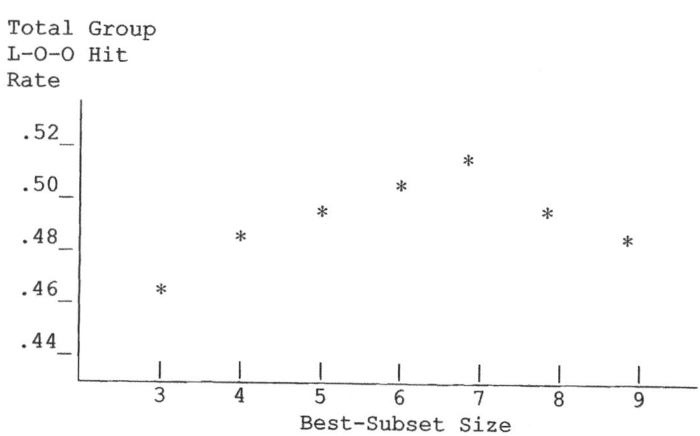

Figure 17.1 Total group L-O-O hit rate versus best-subset size for the 3-group Ethington data.

recognized that an all-possible-subsets analysis may be criticized as one that "milks the data," and that such an analysis "capitalizes on chance." Be that as it may, this is the deletion analysis favored by us.

17.4 PREDICTOR ORDERING

The problem of predictor ordering in the current context pertains to relative contribution, or importance, of the predictor variables to classification accuracy. Therefore, it makes sense to deal with an index that involves a proportion of correct classifications; that is, a hit rate. (If misclassification costs were to be involved, an index of expected loss would be appropriate.) In multiple regression analysis and in descriptive discriminant analysis, a popular (but highly questionable) approach to variable ordering is to use a stepwise analysis.

For predictor ordering (as well as predictor deletion) purposes, the (forward) stepwise analysis programs available through the SAS and SPSS packages are *generally not* appropriate for use with a predictive discriminant analysis problem. The basic reason these programs are generally inappropriate is that the criterion used for entering variables into the analysis is one pertaining to group separation (typically Wilks Λ), *not* classification accuracy. If, however, a normal linear rule with equal base rates is to be used [see cell (2,2) in Table 13.3], the Λ criterion would be equivalent to a classification criterion.

17.4.1 Meaning of Importance

It has been argued (e.g., Huberty, 1989) that a stepwise discriminant analysis should *not* be used for the purpose of determining relative predictor importance. In assessing the relative importance of a predictor, one might ask: How well (in the sense of classification accuracy) can we do without the predictor? Viewing relative variable importance in this manner, one conducts p (the number of predictors) PDAs, each involving $p - 1$ predictors. That is, conduct an analysis with X_1 deleted, then one with X_2 deleted, and so on. Suppose that the analysis conducted with X_1 deleted yields the lowest hit rate of interest (for the total group or for a particular group); then X_1 would be judged to contribute most to predictive accuracy. That is, the poorest classification accuracy would result if X_1 were deleted; therefore, X_1 is considered "most important."

17.4.2 Variable Ranking

Visual Ranking As an example, consider the 3-group Ethington data set (3GED). In Table 17.2 the Group 2 L-O-O hit rates are given for the nine 8-variable analyses

TABLE 17.2 Linear L-O-O Group 2 Hit Rates, Transformed Hit Rates, and Predictor Ranks for the 3-Group Ethington Data

Predictor Deleted	$H_{(i)}$	$Z_{(i)}$	Rank
X_2	.812	6.89	3
X_4	.812	6.89	3
X_9	.812	6.89	3
X_7	.820	7.07	3
X_8	.820	7.07	3
X_6	.828	7.25	6.5
X_5	.836	7.42	6.5
X_1	.852	7.78	8.5
X_3	.861	7.97	8.5
(None)	(.812)		

(ordered). There are at least two approaches that one might use to arrive at a reasonable rank ordering of the predictors based on hit rates as given in Table 17.2.

One approach is simply an "eyeball assessment." By merely looking at the second column in Table 17.2, one might suggest the following ordering:

Predictors	Rank
X_2, X_4, X_9	2
X_7, X_8	4.5
X_5, X_6	6.5
X_1, X_3	8.5

Such an eyeball variable ranking may be questionable because some Group 2 hit rates are quite close; for example, .820 for X_7 and X_8, and .828 for X_6. The reader may very well arrive at yet another variable rank ordering. A problem, then, with an eyeball assessment is that visual perceptions may vary (considerably?) from researcher to researcher. That is, this assessment may be too "subjective" for some researchers. Subjectivity may be greater with a "low" total group (or separate groups) hit rate.

Ranking Transformed Hit Rates The *second* approach to arrive at a variable ranking involves some quantification as a basis for increased "objectivity." [Huberty and Wisenbaker (1992b) suggest a variable ranking approach in the context of group separation. The basic idea of this approach will now be applied in the PDA context.] It seems reasonable to somehow take into consideration the variability of the set of hit rates, such as those in Table 17.2. One way to do this for total-group hit rates is as follows. Find the expected number of total group hits:

$$e = \sum_{j=1}^{J} e_j = \sum_{j=1}^{J} q_j n_j,$$

where q_j is the estimated prior probability of membership in Group j. Let $H_{(i)}$ denote the observed total-group hit rate obtained with predictor X_i deleted. Then $N \cdot H_{(i)}$ would be the observed number of total-group hits. The variance of $N \cdot H_{(i)}$ is given by $e(N - e)/N$. Thus, a reasonable transformation of $H_{(i)}$ that reflects this variability is

$$Z_{(i)} = \frac{N \cdot H_{(i)} - e}{\sqrt{e(N - e)/N}}, \tag{17.1}$$

where $N = \sum n_j$. The $Z_{(i)}$ values are then ordered from lowest to highest. The "best" predictor is the one that is associated with the lowest $Z_{(i)}$ value.

A special case of (17.1) may be used for particular-group hits, rather than for total-group hits. To demonstrate this special application, we will now consider Group 2 of the 3-group Ethington data. For Group 2, the following index is used in place of (17.1):

$$Z_{(i)} = \frac{n_2 \cdot H_{(i)} - e_2}{\sqrt{e_2(n_2 - e_2)/n_2}},$$

where $H_{(i)}$ now denotes the hit rate for Group 2 obtained with predictor X_i deleted. The transformed hit rate for each predictor is in the third column of Table 17.2.

For the 3-group Ethington data and focusing on Group 2, $Z_{(2)} = 6.89$ is lowest. What is needed next is a numerical value (based on a standardized scale) to add to the lowest $Z_{(i)}$ value to determine a cutoff for ranking purposes. This numerical value would, typically, range from .1 to .6. Which value to use is a judgment call on the part of the researcher. Trial and error indicated in the current situation that using .2 is reasonable in terms of sensitivity to hit-rate differences and tied ranks. For this example, $Z_{(6)}(= 7.25)$ is the smallest value that is greater than $Z_{(2)}$. The next step is to find $Z_{(6)} + .2$. Predictors with $Z_{(i)}$ values less than $Z_{(6)} + .2 (= 7.45)$ are assigned the same rank. This procedure is continued until all p predictors are ranked. For this example, ranks of the nine predictors are given in the rightmost column in Table 17.2.

A comparison of the hit rates obtained when the respective predictors are deleted with the all-variable hit rate—.811 for this data set—suggests a simplification of the ranking process. If deleting a variable does not decrease the hit rate, it seems reasonable that such a variable would be judged to be unimportant. Such unimportant variables may be ignored in the ranking process. For example, with this data set (see Table 17.2) only variables X_1, X_3, X_5, X_6, X_7, and X_8 *might* be retained.

This ordering procedure may be used for predictor ordering for each of the J-group hit rates, as well as for the total-group hit rate. Multiple predictor orderings may be substantive interest.

17.5 REANALYSIS

By examining formulations of classification statistics presented in Table 13.3, it may be concluded that correlations (or covariances) among the predictor variables are "taken into consideration." This is the case because involved in all rule forms is a distance index that is a function of a predictor variable covariance matrix. Just what effect these intercorrelations have on classification accuracy is not well known, except in some very special situations.[1]

With more than two predictors and more than two groups, the effects of increasing the number of correlations is not describable in simple terms. The point to be made here is that classification results should be interpreted in light of the predictors involved in the analysis. Therefore, it is desirable to delete some predictors on the basis of methods discussed earlier in this chapter, a *reanalysis* using only the selected predictors should be conducted.

Suppose, from a 12-predictor analysis, it is decided that five predictors should be deleted. If it would be informative to determine an ordering of the seven predictors retained, the ordering should be based on a 7-predictor analysis, *not* on the 12-predictor analysis. It must be remembered that an interpretative statement about a single variable or a subset of the total set of variables in a multivariate analysis must be considered *in the company* of all of the remaining variables analyzed.

The approach to predictor ordering presented in this chapter represents an assessment of relative predictor contribution (to classification accuracy) that depends on the data on hand. That is, a resultant predictor ordering depends on the predictor set and, of course, on the predictor measurements available. So, one must be careful in comparing relative importance of predictors across studies for reasons other than the involvement of a different sample of analysis units.

17.6 COMMENTS

Predictor *deletion* is a very important consideration to be made in the context of predictive discriminant analysis (PDA). Group assignment and assessment of predictive accuracy are essential concerns in PDA, and the accuracy and precision of group membership prediction both may be enhanced with *fewer* predictors than the total number initially included for study. Therefore, it behooves the researcher to consider deleting some predictors in forming a final prediction rule. [It should be noted that such is *not* the concern when it comes to considering the deletion of variables in the context of descriptive discriminant analysis (DDA); see Section 6.2.] When some predictors are deleted, it may be of interest to determine if there are units with changed group assignment relative to that when the deleted predictors are included in the rule.

[1] Cochran (1964) has shown that for a particular two-group bivariate normal situation, a negative correlation enhances classification accuracy, and accuracy is dampened if a positive correlation is too high or too low. Elashoff et al. (1967) have shown that these conclusions are not generally applicable for dichotomous predictors.

Predictor *ordering* is usually done for substantive rather than empirical reasons. It simply seems "natural" for a researcher to want to comment about predictors that are (relatively) "important" and those that are (relatively) "unimportant."

Predictor deletion and predictor ordering may be considered in terms of classification accuracy for the *total group* of units. In some research situations, however, it may very well be the case that predictive accuracy for a particular group is of special concern. In this case, predictor deletion and ordering should be considered in terms of the estimated hit rate for the particular group of interest. For instance, if a researcher is interested in identifying a particular type of disabled child—say, dyslexic as opposed to "normal" children—in a two-group PDA situation, the criterion of correct classification for the smaller group may be considered in deleting a subset of predictors and in ordering the predictors.

All of the discussion in this chapter has been presented under the assumption of equal population covariance matrices. That is, analysis procedures suggested for deleting and ordering predictors were those that call for equal covariance matrices. Of course, in a PDA context, it is the accuracy of classification—separate-group or total-group—that is of concern when deleting or ordering predictors. With equality of covariance matrices being untenable, a quadratic classification rule (see Section 13.3) may be used to generate the necessary classification results. Furthermore, some nonnormal rule (see Section 19.2) may even be used to determine predictor subsets or to assess relative predictor importance. No matter what form of classification rule is used, the important thing to consider is that some estimate of classification accuracy be used as a criterion for predictor deletion and for predictor ordering.

The estimators considered in this chapter are counting estimators. The ideas expressed with respect to predictor deletion and ordering may also be incorporated using posterior probability estimators. (See Section 15.3 for a discussion of counting and posterior probability estimators.)

There are two generalizability problems with which one must come face to face when conducting predictor deletion and predictor ordering analyses (see Huberty, 1989, pp. 62–64). The *first* problem pertains to *model specificity*. It was emphasized in Section 1.5 that consideration of the initial choice of response variables—predictors in a PDA context—is very important. Thorough study and sound judgment are needed in choosing predictors at the outset.

As many relevant predictors (relevance being based on substantive theory) as feasible need to be chosen for inclusion in the initial predictor system, and as many irrelevant predictors as possible need to be excluded. This is an easier-said-than-done situation, of course. Limited knowledge and resources sometimes preclude the researcher from including all relevant predictors and from excluding all irrelevant predictors. Herein lies the problem. That is, the final predictor subset obtained and the ordering of the predictors in this subset should be discussed only with all accompanying predictors (and measurement scales used) in mind. A decided-upon final subset out of an initial set of p predictors may not be a best subset if one or more predictors were to be added to, or deleted from, the initial set. Furthermore, a predictor that is considered very important (by whatever index) in, say, a three-predictor situation, may be relatively unimportant (by the same index) in a four-predictor situation.

The *second* generalizability problem pertains to *sample specificity*. As well as considering the predictors (and how they are measured), one should consider the (design) sample of analysis units in drawing conclusions about predictor deletion and ordering. Strictly speaking, results of deleting and ordering predictors should only be considered descriptive for the sample of units on hand. That is, inferences about best subsets and predictor importance to other units should be made with great caution. The best predictor subset for one sample of units may not be the best for other samples. The greater the ratio of sample size to number of predictor variables, the more reasonable are the implied generalizations. A large such ratio alone, however, does not ensure valid generalizations. Valid generalizations may be obtained only to the extent that the pattern of predictor variable intercorrelations for nondesign sample analysis units follow the pattern present in the design sample. To enhance the validity of one's conclusions regarding predictor deletion and ordering, some types of external analysis should be conducted. One way to accomplish this is to employ a hold-out or test sample. Preferably, one can use a resampling strategy such as jackknifing or bootstrapping.

A *final* word about predictor deletion and predictor ordering is needed. Since the advent of statistical computing packages, the typical approach to variable deletion and ordering taken by researchers has been to employ a stepwise discriminant analysis program. Beyond a pre-PDA and a preliminary exploratory analysis, the advice given here for using such a computer program is simple: Don't do it! See Huberty (1989) for elaboration on this advice, the substance of which was advanced earlier in this chapter.

17.7 SIDE NOTE

There is an eyeball method of deleting a predictor that *may* be attractive to some applied researchers. It was discussed in Section 13.4 that a classification rule may be thought of as a set J linear composites of the p predictor variables. Recall that an analysis unit is assigned to that group with which is associated the largest composite score. Suppose the J composite weights for predictor i are numerically very close. If so, then it might seem that predictor i contributes equally to all J composites. Thus, if this predictor were deleted, the relative magnitudes of the J composites would be nearly the same as when that questionable predictor is retained.

For example, consider the three LCFs given in Table 13.1. The three weights for predictor qelib might be considered to be "equal"; the same for qewrite. Linear L-O-O analysis on the 3-group Ethington data yielded the following total-group hit rates:

Predictor Deleted	Hit Rate
qelib (X_4)	.489
qewrite (X_8)	.485
Both	.496
(None)	.485

So, for this data set, this eyeball method of seeking predictors to delete appears to work well.

Further Reading

Duarte Silva (2001) suggests a rather general algorithm for variable deletion that can be used in contexts of MANOVA, CCA, DDA, and PDA.

Ganeshanandam and Krzanowski (1989) suggest an approach to predictor deletion that is the same as the Smith (1984) forward selection method. Monte Carlo simulations of multivariate binary and normal data in the two-group setting are conducted.

Habbema and Hermans (1977) present a variable deletion algorithm that is very general; because density values are estimated directly from the data on hand, no distribution assumption (e.g., normality) is involved.

Hand (1997, pp. 149–152) presents a readable point of view on predictor deletion.

Krzanowski (1995) applies the leave-one-out method of two- and three-group classification in the context of predictor deletion with a mixture of continuous and binary predictors; real data as well as simulated data were used to compare three deletion methods. A distance measure proposed by the author was the "winner."

Langbehn and Woolson (1997) conduct a Monte Carlo comparison of nine different methods for forward stepwise deletion involving the unweighted sum of binary (predictor) variables (SBV).

McKay and Campbell (1982) provide one of the first reviews of the variable deletion problem in predictive discriminant analysis and discuss how deletion in PDA relates to deletion in descriptive discriminant analysis.

Nath and Jones (1988) suggest an approach to deleting and ordering predictors using a linear programming approach with jackknife methods for the two-group situation.

O'Gorman and Woolson (1993) recommend routine application of rank transformations for (predictor) variable deletion as a preanalysis using PDA methods for studies with $N < 100$.

Pynnönen (1988) discusses the predictor deletion problem in a quadratic PDA context.

Rencher (1998, pp. 250–251) provides four references that discuss the deletion of predictors in the case of heteroscedastic covariance matrices.

Seaman and Young (1990) propose a predictor deletion algorithm as an alternative to an all-possible-subset approach—for two groups only. The criterion of subset effectiveness used is an estimated probability of misclassification using a L-O-O approach.

Snapinn and Knoke (1988) propose an "adjusted" bootstrap method of estimating error rates in the predictor deletion process.

Snapinn and Knoke (1989) review a number of methods to consider in deleting predictors and in estimating error rates in PDA; smoothing error-rate estimators are favored.

Definition ***Standard deviate***: Transvestite in the traditional drag.

EXERCISES

1. Why is the variable deletion problem so important in the context of a PDA?

Computer Applications

2. Using a Morris analysis on the 3-group Ethington data set (3GED), what are the three best subsets of size 4, all of which include X_7, X_8, and X_9? Use a deletion criterion of classification accuracy for Group 2 only.

3. Consider the best subset of size 6 from Table 17.1. Order these six predictors by carrying out six 5-predictor analyses, as discussed in Section 17.3. (Use the linear L-O-O total-group hit rate.)

4. Consider the 5-group Ethington data set (5GED) (using respective group priors of .15, .20, .30, .20, .15).

(a) Obtain all possible predictor subsets using the Morris program for linear L-O-O total-group results.

(b) Consider the best subset of size 5; order the five predictors in terms of total-group predictive accuracy.

5. Consider your personal data set (without categorical predictors).

(a) Determine a good predictor subset of whatever size you want.

(b) Order the predictors in the subset retained.

CHAPTER 18

Two-Group Classification

18.1 INTRODUCTION

A great deal of effort has been devoted to the study of the two-group predictive discriminant analysis problem over the past six decades. Many practical issues in the two-group situation have essentially been resolved, but in the statistical literature, at least, specific aspects of the two-group situation continue to be studied. In the context of two-group classification, current research areas include robustness of classification rules, effects of outliers on classification accuracy, effects of nonnormality and unequal covariance matrices on classification accuracy, classification via density estimation, nonparametric rules, classification with a mixture of variable types, and classification accuracy as a criterion in predictor selection. (See the Further Reading at the end of this chapter.)

Some of the peculiarities of the two-group situation have been mentioned previously. For example, in Section 15.3 a two-group "shrinkage formula" for hit rate estimation was presented. The specific peculiarity of the relationship between two-group classification and multiple regression analysis has been discussed by numerous writers—see, for example, the book by B. Flury (1951–1999) and Riedwyl (1988, pp. 94–96) and the book by Rencher (2002, pp. 275–276). The J-group classification rules will now be reformulated for the two-group case, which will lead into an analogy with regression analysis. This analogy is followed by a discussion of a relationship between multiple regression analysis (MRA) and predictive discriminant analysis (PDA), and a discussion of sample size requirements in a two-group PDA situation. The chapter concludes with a brief discussion of univariate classification.

18.2 TWO-GROUP RULE

As discussed in Chapter 12 [see (12.9) and (12.10)], a form of the Bayesian probability rule may be stated as:

Applied MANOVA and Discriminant Analysis, Second Edition, by Carl J. Huberty and Stephen Olejnik
Copyright © 2006 John Wiley & Sons, Inc.

> Assign unit u to Population j if
> $$\pi_j \cdot f(\mathbf{x}_u|j) > \pi_{j'} \cdot f(\mathbf{x}_u|j')$$
> for $j \neq j'$.

(18.1)

In the two-group case, the rule becomes

> Assign unit u to Population 1 if
> $$\pi_1 \cdot f(\mathbf{x}_u|1) > \pi_2 \cdot f(\mathbf{x}_u|2);$$
> otherwise, assign the unit to Population 2.

(18.2)

According to McLachlan (1992, p. 8), B. L. Welch (1911–1989) showed in 1939 that (18.2) is deducible either from the Bayes Theorem or the Neyman–Pearson Lemma. An alternative expression of the rule, and one that is typically given in the literature, is

> Assign unit u to Population 1 if
> $$\frac{f(\mathbf{x}_u|1)}{f(\mathbf{x}_u|2)} > \frac{\pi_2}{\pi_1};$$
> otherwise, assign the unit to Population 2.

(18.3)

If differential costs of misclassification (see Section 13.7) are to be considered, the ratio on the right in (18.3) becomes

$$\frac{C(1|2)\pi_2}{C(2|1)\pi_1}, \tag{18.4}$$

where $C(1|2)$ is the cost of classifying an analysis unit into Population 1 when it actually belongs to Population 2. Of course, only relative costs [i.e., $C(1|2)/C(2|1)$] need to be specified. For example, if it is three times as consequential to erroneously classify a unit into Population 1, then let $C(1|2)/C(2|1) = 3$.

If misclassification costs and prior probabilities are assumed equal, we have the rule:

> Assign unit u to Population 1 if
> $$\frac{f(\mathbf{x}_u|1)}{f(\mathbf{x}_u|2)} > 1;$$
> otherwise, assign the unit to Population 2.

(18.5)

To make any of these rules functional, we need to impose some probability model to obtain estimates of the densities $f(\mathbf{x}_u|j)$, $j = 1, 2$. Assuming a multivariate normal probability model with equal covariance matrices, employing logarithms (as in Section 13.4), and performing some algebra, it can be shown that the natural logarithm of the ratio of estimated densities in (18.5) is equal to

$$[(\bar{\mathbf{x}}_1 - \bar{\mathbf{x}}_2)'\mathbf{S}_e^{-1}]\mathbf{x}_u - \frac{1}{2}(\bar{\mathbf{x}}_1'\mathbf{S}_e^{-1}\bar{\mathbf{x}}_1 - \bar{\mathbf{x}}_2'\mathbf{S}_e^{-1}\bar{\mathbf{x}}_2). \tag{18.6}$$

Thus rule (18.5) may, therefore, be stated as:

> Assign unit u to Population 1 if
>
> $$\left[(\bar{\mathbf{x}}_1 - \bar{\mathbf{x}}_2)'\mathbf{S}_e^{-1}\right]\mathbf{x}_u - \frac{1}{2}\left(\bar{\mathbf{x}}_1'\mathbf{S}_e^{-1}\bar{\mathbf{x}}_1 - \bar{\mathbf{x}}_2'\mathbf{S}_e^{-1}\bar{\mathbf{x}}_2\right) > 0; \tag{18.7}$$
>
> otherwise, assign the unit to Population 2.

We now turn to the formal relationship between two-group classification (i.e., group membership prediction) and multiple regression analysis.

18.3 REGRESSION ANALOGY

In a two-group situation under the condition of equality of the two population covariance matrices, there are two linear classification functions (LCFs). For Group 1 we write [see (13.21)]

$$L_{u1} = (\bar{\mathbf{x}}_1'\mathbf{S}_e^{-1})\mathbf{x}_u - \frac{1}{2}\bar{\mathbf{x}}_1'\mathbf{S}_e^{-1}\bar{\mathbf{x}}_1 + \ln q_1;$$

and for Group 2,

$$L_{u2} = (\bar{\mathbf{x}}_2'\mathbf{S}_e^{-1})\mathbf{x}_u - \frac{1}{2}\bar{\mathbf{x}}_2'\mathbf{S}_e^{-1}\bar{\mathbf{x}}_2 + \ln q_2,$$

where $\bar{\mathbf{x}}_j$ is the $p \times 1$ vector of predictor means for Group j, $\mathbf{S}_e$ is the $p \times p$ error covariance matrix, $\mathbf{x}_u$ is the $p \times 1$ vector of observations for unit u, q_j is the base rate (or, estimated prior probability) for Group j, and $j = 1, 2$. Now, the difference between L_{u1} and L_{u2} is

$$L_u = \left[(\bar{\mathbf{x}}_1 - \bar{\mathbf{x}}_2)'\mathbf{S}_e^{-1}\right]\mathbf{x}_u - a, \tag{18.8}$$

where $a = \frac{1}{2}(\bar{\mathbf{x}}_1'\mathbf{S}_e^{-1}\bar{\mathbf{x}}_1 - \bar{\mathbf{x}}_2'\mathbf{S}_e^{-1}\bar{\mathbf{x}}_2) + \ln q_2 - \ln q_1$. Noting that $-(\ln q_2 - \ln q_1) = -\ln(q_2/q_1)$, the expression for L_u in (18.8) may be used as a basis for a classification rule:

> Assign unit u to Population 1 if
>
> $$\left[(\overline{\mathbf{x}}_1 - \overline{\mathbf{x}}_2)'\mathbf{S}_e^{-1}\right]\mathbf{x}_u - \frac{1}{2}\left(\overline{\mathbf{x}}_1'\mathbf{S}_e^{-1}\overline{\mathbf{x}}_1 - \overline{\mathbf{x}}_2'\mathbf{S}_e^{-1}\overline{\mathbf{x}}_2\right) > \ln\frac{q_2}{q_1}; \qquad (18.9)$$
>
> otherwise, assign the unit to Population 2.

[For the two-group case, this rule yields results that are identical to those obtained using rule (13.22).]

If the estimated prior probabilities of group membership are equal (i.e., $q_1 = q_2$), rule (18.9) is simplified to rule (18.7). It may be noted that rule (18.7) is based on a linear composite[1] of the predictor scores ($\mathbf{x}_u$); the weights for the p predictors are the elements in the row vector

$$\mathbf{b}' = (\overline{\mathbf{x}}_1 - \overline{\mathbf{x}}_2)'\,\mathbf{S}_e^{-1}. \qquad (18.10)$$

Furthermore, the midpoint of the means of this linear composite scores for the two groups is $\frac{1}{2}(\overline{\mathbf{x}}_1'\mathbf{S}_e^{-1}\overline{\mathbf{x}}_1 - \overline{\mathbf{x}}_2'\mathbf{S}_e^{-1}\overline{\mathbf{x}}_2)$ (see Morrison, 1990, p. 769). So, the assignment procedure stated in (18.7) may be described as: Find the linear composite score for a given analysis unit; if this composite score is closer to the mean composite score for Group 1, assign the unit to Population 1; otherwise, assign the unit to Population 2.

As can be seen from (18.9), for unequal priors, $\ln(q_2/q_1)$ is a "cutoff score" to which values of the linear composite on the left in (18.9) may be compared.

There is a quadratic counterpart of (18.9) for which the same cutoff score is used. The quadratic composite is given by:

$$\frac{1}{2}\ln\frac{|\mathbf{S}_2|}{|\mathbf{S}_1|} - \frac{1}{2}(\mathbf{x}_u - \overline{\mathbf{x}}_1)'\mathbf{S}_1^{-1}(\mathbf{x}_u - \overline{\mathbf{x}}_1) + \frac{1}{2}(\mathbf{x}_u - \overline{\mathbf{x}}_2)'\mathbf{S}_2^{-1}(\mathbf{x}_u - \overline{\mathbf{x}}_2). \qquad (18.11)$$

It may be shown that the $(1 \times p)$ vector of weights, $\mathbf{b}' = (\overline{\mathbf{x}}_1 - \overline{\mathbf{x}}_2)'\mathbf{S}_e^{-1}$, is proportional to the $(1 \times p)$ vector of weights applicable to the p-variable scores in a multiple regression analysis (MRA) with a dichotomous criterion. Because a set of weights proportional to $\mathbf{b}$, say $c\mathbf{b}$, where c is a constant, would not affect assignment decisions, a two-group predictive discriminant analysis could be accomplished using MRA.

An *internal* two-group classification analysis may be completed by using a multiple regression computer program as follows. Assign a criterion score of 0 to each unit in Group 1, and a criterion score of 1 to each unit in Group 2. Submit the $N \times (p + 1)$ data matrix ($N = n_1 + n_2$) to a multiple regression program (e.g., SAS PROC STEPWISE or SPSS REGRESSION). To obtain a (2×2) classification table, then, it is necessary

[1]In two-group *descriptive* discriminant analysis, this composite is called the *linear discriminant function* (LDF). LDFs were discussed in some detail in Section 5.2.

TABLE 18.1 Regression Classification Results for Groups 1 and 2 of the 3-Group Ethington Data

		Predicted Group		
		1	2	Total
Actual	1	42	34	76
Group	2	17	105	122
Total		59	139	198

to count the number of predicted criterion scores closer to 0, and assign these to Group 1; those closer to 1 are assigned to Group 2.

As an example, consider Groups 1 and 2 of the 3-group Ethington data set (3GED). For each unit in the two groups, let Y be defined as follows for each unit:

$$Y = 0 \text{ if unit is from Group 1}$$
$$Y = 1 \text{ if unit is from Group 2}$$

Thus, from a multiple regression viewpoint, we have a set of nine predictors and a dichotomous criterion, Y. This data set was analyzed using SPSS REGRESSION. The results are given in Table 18.1. These results are precisely the same as those obtained if a *linear internal analysis with prior probabilities based on group sizes* was conducted [see Exercise 1(d)].

The italics in the immediately preceding sentence are used to emphasize that doing a multiple regression analysis is equivalent to a two-group PDA *only* under four constraints. These are: (1) a *linear* classification rule is used; (2) results of an *internal* PDA are considered; (3) the PDA is done with *group-size-based priors*; and (4) there are *equal costs* of misclassification. To simply state that a two-group PDA and a regression analysis (with a dichotomous criterion) "are equivalent" or "yield the same results" does not tell the whole story.

18.4 MRA–PDA RELATIONSHIP

When the criterion variable is dichotomous, there is an interesting relationship between the multiple correlation coefficient, R, and the distance between the two sample centroids, D. This relationship may be expressed as:

$$D^2 = \frac{R^2}{1 - R^2} \frac{N(N-2)}{n_1 n_2}, \tag{18.12}$$

where $N = n_1 + n_2$, or as:

$$R^2 = \frac{n_1 n_2 D^2}{N(N-2) + n_1 n_2 D^2}. \tag{18.13}$$

As such, R, which may be viewed as a canonical correlation in this context, may also be regarded as a relative measure of distance.

It is well known that the *true* probability of correct classification under conditions of equal costs and equal prior probabilities (with a linear rule) is the optimal hit rate (see Section 15.3),

$$P^{(o)} = \phi\left(\frac{1}{2}\Delta\right),$$

where ϕ denotes the standard normal distribution function, and [see (12.2)]

$$\Delta^2 = (\boldsymbol{\mu}_1 - \boldsymbol{\mu}_2)'\boldsymbol{\Sigma}^{-1}(\boldsymbol{\mu}_1 - \boldsymbol{\mu}_2).$$

One approach to estimating $P^{(o)}$ is to use the "plug-in" estimator

$$\hat{P}^{(o)} = \phi\left(\frac{1}{2}D\right). \tag{18.14}$$

This estimator is a (negatively) biased one, however, because D^2 is a (positively) biased estimator for Δ^2.

To obtain a less biased estimator for $P^{(o)}$ one might use a less biased estimator for Δ^2. To arrive at a "shrunken D^2" a reasonable approach would be to use the "shrunken R^2" in (18.12). In the context of multiple prediction, as opposed to multiple correlation, an acceptable shrunken R^2 is (see Huberty and Hussein, 2001)

$$\tilde{R}^2 = 1 - \frac{N+p}{N-p}(1 - R^2). \tag{18.15}$$

Equation (18.12) suggests that a shrunken D^2 might be obtained using

$$\tilde{D}^2 = \frac{\tilde{R}^2}{1 - \tilde{R}^2}\frac{N(N-2)}{n_1 n_2};$$

therefore, from (18.13) and (18.15) we get

$$\tilde{D}^2 = D^2\frac{N-p}{N+p} - \frac{2pN(N-2)}{(N+p)n_1 n_2}.$$

However, this estimator is not acceptable because for small D^2 and p large relative to N, $\tilde{D}^2$ may be negative. Another candidate for $\tilde{D}^2$, which is a biased estimator for Δ^2 but less biased than D^2, is one proposed by Lachenbruch and Mickey (1968, p. 763):

$$\tilde{D}^2 = D^2\frac{N-p-3}{N-2} - \frac{pN}{n_1 n_2}. \tag{18.16}$$

[This is the same as (15.1).]

The MRA–PDA relationship presented in this section is "neat" if the discussion is restricted to estimating the optimal hit rate, $P^{(o)}$. As noted in Section 15.3, however, the true hit rate of greater interest in practice is the actual hit rate, $P^{(a)}$. In the two-group context, G. J. McLachlan derived a formula estimator for $P^{(a)}$, $1 - Q$, which was discussed in Section 15.3 [see (15.6)].

A final comment about the relationship between MRA and PDA may be summarized as below.

MRA	PDA
1 group of units	$J \geq 2$ groups of units
1 predictor composite	J predictor composites

18.5 NECESSARY SAMPLE SIZE

A rule of thumb for minimum sample size in predictive discriminant analysis was advanced in Section 15.7. To repeat, it was suggested that the smallest group be comprised of at least $3p$ units. A basis for this guide is presented in this section. [This section is built largely on the work of Lachenbruch (1968).]

As mentioned in Section 18.4, the true or optimum[2] hit rate with equal priors and equal costs in a two-group situation is

$$P^{(o)} = \phi\left(\frac{1}{2}\Delta\right), \qquad (18.17)$$

where ϕ denotes the standard normal distribution function and Δ is the true inter-centroid distance. For the purpose of developing a sample size table, we consider estimating the *expected hit rate* over all samples of size n $(= n_1 = n_2)$, $\tilde{P}$[3]. The sample sizes sought are those required for $\tilde{P}$ to be within some specified value, γ, of the optimal hit rate, $P^{(o)}$. The sample size, n $(= n_1 = n_2)$, is a function of γ, Δ, and p, the number of predictors. The values used in developing the table are $\gamma = .05$ and .01, $\Delta = 1, 2$, and 3, and $p = 4, 8, 12, 16$, and 20. Minimum required sample sizes are given in Table 18.2.

Sample size requirements may be summarized as follows:

1. The greater the expected group separation, the smaller the sample size needed.
2. Smaller tolerance requires larger sample.
3. As the number of predictors increases, the required sample size increases, but the ratio of n to p decreases.

[2]A hit rate is optimum in the sense that it would be the hit rate if all parameters were known (see Section 15.2).

[3]
$$\tilde{P} = \phi\left[\frac{\Delta}{2}\sqrt{\frac{(N-p-2)(N-p-5)}{(N-p-3)(N-3)\left(1+\frac{2p}{\Delta^2 n}\right)}}\right].$$

TABLE 18.2 Minimum Sample Size, $n(= n_1 = n_2)$, in Each Group Required for $\overline{P}$ to be Within γ of $P^{(o)a}$

p	Δ	γ	$\overline{P}$	n	p	Δ	γ	$\overline{P}$	n
4	1	.05	.642	15	16	1	.05	.642	54
	2		.794	13		2		.791	41
	3		.884	10		3		.885	31
	1	.01	.682	89		1	.01	.682	350
	2		.831	56		2		.831	202
	3		.923	37		3		.923	125
8	1	.05	.642	28	20	1	.05	.642	67
	2		.792	22		2		.791	51
	3		.885	17		3		.884	38
	1	.01	.681	172		1	.01	.681	421
	2		.831	104		2		.831	250
	3		.923	66		3		.923	154
12	1	.05	.642	41					
	2		.792	32					
	3		.885	24					
	1	.01	.683	300					
	2		.831	152					
	3		.923	96					

[a] $P^{(o)} \doteq .691$ if $\Delta = 1$, .841 if $\Delta = 2$, and .933 if $\Delta = 3$.

4. Sample sizes for intermediate p values may be approximated using linear interpolation.

From Table 18.2 it may be observed that for the more reasonable values, $\gamma = .05$ and $\Delta = 2$, the required sample is about $3p$. This was the sample size requirement proposed in Section 15.7.

18.6 UNIVARIATE CLASSIFICATION

In many fields of study, the classification of analysis units into one of two groups or categories based on a single predictor variable has a fairly long history. Over the past five or six decades, univariate classification results have been utilized for two purposes: (1) to assess the magnitude of the effect for the grouping variable—when the two groups are pre-established and well defined—as it relates to the response variable, and (2) to establish a "good" cutoff response variable score for setting a standard so that the sample of units may be split into two groups to identify or label the units appropriately.

The notion underlying an index of *effect size* may be related to the measurement of overlap of two groups, an idea proposed over 65 years ago by J. W. Tilton (1891–1955) in 1937. This idea was explicitly tied to the notion of effect size by Alf and Abrahams (1968), Huberty and Holmes (1983), Huberty and Lowman (2000), Levy (1967), and Oakes (1986, pp. 53–55). The effect-size index used is a proportion of total-group correct classifications. The relationship between the proportion of correct classification and the squared point-biserial coefficient—an effect size index often used in a two-group t-test setting—is discussed by Alf and Abrahams (1968) and Levy (1967). The two-group univariate classification problem with a nonparametric solution is discussed by Soltysik and Yarnold (1994a) and Yarnold and Soltysik (1991).

When using a hit rate as an effect-size index for the two-group comparison problem, the roles of the two variables involved are reversed from the roles in a two-group t test. In the latter situation, the response variable is the outcome variable, while the grouping variable is the predictor. It is just the reverse in a two-group classification situation. For a researcher to use a hit rate as an effect-size indicator, it must "make sense" for the response variable to play the role of a predictor.

The second purpose of two-group univariate classification pertains to determining cutoff scores. In some contexts this amounts to the setting of standards. Serious consideration of prior probabilities of group membership (or base rates) in determining cutoffs was given about 50 years ago by P. E. Meehl (1920–2003) and Rosen (1955). The problem of determining optimal—in the sense of maximizing hit rates of interest—cutoff scores is discussed by Koffler (1980) and Rorer et al. (1966). Gottesman and Prescott (1989) review the use and abuse of a particular assessment instrument to classify individuals into one of two categories with consideration of prior probabilities, separate group hit rates, and misclassification costs. Classification rules considered are simply based on univariate cutoff scores. Hayes and Martin (1986) address the usefulness of a test as a screening device for the placement of gifted children. Various test cutoff scores are assessed via two-group classification analyses where a classification result is determined simply by whether a child's test score is above or below a specified cutoff value.

Further Reading

The two-group situation in PDA has been studied extensively by statisticians and methodologists. There are many writings devoted to both the theoretical and applied aspects of the two-group problem. Below are some readings pertaining mostly to the latter.

Aeberhard et al. (1993) propose a modification of the original regularized discriminant analysis two-group classification procedure that performs pretty well, even for a "small" n/p ratio.

Bedrick et al. (2000) give a method for estimating the Mahalanobis distance between two multivariate normal populations when a subset of the predictor variables is measured via ordered categorical responses.

Cawley and Talbot (2003) propose a two-group L-O-O analysis for the kernel function approach of linear PDA.

Chouvarda et al. (2003) give a very thorough discussion of a two-group PDA. They discuss predictor deletion and bootstrapping using "artificial" data sets as well as real data sets.

Cox and Pearce (1997) give an update on logistic classification and propose a robust adjustment to the approach for the two-group situation.

Dorans (1988) proposes a shrunken generalized distance estimator for D^2. Yet another formula estimator for the two-group total error rate is presented.

Duarte Silva et al. (2002) find that when misclassification costs differ widely in the two-group context, this may have a "major impact" on classification results.

Everitt and Der (1996, pp. 128–136) give a detailed discussion of a two-group PDA via SAS.

Flury and Riedwyl (1988) cover the two-group situation extensively, using the analogy between multiple regression and two-group PDA. In Chapter 8 they discuss "identification analysis," a sort of special case of two-group PDA with $n_1 = n$ and $n_2 = 1$. In Chapter 11 they present a novel approach to comparing the covariance structures of two groups.

Fung (1995a) proposes two measures of influence on individual misclassification probabilities in the normal-based linear two-group PDA.

Fung (1998) shows that two diagnostic measures he had proposed earlier for a PDA data set are asymptotically equivalent and illustrated the equivalence with two real two-group data sets.

Joachimsthaler and Stam (1989, 1990) review mathematical programming approaches for the classification problem in two-group discriminant analysis.

Koolaard and Lawoko (1993) report results of an extensive simulation study that compared eight methods of estimating error rates in a two-group linear PDA context involving "correlated training data"; very briefly, L-O-O is one of the recommended methods.

Koolaard and Lawako (1996) compare simulated two-group linear normal-based PDA results with those based on a "Euclidian distance function." Comparative results depend on predictor intercorrelations.

Lei and Koehly (2003) simulate two-group data to compare predictive accuracy of PDA versus logistic regression; no appreciable differences were found for a number of imposed conditions.

Loh (1995) studies the efficiency of a "new adaptive ridge (linear) classification rule" in a two-group context and concluded that it "performs reasonably well."

Meshbane and Morris (1995) review the issue of equal versus unequal priors in a two-group PDA; a simulation comparison indicated that use of equal priors yielded good results when sample group sizes are similar.

Rubin (1990) finds, through simulation, that a quadratic rule was superior to 15 linear programming rules in the two-group case.

Rudolph and Karson (1988) discuss the effect of unequal priors and unequal misclassification costs. Even though the expression "MDA" is used, the authors deal only with the two-group case.

Snapinn and Knoke (1988) review bootstrapped and smoothed classification error estimators. The focus for comparing four error-rate estimators is on an external analysis.

Soltysik and Yarnold (1994b) present the multivariate optimal discriminant analysis (MultiODA) approach with $J = 2$.

Vlachonikolis (1990) discusses two-group L-O-O PDA with equal priors using a mixture of binary and continuous predictors; three proposed error-rate estimates yielded decent results.

Xiao (1994) presents a modification—"regularization"—of the nonlinear mathematical programming approach to two-group PDA; some computational intensity is reduced.

Yarnold and Soltysik (1991) apply the "optimal linear discriminant analysis" in a single predictor context; a maximum classification accuracy (MCA) method is presented to determine the MCA for a maximum of $n_j = 30$.

Yarnold et al. (1994) briefly review six classification methods, and focus on one of them, optimal discriminant analysis (ODA), in a two-group, three-variable medical example; the MultiODA method that involved L-O-O mathematical/linear programming (which does not require normality or equal covariance matrices, but is very computationally intensive) outperformed the other five methods with $n_1 = 23$ and $n_2 = 22$.

Professor ***"Where is infinity?"***

Student ***"Don't know, but I drove it!"***

EXERCISES

Computer Applications

1. Using the SAS and SPSS programs and the first two groups of the 3-group Ethington data set (3GED), consider the nine variables as predictors and conduct two analyses: (i) a multiple regression analysis using the grouping variable (with two levels) as the criterion variable; and (ii) an internal PDA with a pooled covariance matrix, proportional priors, and equal misclassification costs.

(a) Determine the set of nine regression weights.

(b) Find the difference of the nine corresponding LCF weights for the two groups.

(c) Verify that the weights from part (a) are proportional to the differences in part (b) (i.e., the nine ratios are approximately the same).

(d) Verify that the results in Table 18.1 are the same as the results of your PDA.

(e) Find R^2 from your regression analysis and D^2 from your PDA. Then verify (18.12) or (18.13).

2. Using Table 18.2 as a reference, are the two group sizes ($n_1 = 76, n_2 = 122$) for the 3-group Ethington data adequate for accurate—however you define "accurate"—classification?

3. Suppose it is somehow determined that it is four times as serious to misclassify a Group 1 into Group 2 as it is to err the other way. That is, $C(1|2)/C(2|1) = \frac{1}{4}$. To take these relative misclassification errors into consideration [see (18.4)], let $q_1 = \frac{2}{3}$ and $q_2 = \frac{1}{3}$—ignoring classification costs, $q_1 = \frac{1}{3}$ and $q_2 = \frac{2}{3}$—and conduct both an internal analysis and an external analysis using the first two groups of the 3-group Ethington data set. (SAS DISCRIM may be used to accomplish this.) Note that $C(1|2)\pi_2/C(2|1)\pi_1 = \pi_2/4\pi_1$; this might imply the use of $q_1 = 4(\frac{1}{3}) = \frac{4}{3}$ and $q_2 = \frac{2}{3}$. But the package programs require the sum of the priors to be 1.0; therefore, $q_1 = \frac{2}{3}$ and $q_2 = \frac{1}{3}$ may be used. With $C(1|2)/C(2|1) = \frac{1}{4}$, conduct a PDA using Groups 1 and 2 of the 3-group Ethington data set. Compare your results with those obtained in Exercise 1(d).

4. For this exercise consider only two groups of your personal data set—whichever you want, assuming that you have more than two groups. (Delete categorical predictors, if any.)
   (a) Specify the two priors you would use.
   (b) Is it reasonable to consider differential costs of misclassification? If so, do an analysis with the unequal costs incorporated into your priors. How do your results compare with those for equal costs?
   (c) Check your group sizes with those in Table 18.2.

CHAPTER 19

Nonnormal Rules

19.1 INTRODUCTION

When approaching the problem of classification with measures on nonnormal variables, the door is open to a number of possibilities. The discussion in Chapter 13 dealt with classification involving a set of continuous predictor variables whose theoretical joint distribution was assumed to be multivariate normal. In this chapter we deal with nonnormal predictor variables. Two types of variables are considered: continuous nonnormal variables and categorical variables. The latter variables have also been labeled "qualitative" and are variables that are typically measured using a *nominal* scale.

Continuous nonnormal variables may come into play in a number of situations. One situation is when there is a "ceiling" (or "floor") effect that results when, for example, a performance test is administered. The reader can undoubtedly think of other situations that would yield asymmetric score distributions. Another situation may result when a distribution of scores for one or more groups on one or more variables is clearly bimodal. These are, perhaps, examples of hypothetical situations; examples of classification using readily known and recognized continuous nonnormal multivariate distributions may be somewhat rare.

Categorical response variables are, however, fairly common. Examples of categorical variables that have been used in the literature as group membership predictors are: sense modality preference, academic field, geographical residence, race, marital status, ethnic group, gender, academic degree, disease presence, and type of undergraduate training.

The ensuing discussion focuses on two approaches to handling data on nonnormal variables in a prediction problem. The *first* is to transform the data so that the transformed variables have good distributional properties. In particular, the target form of the distribution is typically that of multivariate normality. If approximate normality is obtained, the general form of classification rules discussed in Chapter 13

Applied MANOVA and Discriminant Analysis, Second Edition, by Carl J. Huberty and Stephen Olejnik
Copyright © 2006 John Wiley & Sons, Inc.

is applicable. Transformations appropriate for continuous nonnormal variables and for categorical variables are presented. The *second* approach discussed is that of obtaining estimates of posterior probabilities of group membership directly from the data rather than assuming any particular distribution form. Here, too, both continuous nonnormal and categorical variable situations are covered.

The interested reader may refer to Section 22.2 for more discussion of nonnormal rules.

19.2 CONTINUOUS VARIABLES

The three analyses on which we now focus are used when no knowledge of underlying predictor variable distribution form is available or assumed.

19.2.1 Rank Transformation Analysis

One data transformation that is generally applicable to nonnormal continuous distributions is the rank transformation. For a given predictor variable, all the $N = \Sigma n_j$ analysis unit observations in the J groups are pooled and then ranked from 1 for the smallest observation to N for the largest. Midranks are used for tied observations. This procedure is repeated across all p predictors. Then the usual computer package programs are applied as if the data samples were taken from normal distributions.

For a new analysis unit to be classified using the rule determined from the original data, the unit's observation must be assigned a rank. Let $X_{(l)}$ denote the lth ordered observation on a given variable in the original data set; let X_{N+1} denote the observation for the new unit, and let $R(X_{N+1})$ denote its rank to be assigned as follows:

1. If $X_{N+1} < X_{(1)}$, then $R(X_{N+1}) = R(X_{(1)}) = 1$.
2. If $X_{N+1} > X_{(N)}$, then $R(X_{N+1}) = R(X_{(N)}) = N$.
3. If $X_{N+1} = X_{(l)}$, then $R(X_{N+1}) = R(X_{(l)})$.
4. If $X_{(l)} < X_{N+1} < X_{(l+1)}, l = 1, 2, \ldots, N - 1$, then

$$R(X_{N+1}) = R(X_{(l)}) + [R(X_{(l+1)}) - R(X_{(l)})]\frac{X_{N+1} - X_{(l)}}{X_{(l+1)} - X_{(l)}}.$$

Note that $R(X_{N+1})$ need not be an integer. If cross-validation of classification results is desired (see Section 15.3), this interpolation process is a bit tedious. The process is, however, readily accomplished by using a fairly simply written computer program.

The analysis of the 3-group Ethington data set (3GED) using the rank transformation can be done with the SAS package using the following commands:

SAS SYNTAX FOR TRANSFORMING RAW DATA TO RANKS

```
proc rank data=A2 ties = mean out = two;
var counsum - - qesci;
ranks R1 - R9;

data A2;
set=two;
run;

proc discrim data=two pool=yes crossvalidate crosslist
  posterr;
class grade;
var consum - - qesci;
priors '1'=.25 '2'=.50 '3'=.25;
run;
```

The L-O-O classification results are given in Table 19.1. These results turn out to be the same as the linear L-O-O results obtained via the standard normal-based classification rule—see Table 15.2.

19.2.2 Nearest-Neighbor Analyses

It is reasonable to assume that analysis units that are close together (in the sense of some appropriate metric) will belong to the same group. Thus to classify a unit whose group membership is unknown, it may be desirable to weight the evidence of already classified nearby units most heavily. This is, in effect, what nearest-neighbor (NN) rules do.

Using a single-nearest-neighbor (1-NN) rule, a unit is classified into the group corresponding to the membership of the single unit that is closest. If the number of

TABLE 19.1 Linear L-O-O Rank-Based PDA Results for the 3-Group Ethington Data (3GED)

		Predicted Group			
		1	2	3	Total
Actual Group	1	16 (.211)	55	5	76
	2	12	99 (.811)	11	122
	3	6	47	13 (.197)	66
Total		34	201	29	264

units in each group is large, it makes some sense to consider, instead of the single nearest neighbor, the most frequently represented among M nearest neighbors. Of these M units, let m_j represent the number of units that belong to Group j. The posterior probability of unit u belonging to Group j is estimated by:

$$\hat{P}(j|\mathbf{x}_u) = \frac{q_j \cdot m_j}{\sum_{j'=1}^{J} q_{j'} \cdot m_{j'}}. \tag{19.1}$$

(This is similar to normal-based posterior probability expressions in Section 13.3.) With equal priors, the M-NN rule simply amounts to assigning the unit to the group for which the proportion of the M nearest neighbors is largest, because in this case,

$$\hat{P}(j|\mathbf{x}_u) = \frac{m_j}{\sum_{j'=1}^{J} m_{j'}}. \tag{19.2}$$

An estimate of $P(j|\mathbf{x}_u)$ for a given unit is the proportion of the M units that are in the neighborhood of the $\mathbf{x}_u$ that belong to Group j. The "neighborhood" of $\mathbf{x}_u$ is defined by the distance from $\mathbf{x}_u$ to the Mth nearest unit from each of the N units—adjusted by the priors as in (19.1).

Thus, the M-NN rule can be viewed as an attempt to estimate the posterior probabilities of group membership from the available data. It seems reasonable, then, to use a large value of M to obtain reliable estimates. On the other hand, it is desirable that the M nearest neighbors be very close to the unit under consideration; this forces the choice of a small M. Research on NN rules has not indicated how to choose an optimal M value for a given situation. Cover and Hart (1967) have shown that there exists no "optimum" M, $M \neq 1$, for an M-NN rule. In a given situation, a researcher might try, say, $M = 1$, 3, and 5, and make a choice by comparing the classification results.

An NN classification is easily obtained through the SAS DISCRIM procedure. This procedure has a number of options in addition to a choice of M—SAS uses K rather than M. Only three are mentioned here. *One* option is that two different metrics may be used to index the distance between units u and u': the basic squared Euclidean distance, $(\mathbf{x}_{u'} - \mathbf{x}_u)'(\mathbf{x}_{u'} - \mathbf{x}_u)$, or the squared Mahalanobis distance, $(\mathbf{x}_{iu'} - \mathbf{x}_u)'\mathbf{S}_e^{-1}(\mathbf{x}_{u'} - \mathbf{x}_u)$, where $\mathbf{S}_e$ is the $(p \times p)$ error covariance matrix. The latter distance index is, by default, used by SAS with the "pool = yes" command. A *second* option is that a minimum acceptable posterior probability of group membership (THRESHOLD=) may be specified. A *third* option is a L-O-O analysis (using CROSSVALIDATE). As pointed out by White (1994), however, there is (or, at least, was) a problem with the SAS NN program in terms of the L-O-O analysis. The problem is that a SAS NN L-O-O analysis yields "an optimistic bias" in the L-O-O results. White proposed a modification of the SAS procedure that removes the bias. (As of mid-2005, the SAS NN L-O-O output has not changed.)

A nearest-neighbor analysis may be conducted on the 3-group Ethington data set using the following SAS syntax.

SAS SYNTAX FOR A L-O-O NN ANALYSIS USING THREE GRADE GROUPS AND NINE PREDICTOR VARIABLES

```
proc discrim pool = yes crossvalidate crosslist K = 3;
class grade;
var counsum - - qesci;
priors '1'=.25 '2'=.50 '3'=.25;
run;
```

K=3 requests that a unit be classified based on the membership of the three units closest to the target unit.

(Note that SAS uses K rather than M to specify the neighborhood size.)

The results of the linear L-O-O 3-NN analysis are given in Table 19.2. The three separate group hit rates are .18, .66, and .18, respectively; the total group hit rate is .41. The hit rates for Group 1 and Group 3 are comparable to those reported in Table 15.2. For this data set, however, the 3-NN Group 2 hit rate of .66 is lower than the normal-based hit rate of .81.

To classify a **new analysis unit** via SAS DISCRIM, the "TESTDATA = dataset name " option is used to identify a SAS data set listing the units to be classified—see Section 16.9.1.

TABLE 19.2 L-O-O Linear 3-NN Results for the 3-Group Ethington Data

		Predicted Group			
		1	2	3	Total
Actual Group	1	14 (.184)	50	12	76
	2	20	81 (.664)	21	122
	3	11	43	12 (.182)	66
Total		45	174	45	264

19.2.3 Another Density Estimation Analysis

An expression for estimates of posterior probabilities of group membership was given in Section 13.3 [see (13.8)] and is repeated here:

$$\hat{P}(j|\mathbf{x}_u) = \frac{q_j \cdot \hat{f}(\mathbf{x}_u|j)}{\sum_{j'=1}^{J} q_{j'} \cdot \hat{f}(\mathbf{x}_u|j')}. \tag{19.3}$$

In Chapter 13, density estimates, $\hat{f}(\mathbf{x}_u|j)$, were obtained under multivariate normal models [see (13.7) and (13.13)], thus giving the maximum probability rules stated in (13.10) and (13.15). Earlier in this chapter, a simple "nonparametric" model was considered in arriving at posterior probability estimates for the M-NN rule [see (19.1)].

A more formal approach to density estimation in the context of predictive discriminant analysis involves the use of something called *kernel functions*. Details of this density estimation approach will not be given here (see, e.g., Hand, 1997, pp. 79–87). It suffices to say that the estimates, $\hat{f}(\mathbf{x}_u|j)$ in (19.3), are based directly and explicitly on the sample n_j vectors of observations. A computer program utilizing kernel estimators has been developed and is distributed by J. D. F. Habbema and associates in The Netherlands [see Habbema and Hermans (1977) for an application]. Five different kernel methods are available via the SAS DISCRIM procedure.

19.2.4 Other Analyses

At least two other ways of handling nonnormal continuous variables have been suggested and studied, to some extent. *First*, an alternative to the rank transformation, namely, the inverse normal scores transformation, was used in a $J = p = 2$ context by Koffler and Penfield (1982). (Only the case where the two predictor variables were uncorrelated was considered.)

Second, rather than transform the data or estimate densities from the data directly, one might assume that the posterior probabilities, $P(j|\mathbf{x}_u)$, have a logistic form. Most of the study of logistic classification has been restricted to the two-group case; Albert and Lesaffre (1986) do, however, discuss the multiple-group case. Albert and Lesaffre give credit to the pioneering work of J. A. Anderson (1939–1983) in the study of logistic classification. Logistic classification is discussed (with references) in Section 21.2.2.

19.3 CATEGORICAL VARIABLES

Just as when dealing with continuous nonnormal predictor variables, two approaches will be discussed for dealing with categorical predictors: (1) obtaining estimates of posterior probabilities directly from the data; and (2) transforming the data so that normal-based rules can be used. One analysis that exemplifies the first approach is discussed briefly, whereas three analyses involving transformations are described in some detail.

The analyses discussed in this section need only be considered with a restricted type of categorical variable. This type is the *unordered* categorical variable having more than two categories. Other types, such as ordered two-category, ordered multicategory, and unordered two-category, are easily handled by transforming them to a form that is adaptable for normal-based analyses. A binary or dichotomous variable can be scaled (or metricized or calibrated) by using a 0–1 assignment. It has been shown (Bryan, 1961, p. 735) that this "scaling" is optimal; Maxwell (1961) suggests the 0–1 scaling for dichotomous variables in the extraction of canonical variates. Thus for a variable such as Gender, the value for male could be 0, and that for female 1.

Another special type is an ordered categorical variable. The way this type can be handled is to use *integer scaling*. For example, with the variable socioeconomic status, it is reasonable to assign 1 for low, 2 for middle, and 3 for high. See Krauth (1986) for another scaling method for ordered categories.

So, then, the only type of categorical predictor that calls for special attention is one with three or more *unordered* categories, the type discussed in the remaining part of this section. Examples of such variables are marital status, college academic major, occupation, and disease diagnosis.

19.3.1 Direct Probability Estimation Analysis

The proportion of analysis units from a category of a variable being in a particular criterion group may be used as an estimate of the probability that units in the category belong to that group. This notion is the basis for estimating posterior probabilities of group membership directly from the sample data. There is an obvious problem with such an analysis. Unless the total number of variable-category "cells" is small relative to the total number of units, such probability estimates will be unreliable because some or many of the estimates may be based on little information. [This problem has become known as Bellman's *curse of dimensionality*; see Hand (1997, p. 80).] For this reason, plus the existence of other acceptable analyses (see the next three subsections), this analysis will not be elaborated upon here. Overall and Klett (1972, Chapter 16) present a detailed discussion, plus listings of computer programs for this analysis. McLachlan (1992, Chapter 7) discusses theoretical aspects of multinomial-based classification.

19.3.2 Dummy Variable Analysis

One way of dealing with a categorical variable is to transform it to binary form by means of defining dummy (or indicator) variables. For example, in place of $X =$ marital status ($1 =$ single, $2 =$ married, $3 =$ divorced or separated, $4 =$ widowed), $X_1 = 1$ if single, 0 otherwise; $X_2 = 1$ if married, 0 otherwise; $X_3 = 1$ if divorced or separated, 0 otherwise, may be used. For this example, the transformation produces three new variables from the one four-category variable. If there are p categorical variables, the ith one having c_i categories, the dummy variable transformation would yield $\Sigma_{i=1}^{p}(c_i - 1)$ different binary variables. Here, too, with this analysis one would be confronted with the curse-of-dimensionality problem.

19.3.3 Overall–Woodward Analysis

A *second* transformation is due to Overall and Woodward (1977). With this analysis, all p categorical variables are analyzed simultaneously. The method of analysis is developed by analogy to that of deriving linear discriminant functions (LDFs). It is essentially a principal component analysis of frequency patterns across the categorical variables. Suppose that there are p categorical variables, c_i categories for variable X_i, yielding a total of $C = \Sigma_{i=1}^{p} c_i$ categories. The $C \times N$ *incidence matrix* containing the category incidences is used to arrive at a $C \times J$ matrix, Z, of proportions of units in each of the J groups that fall in each of the C categories, deviated about the unweighted mean of each category proportion across all J groups. That is, the jth column of Z is defined by the $C \times 1$ vector $\mathbf{p}_j - \bar{\mathbf{p}}$, where $\mathbf{p}_j$ is the vector containing the proportion of units in Group j that fall in each category, and $\bar{\mathbf{p}}$ is the vector containing the unweighted mean of the category proportion vectors across all J groups. The Z matrix is then transformed to $\widetilde{Z} = D^{-1}Z$, where D is the $C \times C$ diagonal matrix containing the positive square roots of the mean (over the J groups) error variances for the 0–1 observations in the C categories.

The r principal components of the $C \times C$ matrix, $\widetilde{Z}\widetilde{Z}'$, are then found. Let $\widetilde{A}$ be the $C \times r$ matrix of the r components (i.e., eigenvectors) of $\widetilde{Z}\widetilde{Z}'$. The elements of $A = D^{-1}\widetilde{A}$ are the raw-score category weights that define the r components. Thus each unit has r values for analysis input. These r values are simply sums of the elements of A corresponding to categories in which the unit belongs. For this transformation, r new response "variables" are produced to represent the p categorical variables.

19.3.4 Fisher–Lancaster Analysis

The *third* and final transformation to be considered is one that yields scale values for each category, separately for each variable. For p categorical variables, there will result p sets of scale values. The procedure was independently developed, from different points of view, by Fisher (1940) and H. O. Lancaster (1913–2001) in 1957. The scale values are found, in a Fisher sense, so that the group discriminatory power of a given variable is maximized; in a Lancaster sense, the leading canonical correlation between the grouping variable and the categorical variable is maximized.

For a categorical variable X_i, a $c_i \times J$ table of frequencies is constructed. Then a $c_i \times c_i$ matrix, M, is formed; entries of M are functions of entries of the $c_i \times J$ incidence matrix:

$$m_{ii'} = \left[\sum_{j=1}^{J} \frac{n_{ij}n_{i'j}}{n_{\cdot j}} - \frac{n_{i\cdot}n_{i'\cdot}}{N} \right] (n_{i\cdot}n_{i'\cdot})^{-1/2},$$

where n_{ij} is the frequency in category i for Group j. The largest eigenvalue of M and the associated eigenvector, $\mathbf{v}$, is obtained. Scale values for category k of x_i are obtained using

$$w_{ik} = \frac{v_{ik}}{\sqrt{n_{k\cdot}}}. \tag{19.4}$$

The w_{ik} values are the optimum category values that are used in *scoring* the given categorical variable. That is, a unit in the kth category of variable X_i is given a *score* of w_{ik}. The process is repeated for all p variables, thus yielding q sets of w values.

What is accomplished via this analysis is a single "optimum" scaling of the categories for each categorical variable. In a canonical analysis, however, there often exists more than one nonzero eigenvalue of interest. In the present context, there are $s = \min(J, c_i)$ nonzero eigenvalues for each categorical variable, one of which is unity; interest is on the remaining $s - 1$ eigenvalues. Therefore, potentially, there are $s - 1$ sets of category values for each categorical variable. There is some question as to whether or not the less-than-optimal scales should be used. In the current book, only the first scale is considered.

The Fisher–Lancaster analysis has been studied extensively by Badarinathi (1983), Bryan (1961), and Kundert (1973). Henschke et al. (1974) present a detailed description of the technique—the radical sign in the formula for the w_{ik} values was inadvertently deleted by these authors on page 110. This transformation has the same goal as that of the computer algorithm discussed by Young et al. (1976); namely, to maximize a canonical correlation.

A listing of a computer program (written by J. M. Wisenbaker, University of Georgia) labeled FLPC that can be used to accomplish the Fisher–Lancaster analysis may be found at the Wiley website.

Of the four categorical variable analyses mentioned in this section, the Fisher–Lancaster approach is preferred; see Huberty et al. (1986) for elaboration of this preference. This approach is illustrated in Section 19.4.

19.3.5 Other Analyses

Other analyses for handling categorical variables, as well as the use of kernel functions, nearest-neighbor methods, and logistic functions, are discussed by Hand (1997, Chapter 5).

19.4 PREDICTOR MIXTURES

In many areas of study it is common to find that a set of predictors includes continuous as well as various types of categorical variables. Examples of such sets appear in medicine (Afifi and Azen, 1979, pp. 16, 23), in pharmacy (Solander, 1978), in social science (Talarico, 1978), in marketing (Dillon et al., 1978) and in education (Bobbitt, 1990), to list a few. It is desirable to have analysis methods to handle mixtures of variable types.

With the analyses discussed in Sections 19.2 and 19.3, one can entertain the use of virtually any mixture of variables. An application of the Fisher–Lancaster method will now be illustrated.

Consider the HSB data set described in Appendix A. With this data set, X_1 (Locus of Control) and X_2 (Self-Concept) are continuous, whereas X_3 (Occupational Aspiration at age 30) and X_4 (Main Activity in the year after high school) are unordered

TABLE 19.3 Category Weights for X_3 and X_4 in the HSB Data

Variable	Category	Weight
X_3	1	−1.625
	2	.636
	3	.952
	4	−.535
	5	1.040
X_4	1	1.123
	2	−1.060
	3	.778
	4	.805

categorical variables. Variable X_3 has five categories while X_4 has four categories. Even though more than one set of Fisher–Lancaster weights [see Eq. (19.4)] is obtainable with the HSB data, only the first or "leading" set is considered for each variable. The two sets of weights are given in Table 19.3. A unit in the, say, second category of X_3 is assigned an X_3 "score" of .636, while an analysis unit in the fourth category of X_4 is assigned a score of .805.

Thus, such a scaling of X_3 and X_4 enables a researcher to analyze data on X_3 and X_4 along with data on X_1 and X_2 with the "standard" PDA (and DDA, too, for that matter) computer programs. Illustrative results are given by Huberty et al. (1986).

19.5 COMMENTS

A researcher could, of course, ignore the concerns of lack of normality when dealing with continuous variables. Of course, too, normality is not a prerequisite to deriving linear (or, even quadratic) classification functions for the purpose of group assignment. LCF values are mere linear composite scores. The optimality of the results, however, might be in question. That is, by using the LCFs for assignment purposes, one cannot be assured that the number or proportion of analysis units correctly classified is a maximum or even close to a maximum. The robustness of the LCF rule to nonnormality has been studied almost exclusively in a two-group setting, with the Baron (1991) three-group trivariate study being an exception. Baron also summarizes a number of two-group studies of LCF robustness to nonnormality. Five factors that may affect the degree of robustness are (1) sample size, (2) covariance structure, (3) group separation, (4) type of nonnormality, and (5) data structure. As may be obvious, robustness of an LCF rule to nonnormality is not a simple issue; the general extent to which optimality is sacrificed using nonnormal predictors has not been well established. Suggested ways of handling nonnormal predictor variables are summarized in Table 19.4.

TABLE 19.4 Suggested Ways of Handling Nonnormal Predictors

Nonnormal Type	Way of Handling
Continuous	Rank transformation
Bernoulli	0–1 scoring
Ordered categorical	Integer scaling
Unordered categorical	Fisher–Lancaster scaling

Because of lack of reliability of predictor measures, ordered categories may be imposed on continuous predictors for the purpose of classification, a suggestion made by Lykken and Rose (1963, p. 140).

A final recommendation might be considered. Suppose scores on predictor variables of interest are obtained via some type of questionnaire that consists of a set of items. It is recommended that item scores should *not* be considered as *variable* scores. Rather, what should be used to obtain variable scores is some type of *dimension reduction*. An analysis that may be used is a principal component analysis (PCA). With such an analysis, one obtains a set of components that are linear composites of the item scores. The number of components to retain to identify more meaningful variables is, again, a judgment call. See Khattree and Naik (2000, Chapter 2) for PCA details.

Further Reading

Readings in four areas of nonnormal classification are provided.

Classification with Discrete Predictors

Coste et al. (1997) propose the optimal discriminant analysis for ordinal responses (ODAO) method for classifying analysis units using predictor variables that are scored using an ordinal scale.

Dillon and Westin (1982) report the study of the effect of the number of levels of an unordered categorical variable on the classification performance when dummy coding is followed by the use of a linear rule versus four discrete methods in a two-group setting. Monte Carlo simulated data are used.

Gitlow (1979) compares the dummy variable approach with a multivariate nominal scaling approach using real data with three groups.

Hand (1983) illustrates the application of a linear classification rule and a kernel rule with six real data sets.

Krauth (1986) proposes cluster analysis and PDA procedures for ordered categorical response variable measures which are based on a natural definition of the neighborhood of configurations.

Yarnold et al. (1998) focus on the use of optimal discriminant analysis (ODA) with binary data.

Classification with Variable Mixtures

Choi (1986) includes articles by W. J. Krzanowski and by I. G. Vlachonikolis that cover PDA with a mixture of continuous and categorical variables and with a mixture of continuous and binary variables, respectively.

Krzanowski (1980) uses a classification analysis based on a "location model" to illustrate the handling of a two-group real data set consisting of measures on seven continuous variables, two binary variables, and two three-level categorical variables.

Krzanowski (1994) discusses a quadratic PDA approach when some of the predictors are continuous and some are categorical (binary or with three or more unordered categories).

Kumar and Sahai (1993) illustrate the prediction of 1 of 4 family planning devices using 12 predictors, 9 of which are continuous, 2 unordered categorical, and 1 dichotomous; the 4 group hit rates ranged from .77 to .83 (with unspecified priors).

Comparison of Classification Methods

Baron (1991) reports on the robustness of linear PDA to nonnormality and compares the (internal) classification accuracy of linear PDA, kernel density estimation, rank PDA, and logistic PDA using simulated three-group trivariate data. For normal and nonnormal data, logistic PDA performs quite well. Many two-group comparison studies are cited.

Boothby and Brewer (1990) compare real-data two-group classification results yielded by a logit analysis, a discrete analysis, and a linear PDA; a hold-out sample validation approach is used with varying misclassification cost functions.

Hand (1992) briefly reviews nine formulations of classification rules applicable with continuous and categorical variables and refers to Campbell et al. (1991), who concluded that classifications based on ordinal models confer no advantage over some other approaches.

Joachimsthaler and Stam (1988) compare the use of the LCF, QCF, logistic model, and a model based on linear programming on a variety of normal and nonnormal predictor distributions in a two-group setting. Simulated data are used.

Johnston and Seshia (1992) compare a nonparametric PDA with a normal-based linear PDA, linear regression, and logistic regression when all predictors and the grouping variable are categorical variables with ordered categories.

Long (1997) points out some potential problems of using integer scaling with ordered categorical variables.

McLachlan (1992) devotes a chapter to a discussion of logistic discrimination, reviews comparisons of LCF and logistic approaches to classification (Chapter 8), and reviews a number of studies in which performance of various nonparametric rules are compared (Chapter 9).

Munakata-Marr et al. (2003) use real data to compare the k-nearest-neighbor method with what they term "penalized discriminant analysis" and "GelCompar 11" methods.

Nath et al. (1992) compare linear, quadratic, and mathematical programming rules for the two-group situation. Normal distributions as well as four non-normal distributions are simulated. Comparisons are mixed across various data conditions.

Schmitz et al. (1983) compare the performance of the LCF, QCF, a logistic model, and a kernel model on a mixture of continuous and discrete predictors in a two-group setting. Simulated data are used.

Srinivasan and Kim (1987) compare six different classification rules (linear and quadratic, unordered logit analysis, goal programming, recursive partitioning algorithm, and an analytic hierarchy process) using risk (high risk versus non-high-risk) as the grouping variable and six financial variables plus two categorical variables as predictors. Real data ($n_1 = 39$, $n_2 = 176$) are used.

Other

Steel and Louw (2001) present formulas for exact calculation of bootstrap estimates of expected prediction error for k-nearest-neighbor classification, and propose a "weighted k-NN classifier" based on resampling ideas.

Velilla and Barrio (1994) propose a data transformation for nonnormal distributions in a linear or quadratic PDA context.

Definition ***Ordinal scale*:** Device for taking weights in the Vatican.

EXERCISES

1. Identify at least two continuous variables of interest in your area of study whose score distributions are clearly nonnormal.

2. Identify five categorical variables of interest in your area of study, three of which at least have more than two unordered categories.

3. For each of the variables identified in Exercises 1 and 2, specify a means of transforming the "scores" so that the transformed scores may be used as input for a normal-based analysis.

4. Specify how your would numerically scale (i.e., "measure") each of the following predictors:

(a) Satisfaction with salary (Excellent, Good, Fair, Poor)

(b) Disease stage (initial, advanced)

(c) Product preference (Brand A, Brand B, Brand C)

5. Suppose you have a categorical predictor variable—three unordered categories—and want to conduct a PDA with the predictor along with some other continuous predictors. Briefly discuss how you would proceed.

Computer Applications

6. Using the SAS package, run two or three nonparametric analyses on the 3-group Ethington data set (3GED) via the DISCRIM procedure. Consider some combinations of options available through DISCRIM: THRESHOLD, CROSSVALIDATE, POSTERR, POOL, and perhaps others (you decide). For each analysis, obtain classification tables using any option combination(s). Based on the resulting tables, make some rough (eyeball) comparisons among the analyses (on the Ethington data). ("Rough" is used because bona fide comparisons can only be made using methods discussed in Section 17.5.)

7. Consider your personal data set:
   - **(a)** If you included some categorical predictor(s), transform it (them) using a method described in this chapter.
   - **(b)** Conduct a PDA with all of your predictors.

CHAPTER 20

Reporting PDA Results

20.1 INTRODUCTION

Prior to reporting results of a predictive discriminant analysis (PDA), an indication of the purpose of using a PDA should be clearly stated. Is the analysis used to verify some theory? To establish a prediction rule to be used in the future? To validate some categorization method? To determine some type of "cutoff" score(s)? To support results of another analysis? To identify specific types of analysis units?

In addition to addressing the purpose question, there are other reporting basics to which the writer should attend. Suggestions are made in this chapter for describing the study, design, context, and analysis (group definitions, variables and their measures, computer software used, classification rule used, and descriptives) as well as for the classification results themselves. [An excellent discussion pertaining to the reporting of quantitative results is given by Bailar and Mosteller (1992).]

Using the 3-group Ethington data set (3GED), a brief discussion of reporting PDA results will be given.

20.2 EXAMPLE OF REPORTING PDA RESULTS

Introduction Just as in reporting DDA results (see Section 7.2), a PDA study introduction should include a very clear purpose statement, as well as a review of related literature. The purpose of the current study is to assess the accuracy of predicting academic grades of 264 community college students using nine student characteristics as predictors.

Study Design For our data set, the three criterion groups are defined as A students ($n_1 = 76$), B students ($n_2 = 122$), and students who earned C or C− ($n_3 = 66$). Of the large number of "variables" obtained on the basis of responses to items on the CCSEQ, nine predictors were identified. Predictor identification was based on a

Applied MANOVA and Discriminant Analysis, Second Edition, by Carl J. Huberty and Stephen Olejnik
Copyright © 2006 John Wiley & Sons, Inc.

minimum of the sum of six item scores. Reliability and validity information for the last six predictors is provided by Ethington and Polizzi (1996). The nine predictors were selected on the basis of professional judgment of their relevance to predicting student grades.

No missing data were found in the 264×9 data matrix; also, no aberrant predictor scores were present.

Computer Software Used The latest versions of SPSS DISCRIMINANT and SAS DISCRIM were used for the data analyses. [In some PDA applications, one may refer to Fraley and Raftery (2003) for the use of another software package, MCLUST.]

Descriptives The reporting of variable descriptives for a PDA study would be the same as the reporting for a DDA study—see Section 7.2.

Requisite Classification Rule Information We assumed that the joint distribution of each of the nine predictor variables is approximately normal in each of the three populations. Two tests of the equality of the three covariance matrices provided no evidence to conclude that they are different. The SPSS test results are $F(90, 125132) \doteq 1.191$, $P \doteq .105$; $\chi^2(90) \doteq 107.25$, $P \doteq .104$. Also, logarithms of the three covariance matrix determinants are 24.6, 23.1, and 22.5. This information indicated that there is support for the use of a *linear* classification rule.

Deciding on the prior probabilities of grade prediction to be used is based on researcher judgment. We arrived at the following priors: .25, .50, and .25.

An *external* classification rule is generally recommended to estimate group hit rates. Therefore, a linear leave-one-out (L-O-O) rule was used.

A final data examination was made. What was sought was the possibility of deleting one or more predictors before performing our final PDA. We judged that all nine predictors should be retained. (If predictor deletion is done, an all-subsets program should be used.)

Classification Results

Individual Student Prediction Results To list the three posterior probabilities of group membership for each student would probably go against the wishes of most journal editors. [For an exception, see Seshia et al. (1983), who reported the largest posterior probability for each of 104 units.] This information should, however, be made available upon request. In addition to the individual student posterior probabilities, student typicality probabilities associated with the predicted Grade may also be made available. The latter probabilities (obtained from SPSS output) may be examined to determine possible outliers and in-doubt students—see Section 15.6.

Grade Prediction Results The 3×3 classification table is given in Table 20.1. The three separate group hit rates are given in parentheses on the main diagonal. The total-group hit rate is $(16 + 99 + 13)/264 \doteq .485$. It may be noted that the total-group hit rate is only approximately 19 percent better than what may be expected by chance.

TABLE 20.1 Linear L-O-O Group Classification Results

		Predicted Group			
		1	2	3	Total
Actual Group	1	16 (.211)	55	5	76
	2	12	99 (.811)	11	122
	3	6	47	13 (.197)	66
Total		34	201	29	264

The Group 1 and Group 3 hit rates are no better than what may be expected by chance. The Group 2 hit rate (.811) is, however, about 62 percent better than what may be expected by chance. That is, predicting a grade of B is about 62 percent better than chance by using the derived classification rule. [The author(s) of a manuscript that reports chance results should give a reference.] It may also be informative to report classification results using some THRESHOLD value—see Section 15.6.

Classification Rule for New Students The (linear) classification rule developed with the sample of 264 community college students may be used with new students (assuming measures on the nine predictor variables may be obtained). The (linear) rule to be used is in the form of three linear composites of the nine predictors. The three sets of weights (and constants) are given in Table 20.2. Given a set of nine predictor scores for a new student, a linear composite score for Group 1 is found by multiplying each predictor score by the respective weight, summing these nine products, and adding the constant. Three such composite scores are found, and the new student is assigned to the group for which the largest composite score is determined.

TABLE 20.2 Classification Rule Weights (and Constants)

Variable	G_1	G_2	G_3
X_1	−.0496	−0.320	−0.241
X_2	0.039	0.064	−0.012
X_3	0.442	0.486	0.567
X_4	0.055	0.069	0.058
X_5	0.354	0.242	0.318
X_6	0.107	0.115	0.094
X_7	0.529	0.554	0.436
X_8	0.348	0.371	0.350
X_9	0.218	0.164	0.125
Constant	−16.121	−15.785	−15.798

As an example of applying a classification rule derived from a given data set to a new student will now be illustrated. Suppose the nine predictor scores for a new student are: $X_1 = 3$, $X_2 = 12$, $X_3 = 13$, $X_4 = 9$, $X_5 = 11$, $X_6 = 13$, $X_7 = 10$, $X_8 = 14$, and $X_9 = 14$. For this student, the three LCF scores are $\text{LCF}_1 \doteq 7.60$, $\text{LCF}_2 \doteq 8.15$, and $\text{LCF}_3 \doteq 19.33$. Because the LCF_3 score is highest, the new student would be identified with Group 3. That is, with this student's predictor score vector, he/she would be predicted to earn a C or C−. It should be noted that such a procedure to classify/identify a new student may very well suggest that the predicted grade for a new student may be doubtful (e.g., A versus B). This *may* suggest some type of guidance for the new student.

In a research situation where it is clear that a quadratic classification rule is appropriate, calculation of respective group composite scores is much more complicated. An example of applying a quadratic rule is given in Section 16.9.

Conclusions/Discussion The purpose of this study was to determine how well three academic grades of community college students may be predicted, using predictor scores based on college experience questionnaire responses. Of the three grades considered (A, B, and C or C−), it was found that for only the B group was the prediction accuracy of some note. The B group hit rate was approximately 81 percent, which is about 62 percent better than a chance hit rate.

A classification rule for use with new students was developed. Group assignment/identification for new students may be helpful for student guidance purposed.

The results of this study should be presented in connection with previous research—discussed in the Introduction. The comments at the very end of Section 7.2 pertaining to giving references in reporting DDA results apply to reporting PDA results as well.

20.3 SOME ADDITIONAL SPECIFIC PDA INFORMATION

Typical reporting of PDA results in journal articles focuses on the classification based on the computer data set. That is, it appears that most interest is on the total group hit rate and, rarely, on separate group hit rates. In a "practical" study situation, one would want information in addition to hit rates. For example, it might be of some practical value to "profile" certain community college students. Arriving at a profile for students who are "clearly" (using a THRESHOLD of, say, .45) predicted to earn a C or C−, might suggest some advice that could be given to new students. A profile for such a student may be developed by examining the 9-element mean vector for the 10 students (see Table 15.4) who were "clearly" predicted to get a C or C−—such a profile would reflect a *typical* C or C− student. Similarly, for advisement purposes, profiles of B students who were predicted to earn a C or C− may be found—for the Ethington data, 8 such students were "clearly" determined (see Table 15.4). Profiles of in-doubt students might also be examined. After predictor deletion has been examined (discussed in Section 17.2), profiles of typical students may be a much easier task.

A final comment is in order here. The above profiling should only be considered for analysis units that belong to groups for which classification accuracy is better than chance—see Chapter 16.

20.4 COMPUTER PACKAGE INFORMATION

Computational results for a PDA may be obtained via SAS or SPSS. A summary of such information, in addition to the usual descriptives, is given in Table 20.3.

20.5 REPORTING TERMS

Following are some terms that might be used in reporting results of a study in which a predictive discriminant analysis is conducted. For some parts of the write-up there are alternative terms, the choice between which would depend on the research situation and on the research questions of interest.

- Grouping variable
- Predictor variables
- Group covariance homogeneity
- Linear/quadratic classification rule

TABLE 20.3 DA Printout Information

	SAS DISCRIM	SPSS DISCRIMINANT
Preliminary = $\boldsymbol{\Sigma}$s	Yes[a]	Yes[b]
PDA		
Internal		
Linear	Yes	Yes
Quadratic	Yes	Yes
External		
Holdout	Yes	Yes
L-O-O		
Linear	Yes	Yes
Quadratic	Yes	Yes[c]
LCFs	Yes	Yes
Posterior Probability	Yes	Yes[d]
Typicality Probability	No	Yes

[a] Chi-squared test.
[b] F test.
[c] Results are incorrect.
[d] Only two are reported.

- Prior probabilities of group membership
- Misclassification costs
- Predictor deletion
- Predictor ordering
- Categorical variable scaling
- In-doubt units (use specific term)
- Outlying units
- Internal classification
- Leave-one-out classification
- Separate group hit rate (or classification accuracy)
- Total group hit rate
- Proportional/maximum chance criterion
- Better than chance
- Improvement over chance
- Judgment

While reading reports or journal articles in which group membership prediction (discussed in Part IV) or in which group separation (discussed in Part II) is studied, it is not uncommon to find general descriptive terms used, such as *discriminant analysis* or *multiple discriminant analysis* or *discriminant function analysis*. Unless a reference (with specific page numbers for a book) is given, such terms have little informative value and may even be misleading—this may be the case even if a reference is given. The first term tells the reader little. Does it imply a predictive discriminant analysis (PDA)? A descriptive discriminant analysis (DDA)? Or, a mixture of the two? Sometimes the second term is used to mean multiple response variables and sometimes to mean three or more groups. Like the first term, *multiple discriminant analysis* is a nondescript term. Although more specific, the third term (*discriminant function analysis*) can also be misleading. As pointed out by Tatsuoka (1988, p. 369), "The term 'discriminant function' has come to be used in two different senses in the literature." One is the sense in which it was used in Section 13.4 (as a classification function), and the other, as it is used in Section 5.2 (as a linear discriminant function). These two composites, or "functions," are quite different. It is recommended that none of the three expressions given above be used in reporting either a PDA or a DDA.

It would be much more informative if the writer would use specific, well-defined terms. Meanings and even definitions of terms need not be given in a report; rather, reference may be made to a book (giving specific page numbers). In Section 7.7, there is a list of nine "expressions" we suggest that a researcher "shy away from."

For tips on writing in general, and tips on reading quantitative material, the interested reader might refer to Abelson (1995), Best (2001), Cohn (1989), Crossen (1994), Huff (1993), Lederer and Downs (1995), Moroney (1951), and Murray et al. (2001). These eight books contain many helpful—some being humorous—suggestions. The same may be said regarding at least five books written by a mathematician, J. A. Paulos (1985, 1988, 1991, 1995, 2000).

20.6 SOURCES OF PDA APPLICATIONS

As mentioned in Section 7.5, examples of applications of PDA may be obtained by conducting a Web search using the key word "discriminant analysis." (A warning about using a published application as a "model" is given in Section 7.5.)

Five specific sources for applications of PDA follow.

Baron (1991) refers to a number of applications of PDA in medical research.

Devillers and Karcher (1991) refer to numerous applications of PDA in toxicology, ecology, and related fields; this edited volume includes one study (by Gombar and Enslein) that is a detailed application of a two-group PDA of a structure–activity relationship.

Hand (1997, Chapter 10) discusses applications of PDA in four different contexts: chromosome analysis, credit scoring, speech recognition, and character recognition.

Jurs (1986) refers to a number of applications of PDA in the area of analytical chemistry.

McLachlan (1992, pp. 201–211) presents two very detailed discussions of two PDA application contexts: genetic counseling and diabetic diagnosis.

20.7 CONCERNS

Applications of predictive discriminant analysis have been quite widespread during the past two decades. The increase in the use of PDA is due, in part, to an increase in teachings (formal coursework in academe and professional workshops) and in writings (textbooks and journal articles). Teaching and writing are media through which translations from statistical theory and formal presentations are often made to practicing researchers. More such translations will mean more and (hopefully) better applications. One of the problems with the practice of scientific inquiry pertains to the delay (in years) that occurs between translation and application. In the interim we often find misuses (abuses?), misinterpretations, and misapplications of some techniques/procedures and reporting of questionable results by some practicing researchers. So we progress in growth!

There are four concerns that pertain to applications of PDA and the reporting of PDA results. The *first* concern with many published applications is the focus on internal rather than external classification results (see Section 15.3). A *second* concern is with the assessment of the goodness of the classification results. This concern pertains to comparing the obtained results to results that could have been obtained "by chance." Willis (1984) points out some misleading conclusions that can be reached by ignoring the maximum-chance criterion (see Section 16.3) when criterion group sizes differ widely. A *third* concern is with criteria used to delete and order predictor variables in a PDA (see Chapter 17). The *fourth* concern is the "mix" of predictive and descriptive discriminant analysis, along with the (usually questionable) use of "stepwise discriminant analysis" that are still fairly prevalent in the applied literature.

Huberty and Hussein (2003) reviewed 20 journal articles (1998–2000) that reported applications of discriminant analysis. Generally, the reporting was found to be somewhat lacking. One problem found was the "mixing" of PDA and DDA. One view of PDA versus DDA is given in Table 7.8.

20.8 OVERVIEW

An overview of predictive discriminant analysis in the form of a flowchart is given in Figure 20.1.

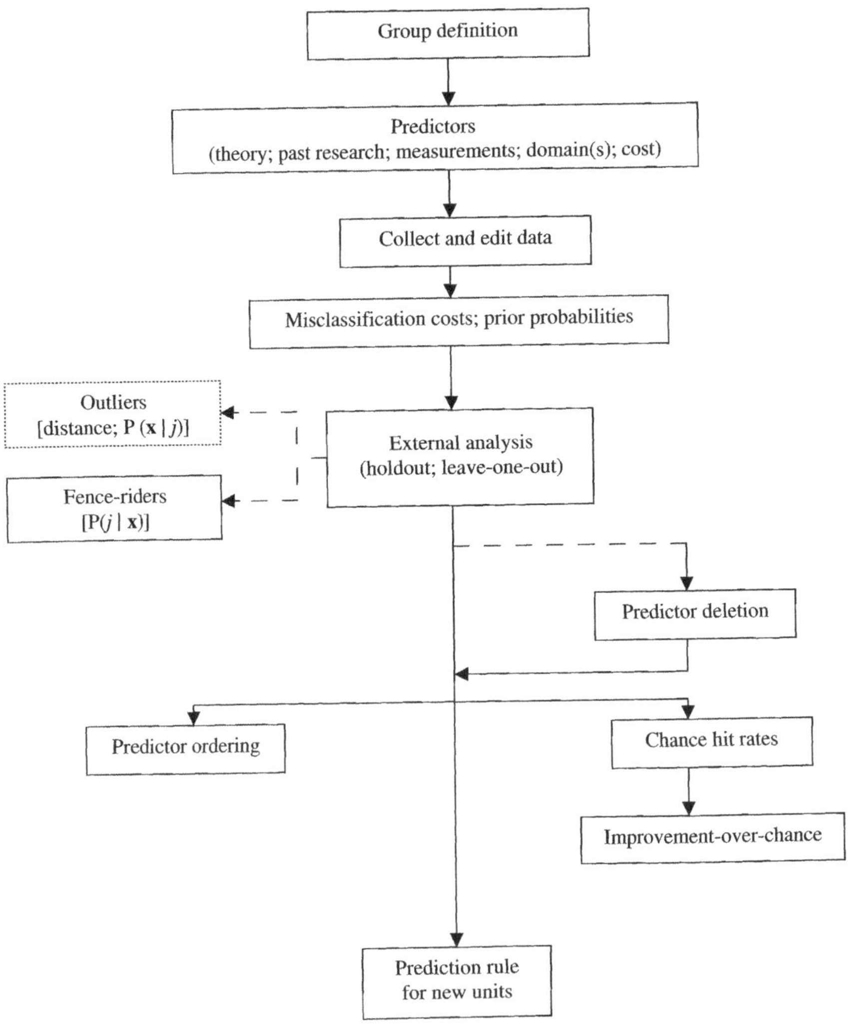

Figure 20.1 Predictive discriminant analysis.

Further Reading

Finn et al. (1997) apply both linear and quadratic PDAs to six sites of sockeye salmon in Alaska using 11 to 17 different site salmon amplitudes as predictors; with site sizes from 52 to 157, linear and quadratic "cross-validation" results were mixed.

Harris and Kaine (1994) apply cluster analysis and discriminant analysis (both DDA and PDA!) to a five-variable data set.

Huberty and Hussein (2003) describe the reporting of discriminant analysis in 20 behavioral science journal articles (1998–2000). Both a DDA and a PDA were conducted on the same data set in nine of the articles. In general, the reporting of discriminant analysis results was very disappointing.

Mason (1998) constructs and validates a quadratic PDA method for forecasting rainfall in South Africa.

Parker and Leinhardt (1995) point out the distinction between the use of "percent"and the use of "percentage."

Definition ***t test***: Admission procedure used in English public schools.

EXERCISES

1. From journals in your substantive area that publish empirical studies, select an article that reports an application of PDA. Comment on the extent to which each of the following is discussed.

(a) Purpose

(b) Descriptions of group definitions, predictor variables, predictor measures, and sampling procedure

(c) Computer program used

(d) Classification rule used

(e) Variable deletion/ordering

(f) Individual unit results

(g) Group results (internal or external? relative to chance?)

2. Propose a study in your discipline that would call for a PDA. Indicate some design and analysis specifics for your research study.

3. Consider a data set specified by you or your instructor. Carry out a PDA on these data. Giving your data set some "context," write up a report of your analysis.

4. This exercise pertains to your personal data set (Exercise 2 in Chapter 1). Draft a brief introduction, sections describing your methods (analysis units, sampling, variables, variable measures), and sections pertaining to results and interpretation. All computer output should be retained. You might even include some references on substantive related material as well as on analyses. Give it a try!

CHAPTER 21

PDA-Related Analyses

21.1 INTRODUCTION

A few multivariate analyses "related" to normal-based predictive discriminant analysis (PDA) were briefly mentioned earlier in Part IV. In this chapter, we define and briefly discuss seven PDA-related analyses. Two fairly nonspecific categories are used: nonlinear methods and other methods. All of the stated definitions are cited from Dodge (2003). To conclude this chapter, additional readings are given for PDA-related methods.

21.2 NONLINEAR METHODS

21.2.1 Classification and Regression Trees (CART)

Definition Classification algorithms based on binary splits of variables, often followed by pruning to reduce the complexity of the resulting decision tree (Dodge 2003, p. 68).

Notes Breiman and co-workers (1984) provide a detailed discussion of CART methods. These nonparametric methods are also discussed by McLachlan (1992, pp. 324–332), Webb (2002, Chapter 7), and Yarnold (1996).

21.2.2 Logistic Regression

Definition A model of the dependence of Bernoulli random variables of explanatory variables. The logit of the expectation is explained as a linear form of explanatory variables. ... The model is especially useful in the case-control studies and leads to the effect of risk factors by odds ratios (Dodge, 2003, pp. 243–244).

Applied MANOVA and Discriminant Analysis, Second Edition, by Carl J. Huberty and Stephen Olejnik
Copyright © 2006 John Wiley & Sons, Inc.

Notes A number of studies have appeared that pertain to a comparison of logistic regression and (usually, two-group) PDA. The conclusion reached by Dattalo (1994), Fan and Wang (1999), Lei and Koehly (2003), Meshbane and Morris (1996), and Todd et al. (1995) is that classification accuracy for the two methods was "comparable." Lesaffre et al. (1989) conclude that "the literature reporting logistic regression models is full of over-optimistic results" (p. 3005). McLachlan (1992, pp. 276–282) presents a detailed discussion of the comparison of logistic regression with linear PDA. His conclusion is that, with respect to hit/error rates, there is no strong preference for one over the other. See, also, Efron (1975). Other discussions of logistic regression may be found in Hand (1997, pp. 42–44), McLachlan (1992, pp. 255–282), and Webb (2002, pp. 124–128). Rencher (1998, pp. 254–261) provides a discussion of logistic and probit classification.

21.2.3 Neural Networks

Definition A regression model in which the responses are nonlinear functions of inputs through layers of connected hidden variables, originally by treating biological neurons as binary thresholding devices. They are flexible models useful for discrimination and classification . . . and are implanted by a computerized "black-box" trained by a training data set (Dodge, 2003, p. 282).

Notes Chatfield (1993) questions the validity of neural networking as a viable forecasting tool. Marzban and Stumpf (1996) apply a neural network method of analysis to predict whether or not a tornado is detected; the method depends on less restrictive data conditions and are, thus, claimed to be more widely applicable than the usual PDA methods. Peng et al. (2003) compare the neural networks classification with linear PDA using real data. Yoon et al. (1993) compare results of an artificial neural network classification method with two-group quadratic PDA results; with the former, results are more impressive when binary predictors are included. Discussions of neural networks may also be found in Gordon (1999, pp. 170–172), Hand (1997, pp. 44–56), and Webb (2002, pp. 133–173).

21.3 OTHER METHODS

21.3.1 Cluster Analysis

Definition A general approach to multivariate problems in which the aim is to see whether the individuals fall into groups or clusters. There are several methods of procedure; most depend on setting up a metric to define the "closeness" of individuals (Dodge, 2003, p. 69).

Notes As mentioned earlier in the current book, the term *classification* has been used in a cluster analysis context as well as in a PDA context. Webb (2002, p. 5) considers cluster analysis as an "unsupervided classification" method. Furthermore, cluster analysis has, interestingly, been "connected" to pattern recognition, as indicated by

Fukunaga (1990, Chapter 11) and Webb (2002, Chapter 9). See, also, Diday et al. (1994, Sections 1.2 and 1.4).

21.3.2 Image Analysis

Definition The extraction of the underlying scene from an image, usually pixellated and subject to degradation by noise. Also known as image enhancement or image reconstruction (Dodge, 2003, p. 193).

Note Two readings regarding image analysis, which is used in the contexts of medicine and astronomy, for example, are Diday et al. (1994, Section 4.2) and McLachlan (1992, Chapter 13).

21.3.3 Optimal Allocation

Definition In general, the allocation of numbers of sample units to various strata so as to maximize some desirable quantity such as precision for fixed cost. Secondarily, allocation of numbers of sample units to individual strata is an optimum allocation for a given size of sample if it affords the smallest value of the variance of the mean value of the characteristic under consideration. Optimum allocation in this sense for unbiased estimators requires that the number of observations from every stratum should be proportional to the standard deviation in the stratum as well as to the stratum number (Dodge, 2003, p. 293).

Note The relationship between optical allocation/scaling and PDA is discussed by Webb (2002, pp. 120–122).

21.3.4 Pattern Recognition

Definition A branch of computer science concerned with identification of objects of known classes, or grouping of objects. The first stage gives the pattern a digital code; the second stage is closely analogous to discriminant analysis or cluster analysis. The techniques, however, are instrumental and do not, as a rule, involve any consideration of underlying distributions (Dodge, 2003, p. 303).

Notes The "connection" between pattern recognition and PDA is implied by the title of the book by McLachlan (1992). PDA and pattern recognition are used for the lone basic purpose: Develop a rule for assigning an analysis unit (or a pattern) to one of two or more defined groups. In normal-based PDA (as discussed in the current book), a distribution form—typically multivariate normality—is assumed for the data on hand. In pattern recognition, a particular distribution form for the patterns is not assumed. According to Marriott (1990, p. 153) the initial stage involves digitally coding each pattern; this is followed by an analysis that is akin to a PDA if there are well-defined groups of patterns, or akin to a cluster analysis. The two books by Webb (2002) and Fukunaga (1990) have extensive discussions of both pattern recognition and PDA. Webb (2002) considers pattern recognition as a "general" approach to multivariate classification analysis. Also, the PDA book by Hand (1997) mentions pattern recognition on 11 different pages and provides many references for the reader.

Further Reading

Cheng and Titterington (1994) give a rather detailed discussion of artificial neural networks (ANNs) in relation to a number of "standard" statistical data analysis procedures such as a PDA.

Dudoit et al. (2002) compare nearest-neighbor, classical PDA, and classification trees using real data. They conclude that the first two methods "performed remarkably well compared" to the third method.

Faraj (1994) is one refernce for an analysis method called "generalized discriminant analysis."

Flitman (1997) comparatively evaluates neural networking, logistic regression, and two-group PDA in the context of predicting college student success/failure.

Friedman (1989) proposes a classification approach called "regularized discriminant analysis" as a compromise between the use of linear and quadratic classification functions.

Hadorn et al. (1992) compare seven "statistical models" (including logistic regression and CART) in predicting patient mortality (yes–no) using STATA software on real data.

Hand (1997, pp. 34–36) provides a brief discussion of regularized discriminnat analysis (as a "regularization" method).

Hussein (2001) reports an extensive simulation study involving a mixture of continuous and categorical predictors and linear and quadratic classification rules. His conclusion is that logistic classification is a little more robust than normal-based PDA to unequal group dispersion.

Jain et al. (2000) provide a very thorough review of pattern recognition, including neural networks, logistic analysis, decision trees, nearest-neighbor rules, linear PDA, and cluster analysis, giving a total of 176 references.

Koolaard et al. (1998) find, via simulation, that when the $\min(n/p)$ ratio is "small," the "regularized approach" to quadratic classification outperforms the normal-based PDA method.

Koolaard et al. (2002) incorporate the use of the Bhattacharyya distance to simplify computations in the standard regularized approach; a small loss of classification accuracy results.

Kumar and Haynes (2003) compare (artificial) neural networks with classical linear PDA. Using real data, they conclude that their neural networks method performs better than classical PDA.

Loucopoulos and Pavur (1997) present a new mathematical programming (MP) approach for the three-group classification problem; some data simulation indicated greater computational efficiency than that of earlier MP approaches.

Lu et al. (2003) find that regularized discriminant analysis outperforms linear and quadratic PDA with "small" group sizes.

Mkhadri et al. (1997) present an overview of "regularized discriminant analysis" in a small-sample setting, and an alternative approach based on spectral decomposition of group covariance matrices, dependence trees, and logistic methods.

Nelson et al. (2003) perform a comparison study involving four factors (five classification methods, four numbers of analysis units, two numbers of groups, and three numbers of predictor variables). The five methods are neural networks, linear and quadratic PDA, and two decision tree methods. The conclusion is that the linear PDA "worked best for almost all considered conditions" (p. 565), but the neural network method worked well with a large number of units.

Renaud (2002) explores the properties of yet another alternative to PDA, "projection pursuit discriminant analysis," which involves the consideration of "wavelets."

Rencher (1998) provides a brief discussion (with references) of at least nine topics related to PDA (ridge-type adjustment for LCF weights, random effects, categorical predictors, linear programming, influence of individual observations, PCA and PDA, data imputation, neural networks, and nonparametric classification).

Rencher (2002) has a chapter on cluster analysis (Chapter 14); he also discusses multidimensional scaling (MDS) and correspondence analysis (Chapter 15).

Soltysik and Yarnold (1994a) describe a linear programming algorithm that dramatically reduces computer resources necessary to conduct optimal discriminant analysis (ODA).

Villarroya et al. (1995) propose a new classification method, minimum distance probability (MDP), which is based on a distance for continuous, discrete, or mixed variables with known or unknown distributions, and can be used as an alternative to the "standard" PDA.

Webb (2002) intermingles many analysis methods: discriminant analysis (i.e., PDA), regularized discriminant analysis, pattern recognition, neural networking, multidimensional scaling, optimal scaling, Bayesian networks, cluster analysis, factor analysis, principal component analysis, and logistic discrimination. Many, many references are given for these methods.

Williams et al. (1999) compare five methods (quadratic PDA, linear PDA, logistic regression, neural networks, and classification trees) for biochemical-based Down syndrome risk prediction (yes–no with very unequal group sizes); resampling is used to conclude that the quadratic method did not, comparatively, yield very good results—priors used is not mentioned.

(Additional references for PDA-related analyses are given in the Further Reading for Chapter 19.)

Definition ***Typology***: Apologizing for your neckwear.

PART V

Issues and Problems

With nearly all collections of data analysis techniques there are issues pertaining to statistic choice, analysis strategies, and interpretation of results, as well as to some philosophical points of view. Eight issues involved in discriminant analysis are briefly reviewed in Chapter 22. All eight issues have at least been mentioned in earlier chapters.

Also associated with many data analytic methods are some special problems. The problems are associated with the use of discriminant analysis techniques. These are basically unresolved problems. Statistical theory has not, to date, suggested clear-cut resolutions. Only some of the eight problems briefly reviewed in Chapter 23 have been mentioned in previous chapters.

The resolution of data analysis issues and problems for the applied researcher is a difficult task, indeed. When confronted with issues and/or problems, both the empirical researcher and the statistical methodologist may take any of, for example, five options: ignore them, refer to an "authority," carry out the analysis with the admittance of there being an issue or a problem, report multiple sets of results, or use a single approach and defend it. The reader may think of additional options. In Chapters 22 and 23, issues and problems are presented with a preferred resolution stated with respect to some; others are left "open" for judgment on the part of the reader.

Applied MANOVA and Discriminant Analysis, Second Edition, by Carl J. Huberty and Stephen Olejnik
Copyright © 2006 John Wiley & Sons, Inc.

CHAPTER 22

Issues in PDA and DDA

22.1 INTRODUCTION

There are some issues pertaining to discriminant analysis that are sometimes ignored by practicing researchers, about which methodologists disagree, and for which there are no theoretical solutions. When confronted by some issues, or even all issues for that matter, many methodologists would begin their response with: "It depends." The dependence for some would relate to statistical background, to positions taken by "authorities," or to philosophical beliefs. The reader should not necessarily expect to find clear-cut resolutions in this chapter. Rather, the intent of this chapter is to point out, as was done in earlier chapters, that some, at least, of the issues are real. Four issues are briefly discussed in this chapter: (1) choices in PDA, (2) stepwise analyses, (3) standardized LDF weights versus structure r's, and (4) data-based structure.

22.2 FIVE CHOICES IN PDA

22.2.1 Linear Versus Quadratic Rules

As mentioned in Section 13.6, the general recommendation is to use a linear rule. Analyses of 31 real data sets by Meshbane and Morris (1995) led them to that recommendation. If covariance matrix heterogeneity is suspected at the outset of the analysis, a large number of analysis units in the smallest group should be seriously considered [i.e., at least $\min_j(n_j) \geq 5p$]—see Huberty and Lowman (1997) and Jain et al. (2000, p. 11). The linear rule is suggested because of presumed greater stability than that for the quadratic rule, in general (see, e.g., Flury et al., 1994). In simulation results, however, Connally (2004) found that the (internal) quadratic rule outperformed the linear rule.

Applied MANOVA and Discriminant Analysis, Second Edition, by Carl J. Huberty and Stephen Olejnik
Copyright © 2006 John Wiley & Sons, Inc.

22.2.2 Nonnormal Classification Rules

What if multivariate normality (in each population) of continuous X variables cannot be assumed? Practically speaking, this is a very difficult issue to address for at least three reasons. *First*, assessing this condition is not a trivial exercise for the practicing researcher. (For a brief discussion of statistical tests, see Section 13.6.) *Second*, the effect of nonnormality on classification results from normal-based rules has not been thoroughly studied. *Third*, the choice of a nonnormal rule (see Chapter 19) has been studied only to a limited extent, and most of the studies have been, understandably, restricted to the two-group case.

A choice of rule form must also be made when the X variables are categorical. Again, study of such rules has been restricted largely to the two-group situation. The most serious analysis issues may be claimed to arise with predictors having more than two nonordered categories. A preference for the use of a particular scaling method was advanced in Section 19.3.

22.2.3 Prior Probabilities

There are formal (in a statistical sense) ways of estimating the priors (see McLachlan, 1992, Chapter 2). The approach we take to deciding on group priors to use is, at least in part, a judgmental one. Suppose a researcher has J well-defined populations in mind, and a proportional sampling method is used. In this situation, "estimating" the priors is not too difficult. But, what if proportional sampling is not used? We suggest, then, that the relative population sizes determine the priors, regardless of the group sizes used.

Now suppose that the populations suggested by the researcher are not "well-defined." What we suggest, then, is to consult with some "experts" in the area of study regarding the relative population sizes. The estimated priors to be used can then be determined using the "average" of the suggested size proportions for each population. [This process was actually used to arrive at priors used by Huberty et al. (1997).]

Another judgment call may be made that is related to unequal priors. Suppose one population is pretty well known to be quite "small." For example, let one population proportion be .10—this would be one prior to be used. In a PDA situation, it is advised that the corresponding group size be greater than 10 percent of the total sample size. This advice is particularly suggested when the smaller(est) group hit-rate is of greater(est) interest.

22.2.4 Misclassification Costs

Misclassification costs are virtually ignored in the PDA applications that we have found. The main reason for such an omission is, perhaps, that the costs are not explicitly mentioned in the SAS and SPSS manuals. See McLachlan (1992, pp. 7–9) for a discussion of misclassification costs. For the two-group situation, costs may be considered jointly with the prior probabilities (see Sections 13.7 and 18.2). Rudolph and Karson (1988) discuss the effect of unequal misclassification costs.

22.2.5 Hit-Rate Estimation

In connection with rule form, there is a fifth choice to be made in conducting a PDA: the choice of a hit-rate estimation method. The estimation of classification hit rates or, equivalently, error rates has been a topic of interest to statisticians and methodologists for some time. McLachlan (1992) devotes an entire chapter (Chapter 10: Estimation of Error Rates) to this issue; a statistical theory perspective is taken to a large extent. To the applied researcher the issue is two-sided: (1) accuracy versus precision of estimators and (2) actual computation of an estimate. Of course, a researcher wants a "good" hit-rate estimate. In other areas of data analysis that involve parameter estimation, the general impression that the nonmethodologist gets is to focus on *accuracy* of estimators; that is, on lack of estimation bias. True, estimation bias, or accuracy, is a legitimate and important concern. But another important concern regarding estimator quality is variability or *precision*. After all, isn't estimation precision a concern when making generalizations, just as is estimation bias? It would be nearly ideal to use an estimator that has little bias (i.e., is highly accurate) and that has little variability (i.e., is highly precise).

Of course, ideals are seldom attained, especially in the case of hit-rate estimation with more than two groups. Resampling estimation methods—leave-one-out, bootstrap, and jackknife—have been suggested by many (e.g., McLachlan, 1992, Chapter 10). See, also, Hand (1997, Chapter 7) who discusses resampling methods along with many references for review. These methods generally yield good estimators in the sense of accuracy. But, as mentioned in Section 15.3, some simulation studies have indicated a general lack of precision for these estimators. So what might be recommended for the practicing researcher? There appear to be two routes one might take. One is to use large samples—"large" may be defined as $\min_j(n_j) > 5p$. Another route might be to use another estimator, such as the maximum-posterior-probability estimator discussed in Section 15.3.4. The suggestion given here is to use a combination of a *linear* M-P-P estimator and a L-O-O estimator when multivariate normality is a reasonable assumption. This combination was denoted as a linear M-P-P/L-O-O estimator in Section 15.3.4. If multivariate normality is very doubtful, the suggestion here is to use the linear L-O-O estimator.

22.3 STEPWISE ANALYSES

It is quite common to find the use of "stepwise analyses" reported in empirically based journal articles. Two very popular analyses are stepwise regression analysis and stepwise discriminant analysis, results for both of which are available via the two statistical computer packages, SAS and SPSS. With these packages one can conduct a forward selection, backward elimination, forward stepwise, and backward stepwise analysis. When it is claimed that a stepwise analysis was run, more likely than not it was a forward stepwise analysis using default values for variable deletion, which simply results in a forward selection analysis.

The stepwise discriminant analysis programs in the two packages have built-in criteria for stepping that relate to *group separation*. That is, in a forward analysis, for a given step the next variable added is the one that increases group separation the most for all the remaining candidates. (It is usually assumed that the addition of a variable cannot decrease group separation.) So, a stepwise discriminant analysis *might* be considered appropriate in a MANOVA/DDA context in which group separation is of interest. But, only in very restrictive situations would such an analysis be tentatively considered in a PDA context in which group membership prediction is of interest (see Section 17.2).

There appear to be two reasons why researchers want to examine the results of the "steps" from a stepwise analysis. (As indicated in Chapters 6 and 17, some "nonstep" results of a stepwise analysis can be informative.) The two reasons pertain to variable deletion and variable ordering. How typical it is to see researchers view the first, say, five variables entered into the (forward) analysis as constituting the *best* subset of five variables. But, if challenged, most applied researchers may be hard put to explain their sense of "best" and to defend their subset as being the best subset of a given size. (The writers of the stepwise algorithms themselves would not claim that the first five variables entered constitute the best, in any sense, subset of size 5.) So, if a researcher is seeking the best subset of size 5, something other than a stepwise analysis should be utilized. This is the recommendation if one is researching in a MANOVA/DDA context (see Section 6.2) or in a PDA context (see Section 17.2).

If a variable is the third one entered into a stepwise discriminant analysis, it is often judged to be the third "most important" variable. That is, a stepwise discriminant analysis is often used by applied researchers to solve the variable ordering problem. In many reports it is usually not made clear what the basis of importance is. That is, important with respect to what? Because the stepwise analysis programs in the two packages focus on group separation, one might be tempted to utilize one of the programs to order the outcome variables with respect to separation (i.e., in a MANOVA/DDA context). In a PDA context where group membership prediction is of interest, it is dubious even to consider using one of those programs. In either context there are clearly better ways of ordering outcome variables (see Section 6.3) or predictor variables (see Section 17.4). A much more detailed critique of the use of stepwise regression and discriminant analyses is given by Huberty (1989).

22.4 STANDARDIZED WEIGHTS VERSUS STRUCTURE *r*'s

The notion of a linear composite or combination of response variables is very basic to many multivariate analyses (see Section 2.7). For some analyses (e.g., multiple regression, principal components, predictive discriminant analysis), composites are formed for the primary purpose of obtaining composite scores for the units being studied. For other analyses (e.g., multiple correlation, canonical correlation, common factor analysis, descriptive discriminant analysis), there is interest usually in identifying the *constructs* represented by the composites. That is, a researcher often desires to give substantive meaning to the underlying unobservable (i.e., latent) constructs represented by the composites to "interpret" the composites.

The reason quotes were used in the preceding sentence is because different views of what it means to interpret a composite have been expressed. In some instances, how to interpret "interpret" is not clear. In the case of linear discriminant functions (LDFs), one may interpret a function by ordering the outcome variables with respect to importance or contribution, or by naming the function. To some, these are quite different purposes. Rencher (1992, p. 218), however, states that "no attempt will be made to distinguish between these two modes of interpretation." In the case of linear classification functions (LCFs)—as well as linear multiple regression equations—seldom do methodologists discuss naming the linear composites (see, however, Huberty, 1994a). An ordering of predictors in PDA may be accomplished legitimately without examining the multiple LCFs (see Section 17.4).

The present issue of concern, then, pertains to the interpretation of LDFs. If, to the reader, this includes assessing relative outcome variable importance, it may be claimed that the issue has been, to some extent at least, resolved (see Section 6.3). This is not to say that all methodologists would agree with the ordering method suggested. In fact, some methodologists prefer to examine standardized LDF weights to determine an outcome variable ordering. Problems with this approach and a discussion of problems with variable ordering in general are reviewed by Huberty (1989), Huberty and Wisenbaker (1992b), Thomas (1997), and Thomas and Zumbo (1996).

The real issue, then, as far as this chapter is concerned, pertains to the role of standardized LDF weights versus variable-LDF correlations—the latter are usually termed structure r's—in naming the construct associated with an LDF. It is clear that during the early history of factor analysis, the index most commonly advocated and used in defining constructs was the structure r. The reasoning behind the use of this index pertains to shared variation and, of course, correlation. If, say, three out of a set of response variables correlated highly (positive or negative) with a factor "composed" of all of the response variables, the factor would essentially be "made up" of whatever those three variables represented. The actual naming of the factor, then, would be made by subjectively coalescing the attributes or traits (of the measured analysis units) represented by the three variables.

This approach to data-based construct identification was very commonly used in factor analysis well into the 1990s. The structure r index has also been commonly used in canonical correlation analysis and in what is termed in this book "descriptive discriminant analysis." In recent years, however, some quantitative methodologists have proposed that standardized LDF weights, *not* variable-LDF correlations, should be utilized in identifying or labeling the constructs underlying grouping variable effects; that is, the weights should be examined to see if a substantive interpretation can be given to the LDF. This position has been taken by Harris (1989), Mulaik (1994), Rencher (2002, pp. 288–291), and Tatsuoka (1988, p. 521). This issue has not arisen in many writings in which descriptive discriminant analysis results are reported because of the *minimal* expressed interest in "interpreting" LDFs (see, e.g., Huberty and Hussein, 2003).

There is a subissue with respect to standardized LDF weights that pertains to the method of *standardizing* (see Section 5.4). The weights may be standardized in terms of error variance or in terms of total variance. This issue is discussed and debated by Mueller and Cozad (1988, 1993) and by Nordlund and Nagel (1991).

Another comment about the weight-versus-structure r issue pertains to interpretation consistency. If a researcher is convinced that the use of structure r's makes sense in, say, a canonical correlation context, he or she would also advocate the use of structure r's in the contexts of multiple correlation, common factor analysis, and descriptive discriminant analysis.

Finally, even if structure r's are used in LDF interpretation, there remains the issue of the minimum r magnitude to be used in making the interpretation. Because the sampling distribution of a structure r is not known, formulas for estimated standard errors are not available, thus precluding direct testing of an r for statistical significance. It makes sense, as Dalgleish (1994) points out, that if an r is not statistically significantly different from zero, the associated outcome variable should not be considered in interpreting the given LDF. Dalgleish concluded that a reasonable approach to assessing the precision of a structure r is via the bootstrap, a method of resampling. Although bootstrapping itself may not be too conceptually complex, two complexities do enter into its application to the structure r precision issue. *One* complexity is that of "aligning" bootstrap sample structure with the structure of the base sample. The alignment issue has been discussed by Clarkson (1979). The *second* complexity that arises pertains to bias in estimating the structure r's. Dalgleish (1994) briefly reviews some of the related literature on bias correction with the bootstrap.

22.5 DATA-BASED STRUCTURE

What is this notion of *structure*? When posed with this question, the proverbial person on the street would, perhaps, think of the *construction* of a building. So much for this metaphorical example. In short, a structure underlying a system of variables is comprised of one or more constructs. So, then, what is a construct? Simply put, but of somewhat limited definitional help, a construct may be thought of as a concept. Some help on the meaning of construct is provided by Liebert and Spiegler (1982, p. 764):

> *Theoretical constructs* identify phenomena considered important to the theory. Energy is a construct from physics, oxidization is a construct from chemistry, and natural selection is a construct from biology. Personality theorists have used many constructs; among the more familiar ones are ego, anxiety, conditioning, and self-awareness. Theoretical constructs do not actually exist, nor can they be seen or touched. They are merely useful inventions which help to give order to observed phenomena. Theoretical constructs are often shorthand summaries of relationships among many different variables, and they therefore serve to facilitate communication about these relationships.

Without a doubt, the multivariate analysis most commonly used to identify structure underlying a set (or multiple sets) of variables is exploratory factor analysis. This analysis is often utilized in the study of measurement instrument validation; in particular, construct validation. In the behavioral science journal *Educational and*

Psychological Measurement, there is a section called Validity Studies. A common analysis method used in this section across journal issues is exploratory factor analysis. This analysis is used to search for underlying constructs, constructs determined by the data. Is this a legitimate pursuit of knowledge?

If that question had been addressed in the 1940s (e.g., MacCorquodale and Meehl, 1948), 1950s (e.g., Fruchter, 1954), 1960s (e.g., Horst, 1965; Kaplan, 1964, p. 759), 1970s (e.g., Rummel, 1970), 1980s (e.g., McDonald, 1985), and in the 1990s (e.g., Comrey and Lee, 1992), the answer would have been a resounding "yes." But in the more recent years, the thinking of some methodologists (e.g., Mulaik, 1994) about basing construct identification on data has gone in another direction. These writers posit that constructs should be theory based, not data based, that suggested constructs should evolve from substantive theory, and that data should only be used to *verify* the existence of hypothesized constructs. This would be well and good if the substantive theory of interest is refined and has been studied to the extent of suggesting meaningful and interpretable constructs that may be identifiable in such a way that observable variables may be chosen that may, in turn, lead to data verification. In the behavioral sciences, at least, most theories are not so well defined.

One might conclude that we do not have a *clear* two-sided issue here. Although some may have a preference for one side or the other, the position taken by Velicer and Jackson (1990, p. 782) makes sense: "Exploratory analytic approaches (or result-centered research strategies) should be preferred except for those cases where a well-defined theory exists." The exploratory versus confirmatory analyses is not a clear two-sided issue according to Tukey (1980, p. 782): "Neither exploratory nor confirmatory is sufficient alone. To try to replace either by the other is madness. We need them both."

Using data to discover and/or define structure underlying the variable system on hand is common, as mentioned above, by users of factor analysis or component analysis. Even though structure may be discovered in a multivariate group difference study in which MANOVA may be employed, such discovery does *not* appear to be of routine interest to applied researchers (Huberty and Hussein, 2003). The analysis in the study of group differences has traditionally been simply that of statistical testing (as in Chapter 3). One reason for this, perhaps, has been a lack of emphasis in textbooks on seeking structure in a group comparison context. For example, Part Two of the Chatfield and Collins (1980) text is entitled, Finding New Underlying Variables, and the two topics covered are principal component analysis and factor analysis. MANOVA and descriptive discriminant analysis (including "canonical variates") are in another part of that text in which structure and constructs are not discussed. More encouragement is needed.

A view of data-based structure was expressed by Cole et al. (1993). In this article, the authors discuss two competing strategies for identifying structure that underlies multivariate group differences. *First*, they differentiate a *latent variable* system (where the observed outcome variables serve as indicators or manifestations of an underlying construct) and an *emergent variable* system (where the construct is the resultant composition of the outcome variables). *Second*, they propose that MANOVA and DDA would be the appropriate analysis strategy in the context of an emergent variable system, while structural equation modeling (SEM) would be the appropriate analysis strategy in the context of a latent variable system. Discussion was limited to the comparison of two groups. It was recognized that for some data sets, results from MANOVA/DDA and from SEM would yield similar substantive conclusions, but not so in general.

Further Reading

Connally (2004) reviews the effect of covariance matrix inequality on the comparison of linear versus quadratic classification. The simulation results indicated that the internal quadratic rule outperformed the internal linear rule in nearly all conditions studied.

Flury et al. (1994) review the linear versus quadratic PDA comparison; fairly extensive simulation comparisons were made with constraints on covariance matrices. One conclusion drawn was: "Ordinary quadratic discrimination should be avoided whenever possible" (p. 118).

Schiavo and Hand (2000) conclude that although the L-O-O hit rate extimator is almost unbiased, "it suffers from having a large variance" (p. 298).

Schott (1993) develops a "test of dimensionality which is appropriate when the covariance matrices are heterogeneous" (p. 163). The effect of the dimensionality reduction on misclassification probabilities of a quadratic PDA is studied via a simulation.

Definition ***Bernoulli***: An Italian dish made with meat balls.

CHAPTER 23

Problems in PDA and DDA

23.1 INTRODUCTION

Many of the writings on problems in discriminant analyis are based on statistical theory. Five problems related to discriminant analysis in practice were selected for review in this chapter: (1) missing data, (2) outliers and influential units, (3) initial group misclassification, (4) misclassification costs, and (5) statistical versus clinical prediction. Much of the discussion in these five sections consists of references to the writings of statisticians. Specific solutions of some of the problems are not very straightforward to implement for most practicing researchers. Therefore, some suboptimal solutions that are relatively easy to implement are discussed.

23.2 MISSING DATA

23.2.1 Data Inspection

For any data analysis situation one can construct, or at least conceptualize, a data matrix. For a situation involving one grouping variable, G (with J levels), and p response variables, a data matrix may be presented as in the following schematic. (An alternative data schematic may be preferred by some. Such a schematic would have G as a variable across the top—a nominal scale of measurement would be used to measure G—and on the left would be N rows simply indicated by $u_1, u_2, \ldots, u_N$.)

		X_1	X_2	$\cdots$	X_p
	u_1				
G_1	u_2				
	$\vdots$				
	u_{n_1}				
	....				
	u_{n_1+1}				

Applied MANOVA and Discriminant Analysis, Second Edition, by Carl J. Huberty and Stephen Olejnik
Copyright © 2006 John Wiley & Sons, Inc.

		X_1	X_2	$\cdots$	X_p
G_2	u_{n_1+2}				
	$\vdots$				
	$u_{n_1+n_2}$				
	$\cdots\cdots\cdots\cdots$				
	$\vdots$				
	$\cdots\cdots\cdots\cdots$				
G_J	$\vdots$				
	u_N				
	$\cdots$				

If this were a complete data matrix, there would be $N \cdot p$ observations. For various reasons, a researcher may, after all data are collected and filed, end up with fewer than $N \cdot p$ observations.

It is strongly recommended that a researcher carefully examine a listing of his or her data set. For what should the researcher look? Basically, two questions should be kept in mind: (1) Are there any missing data? and (2) are there any aberrant observations? Aberrant measures may be the result of errant recording or errant data entry. Such errors may easily be corrected. Other aberrant measures call for special attention (see Section 23.3).

23.2.2 Data Imputation

Let us now deal with the missing data problem. (It is assumed that second or even third attempts, if reasonable, have been made to complete the data matrix, and that not all attempts were successful.) There are at least three "patterns" that may result with missing data. *One* pattern results when there are many measures missing for one or more analysis units. If there are, say, $p/2$ or more measures missing for a given unit, it might be reasonable to delete the corresponding row from the data matrix; that is, delete that unit. A *second* pattern may result when there are many measures missing for one or more response variables. If a given variable is measured on fewer than, say, $N/2$ units, it might be reasonable to delete the corresponding column from the data matrix. It may also be reasonable to drop a variable if it is not measured on an appreciable number of units in a given group. To no surprise, some judgment calls will have to be made in deciding whether to drop a unit or drop a variable. It must be recognized that in a predictive discriminant analysis (PDA) context, deleting a unit with some missing variable measures precludes one from predicting group membership for future units with comparable profiles.

As one might expect, a *third* pattern—which is really not a pattern at all—of missing data may result when response variable measures are missing for a "few" units and/or a "few" variables. With such a pattern it may not be reasonable to drop all associated units or variables because such decisions may considerably reduce the size of the data matrix. Therefore, we have to either deal with the data we have, or somehow fill in the "holes" in our data matrix; that is, use a data imputation method.

(It is assumed in this discussion that all response variables are continuous, each with measures that do not involve category assignments.) Now with the data on hand we can easily estimate all needed mean vectors. Arriving at the appropriate covariance and SSCP matrices, and inverting those that need to be inverted, using only the data on hand is messy; see Hand (1977, pp. 186–188) for a discussion of, and references for, this problem in the PDA context. An analysis involving all unit measures with or without missing variable measures is sometimes called an *available-case analysis*.

For the practicing researcher, a more reasonable analysis may be one that uses an *imputation* method for the missing data. Now the question becomes: How do we impute the missing observations? Without a doubt, the most popular imputation method used in most data analysis contexts is to replace a missing observation with an arithmetic mean. Consider this approach in our current situation. Suppose that the measure on V_2 is missing for unit 7 in G_3. There are at least two different means that might be considered as replacements for the missing observation. One is the mean of available measures on V_2 in G_3 only. The other mean is that of the available measures on V_2 across *all* the groups. Which one is more appropriate? From a substantive viewpoint, it may make more sense to use the former mean because the imputed measure (i.e., the mean) is based on V_2 measures for units from the *same* group as unit 7, and therefore should more closely approximate the "real" measure on V_2 for unit 7. On the other hand, the all-group mean may be preferred by some methodologists. The reason for this preference is that the imputed value would tend to support null group separation or chance classification. The use of the all-group mean would protect the researcher from the potential criticism of "stacking the deck" in favor of a nonnull condition. There is some evidence in the two-group situation (Hufnagel, 1988) that use of a mean based on a single group will yield better hit-rate estimates for some predictor variable characteristics, some proportions of missing observations, and some intergroup distances. The relative merits of the two mean types in a MANOVA context have not been studied to any known extent.

There is another general imputation method that has been studied a fair bit by statisticians during the past 35 years or so [see, e.g., Chan et al. (1976), Hufnagel (1988), and Jackson (1968)]. This method involves the use of multiple regression. To start, all missing response variable values are replaced with the variable means for a given group. Then each variable on which there are missing values is regressed on all others using the completed group data to develop regression equations. These equations—one for each response variable—yield estimates that replace the values originally missing. (Variations in the method may be used at this point that involve standardizing the data, and the use of a principal components analysis.) An iteration process then follows. New equations are built using the previous estimates, which, in turn, yield new estimates for the missing values. And so on, until successive iterations fail to substantially change the estimated values obtained. Chang et al. (1976) and Hufnagel (1988) found this method to work quite well under some limited conditions, while Jackson (1968) concluded that the simple mean substitution method produced results quite comparable to those of a regression estimation method. [It is not clear from the writings of Chan et al. (1976) and Jackson (1968) as well as earlier writings of Chan (and colleagues) which of the two means—separate-group or total-group—were being substituted for missing values.] This problem clearly needs further study.

The SPSS DISCRIMINANT procedure has an option that may be reasonable if the number of units on which there is at least one missing observation is not too large. The CLASSIFY subcommand is used with the keyword MEANSUB. With this, a classification rule is built using a *complete-case* analysis—an analysis using only those rows in the data matrix that are complete. Next, *total*-group means for the respective response variables are substituted for the missing data values. Then the rule developed earlier is applied to *all* cases. The SAS CANDISC, DISCRIM, and STEPDISC procedures will include in an analysis only those units for which there are complete data vectors.

23.2.3 Missing G Values

An alternative data schematic was mentioned early in this chapter. With this schematic, one column would be used for the grouping variable, G. Now, what if some G values for particular units are missing? This situation might arise when groups of units are determined from a written report or from survey information. Whereas the missing data problem discussed earlier in this section pertains to missing values on the X's (or Y's), now the problem pertains to not knowing group membership for some units. [The context here may be either that of predictive discriminant analysis (PDA) or of descriptive discriminant analysis (DDA).] How might this problem be approached?

One possible approach is to construct a classification rule using those units with known G values. The rule may then be used for predicting group membership of the unit(s) with missing G values—SAS DISCRIM does this. Group membership prediction may be quite clear for some units—"high" posterior probabilities—and not so clear for other units—"low" posterior probabilities. If so, an iterative process may be employed. Units (with missing G values) for which predicted group membership is clear may be put in with units with known G values and a new classification rule may be built. This new rule may then be used for predicting group membership of the remaining units with missing G values. And so on.

A little more complicated problem may arise. How does one proceed if there are missing X (or Y) values as well as missing G values on the same units? On different units? Whatever the case may be, multiple analyses would be needed, with judgments made along the way. In other words, do what makes sense!

23.2.4 Ad Hoc Strategy

Suppose one has an incomplete data matrix of N rows and p columns—that is, there is a total of N units—but on only $N^*(<N)$ units are there *complete* p-dimensional observation vectors. One could conduct an analysis (PDA or DDA) using only the N^* units. Then it may be reasonable to conclude that, say, two of the p response variables may be deleted with little or no loss in the effect of interest (predictive accuracy or group separation). One could then return to the original data matrix and determine a new data matrix of N^{**} (where $N^* < N^{**} \leq N$) rows and $p - 2$ columns—that is, there may be $N^{**} - N^*$ units that had missing data on the two deleted variables but complete data on the other $p - 2$ variables. Then an analysis could be conducted using the $p - 2$ variables and the N^{**} units to determine if more variables might be deleted with little or no loss in the effect of interest. If so, a new data matrix with, presumably, a greater number of rows than N^{**} may then be analyzed. Again, "weak" variables may be deleted, and so on. It is recognized that multiple decisions would need to be made *on different data matrices*. Needless to say, judgment and reasonableness will need to be exercised. [This ad hoc analysis strategy is illustrated in a PDA context by Huberty and Julian (1995).]

23.3 OUTLIERS AND INFLUENTIAL OBSERVATIONS

Data analysis diagnostics have recently become very commonplace in multiple correlation/regression textbooks and computer programs. Multiple regression diagnostics pertain to model fit, data conditions (e.g., collinearity), and identification of outlying observations and influential observations. Diagnostics in discriminant analysis pertain mostly to the two latter identifications. There is, however, another aspect of discriminant analysis diagnostics, namely, the identification of in-doubt observations, or fence riders (see Section 15.6). The focus in the current section is on outliers and influential observations.

23.3.1 Outlier Identification

The problem of identifying outliers in multivariate data is a challenging one, indeed, particularly in a multigroup context. The outlier problem in discriminant analysis has been studied to a very limited extent; see McLachlan (1992, pp. 181–185) for some references. The SPSS computer package yields information that may be helpful in identifying potential outliers in a PDA context. An outlier index that was discussed in Section 14.4.2 and in the Chapter 14 Technical Note is a *typicality probability*. This is a tail area of a probability distribution, a "small" value indicating that a unit is a "great" distance from a centroid. With the SPSS DISCRIMINANT procedure, a chi-squared distribution is used to determine this probability, denoted as $P(D/G)$, associated with the group to which the unit is assigned. The $P(D/G)$ values—denoted by $\hat{P}(\mathbf{x}_u | j)$ in Section 14.4—may be obtained under the condition of covariance matrix homogeneity (when the sample error covariance matrix is used) or under covariance heterogeneity (when separate sample covariance matrices are used). The latter typicality values are obtained by using the subcommand CLASSIFY = SEPARATE/. When the group covariance matrices are clearly not equal, the unit typicality probabilities are difficult to interpret because different distance metrics are used in the calculations.

For a two-group situation under normal homoscedasticity, McDonald et al. (1976) present a test to determine if an observation vector belongs to one of the two populations or to a third population. The test statistic for $\mathbf{x}_u$ and Group j yields a P value that is a typicality probability estimate, $\hat{P}(\mathbf{x}_u | j)$.

There is a type of outlier that may not be identified via the SPSS typicality index. If one type identified is considered an "external" outlier, the other type is an "internal" outlier. An internal outlier is located "in-between" two groups, as opposed to being located "outside" the groups. Such an outlier is most difficult to identify unless the groups are separated to an extreme extent. The location of an outlier, internal or external, may be determined (in the at most two-dimensional discriminant space) by examining discriminant space plots yielded by the SPSS DISCRIMINANT and SAS CANDISC procedures.

23.3.2 Influential Observations

So, what is to be done with units identified as outliers? One thing to check is to see if there is a "cluster" of outliers that might suggest the definition of an additional group. Discriminant space plots may be helpful here. Another thing to check is the *influence* of the identified outliers. Influence on what? If, in a descriptive discriminant analysis context, the concern would be with influence on group separation in general, and construct definition (via LDFs) in particular. If one is in a predictive discriminant analysis context, the concern would be with influence on classification accuracy—separate-group or total-group accuracy. One way to assess the influence of outliers is to conduct repeated analyses. If there is only one outlier, there would be two analyses to conduct, one with the outlier in the analysis and the other with it deleted. With multiple outliers, multiple analyses would need to be conducted. Just how the multiple analyses are to be done is potentially problematic. Do we delete outliers one at a time, two at a time, and so on? Another judgment call must be made.

The study of influential observations in a predictive discriminant analysis context has been restricted to the two-group situation. Three studies by Critchley and Vitiello (1991), Gomez et al. (1990), and Sadek (1992) indicate that the influence problem in PDA is quite complicated. No studies of the influence problem in descriptive discriminant analysis are known.

23.4 INITIAL GROUP MISCLASSIFICATION

The assumption was made in Part IV that for the most part, the criterion groups were *well* defined; that is, group membership of each unit considered was assumed to be correct at the start of the analysis. For an example, consider the 3-group Ethington data set (3GED). Group 1 students were those who were given an A, Group 2 students were given a B, and Group 3 students were given a C or C−. For some reasons, however, some students may have been overrated or underrated. That is, there are research situations in which the initial group membership of units may be in question. This is particularly the case when predicting membership of children in special education groups, patients with particular medical diagnoses (e.g., psychiatric disorders), corporations grouped by Standard Industrial Codes, voters who had voted for election candidates, excavated archaeological pottery pieces grouped by place of manufacture, and so on. Initial grouping of subjects may also be questioned when groupings are based on responses to a questionnaire or survey form.

The influence of initially misclassified analysis units on hit-rate estimation may be nonconsequential, may be considerable, or may be a problem of unknown consequence. The moral of the story, so to speak, is to verify, as much as is feasible, the initial group membership of all units. Often, it is conjectured, initially misclassified units are those in "in-between" groups, those labeled in Section 14.4 as in-doubt cases or fence-riders. Reassigning units is problematic; reassignment of some or all units may be difficult to defend in some research situations. For study of this problem in a two-group context, the reader may refer to Hand (1981, pp. 194–197) and McLachlan (1992, pp. 35–37).

Suppose one has a classification problem where the goal is to estimate group classification hit rates—as opposed to determining a specific classification rule that might be used with follow-up samples. In this situation, initial misclassification may not be a concern because there is a way to estimate hit rates when group membership of some units is not clear. A hit-rate estimator that may be employed is the maximum-posterior-probability (M-P-P) estimator discussed in Section 15.3.4.

It may be farfetched, but if it is suspected that there are many initially misclassified units, it might be thought that some kind of cluster analysis (Aldenderfer and Blashfield, 1984; Kaufman and Rousseeuw, 1990) should be conducted. If a cluster analysis were conducted and reasonable clusters were identifiable, what would the finding of a reasonably good prediction rule indicate? Would the rule have any utility with the data-defined groups? Is it appropriate to develop a rule using the same data that were used to define the criterion groups? It may not make much sense, in practice, to develop a classification rule for criterion groups that are not initially well defined.

23.5 MISCLASSIFICATION COSTS

As mentioned in Sections 13.7 and 18.2, the consideration of misclassification costs when $J = 2$ can be made in a fairly simple manner by modifying the initial priors. When $J > 2$, the use of misclassification costs becomes much more complicated. Grouven et al. (1996) introduce a "menu-driven, user-friendly" computer program that allows for the incorporation of misclassification costs when $J > 2$ in a linear or quadratic PDA context.

23.6 STATISTICAL VERSUS CLINICAL PREDICTION

Clinicians or clinician-like professionals in artificial intelligence, criminal justice, education, medicine, neuropsychology, psychiatry, and psychology are regularly faced with the problem of making predictions. Criteria for these predictions may be diagnostic condition, recovery time, recovery type, disease type, survival time, treatment response, relapse, personality type, graduate school success, mental/behavioral disorder type, and so on. The list is virtually endless. Predictor information utilized may involve personality measures, intelligence scores, biopsy slide ratings, psychological test scores, achievement scores, and so on. Many such predictions are accomplished via professional judgment. It is not known if many clinicians use empirical rules in practice to make their predictions. Such rules would generally be based on multiple regression analysis (MRA) or on predictive discriminant analysis (PDA).

In his review of clinical versus actuarial prediction, Marchese (1992) focuses almost exclusively on MRA. (The criterion variable used in some reviewed studies was, however, dichotomous.)

In clinical practice, prediction[1] is sometimes made with respect to a categorical criterion variable often with more than two categories. Such prediction would call for the use of predictive discriminant analysis (PDA). Studies of how PDA results compare with clinical prediction are few and far between. Some preliminary studies using PDA have been made by Moras et al. (1992), Seshia et al. (1983), Sexton et al. (1987), Shrout et al. (1986), Swiercinsky and Warnock (1977), and Wedding (1983). Willis (1986) reviews some issues in the use of PDA is making clinical predictions.

The details of clinical prediction or of statistical/actuarial prediction will not be given here. Dawes et al. (1989) discuss some of the general notions and interpretations of the two methods of prediction. The general impression of reviews of the comparison of clinical and statistical prediction (Dawes et al., 1989; Marchese, 1992) is that there is overwhelming evidence of the superiority in predictive accuracy with the latter approach. This conclusion is based mostly on syntheses of comparisons where the statistical method involves multiple regression. What is sorely needed are some comparison studies where the statistical method involves PDA. With a categorical criterion variable, PDA results would enable the clinician to obtain such information as the following:

- People who clearly belong to a particular group
- People whose group membership is not clear
- Estimate of predictive accuracy for each group
- Identification of outlying individuals
- Relative importance of predictors

A prediction system might be set up so that when predicted group membership (via PDA) is not clear-cut, the empirical information may be supplemented with the professional judgment of the clinician. Such a statistical-clinical prediction process needs to be studied. (Of course, initial choice of predictors is extremely important; see Section 1.5.) There obviously is much empirical research needed in this area of study.

As Marchese (1992, p. 766) mentions, the use of statistical prediction in some clinical settings has not been well received. It is the case that in some clinical settings, empirical predictions may be extremely difficult—at least perceptually so—to make because of the difficulty of identifying reasonable prediction rules. In particular, it may be the case that nonlinear prediction rules ought to be considered. As has been demonstrated, however, nonlinear rules used with particular score ranges of some predictors can lead to chaos (Paulos, 1991, pp. 32–37). Would chaos theory be applicable to some clinical prediction situations?

[1] Rather than a prediction problem, this might more appropriately be considered an *identification* problem. That is, it is to be determined if a person is to be identified with one group or another. This use of the word "identification" is different from the use found in Manski (1995).

23.7 OTHER PROBLEMS

Three additional problems in discriminant analysis, mostly pertaining to PDA, are reviewed in the comprehensive treatise by McLachlan (1992):

- PDA with repeated measures (pp. 78–86)
- Assessment of assumptions (pp. 169–178)
- Partially classified data (pp. 37–46)

Further Reading

Bar-Hen and Darcdin (1997) propose a test of the hypothesis that a vector of scores for an analysis unit is in one of the two populations against the hypothesis of membership in a third (not initially considered) population; simulations were performed.

Barnett and Lewis (1994) discuss the presence of outliers in a variety of multivariate contexts including PDA.

Bello (1993) simulates the comparative performance of five deterministic imputation techniques when using linear, quadratic, and kernel two-group PDA methods; it is recommended that three iterative imputation techniques (maximum-likelihood estimate, general iterative principal component, single-value decomposition) be tried and compared.

Bello (1995) addresses the effects of imputed values on linear, two-group PDA error rate estimates; eight estimators and five imputation algorithms were considered, with the EM algorithm judged to perform well.

Dawes et al. (1993) review the general issue of statistical versus clinical prediction; some 65 references are cited that relate to the issue itself or to applications, some of which utilize predictive discriminant analysis techniques.

Fung (1992) suggests two indices useful for studying outliers and influential observations in a two-group PDA context.

Fung (1995b) proposes four measures of influence on the (posterior) probability of group membership for an analysis unit not considered in developing the linear normal-based classification rule for the initial two groups; a simulation yielded satisfactory results.

Fung (1996a) proposes eight measures for diagnosing influential observations in a two-group quadratic PDA context and applied them to a data set of two species of biting flies.

Fung (1996b) extends his previous study of two-group linear PDA diagnostic measures to multiple-group measures with a reduced computational load.

Hand (1992, pp. 50–51) discusses statistical versus traditional approaches to classification in medicine.

Hand (1997, Chapter 9) reviews seven different "special" problems related to PDA; included are variable deletion methods, group formation based on partitioning a continuum, and categorical predictors.

Hawkins and McLachlan (1997) suggest a "high-breakdown criterion" to detect outliers in a linear PDA context.

Krusinska et al. (1993) illustrate the influence of outliers in a PDA involving three main types of pathological changes and 26 predictors; deletion of determined outliers resulted in better classification results.

Lachenbruch (2001) proposes a regression-like index of leverage to detect outliers for the two-group linear PDA situation; a plot of ordered values of the index against centiles of a chi-squared distribution may suggest the presence of outliers.

Little and Rubin (2002) discuss new missing-data methods and apply current software to real data.

Naes and Indahl (1998) describe a method of "handling" high predictor collinearity in three multiple-group PDA contexts.

Rocke and Woodruff (1996) demonstrate a "hybrid method" with simulated data to identify outliers. Originated computer software is used.

Schafer (1997) provides an extensive discussion of missing-data situations involving continuous and categorical variables; in particular, missing data in a PDA context is covered.

Steel and Louw (2001) propose a new measure of influence of an analysis unit in the predictor variable selection/deletion process.

Twedt and Gill (1992) compare three data imputation methods and conclude that the three yield comparable two-group hit-rate estimates.

Verboon and van der Lans (1994) propose a method for robust "canonical discriminant analysis" that is useful to reduce the influence of outliers in the context of PDA.

Viragoontavan (2000) compares six missing-data methods in a PDA context. His simulation results indicate that two multiple-imputation methods (using SOLAS and NORM) were "uniformly the most effective." The hot-deck method was third best, followed by the group-mean and regression-based methods. Listwise deletion was least effective.

Whitcomb and Lahiff (1993) conclude that both sample size and distance between two population mean vectors are important in assessing the impact of spurious observations in PDA.

Definition Paradigm: $0.20.

APPENDIX A

Data Set Descriptions

The following five data sets are available at the Wiley website.

Data Set A1 (5GED)

This data set is based on a sample of community college students. Over 700 students responded to 150 (or fewer) items on the Community College Student Experience Questionnaire (CCSEQ) (Ethington et al. 2001). Items of interest to us were those that were scored with numerical values for two to four categories. Nine response variables were defined for our use; see Table A.1. CCSEQ validity and reliability information for the last six effort scales is reported by Ethington and Polizzi (1996); validity and reliability index values are judged to be respectable.

An inspection of the 700 × 9 data matrix was made. We ended up with a 545 × 9 complete data matrix. The 545 (squared) Euclidean distances for each student (represented by a vector of 9 variable scores) to the "typical student" (represented by variable means) were calculated (via SAS). The 545 distances ranged from 5.60 to 37.92 with no appreciable gaps; therefore, it was judged that no outliers were present.

Two grouping variables are considered. These are Race with three levels (Black, Hispanic, White) and Grade with five levels (A, A− or B+, B, B− or C+, and C or C−). To obtain the Grade variable, students were asked to report the grade they typically earned in their classes. The number of students in each Race–Grade combination are given in Table A.2.

In the covariance analysis context, a hypothetical covariate termed Time is included.

Data Set A2 (3GED)

This data set is a subset of the 5GED data set. Here, we are using three Grade levels A, B, and C or C−, and does *not* include Race.

Applied MANOVA and Discriminant Analysis, Second Edition, by Carl J. Huberty and Stephen Olejnik
Copyright © 2006 John Wiley & Sons, Inc.

TABLE A.1 Variables Selected from the CCSEQ

Variable	No. Items	Score Range
1. Counselor Interaction (counsum)	7	0–7
2. Self Understanding (gainsum)	7	7–28
3. Instruction Received (learnsum)	9	9–27
4. Library Effort (qelib)	7	7–28
5. Student-Faculty Effort (qefac)	8	8–32
6. Interstudent Effort (qestacq)	6	6–24
7. Art/music/theater Effort (qeamt)	6	6–24
8. Writing Effort (qewrite)	8	8–32
9. Science Effort (qesci)	9	9–36

TABLE A.2 Cell Sizes for the Race-by-Grade Design

Race	A	A− or B+	B	B− or C+	C or C−	$\eta_{k\cdot}$
Black	20	36	35	57	34	182
Hispanic	15	36	45	54	19	169
White	41	56	42	42	13	194
$\eta_{\cdot j}$	76	128	122	153	66	545

Data Set A3 (FLEX)

A gerontologist working in a senior center was interested in studying the benefits of a flexibility program she had developed. Twenty seniors agreed to participate in a 5-week program. These individuals had been physically active at the senior center for the past year. At the end of each week a score on a flexibility scale was obtained. Scores on this scale could range between 0 and 100. The data file labeled FLEX contains the flexibility scores of the 20 participants.

Data Set A4 (FLEX2)

Two additional groups of seniors participated in the flexibility study described in data set A3. Twenty seniors who had been socially active but not physically active (Group 2) and 20 seniors who had been neither physically or socially active (Group 3) also volunteered to participate in the study. The data file labeled FLEX2 contains the flexibility scores of all 60 participating seniors.

Data Set A5 (HSB)

This data set is a subset of a High School and Beyond (HSB) data set based on a nationwide survey of 1980 high school seniors. The survey was conducted by the National Center for Education Statistics. The grouping variable we consider is Participation in Church Activities with three levels: No Participation ($n_1 = 99$),

Participation ($n_2 = 60$), and Participation as a Leader ($n_3 = 40$)—$N = 199$. The HSB data set involves two response measures with underlying continua: Locus of Control composite scale score (X_1), and Self-Concept composite scale score (X_2). Two additional response variables considered are categorical with unordered categories: Occupational Aspiration at Age 30 with five categories (X_3), and Main Activity in the Year after High School with four categories (X_4). In the text we focus on X_3 and X_4, information which is summarized in Table A.3.

TABLE A.3 Categorical Response Variables

Variable	Category	n
X_3	1: Education	37
	2: Humanities	29
	3: Professional	42
	4: Medical	59
	5: Other	32
X_4	1: Working	45
	2: College/University	93
	3: Technical School, Junior College	41
	4: Other	20

APPENDIX B

Some DA-Related Originators

We realize that there may very well be additional "originators." We can only apologize for omissions; and for any errors. Page indicates where the name is first mentioned in this text. See McLachlan (1992) for many other originators.

Originator	Page
Anderson, J. A. (1939–1983)	307
Anderson, T. W. (1915–)	3
Bargman, R. E. (1921–2004)	84
Barnard, M. M.	3
Bartlett, M. S. (1910–2002)	9
Bayes, T. (1701–1761)	263
Bonferroni, C. E. (1892–1960)	30
Box, G. E. P. (1919–2002)	41
Campbell, N. A.	9
Cochran, W. G. (1909–1980)	323
Cohen, J. (1923–1998)	27
Cooley, W. W. (–2002)	4
Cover, T. M.	364
DasGupta, S. (1925–2001)	3
Dillon, W. R.	297
Efron, B.	386
Elashoff, J. D.	343
Euclid of Alexandria (c.325 B.C.–c.265 B.C.)	26
Ezekiel, M. J. B. (1899–1974)	66
Fisher, R. A. (1890–1962)	3
Flury, B. (1951–1999)	11
Friedman, J. H.	388

Applied MANOVA and Discriminant Analysis, Second Edition, by Carl J. Huberty and Stephen Olejnik
Copyright © 2006 John Wiley & Sons, Inc.

Originator	Page
Fukunaga, K.	303
Fung, W. K.	358
Ganeshanandam, S.	346
Geisser, S. (1929–2004)	208
Glick, N.	297
Habbema, J. D. F.	306
Hand, D. J. (1950–)	4
Hecker, R.	309
Hilden, J.	311
Hills, M.	297
Hodges, J. L. (1922–)	3
Hora, S. C.	302
Hotelling, H. (1895–1973)	39
Jain, A. K.	309
Kempthorne, O. (1919–2000)	10
Klecka, W. R.	307
Krzanowski, W. J.	99
Lachenbruch, P. A.	4
Lancaster, H. O. (1913–2001)	368
Lesaffre, E.	311
Lohnes, P. R. (1929–2001)	4
Lord, F. M. (1912–2000)	299
Mahalanobis, P. C. (1893–1972)	3
McCabe, G. P.	104
McHenry, C. E.	111
McKay, R. J.	346
McLachlan, G. J.	4
McNemar, Q. (1900–1986)	321
Meehl. P. E. (1920–2003)	357
Michaelis, J.	280
Morant, G. M. (1899–1964)	3
Morris, J. D.	13
Mosteller, F. (1909–1990)	12
Neyman, J. N. (1894–1981)	29
Pearson, E. S. (1895–1980)	29
Pearson, K. (1857–1936)	3
Pillai, K. C. S. (1920–1985)	51
Rao, C. R. (1920–)	3
Roy, S. N. (1906–1964)	51
Rulon, P. J.	4
Schaafsma, W.	30
Smith, C. A. B. (1917–2002)	284

Originator	Page
Snapinn, S. M.	346
Sorum, M. J.	296
Tatsuoka, M. M. (1920–1996)	3
Tiedeman, D. V.	3
Tilton, J. W. (1891–1995)	356
Toussaint, G. T.	4
Tukey, J. W. (1915–2000)	12
Wahl, P. W.	281
Welch, B. L. (1911–1989)	350
Wherry, R. J. (1904–1981)	298
Wilks, S. S. (1906–1964)	48
Wrinch, D. (1894–1976)	262
Yao, Y. (–1995)	43

APPENDIX C

List of Computer Syntax

Software	Title	Section
SPSS	Syntax for Descriptive Statistics, Multivariate and Unvariate Analyses	3.6
SPSS	Syntax for Computing Effect Size	4.3
SPSS	Syntax for MANOVA Contrast Analyses	4.5
SPSS	Syntax for Computing Eigenvalues and Structures r's	5.3
SPSS	Syntax for Dimensionality Tests	5.6
SPSS	Syntax to Obtain an LDF Plot	5.7
SPSS	Syntax for Contrast Raw Discriminant Function Weights and Structure r's	5.9
McCabe	Syntax for Finding the Best Subset of Outcome Variables	6.2
SPSS	Syntax for Stepwise Descriptive Discriminant Analysis	6.3
SAS	Syntax for Stepwise Descriptive Discriminant Analysis	6.3
McHenry	Syntax for Contrast Analysis	6.5
SPSS	Syntax for Covariance Equality, Omnibus Test Statistics, Dimensionality Analysis, and Linear Discriminant Functions	8.5
SPSS	Syntax for Computing Main Effect Contrasts in a Factorial Design	8.6
SPSS	Syntax for Simple Effects	T.N.8
SPSS	Syntax for Cell Contrasts using the General Linear Model (GLM) Program	T.N.8
SPSS	Syntax for Regression Slope Vector Equality	9.5
SPSS	Syntax for Analysis of Covariance Omnibus Test and Construct Definition	9.7
SPSS	Syntax for MANCOVA Contrast Analyses	9.9

Applied MANOVA and Discriminant Analysis, Second Edition, by Carl J. Huberty and Stephen Olejnik
Copyright © 2006 John Wiley & Sons, Inc.

Software	Title	Section
SPSS	Syntax for Testing Equality of Regression Planes with Two Covariates	T.N.9
SPSS	Syntax for Testing Equality of Adjusted Centroids with Two Covariates	T.N.9
SPSS	Syntax for Testing Pairwise Contrasts of Adjusted Centroids	T.N.9
SPSS	Syntax for Omnibus Test for Repeated Measures	10.5
SPSS	Syntax for Obtaining Three Estimates of (ε)	10.8
SPSS	Syntax for Polynomial Trend Contrasts	10.10
SPSS	Syntax for Factorial Design with Repeated Measures	T.N.10
SPSS	Syntax for Repeated-Measures Simple Effect	T.N.10
SPSS	Syntax for a Third Trimester Simple Effect	T.N.10
SPSS	Syntax for a Mixed-Model Design	11.5
SPSS	Syntax for Between-Groups and Within-Subjects Contrasts	11.7
SPSS	Syntax for MANCOVA with a Fixed Covariate in a Mixed-Model Design	T.N.11
SPSS	Syntax for a Linear PDA using 3 Grade Groups and 9 Predictor Variables	14.2
SAS	Syntax for a Linear PDA Using 3 Grade Groups and 9 Predictor Variables	14.3
SAS	Syntax for Comparing Linear and Quadratic Internal Hit Rates	16.6
SAS	Syntax for Classifying New Students	16.9
SAS	Syntax for a Linear PDA Using 3 Grade Groups and 9 Predictor Variables	16.9
SAS	Syntax to Obtain Quadratic Classification Function Weights Using 3 Grade Groups and 3 Predictors	16.9
Morris	Prompts for All Possible-Subset Analyses	17.3
SAS	Syntax for Transforming Raw Data to Ranks	19.2
SAS	Syntax for a L-O-O NN Analysis Using 3 Grade Groups and 9 Predictor Variables	19.2

APPENDIX D

Contents of Wiley Website

COMPUTER PROGRAMS

The Wiley website contains nonpackage computer programs corresponding to discussions of six analyses within the text. The six programs are found at the website in the order in which they are mentioned in the text:

Program	**Description**
YAO2	Yao test for two groups
YAOPC	Yao test for contrasts
CLASSVS	Morris program
FLPC	Fisher–Lancaster program
MCCABEPC	McCabe program
MCHENPC	McHenry program

It should be noted that CLASSVS, MCCABEPC, and MCHENPC are FORTRAN source code files from which the executable files, CLASSVS.EXE, MCCABEPC.EXE, and MCHENPC.EXE, respectively, are created. Furthermore, FLPC, YAO2, and YAOPC are *sample* SAS programs that illustrate the use of the SAS macro for conducting the Fisher–Lancaster analysis, Yao two-group test, and Yao contrasts, respectively.

DATA SETS

Twelve data sets are available at the Wiley website. The following is a list of the data files and the location in the text where the data are first cited or described.

Applied MANOVA and Discriminant Analysis, Second Edition, by Carl J. Huberty and Stephen Olejnik
Copyright © 2006 John Wiley & Sons, Inc.

Program	Location
BAUMANN2G2V.sav	Table 2.1
BAUMANN3G2V.sav	Table 3.1
STRESS.sav	Table 8.1
BAUMANN2.sav	Table 9.1
SELFESTEEM.sav	Table 10.1
SELFESTEEM2.sav	Table 11.1
SELFESTEEM3.sav	Chapter 11 Technical Note
5GED.sav	Appendix A Data Set A1
3GED.sav	Appendix A Data Set A2
FLEX.sav	Appendix A Data Set A3
FLEX2.sav	Appendix A Data Set A4
HSB.dat	Appendix A Data Set A5

The Fisher–Lancaster program yields scale values for categories of a categorical response variable with more than two unordered categories (see Section 19.3.4). The scale values may be used in either a predictive discriminant analysis or a descriptive discriminant analysis. This analysis can be conducted using a SAS macro called FL.MAC, written by J. M. Wisenbaker. It can be used for PC versions of SAS having IML capabilities.

To conduct the Fisher–Lancaster scaling, the macro must be used in conjunction with (1) the SAS DATA step that accesses the data to be analyzed, (2) the SAS statement DUM = 1 used for merging purposes, and (3) the statements (one for each categorical variable) that actually invoke the FL macro.

The following is an excerpt from the program:

```
*************************************************************
* Program Name = FLPC                                       *
* Program Purpose = To perform Fisher–Lancaster analysis    *
*    on HSB.DAT which uses the FL.MAC macro                 *
*************************************************************
options        mprint ;                    *******************
DATA ONE;                                  *  Call data set and  *
INFILE 'a:HSB.DAT' MISSOVER;               *  define variables   *
   ** if running on hard drive – >         *******************
      change INFILE statement
      to reflect current directory **;
   INPUT CHURCH X1 X2 X3 X4;
   DUM = 1;

%MACRO FL(NUM, DATA=, GROUP=, VAR=,
             NGROUPS=, NCATS2=, NCATS=);
                                           ********************
   .                                       * Macro for the Fisher- *
   .                                       * Lancaster procedure  *
   .                                       ********************
```

```
%mend FL;
  .
  .
  .
%FL(1, DATA=ONE, GROUP=CHURCH, VAR=X3,   ***********************
  NGROUPS=3, NCATS2=10, NCATS=5)          * Invoke macro for      *
%FL(2, DATA=ONE, GROUP=CHURCH, VAR=X4,   * scaling variables X3  *
  NGROUPS=3, NCATS2=8, NCATS=4)           * and X4                *
                                          ***********************
PROC PRINT;
RUN;
```

The first four lines access the data and define the variables of interest. The statement DUM = 1 must be included so that the new scaled variables can be merged with the original dataset. The next block of code is the actual SAS macro. It can be identified by the first statement, namely %MACRO FL, and the last statement, namely %MEND FL. The parameters to be passed within the macro are defined as follows:

1. NUM starts out with a value of 1 and must be incremented by 1 for each additional categorical variable.
2. DATA designates the SAS dataset containing the data.
3. GROUP identifies the variable identifying group membership.
4. VAR identifies the categorical variable to be scaled.
5. NGROUPS is the value for the number of criterion groups.
6. NCATS2 is the value of twice the number of categories in the variable to be scaled.
7. NCATS is the value for the number of categories in the variable to be scaled.

The next two statements invoke the macro for scaling variables X_3 and X_4, respectively. In other words, it must be called once per categorical variable. The first invocation of the macro defines NUM = 1, indicating that the variable X_3 is to be scaled first, the data can be found in the dataset ONE, the grouping variable is CHURCH, the number of levels of the grouping variable is 3 (NGROUPS), the number of categories in the variable X_3 is 5 (NCATS), and twice the number of categories is 10 (NCATS2). The second invocation of the macro defines NUM = 2 indicating that the variable X_4 is to be scaled second, the data can be found in data set ONE, the grouping variable is CHURCH, the number of levels of the grouping variable is 3 (NGROUPS), the number of categories in the variable X_4 is 4 (NCATS), and twice the number of categories is 8 (NCATS2).

For each macro, the output consists of a cross-classification table for each categorical variable against the grouping variable, the scaling coefficients for each categorical valued variable (WP), and the eigenvalues (LAM). Also on each call, the macro generates Fisher–Lancaster scaled values for the designated categorical variable for the first two eigenvalues (SCALEAi and SCALEBi, respectively). These are appended to the original data set. Once these values are calculated and appended to the original data set, these values can be used for subsequent statistical analyses.

References*

Abelson, R. P. (1995). *Statistics as Principled Argument*. Hillsdale, NJ: Erlbaum. (380)

Aeberhard, S., Coomans, D., and DeVel, O. (1993). Improvements to the classification performance of RDA. *Journal of Chemometrics*, **7**, 99–115. (357)

Afifi, A. A., and Azen, S. P. (1979). *Statistical Analysis*. New York: Academic. (289,297,369)

Agresti, A. (2002). *Categorical Data Analysis*, 2nd ed. New York: Wiley. (321,323)

Albert, A., and Lesaffre, E. (1986). Multiple group logistic discrimination. *Computers and Mathematics with Applications*, **12A**, 209–224. (366)

Aldenderfer, M. S., and Blashfield, R. K. (1984). *Cluster Analysis*. Beverly Hills, CA: Sage. (407)

Alf, E., and Abrahams, N. M. (1968). Relationship between per cent overlap and measures of correlation. *Educational and Psychological Measurement*, **28**, 779–792. (357)

Allen, D. A. (1971). *The prediction of sum of squares as a criterion for selecting predictor variables* (Tech. Rep. 23). Lexington, KY: University of Kentucky, Department of Statistics. (302)

Altman, H., Evenson, R. C., and Cho, D. (1976). New discriminant functions for computer diagnosis. *Multivariate Behavioral Research*, **11**, 367–376. (331)

Anderson, J. A., and Boyle, J. A. (1968). Computer diagnosis statistical aspects. *British Medical Bulletin*, **24**, 230–235. (307)

Anderson, T. W. (2003). *An Introduction to Multivariate Statistical Analysis*, 3rd ed. New York: Wiley. (265)

Anderson, T. W., Das Gupta, S., and Styan, G. P. (1972). *A Bibliography of Multivariate Statistical Analysis*. Edinburgh: Oliver & Boyd. (3)

Andreev, S. (2003). Estimation of similarity between poetic texts and their translations by means of discriminant analysis. *Journal of Quantitative Linguistics*, **10**, 159–176. (8)

Aversa, E. S. (1985). Citation patterns of highly cited papers and their relationship to literature aging: A study of working literature. *Scientometrics*, **7**, 383–389. (99)

*Numbers in parentheses following each entry indicate pages on which the reference is cited.

Applied MANOVA and Discriminant Analysis, Second Edition, by Carl J. Huberty and Stephen Olejnik
Copyright © 2006 John Wiley & Sons, Inc.

Badarinathi, R. (1983). Multivariate generalized classification statistics. Doctoral dissertation, University of Georgia, Athens. (369)

Bailar, J. C., and Mosteller, F. (1992). Guidelines for statistical reporting in articles for medical journals: Amplifications and explanations. In J. C. Bailar and F. Mosteller (Eds.). *Medical Uses of Statistics* (pp. 313–331). Boston: NEJM Books. (117,375)

Bargmann, R. E. (1969). Exploratory techniques involving artificial variables. In P. R. Krishnaiah (Ed.). *Multivariate Analysis II* (pp. 567–580). New York: Academic. (84)

Bar-Hen, A. (1996). A preliminary test in discriminant analysis. *Journal of Multivariate Analysis*, **57**, 266–276. (283)

Bar-Hen, A., and Darcdin, J. J. (1997). A test of a special case of typicality in linear discriminant analysis. *Biometrics*, **53**, 39–48. (409)

Barnard, M. M. (1935). The secular variations of skull characters in four series of Egyptian skulls. *Annals of Eugenics*, **6**, 352–371. (3)

Barnett, S. (1990). *Matrices: Methods and Applications*. New York: Clarendon. (31)

Barnett, V., and Lewis, T. (1994). *Outliers in Statistical Data*. New York: Wiley. (409)

Bartlett, M. S. (1939). A note on tests of significance in multivariate analysis. *Proceedings of the Cambridge Philosophical Society*, **35**, 180–185. (9)

Baron, A. E. (1991). Misclassification among methods used for multiple group discrimination: The effects of distributional properties. *Statistics in Medicine*, **10**, 757–766. (370,372,381)

Baumann, J. F., Edwards, E. C., Boland, E. C., Olejnik, S., and Kame'enui, E. J. (2003). Vocabulary tricks: Effects of instruction in morphology and context on fifth-grade students' ability to derive and infer word meanings. *American Educational Research Journal*, **40**, 447–494. (164)

Baumann, J. F., Seifert-Kessell, N., and Jones, L. A. (1992). Effect of think-aloud instruction on elementary students' comprehension monitoring abilities. *Journal of Reading*, **24**, 143–172. (15)

Bedrick, E. J., Lapidus, J., and Powell, J. F. (2000). Estimating the Mahalanobis distance from mixed continuous and discrete data. *Biometrics*, **56**, 394–401. (357)

Bello, A. L. (1993). A simulation study of imputation techniques in linear, quadratic, and kernel dicriminant analyses. *Journal of Statistical Computation and Simulation*, **48**, 167–180. (409)

Bello, A. L. (1995). Missing values, imputation, and error rate estimators in discriminant analysis. *Australian Journal of Statistics*, **37**, 95–104. (409)

Bernstein, I. H. (1988). *Applied Multivariate Analysis*. New York: Springer. (9)

Best, J. (2001). *Damned Lies and Statistics*. Berkeley: University of California. (380)

Blashfield, R. K., and Aldenderfer, M. S. (1988). The methods and problems of cluster analysis. In J. R. Nesselroade and R. B. Cattell (Eds.). *Handbook of Multivariate Experimental Psychology* (pp. 447–473). New York: Plenum. (99)

Bobbitt, S. A. (1990). *Key Statistics for Public Elementary and Secondary Education: School Year 1990–91, Early Estimates*. Washington, DC: U.S. Department of Education. (369)

Boothby, L. H., and Brewer, J. K. (1990). Predicting early educational program placement with discrete discriminant analysis. *Communications in Statistics: Theory and Methods*, **19**, 4049–4060. (372)

Box, G. E. P. (1949). A general distribution theory for a class of likelihood criteria. *Biometrida*, **36**, 317–346. (41,55,139)

Bray, J. H., and Maxwell S. E. (1982). Analyzing and interpreting significant MANOVAs. *Review of Educational Research*, **52**, 340–367. (85)

Breiman, L., Friedman, J., Olshen, R., and Stone, C. (1984). *Classification and Regression Trees*. New York: Chapman & Hall. (385)

Browne, M. W. (1975). Predictive validity of a linear regression equation. *British Journal of Mathematical and Statistical Psychology*, **28**, 79–87. (296)

Bryan, J. G. (1961, January). Calibration of qualitative and quantitative variables for use in multiple-group discriminant analysis. [Scientific Report 2, Contract AF19(604)–5207]. Hartford, CT: Travellers Insurance Companies. (367,369)

Cacoullos, T. (Ed.) (1973). *Discriminant Analysis and Applications*. New York: Academic. (3)

Campbell, M. K., Donner, A., and Webster, K. M. (1991). Are ordinal models useful for classification? *Statistics in Medicine*, **10**, 383–394. (372)

Campbell, N. A. (1980). Shrunken estimators in discriminant and canonical variate analyses. *Applied Statistics*, **29**, 5–14. (100)

Campbell, N. A. (1984). Some aspects of allocation and discrimination. In G. N. Van Vark and W. W. Howells (Eds.). *Multivariate Statistical Methods in Physical Anthropology* (pp. 177–192). Boston: Reidel. (99)

Carroll, R. M., and Nordholm, L. A. (1975). Sampling charatceristics of Kelley's ϵ^2 and Hay's ω^2. *Educational and Psychological Measurement*, **35**, 541–554. (62)

Cawley, G. C., and Talbot, N. L. C. (2003). Efficient leave-one-out cross-validation of kernel Fisher discriminant classifiers. *Pattern Recognition*, **36**, 2585–2582. (357)

Chan, L. S., Gilman, J. A., and Dunn, O. J. (1976). Alternative approaches to missing values in discriminant analysis. *Journal of the American Statistical Association*, **71**,842–844. (403)

Chapman, D. W., Holloway, R., and Kelly, E. (1977). Using ratings by students to predict high and low achievers in a PSI course: A discriminant analysis. *AV Communication Review*, **25**, 411–422. (301)

Chatfield, C. (1993). Neural networks: Forecasting breakthrough or passing fad? *International Journal of Forecasting*, **9**, 1–3. (386)

Chatfield, C., and Collins, A. J. (1980). *Introduction to Multivariate Analysis*. New York: Chapman & Hall. (399)

Cheng, B., and Titterington, D. M. (1994). Neural networks: A review from a statistical perspective. *Statistical Science*, **9**, 2–54. (388)

Choi, S. C. (1986). Discriminant and classification: Overview. *Computers and Mathematics with Applications*, **12A**, 173–177. (311,372)

Chouvarda, I., Maglaveras, N., Boufidou, A., Mohlas, S., and Louridas, G. (2003). Wigner-Ville analysis and classification of electrocardiograms during thrombolysis. *Medical & Biological Engineering and Computing*, **41**, 609–617. (358)

Clarkson, D. B. (1979). Estimating the standard errors of rotated factor loadings. *Psychometrika*, **44**, 297–314. (398)

Cochran, W. G. (1964). On the performance of the linear discriminant function. *Technometrics*, **6**, 179–190. (343)

Cohen, J. (1968). Weighted kappa: Nominal scale agreement with provisions for scaled disagreement or partial credit. *Psychological Bulletin*, **70**, 213–220. (331)

Cohen, J. (1972). Weighted chi square: An extension of the kappa method. *Educational and Psychological Measurement*, **32**, 61–74. (331)

Cohen, J. (1988). *Statistical Power Analysis for the Behavioral Sciences*. New York: Academic. (27,38,74)

Cohn, V. (1989). *News and Numbers*. Ames, IA: Iowa State University. (380)

Cole, D. A., Maxwell, S. E., Arvey, R., and Salas, E. (1993). Multivariate group comparisons of variable systems: MANOVA and structural equation modeling. *Psychological Bulletin*, **114**, 174–184. (393)

Comrey, A. L., and Lee, H. B. (1992). *A First Course in Factor Analysis*. Hillsdale, NJ: Erlbaum. (399)

Connally, M. (2004). Identifying covariance differences in comparisons of linear versus quadratic classification rule. Doctoral dissertation, University of Georgia, Athens. (393,400).

Cook, T. D., and Campbell, D. T. (1979). *Experimental and Quasi-experimental Designs for Research*. Chicago: McNally. (9)

Cooley, W. W., and Lohnes, P. R. (1962). *Multivariate Procedures for the Behavioral Sciences*. New York: Wiley. (4)

Cooley, W. W., and Lohnes, P. R. (1968). Predicting development of young adults. Project Talent Interim Report 5. (4)

Cooley, W. W., and Lohnes, P. R. (1985). *Multivariate Data Analysis*. Melbourne, FL: Krieger. (84)

Cornell, J. E., Young, D. M., Seaman, S. L., and Kirk, R. E. (1992). Power comparisons of eight tests for sphericity in repeated measures designs. *Journal of Educational Statistics*, **17**, 233–249. (208)

Coste, J., Walter, E., Wasserman, D., and Venot, A. (1997). Optimal discriminant analysis for ordinal responses. *Statistics in Medicine*, **16**, 561–569. (371)

Costner, H. L. (1965). Criteria for measures of association. *American Sociological Review*, **30**, 341–353. (332)

Cover, T. M., and Hart, P. (1967). Nearest neighbor pattern classification. *IEEE Transactions on Information Theory*, **IT-11**, 21–27. (364)

Cox, T. F., and Pearce, K. F. (1997). A robust logistic discrimination model. *Statistics and Computing*, **7**, 155–161. (358)

Cramer, E. M., and Nicewander, W. A. (1979). Some symmetric invariant measures of multivariate association. *Psychometrika*, **44**, 43–54. (63,77)

Critchley, F., and Vitiello, C. (1991). The influence of observations on misclassification probability estimates in linear discriminant analysis. *Biometrics*, **78**, 677–690. (406)

Crossen, C. (1994). *Tainted Truth*. New York: Simon and Schuster. (380)

Cruz-Castillo, J. G., Ganeshanandam, B. R., MacKay, B. R., Lawes, G. S., Lawoko, C. R. O., and Woolley, D. I. (1994). Applications of canonical discriminant analysis in horticultural research. *Horticultural Science*, **29**, 1115–1119. (124)

Dalgleish, L. I. (1994). Discriminant analysis: Statistical inference using the jackknife and bootstrap procedures. *Psychological Bulletin*, **116**, 498–508. (398)

Das Gupta, S. (1973). Theories and methods in classification: A review. In T. Cacoullos (Ed.). *Discriminant Analysis and Applications* (pp. 77–137). New York: Academic. (3)

Dattalo, P. (1994). A comparison of discriminant analysis and logistic regression. *Journal of Social Science*, **19**, 121–144. (386)

David, H. A. (1995). First (?) occurrence of common terms in mathematical statistics. *American Statistician*, **49**, 121–133. (262)

David, H. A. (1998). First (?) occurrence of common terms in probablilty and statistics—A second list, with corections. *American Statistician*, **52**, 36–40. (9)

David, H. A., and Edwards, A. W. F. (2001). *Annotated Readings in the History of Statistics*. New York: Springer. (265)

Dawes, R. M., Faust, D., and Meehl, P. E. (1989). Clinical versus actuarial judgment. *Science*, **243**, 1668–1674. (408)

Dawes, R. M., Faust, D., and Meehl, P. E. (1993). Statistical prediction versus clinical prediction: Improving what works. In G. Keren and C. Lewis (Eds.). *A Handbook for Data Analysis in the Behavioral Sciences: Methodological Issues* (pp. 351–367). Hillsdale, NJ: Erlbaum. (409)

Devillers, J., and Karcher, W. (Eds.) (1991). *Applied Multivariate Analysis in SAR and Environmental Studies*. Boston: Kluwer Academic. (381)

Diday, E., Lechevallier, Y., Schader, M., Bertrand, P., and Burtschy, B. (Eds.) (1994). *New Approaches in Classification and Data Analysis*. New York: Springer. (387)

Dillon, W. R. (1979). The performance of the linear discriminant function in nonoptimal situations and the estimation of classification error rates: A review of recent findings. *Journal of Marketing Research*, **16**, 370–381. (297)

Dillon, W. R., and Goldstein, M. (1984). *Multivariate Analysis: Methods and Applications*. New York: Wiley. (303)

Dillon, W. R., Goldstein, M., and Schiffman, L. G. (1978). Appropriateness of linear discriminant and multinomial classification analysis in marketing research. *Journal of Marketing Research*, **15**, 103–112. (369)

Dillon, W. R., and Westin, S. (1982). Scoring frequency data for discriminant analysis: Perhaps discrete procedures can be avoided. *Journal of Marketing Research*, **19**, 44–56. (371)

Dodge, Y. (2003). *The Oxford Dictionary of Statistical Terms*. Oxford: Oxford University. (12,385,386,387)

Dorans, N. J. (1988). The shrunken generalized distance estimator of the actual error rate in discriminant analysis. *Journal of Educational Statistics*, **13**, 63–74. (298,358)

Duarte Silva, A. P. (2001). Efficient variable screening for multivariate analysis. *Journal of Multivariate Analysis*, **76**, 35–62. (346)

Duarte Silva, A. P., Stam, A., and Neter, J. (2002). The effects of misclassification costs and skewed distributions in two-group classification. *Communications in Statistics—Simulation and Computation*, **31**, 401–423. (358)

Dudoit, S., Fridlyand, J., and Speed, T. P. (2002). Comparison of discrimination methods for the classification of tumors using gene expression data. *Journal of the American Statistical Association*, **97**, 77–87. (388)

Efron, B. (1975). The efficiency of logistic regression compared to normal discriminant analysis. *Journal of the American Statistical Association*, **70**, 892–898. (386)

Ehrenberg, A. S. C. (1990). The unimportance of relative importance [Letter to the editor]. *American Statistician*, **44**, 260. (114)

Elashoff, J. D., Elashoff, R. M., and Goldman, G. E. (1967). On the choice of variables in classification problems with dichotomous variables. *Biometrika*, **54**, 668–670. (343)

Ellis, D. G., and Armstrong, G. B. (1989). Class, gender, and code on prime-time television. *Communication Quarterly*, **3**, 157–169. (118)

Ethington, C. A., Guthrie, A. M., and Lehman, P. W. (2001). *CCSEQ: Test Manual and Comparative Data*, 3rd ed. Memphis, TN: Center for the Study of Higher Education. (411)

Ethington, C. A., and Polizzi, T. B. (1996). An assessment of the construct validity of the CCSEQ quality of effort scales. *Research in Higher Education*, **37**, 711–730. (118,376,411)

Everitt, B. S., and Der, G. (1996). *A Handbook of Statistical Analyses Using SAS*. Boca Raton, FL: Chapman & Hall. (358)

Ezekiel, M. (1930). *Methods of Correlational Analysis*. New York: Wiley. (66)

Fan, X. (1996). An SAS program for assessing multivariate normality. *Educational and Psychological Measurement*, **56**, 668–674. (279)

Fan, X., and Wang, L. (1999). Comparing linear discriminant function with logistic regression for the two-group classification problem. *Journal of Experimental Education*, **67**, 265–286. (386)

Faraj, A. (1994). Interpretation tools for generalized discriminant analysis. In E. Diday et al. (Eds.). *New Approaches in Classification and Data Analysis* (pp. 285–291). New York: Springer. (388)

Fatti, L. P., Hawkins, D. M., and Raath, E. L. (1982). Discriminant analysis. In D. Hawkins (Ed.). *Topics in Applied Multivariate Analysis* (pp. 1–71). Cambridge, MA: Cambridge University. (297)

Finn, J. E., Burger, C. V., and Holland-Bartels, L. (1997). Discrimination among populations of sockeye salmon fry with Fourier analysis of otolith banding patterns formed during incubation. *Transactions of the American Fisheries Society*, **126**, 559–578. (383)

Fisher, R. A. (1936). The use of multiple measurements in taxonomic problems. *Annals of Eugenics*, **7**, 179–188. (3)

Fisher, R. A. (1940). The precision of discriminant functions. *Annals of Eugenics*, **10**, 422–429. (368)

Fleiss, J. L., Levin, B., and Paik, M. C. (2003). *Statistical Methods for Rates and Proportions*, 3rd ed. New York: Wiley. (323,332)

Flitman, A. M. (1997). Towards analysing student failures: Neural networks compared with regression analysis and multiple discriminant analysis. *Computers and Operations Research*, **24**, 367–377. (388)

Flury, B. (1995). Developments in principle component analysis. In W. J. Krzanowski (Ed.). *Recent Advances in Descriptive Multivariate Analysis* (pp. 14–33). Oxford: Clarendon. (11,283)

Flury, B., and Riedwyl, H. (1988). *Multivariate Statistics: A Practical Approach*. New York: Chapman & Hall. (349,358)

Flury, B., Schmid, M. J., and Narayanan, A. (1994). Error rates in quadratic discrimination with constraints on the covariance matrices. *Journal of Classification*, **11**, 101–120. (393,400)

Foley, D. H. (1972). Consideration of sample and feature size. *IEEE Transactions on Information Theory*, **IT-18**, 618–626. (309)

Fraley, C., and Raftery, A. E. (2003). Enhanced model-based clustering, densiy estimation, and discriminant analysis software: MCLUST. *Journal of Classification*, **20**, 263–286. (376)

Frank, R. E., Massy, W. F., and Morrison, D. G. (1965). Bias in multiple discriminant analysis. *Journal of Marketing Research*, **2**, 250–258. (300)

Friedman, J. H. (1989). Regularized discriminant analysis. *Journal of the American Statistical Association*, **82**, 249–266. (388)

Fruchter, B. (1954). *Introduction to Factor Analysis*. Princeton, NJ: Van Nostrand. (399)

Fukunaga, K. (1990). *Introduction to Statistical Pattern Recognition*. Boston: Academic. (387)

Fung, W. K. (1992). Some diagnostic measures in discriminant analysis. *Statistics & Probability Letters*, **13**, 279–285. (409)

Fung, W. K. (1995a). Influence on classification and probability of misclassification. *Sankhya*, **57**, 377–384. (358)

Fung, W. K. (1995b). Detecting influential observations for estimated probabilities in multiple discriminant analysis. *Computational Statistics and Data Analysis*, **20**, 557–568. (409)

Fung, W. K. (1996a). Diagnosing influential observations in quadratic discriminant analysis. *Biometrics*, **52**, 1235–1241. (409)

Fung, W. K. (1996b). The influence of observations on misclassification probability in multiple discriminant analysis. *Communications in Statistics—Theory and Methods*, **25**, 1917–1930. (409)

Fung, W. K. (1998). On the equivalence of two diagnostic measures in discriminant analysis. *Communications in Statistics—Theory and Methods*, **27**, 1915–1922. (358)

Ganeshanandam, S., and Krzanowski, W. J. (1989). On selecting variables and assessing their performance in linear discriminant analysis. *Australian Journal of Statistics*, **31**, 433–447. (346)

Geisser, S., and Greenhouse, S. W. (1958). An extension of Box's results on the use of the F distribution in multivariate analysis. *Annals of Mathematical Statistics*, **29**, 885–891. (208)

Giri, N. C. (2004). *Multivariate Statistical Analysis*. New York: Dekker. (4)

Gitlow, H. S. (1979). Discrimination procedures for the analysis of nominally scaled data sets. *Journal of Marketing Research*, **16**, 387–393. (371)

Glick, N. (1978). Additive estimators for probabilities of correct classification. *Pattern Recognition*, **10**, 211–227. (297,302,303)

Glover, F. (1990). Improved linear programming models for discriminant analysis. *Decision Sciences*, **2**, 771–785. (283)

Gollob, H. F. (1967, September). Cross-validation using samples of size one. Paper presented at the annual meeting of the American Psychological Association, Washington, DC. (302)

Gomez, M. J., DeBenzo, Z., Gomez, C., Marcano, E., Torres, H., and Ramirez, M. (1990). Comparison of methods for outlier detection and their effects on the classification results for a particular data base. *Analytica Chimica Acta*, **239**, 229–243. (406)

Gordon, A. D. (1999). *Classification*. New York: Chapman & Hall. (386)

Gottesman, I. I., and Prescott, C. A. (1989). Abuses of the MacAndrew MMPI Alcoholism Scale: A critical review. *Clinical Psychology Review*, **9**, 223–242. (357)

Greene, T., and Rayens, W. S. (1989). Partially pooled covariance matrix estimation in discriminant analysis. *Communications in Statistics: Theory and Methods*, **18**, 3679–3702. (281)

Gribbons, W. D., and Lohnes, P. R. (1969). *Career Development from Age 13 to 25* (Project 6-2151). Washington, DC: Bureau of Research. (4)

Grouven, U., Bergel, F., and Schultz, A. (1996). Implementation of linear and quadratic discriminant analysis incorporating costs of misclassification. *Computer Methods and Programs in Biomedicine*, **49**, 55–60. (283,407)

Haase, R. F. (1991). Computational formulas for multivariate strength of association from approximate F and χ^2 tests. *Multivariate Behavioral Research*, **26**, 227–245. (77)

Habbema, J. D. F., and Hermans, J. (1977). Selection of variables in discriminant analysis by F-statistic and error rate. *Technometrics*, **19**, 487–493. (346,366)

Habbema, J. D. F., and Hilden, J. (1981). The measurement of performance in probabilistic diagnosis. IV. Utility considerations in therapeutics and prognostics. *Methods of Information in Medicine*, **20**, 80–96. (311)

Habbema, J. D. F., Hilden, J., and Bjerregaard, B. (1978). The measurement of performance in probabilistic diagnosis. I. The problem, descriptive tools, and measures based on classification matrices. *Journal of Methodology in Medical Research Information and Documentation*, **17**, 217–226. (306,307,311)

Habbema, J. D. F., Hilden, J., and Bjerregaard, B. (1981). The measurement of performance in probabilistic diagnosis. V. General recommendations. *Methods of Information in Medicine*, **20**, 97–100. (311)

Hadorn, D. C., Draper, D., Rogers, W. H., Keeler, E. B., and Brook, R. H. (1992). Cross-validation performance of mortality prediction models. *Statistics in Medicine*, **11**, 475–489. (388)

Hakstian, A. R., Roed, J. C., and Lind, J. C. (1979). Two sample T^2 procedures and the assumption of homogeneous covariance matrices. *Psychological Bulletin*, **86**, 1255–1263. (42,43,74)

Hand, D. J. (1981). *Discrimination and Classification*. New York: Wiley. (4,311,407)

Hand, D. J. (1983). A comparison of two methods of discriminant analysis to binary data. *Biometrics*, **39**, 683–694. (371)

Hand, D. J. (1986). Recent advances in error rate estimation. *Pattern Recognition Letters*, **4**, 335–346. (311)

Hand, D. J. (1987). A shrunken leaving-one-out estimator of error rate. *Computers and Mathematics with Applications*, **14**, 161–167. (311)

Hand, D. J. (1992). Statistical methods in diagnosis. *Statistical Methods in Medical Research*, **1**, 49–67. (372,409)

Hand, D. J. (1994). Assessing classification rules. *Journal of Applied Statistics*, **21**(3), 3–16. (311)

Hand, D. J. (1997). *Construction and Assessment of Classification Rules*. Chicester, UK: Wiley. (12,300,302,311,336,346,366,369,381,386,395,403,409)

Harris, G., and Kaine, G. (1994). The determinants of research performance: A study of Australian university economists. *Higher Education*, **26**, 191–201. (383)

Harris, R. J. (1989). A canonical cautionary. *Multivariate Behavioral Research*, **24**, 17–39. (397)

Harris, R. J. (2001). *A Primer of Multivariate Statistics*. Mahwah, NJ: Erlbaum. (51,91)

Hawkins, D. M. (1981). A new test for multivariate normality and homoscedasticity. *Technometrics*, **23**, 105–110. (280)

Hawkins, D. M. (1993). Review of *Discriminant Analysis and Statistical Pattern Recognition* by G. J. McLachlan (1992). *Journal of the American Statistical Association*, **88**, 695–696. (12)

Hawkins, D. M., and McLachlan, G. J. (1997). High-breakdown linear discriminant analysis. *Journal of American Statistical Association*, **92**, 136–143. (410)

Hayes, F. B., and Martin, R. P. (1986). Effectiveness of the PPVT-R in the screening of young gifted children. *Journal of Psychoeducational Assessment*, **4**, 27–33. (357)

Hays, W. L. (1963). *Statistics*. New York: Holt, Rinehart, and Winston. (62)

Healy, M. J. R. (1986). *Matrices for Statistics*. Oxford: Clarendon. (31)

Hecker, R., and Wagner, H. (1978). The validation of classification rates in stepwise discriminant analyses. *Biometrical Journal*, **20**, 713–727. (309)

Henschke, C. I., and Chen, M. M. (1974). Variable selection technique for classification problems. *Educational and Psychological Measurement*, **34**, 11–18. (336)

Henschke, C. I., Kossack, C. F., and Lissitz, R. W. (1974). A generalized classification technique using interval, ordinal, and nominal variables. *Multivariate Behavioral Research*, **9**, 105–118. (369)

Herzberg, P. A. (1969). The parameters of cross-validation. *Psychometrika*, Monograph supplement 16. (296)

Heyck, T. W., and Klecka, W. (1973). British radical M.P.'s 1874–1895: New evidence from discriminant analysis. *Journal of Interdisciplinary History*, **4**, 161–184. (307)

Hilden, J., Habbema, J. D. F., and Bjerregaard, B. (1978a). The measurement of performance in probabilistic diagnosis. II. Trustworthiness of the exact values of the diagnostic probabilities. *Methods of Information in Medicine*, **17**, 227–237. (311)

Hilden, J., Habbema, J. D. F., and Bjerregaard, B. (1978b). The measurement of performance in probabilistic diagnosis. III. Methods based on continuous functions of the diagnosis probabilities. *Methods of Information in Medicine*, **17**, 238–246. (311)

Hills, M. (1966). Allocation rules and their error rates. *Journal of the Royal Statistical Society (B)*, **28**, 1–31. (297)

Hockersmith, F. B. (1969). *Evaluation of New Approaches to Classification Error Rates* (Project 6-2151). Washington, DC: Bureau of Research. (4)

Hodges, J. L. (1950). *Discriminant Analysis. I. Survey of Discriminatory Analysis* (Report 1, Project 21-49-004). Randolph Air Force Base, TX: USAF School of Aviation Medicine. (NTIS PB 102-690.) (3)

Hoffmann, N. G., and Overall, J. E. (1976). *Problems in the Interpretation of Indices of Diagnostic Agreement* (Tech. Rep. 59). Galveston, TX: University of Texas Medical Branch, Psychometric Laboratory. (331)

Hora, S. C., and Wilcox, J. B. (1982). Estimation of error rates in several population discriminant analysis. *Journal of Marketing Research*, **19**, 57–61. (302,303,304,336)

Hornsby-Smith, M. P., Procter, M., Rajan, L., and Brown J. (1987). A typology of progressive Catholics: A study of the delegates to the National Pastoral Congress. *Journal for the Scientific Study of Religion*, **26**, 234–248. (99)

Horst, P. (1965). *Factor Analysis of Data Matrices*. New York: Holt, Rinehart and Winston. (399)

Hubert, L. J. (1978). A general formula for the variance of Cohen's weighted kappa. *Psychological Bulletin*, **85**, 183–184. (331)

Huberty, C. J. (1975). Discriminant analysis. *Review of Educational Research*, **45**, 543–598. (4)

Huberty, C. J. (1983). Some univariate–multivariate generalizations. *Educational and Psychological Measurement*, **43**, 705–721. (283)

Huberty, C. J. (1984). Issues in the use and interpretation of discriminant analysis. *Psychological Bulletin*, **95**, 156–171. (287,331)

Huberty, C. J. (1987). On statistical testing. *Educational Researcher*, **16**, 4–9. (29)

Huberty, C. J. (1989). Problems with stepwise methods: Better alternatives. In B. Thompson (Ed.). *Advances in Social Science Methodology*, Vol. 1 (pp. 43–70). Greenwich,CT: JAI. (103,126,340,344,345,396,397)

Huberty, C. J. (1994a). Why multi-variable analyses? *Educational and Psychological Measurement*, **54**, 620–627. (10,13,397)

Huberty, C. J. (1994b). A note on interpreting an R^2 value. *Journal of Educational and Behavioral Statistics*, **19**, 351–356. (330)

Huberty, C. J. (1994c). *Applied Discriminant Analysis*. New York: Wiley. (282,306,337)

Huberty, C. J. (2000a). Review of *Construction and Assessment of Classification Rules* by D. J. Hand (1997). *Psychometrika*, **60**, 135–137. (12)

Huberty, C. J. (2000b). In focus: Judgment in quantitative research. *Mathematics Educator*, **10**, 5–10. (31)

Huberty, C. J. (2002). Discriminant analysis. In J. Meij (Ed.). *Dealing with the Data Flood* (pp. 585–600). The Hague, Netherlands: Study Centre for Technology Trends. (13)

Huberty, C. J. (2003). Multiple correlation versus multiple regression. *Educational and Psychological Measurement*, **63**, 271–278. (296)

Huberty, C. J. (2005). Mahalanobis distance. In B. Everitt and D. C. Howell (Eds.). *Encyclopedia of Statistics in Behavioural Science*, Vol. 3 (pp. 1110–1111). London: Wiley. (259)

Huberty, C. J., and Curry, A. R. (1978). Linear versus quadratic multivariate classification. *Multivariate Behavioral Research*, **13**, 237–245. (280)

Huberty, C. J., DiStefano, C., and Kamphaus, R. W. (1997). Behavioral clustering of school children. *Multivariate Behavioral Research*, **32**, 105–134. (394)

Huberty, C. J., and Holmes, S. E. (1983). Two-group comparisons and univariate classification. *Educational and Psychological Measurement*, **43**, 15–26. (357)

Huberty, C. J., and Hussein, M. H. (2001). Reporting information in multiple correlation and multiple regression studies. In E. I. Farmer and J. W. Rojewski (Eds.). *Research Pathways* (pp. 325–348). Lanham, MD: University Press of America. (296,330,354)

Huberty, C. J., and Hussein, M. H. (2003). Some problems in reporting use of discriminant analyses. *Journal of Experimental Education*, **71**, 177–191. (117,124,315,382,383,397,399)

Huberty, C. J., Jordan, E. M., and Brandt, W. C. (2005). Cluster analysis in higher education research. In J. C. Smart (Ed.). *Higher Education: Handbook of Theory and Research*, Vol. 20 (pp. 437–457). New York: Kluwer. (99,291)

Huberty, C. J., and Julian, M. W. (1995). An ad hoc analysis strategy with missing data. *Journal of Experimental Education*, **63**, 333–342. (405)

Huberty, C. J., and Lowman, L. L. (1997). Discriminant analysis via statistical packages. *Educational and Psychological Measurement*, **57**, 759–784. (76,291,393)

Huberty, C. J., and Lowman, L. L. (1998). Discriminant analysis in higher education research. In J. C. Smart (Ed.). *Higher Education: Handbook of Theory and Research*, Vol. 13 (pp. 181–234). New York: Agathon. (13)

Huberty, C. J., and Lowman, L. L. (2000). Group overlap as an effect-size basis. *Educational and Psychological Measurement*, **60**, 543–563. (357)

Huberty, C. J., and Morris, J. D. (1989). Multivariate analysis versus multiple univariate analyses. *Psychological Bulletin*, **105**, 302–308. [Reprinted in A. E. Kazdin (Ed.), *Methodological Issues and Strategies in Clinical Research* (1992, pp. 351–365). Washington, DC: APA.] (13)

Huberty, C. J., and Mourad, S. A. (1980). Estimation in multiple correlation/prediction. *Educational and Psychological Measurement*, **40**, 101–112. (299,302)

Huberty, C. J., and Wisenbaker, J. M. (1992a). Discriminant analysis: Potential improvements in typical practice. In B. Thompson (Ed.). *Advances in Social Science Methodology*, Vol. 2 (pp. 169–208). Greenwich, CT: JAI. (306)

Huberty, C. J., and Wisenbaker, J. M. (1992b). Variable importance in multivariate group comparisons. *Journal of Educational Statistics*, **17**, 75–91. (88,107,114,341,397)

Huberty, C. J., Wisenbaker, J. M., Smith, J. D., and Smith, J. C. (1986). Using categorical variables in discriminant analysis. *Multivariate Behavioral Research*, **21**, 479–496. (369,370)

Huff, D. (1993). *How to Lie with Statistics*. New York: Norton. (380)

Hufnagel, G. (1988). On estimating missing values in linear discriminant analysis. Part I. *Biometrical Journal*, **30**, 69–75. (403)

Hussein, M. H. (2001). The linear classification function versus the logistic discriminant function in a three-population situation with a mixture of continuous and categorical variables. Doctoral dissertation, University of Georgia, Athens. (388)

Huynh, H., and Feldt, L. S. (1976). Estimation of the Box correction for degrees of freedom from sample data in randomized block and split-plot designs. *Journal of Educational Statistics*, **1**, 69–82. (209)

Hwang, C-H. (1994). On estimating the dimensionality in discriminant analysis. *Communications in Statistics—Theory and Methods*, **23**, 2197–2215. (100)

Jackson, E. C. (1968). Missing values in linear multiple discriminant analysis. *Biometrics*, **24**, 835–844. (403)

Jain, A. K., and Chandrasekaran, B. (1982). Dimensionality and sample size considerations in pattern recognition practice. *Handbook of Statistics*, **2**, 835–855. (309,311)

Jain, A. K., Duin, P. W., and Mao, J. (2000). Statistical pattern recognition: A Review. *IEEE Transactions on Pattern Analysis and Machine Intelligence*, **22**, 4–37. (310,388,393)

Jain, S., and Jain, A. K. (1994). Discriminant analysis and its application to medical data. *Biometrical Journal*, **36**, 147–151. (325)

James, M. (1985). *Classification Algorithms*. New York: Wiley. (311)

Joachimsthaler, E. A., and Stam, A. (1988). Four approaches to the classification problem in discriminant analysis: An experimental study. *Decision Sciences*, **19**, 322–333. (372)

Joachimsthaler, E. A., and Stam, A. (1989). Solving the classification problem in discriminant analysis via linear and nonlinear programming methods. *Decision Sciences*, **20**, 285–293. (358)

Joachimsthaler, E. A., and Stam, A. (1990). Mathematical programming approaches for the classification problem in two-group discriminant analysis. *Multivariate Behavioral Research*, **25**, 427–454. (358)

Johnson, R. A., and Wichern, D. W. (2002). *Applied Multivariate Statistical Analysis*, 5th ed. Englewood Cliffs, NJ: Prentice Hall. (25,281)

Johnston, B., and Seshia, S. S. (1992). Discriminant analysis when all variables are ordered. *Statistics in Medicine*, **11**, 1023–1032. (372)

Jones, L. V., and Bock, R. D. (1960). Multiple discriminant analysis applied to "ways to live" ratings from six cultural groups. *Sociometry*, **23**, 162–177. (4)

Joy, O. M., and Tollefson, J. O. (1975). On the financial applications of discriminant analysis. *Journal of Financial and Quantitative Analysis*, **10**, 723–739. (88)

Jurs, P. C. (1986). Pattern recognition used to investigate multivariate data in analytic chemistry. *Science*, **232**, 1219–1224. (381)

Kaplan, A. (1964). *The Conduct of Inquiry: Methodology for Behavioral Science*. San Francisco: Chandler. (399)

Kaufman, L., and Rousseeuw, P. J. (1990). *Finding Groups in Data: An Introduction to Cluster Analysis*. New York: Wiley. (407)

Kempthorne, O. (1971). Probability, statistics and the knowledge business. In V. P. Godambe and D. A. Sprutt (Eds.). *Foundations of Statistical Inference* (pp. 470–499, with discussion). Toronto: Holt, Rinehart and Winston. (10)

Kempthorne, O. (1977). Theories of inference and data analysis. In R. O. Collier and T. J. Hummel (Eds.). *Experimental Design and Interpretation* (pp. 376–399). Berkeley, CA: McCutchan. (31,32)

Kendall, M. G. (1957). *A Course in Multivariate Analysis*. London: Charles Griffin. (4)

Keppel, G., and Wickens, T. D. (2004). *Design and Analysis: A Researcher's Handbook*, 4th ed. Upper Saddle River, NJ: Prentice Hall. (195,198,228)

Keselman, H. J. (1975). A Monte Carlo investigation of three estimates of treatment magnitude: Epsilon squared, eta squared, and omega squared. *Canadian Psychology Review*, **16**, 44–48. (62)

Khattree, R., and Naik, D. N. (2000). *Multivariate Data Reduction and Discrimination with SAS Software*. Cary, NC: SAS Institute, Inc. (371)

Kim, S., and Olejnik, S. (2005). Bias and precision of measures of association for a fixed-effect multivariate analysis of variance model *Multivariate Behavioral Research*, **40**, 401–421. (66)

Kirk, R. (1995). *Experimental Design: Procedures for the Behavioral Sciences*, 3rd ed. Pacific Grove, CA: Brooks/Cole. (228, 231)

Klecka, W. R. (1980). *Discriminant Analysis*. Beverly Hills, CA: Sage. (332)

Koffler, S. L. (1980). A comparison of approaches for setting proficiency standards. *Journal of Educational Measurement*, **17**, 167–178. (357)

Koffler, S. L., and Penfield, D. A. (1982). Nonparametric classification based upon inverse normal scores and rank transformations. *Journal of Statistical Computation and Simulation*, **15**, 51–68. (366)

Koolaard, J. P., Ganesalingam, S., and Lawoko, C. R. O. (1998). Comparison of regularized discriminant analysis with standard discrimination methods. *Computational Statistics*, **13**, 495–510. (388)

Koolaard, J. P., Ganesalingam, S., and Lawoko, C. R. O. (2002). The use of distance measure in regularized discriminant analysis. *Computational Statistics*, **17**, 185–202. (388)

Koolaard, J. P., and Lawoko, C. R. O. (1993). Estimating error rates in discriminant analysis with correlated training observations: A simulation study. *Journal of Statistical Computation and Simulation*, **48**, 81–99. (358)

Koolaard, J. P., and Lawoko, C. R. O. (1996). The linear and Euclidean discriminant functions: A comparison via asymptotic expansions and simulation study. *Communications in Statistics—Theory and Methods*, **25**, 2989–3011. (358)

Kraemer, H. C., and Thiemann, S. (1987). *How Many Subjects? Statistical Power Analysis in Research*. Newbury Park, CA: Sage. (74)

Krauth, J. (1986). Classification procedures for ordered categorical data. In W. Gaul and M. Schader (Eds.). *Classification as a Tool of Research* (pp. 249–255). New Holland: Elsevier Science. (367,371)

Kres, H. (1983). *Statistical Tables for Multivariate Analysis: A Handbook with References to Applications*. New York: Springer. (75)

Krusinska, E., Mathiesen, U. L., Franzen, L., Bodemar G., and Wigertz, O. (1993). Influence of "outliers" on the association between laboratory data and histopathological findings in liver biopsy. *Methods of Information in Medicine*, **32**, 388–395. (410)

Krzanowski, W. J. (1980). Mixtures of continuous and categorical variables in discriminant analysis. *Biometrics*, **36**, 493–499. (372)

Krzanowski, W. J. (1990). Between-group analysis with heterogeneous covariance matrices: The common principal component model. *Journal of Classification*, **7**, 81–98. (99)

Krzanowski, W. J. (1994). Quadratic location discriminant functions for mixed categorical and continuous data. *Statistics and Probability Letters*, **19**, 91–95. (372)

Krzanowski, W. J. (1995). Selection of variables, and assessment of their performance, in mixed-variable discriminant analysis. *Computational Statistics and Data Analysis*, **19**, 419–431. (346)

Kumar, K., and Haynes, J. D. (2003). Forecasting credit ratings using an ANN and statistical techniques. *International Journal of Business Studies*, **11**, 91–108. (388)

Kumar, K., and Sahai, A. (1993). Application of discriminant analysis to the family planning data. *Biometrical Journal*, **35**, 869–875. (8,372)

Kundert, K. R. (1973). Tools of analysis for pattern recognition. Doctoral dissertation, University of Georgia, Athens. (369)

Lachenbruch, P. A. (1967). An almost unbiased method of obtaining confidence intervals for the probability of misclassification in discriminant analysis. *Biometrics*, **23**, 639–645. (301)

Lachenbruch, P. A. (1968). On expected probabilities of misclassification in discriminant analysis, necessary sample size, and a relation with the multiple correlation coefficient. *Biometrics*, **24**, 823–834. (311,355)

Lachenbruch, P. A. (1975). *Discriminant Analysis*. New York: Hafner. (4)

Lachenbruch, P. A. (2001). Discriminant diagnostics. *Biometrics*, **53**, 1284–1292. (410)

Lachenbruch, P. A., and Mickey, M. R. (1968). Estimation of error rates in discriminant analysis. *Technometrics*, **10**, 1–11. (301,354)

Lad, F. (1996). *Operational Subjective Statistical Methods*. New York: Wiley. (32)

Lancaster, H. O. (1957). Some properties of the bivariate normal distribution considered in the form of a contingency table. *Biometrika*, **44**, 289–292. (368)

Langbehn, D. R., and Woolson, R. F. (1997). Discriminant analysis using the unweighted sum of binary variables: A comparison of model selection methods. *Statistics in Medicine*, **16**, 2679–2700. (346)

LaRocco, J. M., Pugh, W. M., and Gunderson, E. K. (1977). Identifying determinants of retention decisions. *Personnel Psychology*, **30**, 199–215. (301)

Lauter, J. (1978). Sample size requirements for the T^2 test of MANOVA (tables for one-way classification). *Biometrical Journal*, **20**, 389–406. (75)

Lederer, R., and Downs, R. (1995). *The Write Way*. New York: Pocket. (380)

Lei, P-W., and Koehly, L. M. (2003). Linear discriminant analysis versus logistic regression: A comparison of classification errors in the two-group case. *Journal of Experimental Education*, **72**, 25–49. (358,386)

Lesaffre, E., Willems, J. L., and Albert, A. (1989). Estimation of error rate in multiple group logistic discrimination: The approximate leaving-one-out method. *Communications in Statistics: Theory and Methods*, **18**, 2989–3007. (311,386)

Levy, P. (1967). Substantive significance of significant differences between two groups. *Psychological Bulletin*, **67**, 37–40. (357)

Liebert, R. M., and Spiegler, M. D. (1982). *Personality: Strategies and issues*. Homewood, IL: Dorsey. (398)

Light, R. J. (1971). Measures of response agreement for qualitative data: Some generalizations and alternatives. *Psychological Bulletin*, **76**, 365–377. (332)

Lindeman, R. H., Merenda, P. F., and Gold, R. Z. (1980). *Introduction to bivariate and multivariate analysis*. Glenview, IL: Scott, Foresman. (324)

Little, R. J. A., and Rubin, D. B. (2002). *Statistical Analysis with Missing Data*. New York: Wiley. (410)

Loh, W-L. (1995). On linear discriminant analysis with adaptive ridge classification rules. *Journal of Multivariate Analysis*, **53**, 264–278. (358)

Long, J. S. (1997). *Regression Models for Categorical and Limited Dependent Variables*. Thousand Oaks, CA: Sage. (372)

Looney, S. W. (1988). A statistical technique for comparing the accuracies of several classifiers. *Pattern Recognition Letters*, **8**, 5–9. (323)

Loucopoulos, C., and Pavur, R. (1997). Computational characteristics of a new mathematical programming model for the three-group discriminant problem. *Computers and Operations Research*, **24**, 179–191. (388)

Lu, J. W., Plataniotis, K. N., and Venetsanopoulos, A. N. (2003). Regularized discriminant analysis for the small sample size problem in face recognition. *Pattern Recognition Letters*, **24**, 3079–3087. (389)

Lykken, D. T., and Rose, R. (1963). Psychological prediction from actuarial tables. *Journal of Clinical Psychology*, **19**, 139–151. (332,371)

MacCorquodale, K., and Meehl, P. E. (1948). On a distinction between hypothetical constructs and intervening variables. *Psychological Review*, **55**, 95–107. (399)

Manski, C. F. (1995). *Identification problems in the social sciences*. Cambridge, MA: Harvard University. (408)

Marchese, M. C. (1992). Clinical versus actuarial prediction: A review of the literature. *Perceptual and Motor Skills*, **75**, 583–594. (408)

Marriott, F. H. C. (1990). *A Dictionary of Statistical Terms*. New York: Wiley. (387)

Marzban, C., and Stumpf, G. J. (1996). A neural network for tornado prediction based on Doppler radar-derived attributes. *Journal of Applied Meteorology*, **35**, 617–626. (386)

Mason, S. J. (1998). Seasonal forecasting of South African rainfall using a non-linear discriminant analysis model. *International Journal of Climatology*, **18**, 147–164. (383)

Mauchly, J. W. (1940). Significance test for sphericity of a normal n-variate distribution. *Annals of Mathematical Statistics*, **11**, 204–209. (207)

Maxwell, S. E. (1961). Canonical variate analysis when the variates are dichotomous. *Educational and Psychological Measurement*, **21**, 259–271. (367)

Maxwell, S. E., and Cole, D. A. (1995). Tips for writing (and reading) methodological articles. *Psychological Bulletin*, **118**, 193–198. (127)

Maxwell, S. E., and Delaney, H. D. (2000). *Designing Experiments and Analyzing Data*. Mahwah, NJ: Erlbaum. (30,76,151,159,195,198,210,216,246)

McCabe, G. P. (1975). Computations for variable selection in discriminant analysis. *Technometrics*, **17**, 103–109. (104)

McDonald, L. L., Lowe, V. W., Smidt, R. K., and Meister, K. A. (1976). A preliminary test for discriminant analysis based on small samples. *Biometrics*, **32**, 417–422. (311,405)

McDonald, R. P. (1985). *Factor analysis and related methods*. Hillsdale, NJ: Erlbaum. (399)

McHenry, C. E. (1978). An efficient computational procedure for selecting a best subset. *Applied Statistics*, **27**, 291–296. (111)

McKay, R. J., and Campbell, N. A. (1982). Variable selection techniques in discriminant analysis II. Allocation. *British Journal of Mathematical and Statistical Psychology*, **35**, 30–41. (346)

McLachlan, G. J. (1977). The bias of sample based posterior probabilities. *Biometrical Journal*, **19**, 421–426. (300)

McLachlan, G. J. (1992). *Discriminant Analysis and Statistical Pattern Recognition*. New York: Wiley. (4,11,12,13,265,280,283,289,298,304,311,325,336,350,367,372, 381,385,386,387,393,394,395,405,407,409,415)

Meehl, P. E., and Rosen, A. (1955). Antecedent probability and the efficiency of psychometric signs, patterns, or cutting scores. *Psychological Bulletin*, **52**, 194–216. (357)

Meshbane, A., and Morris, J. D. (1995, April). Assuming equal vs. unequal prior probabilities of group membership in discriminant analysis: Effect on predictive accuracy. Paper presented at the annual meeting of the American Educational Research Association, San Fransisco. (280,324,358)

Meshbane, A., and Morris, J. D. (1996). A method for selecting between linear and quadratic classification models in discriminant analysis. *Journal of Experimental Education*, **63**, 263–273. (386)

Meulepas, E. (1990). On a criterion for omitting variables in discriminant analysis. *Biometrics*, **46**, 1181–1183. (115)

Michaelis, J. (1973). Simulation experiments with multiple group linear and quadratic discriminant analysis. In T. Cacoullos (Ed.). *Discriminant Analysis and Applications* (pp. 225–238). New York: Academic. (280,300)

Mkhadri, A., Celeux, G., and Nasroallah, A. (1997). Regularization in discriminant analysis: An overview. *Computational Statistics and Data Analysis*, **23**, 403–423. (389)

Moore, D. S. (2004). *The Basic Practice of Statistics*. New York: Freeman. (12)

Moras, K., Di Nardo, P. A., and Barlow, D. H. (1992). Distinguishing anxiety and depression: Reexamination of the Reconstructed Hamilton Scales. *Psychological Assessment*, **4**, 224–227. (408)

Moroney, M. J. (1951). *Facts from Figures*. New York: Penguin Books. (380)

Morris, J. D., and Meshbane, A. (1995). Selecting predictor variables in two-group classification problems. *Educational and Psychological Measurement*, **55**, 438–441. (323)

Morrison, D. F. (1990). *Multivariate Statistical Methods*, 3rd ed. New York: McGraw-Hill. (51,283,289,335)

Mosteller, F., and Bush, R. R. (1954). Selected quantitative techniques. In G. Lindzey (Ed.). *Handbook of Social Psychology* (pp. 289–334). Reading, MA: Addison-Wesley. (330)

Mosteller, F., and Tukey, J. W. (1977). *Data Analysis and Regression.* Reading, MA: Addison-Wesley. (12,300)

Mueller, R. O., and Cozad, J. B. (1988). Standardized discriminant coefficients: Which variance estimate is appropriate? *Journal of Educational Statistics*, **13**, 313–318. (88,398)

Mueller, R. O., and Cozad, J. B. (1993). Standardized discriminant coefficients: A rejoinder. *Journal of Educational Statistics*, **18**, 108–114. (88,398)

Mulaik, S. A. (1994). The critique of pure statistics: Artifact and objectivity in multivariate statistics. In B. Thompson (Ed.). *Advances in Social Science Methodology*, Vol. 3 (pp. 241–289) Greenwich, CT: JAI. (397,399)

Munakata-Marr, J., Tenorio, L., and Siegrist, R. L. (2003). Statistical evaluation of bacterial source tracking data obtained by rep-PCR DNA fingerprinting of Escherichia coli. *Environmental Science and Technology*, **37**, 4554–4560. (373)

Murray, D., Schwartz, J., and Lichter, S. R. (2001). *It Ain't Necessarily so.* Larham, MD: Rowman & Littlefield. (380)

Myers, J. L., DiCecco, J. V., White, J. B., and Borden, V. M. (1982). Repeated measures on dichotomous variables: *Q*-test and *F*-test. *Psychological Bulletin*, **92**, 517–525. (323)

Myers, R. A. (1990). *Classical and Modern Regression with Application.* Boston: Duxbury. (307)

Naes, T., and Indahl, U. (1998). A unified description of classical classification methods for multicollinear data. *Journal of Chemometrics*, **12**, 205–220. (410)

Nakanishi, H., and Sato, Y. (1985). The performance of the linear and quadratic discriminant functions for three types of non-normal distribution. *Communications in Statistics: Theory and Methods*, **14**, 1181–1200. (281)

Nath, R., Jackson, W. M., and Jones, T. W. (1992). A comparison of the classical and the linear programming approaches to the classification problem in discriminant analysis. *Journal of Statistical Computing and Simulation*, **40**, 73–93. (373)

Nath, R., and Jones, T. W. (1988). A variable selection criterion in the linear programming approaches to discriminant analysis. *Decision Sciences*, **19**, 554–563. (346)

Nel, D. G., and van der Merwe, C. A. (1986). A solution to the multivariate Behrens–Fisher problem. *Communications in Statistics: Theory and Methods*, **15**, 3719–3735. (77)

Nelson, B. J., Runger, G. C., and Si, J. (2003). An error rate comparison of classification methods with continuous explanatory variables. *IIE Transactions*, **35**, 537–566. (389)

Nordlund, D. J., and Nagel, R. (1991). Standardized discriminant coefficients: Which variance estimate is appropriate? *Journal of Educational Statistics*, **13**, 313–318. (88,398)

Oakes, M. (1986). *Statistical Inference: A Commentary for the Social and Behavioral Sciences.* New York: Wiley. (357)

O'Gorman, T. W., and Woolson, R. F. (1993). On the efficacy of the rank transformation in stepwise logistic and discriminant analysis. *Statistics in Medicine*, **12**, 143–151. (346)

Olejnik, S., and Algina, J. (2000). Measures of effect size for comparative studies: Applications, interpretations, and limitations. *Contemporary Educational Psychology*, **25**, 241–286. (135)

Olejnik, S., and Algina, J. (2003). Generalized eta and omega squared statistics: Measures of effect size for some common research designs. *Psychological Methods*, **8**, 434–447. (134,169,197,231)

Olson, C. L. (1974). Comparative robustness of six tests in multivariate analysis of variance. *Journal of the American Statistical Association*, **69**, 894–908. (279)

Overall, J. E., and Klett, C. (1972). *Applied Multivariate Analysis*. New York: McGraw-Hill. (100,367)

Overall, J. E., and Woodward, J. A. (1977). Discriminant analysis with categorical data. *Applied Psychological Measurement*, **1**, 371–384. (368)

Panel on Discriminant Analysis, Classification, and Clustering. (1989). Discriminant analysis and clustering. *Statistical Science*, **4**(1), 34–69. (4)

Park, P. S., and Kshirsagar, A. M. (1996). Chance of misclassification in the case of several normal populations. *Australian Journal of Statistics*, **38**, 75–82. (331)

Parker, M., and Leinhardt, G. (1995). Percent: A priviledged proportion. *Review of Educational Research*, **65**, 421–481. (383)

Paulos, J. A. (1985). *I Think, Therefore I Laugh: An Alternative approach to philosophy*. New York: Vintage Books. (380)

Paulos, J. A. (1988). *Innumercay*. New York: Hill and Wang. (380)

Paulos, J. A. (1991). *Beyond Numeracy*. New York: Vintage Books. (31,380,408)

Paulos, J. A. (1995). *Mathematician Reads the Newspaper*. New York: Basic Books. (380)

Paulos, J. A. (2000) *I Think, Therefore, I Laugh: The Flip Side of Philosophy*. New York: Vintage Books. (380)

Peng, Y., Jiang, Y., Chen, W., Yang, X., Brender, C., and Han, X. (2003). Comparison of artificial neural network and multivariate linear discriminant analysis in the prediction of extra-capoular extension of prostate cancer. *Medical Physics*, **30**, 1368–1369. (386)

Pison, G., Struyf, A., and Rousseeuw, P. I. (1999). Displaying a clustering with CLUSPLOT. *Computational Statistics and Data Analysis*, **30**, 381–392. (93)

Porter, A. C., and Raudenbush, S. W. (1987). Analysis of covariance: Its model and use in psychological research. *Journal of Counseling Psychology*, **34**, 383–392. (164)

Poudevigne, M. S. (2004). Relationships between physical activity and mood in pregnancy. Doctoral dissertation, University of Georgia, Athens. (195,228)

Press, S. J. (1972). *Applied Multivariate Analysis*. New York: Holt, Rinehart and Winston. (331)

Pynnönen, S. (1988). Testing for additional information in variables in multivariate normal classification with unequal covariance matrices. *Acta Wasaensia*, **23**, 3–95. (346)

Ragsdale, C. T., and Stam, A. (1992). Introducing discriminant analysis to the business statistics curriculum. *Decision Sciences*, **23**, 724–745. (13)

Rao, C. R. (1948). The utilization of multiple measurements in problems of biological classification. *Journal of the Royal Statistical Society, Series B*, **10**, 159–193. (3)

Rao, C. R. (1952). *Advanced Statistical Methods in Biometric Research*. New York: Wiley. (4,49)

Rao, C. R. (1973). *Linear Statistical Inference and its Application*. New York: Wiley. (264)

Rayens, W. S. (1993). Review of *Discriminant Analysis and Statistical Pattern Recognition* by G. J. McLachlan (1992). *Technometrics*, **35**, 324–326. (12)

Renaud, O. (2002). The discrimination power of projection pursuit with different density estimators. *Biometrics*, **89**, 129–143. (389)

Rencher, A. C. (1992). Interpretation of canonical discriminant functions, canonical variates, and principal components. *American Statistician*, **46**, 217–225. (397)

Rencher, A. C. (1998). *Multivariate Statistical Inference and Applications*. New York: Wiley. (13,75,312,346,389)

Rencher, A. C. (2002). *Methods of Multivariate Analysis*, 2nd ed. New York: Wiley. (11,13,25,41,51,56,88,283,312,337,349,397)

Rencher, A. C., and Scott, D. T. (1990). Assessing the contribution of individual variables following rejection of a multivariate hypothesis. *Communications in Statistics: Simulation and Computation*, **19**, 535–553. (115)

Rocke, D. M., and Woodruff, D. L. (1996). Identification of outliers in multivariate data. *Journal of the American Statistical Association*, **91**, 1047–1061. (410)

Rorer, L. G., Hoffman, P. J., and Hsieh, K. (1966). Utilities as base rate multipliers in the determination of optimum cutting scores for the discrimination of groups of unequal size and variance. *Journal of Applied Psychology*, **50**, 364–368. (357)

Rothman, K. J. (1990). No adjustments are needed for multiple comparisons. *Epidemiology*, **1**, 43–46. (32)

Rubin, P. A. (1990). A comparison of linear programming and parametric approaches to the two-group discriminant problem. *Decision Sciences*, **21**, 337–386. (281,358)

Rudolph, P. M., and Karson, M. (1988). The effect of unequal priors and unequal misclassification costs on MDA. *Journal of Applied Statistics*, **15**, 69–81. (331,358,394)

Rulon, P. J., Tiedeman, D. V., Tatsuoka, M. M., and Langmuir, C. R. (1967). *Multivariate Statistics for Personnel Classification*. New York: Wiley. (4)

Rummel, R. J. (1970). *Applied Factor Analysis*. Evanston, IL: Northwestern University. (399)

Sadek, R. F. (1992). Influence of outliers in classification analysis. Doctoral dissertation, University of Georgia, Athens. (312,406)

Saville, D. J. (1990). Multiple comparison procedures: The practical solution. *American Statistician*, **44**, 174–180. (32)

Schaafsma, W., and van Vark, G. N. (1979). Classification and discrimination problems with applications. Part IIa. *Statistica Neerlandica*, **33**, 91–126. (30,301)

Schafer, J. L. (1997). *Analysis of Incomplete Multivariate Data*. London: Chapman & Hall. (410)

Schaub, M., and Tokar, D. M. (1999). Patterns of expectations about counseling: Relations to the five-factor model of personality. *Journal of Counseling and Development*, **77**, 177–188. (7)

Schiavo, R. A., and Hand, D. J. (2000). Ten more years of error rate research. *International Statistical Review*, **68**, 295–310. (312,400)

Schmitz, P. I. M., Habbema, J. D. F., Hermans, J., and Raatgever, J. W. (1983). Comparative performance of four discriminant analysis methods for mixtures of continuous and discrete variables. *Communications in Statistics: Simulation and Computation*, **12**, 727–751. (373)

Schott, J. R. (1993). Dimensionality reduction in quadratic discriminant analysis. *Computational Statistics and Data Analysis*, **16**, 161–174. (283,400)

Seaman, S. L., and Young, D. M. (1990). A non-parametric variable selection algorithm for allocatory linear discriminant analysis. *Educational and Psychological Measurement*, **50**, 837–841. (346)

Seo, T., Kanda, T., and Fujikoshi, Y. (1995). The effects of nonnormality on tests for dimensionality in canonical correlation and MANOVA models. *Journal of Multivariate Analysis*, **52**, 325–337. (100)

Serlin, R. C. (1982). A multivariate measure of association based on the Pillai–Bartlett procedure. *Psychological Bulletin*, **91**, 413–417. (66,72,75)

Seshia, S. S., Johnston, B., and Kasian, G. (1983). Non-traumatic coma in childhood: Clinical variables in prediction of outcome. *Developmental Medicine and Child Neurology*, **25**, 493–501. (288,376,408)

Sexton, D. L., McIlwraith, R., Barnes, G., and Dunn, R. (1987). Comparison of the MCMI and MMPI-168 as psychiatric inpatient screening inventories. *Journal of Personality Assessment*, **51**, 388–398. (408)

Shrout, P. E., Dohrenwend, B. P., and Levav, I. (1986). A discriminant rule for screening cases of diverse diagnostic types: Preliminary results. *Journal of Counseling and Clinical Psychology*, **54**, 314–319. (408)

Shulman, L. S. (1988). Disciplines of inquiry in education: An overview. In R. M. Jaeger (Ed.). *Complementary Methods for Research in Education* (pp. 3–17). Washington, DC: American Educational Research Association. (31)

Smart, J. C. (1982). Faculty teaching goals: A test of Holland's theory. *Journal of Educational Psychology*, **74**, 180–188. (75)

Smith, C. A. B. (1947). Some examples of discrimination. *Annals of Eugenics*, **13**, 272–282. (284)

Smith, J. C. (1984). A comparison of four predictor selection methods in linear classification. Doctoral dissertation, University of Georgia, Athens. (337,346)

Snapinn, S. M., and Knoke, J. D. (1988). Bootstrapped and smoothed classification error rate estimators. *Communications in Statistics: Simulation and Computation*, **17**, 1135–1153. (346,359)

Snapinn, S. M., and Knoke, J. D. (1989). Estimation of error rates in discriminant analysis with selection of variables. *Biometrics*, **45**, 289–299. (347)

Solander, L. (1978). Modeling student academic performance: Predicting "success" or "trouble" in a professional pharmacy program. *American Journal of Pharmaceutical Education*, **42**, 326–329. (369)

Soltysik, R. C., and Yarnold, P. R. (1994a). Univariable optional discriminant analysis: One-tailed hypotheses. *Educational and Psychological Measurement*, **54**, 646–653. (357,389)

Soltysik, R. C., and Yarnold, P. R. (1994b). The Warmack-Gonzalez algorithm for linear two-category multivariate optimal discriminant analysis. *Computers and Operations Research*, **21**, 735–745. (359)

Sorum, M. J. (1972a). Three probabilities of misclassification. *Technometrics*, **14**, 309–316. (296)

Sorum, M. J. (1972b). Estimating the expected and the optimal probabilities of misclassification. *Technometrics*, **14**, 935–943. (297,298)

Spatz, B. A., Thombs, D. L., Byrne, T. J., and Page, B. J. (2003). Use of the theory of planned behavior to explain HRT decisions. *American Journal of Health Behavior*, **27**, 445–455. (8)

Srinivasan, V., and Kim, Y. H. (1987). Credit granting: A comparative analysis of classification preocedures. *Journal of Finance*, **42**, 665–683. (373)

Steel, S. J., and Louw, N. (2001). Variable selection in discriminant analysis: Measuring the influence of individual cases. *Computational Statistics and Data Analysis*, **37**, 249–260. (373,410)

Subrahmaniam, K., and Subrahmaniam, K. (1973). *Multivariate Analysis: A Selected and Abstracted Bibliography, 1957–1972*. New York: Dekker. (4)

Suich, R. C., and Turek, R. J. (1989). Prediction versus independence in contingency tables. *Teaching Statistics*, **11**, 42–43. (332)

Swiercinsky, D. P., and Warnock, J. K. (1977). Comparison of the neuropsychological key and discriminant analysis approaches in predicting cerebral damage and localization. *Journal of Consulting and Clinical Psychology*, **45**, 808–814. (408)

Talarico, S. M. (1978). Defining the sentencing decision: Implications of alternate conceptualizations. *Social Science Quarterly*, **59**, 570–577. (369)

Tardif, B., and Hardy, J. (1995). Assessing the relative contribution of variables in canonical discriminant analysis. *Taxon*, **44**, 69–76. (115)

Tate, R. L. (1981). Multivariate ATI analysis. *Multivariate Behavioral Research*, **16**, 243–259. (77)

Tatsuoka, M. M. (1969). Multivariate analysis. *Review of Educational Research*, **39**, 739–743. (3,4)

Tatsuoka, M. M. (1970). *Discriminant Analysis: The Study of Group Differences*. Champaign, IL: Institute for Personality and Ability Testing. (64,72)

Tatsuoka, M. M. (1982). Statistical methods. In H. E. Mitzel (Ed.). *Encyclopedia of Educational Research* (pp. 1780–1808). New York: Free Press. (31)

Tatsuoka, M. M. (1988). *Multivariate Analysis*. New York: Macmillan. (51,83,276,280,323,380,397)

Tatsuoka, M. M., and Tiedeman, D. V. (1954). Discriminant analysis. *Review of Educational Research*, **24**, 402–420. (3,4)

Therrien, C. W. (1989). *Decision Estimation and Classification*. New York: Wiley. (13)

Thomas, D. R. (1992). Interpreting discriminant functions: A data analysis approach. *Multivariate Behavioral Research*, **27**, 335–362. (115)

Thomas, D. R. (1997). A note on Huberty and Wisenbaker's "View of variable importance." *Journal of Educational and Behavioral Statistics*, **22**, 309–322. (107,115,397)

Thomas, D. R., and Zumbo, B. D. (1996). Using a measure of variable importance to investigate the standardization of discriminant coefficients. *Journal of Educational and Behavioral Statistics*, **21**, 110–130. (397)

Tiedeman, D. V., Bryan, J. G., and Rulon, P. J. (1953). *The Utility of the Airman Classification Battery for Assignment of Airmen to Eight Air Force Specialties* (Project 21-03-032). Cambridge, MA: Educational Research Corporation. (4)

Tiedeman, D. V., Rulon, P. J., and Bryan, J. G. (1951). The multiple discriminant function: A symposium. *Harvard Educational Review*, **21**, 71–95. (4)

Tiedeman, D. V., and Sternberg, J. J. (1952). Information appropriate for curriculum guidance. *Harvard Educational Review*, **22**, 257–274. (4)

Tilton, J. W. (1937). The measurement of overlapping. *Journal of Educational Psychology*, **28**, 656–662. (356)

Todd, S., Woodword, M., and Bolton-Smith, C. (1995). An investigation of the relationship between antioxidant vitamin intake and coronary heart disease in men and women using logistic regression analysis. *Journal of Clinical Epidemiology*, **48**, 307–316. (386)

Toussaint, G. T. (1974). Bibliography on estimation of misclassification. *IEEE Transactions on Information Theory*, **IT-20**(4), 472–479. (4,297)

Tukey, J. W. (1980). We need both exploratory and confirmatory. *American Statistician*, **34**, 23–25. (399)

Tukey, J. W. (1991). The philosophy of multiple comparisons. *Statistical Science*, **6**, 100–116. (159)

Turek, R. J., and Suich, R. C. (1983). An exact test on the Goodman–Kruskal λ for prediction on a dichotomy. *Journal of the Royal Statistical Society (B)*, **45**, 373–379. (332)

Twedt, D. J., and Gill, D. S. (1992). Comparison of algorithms for replacing missing data in discriminant analysis. *Communications in Statistics: Theory and Methods*, **21**, 1567–1578. (410)

Velicer, W. F., and Jackson, D. N. (1990). Component analysis versus common factor analysis: Some issues in selecting an appropriate procedure. *Multivariate Behavioral Research*, **25**, 1–28. (399)

Velilla, S., and Barrio, J. A. (1994). A discriminant rule under transformation. *Technometrics*, **36**, 348–353. (373)

Verboon, P., and van der Lans, I. A. (1994). Robust canonical discriminant analysis. *Psychometrika*, **59**, 485–507. (410)

Villarroya, A., Rios, M., and Oller, J. M. (1995). Discriminant analysis algorithm based on a distance function and on a Bayesian decision. *Biometrics*, **51**, 908–919. (389)

Viragoontavan, S. (2000). Comparing six missing data methods within the discriminant analysis context: A Monte-Carlo study. Doctoral dissertation, Ohio State University, Columbus. (410)

Vlachonikolis, I. G. (1990). Predictive discrimination and classification with mixed binary and continuous variables. *Biometrika*, **77**, 657–662. (359)

von Eye, A. (2002). *Configural Frequency Analysis*. Mahwah, NJ: Erlbaum. (100)

Wahl, P. W., and Kronmal, R. A. (1977). Discriminant functions when covariances are unequal and sample sizes are moderate. *Biometrics*, **33**, 479–484. (281)

Wakaki, H. (1990). Comparison of linear and quadratic discriminant functions. *Biometrika*, **77**, 227–229. (281)

Walters, G. D. (1986). Screening for psychopathology in groups of black and white prison inmates by means of MMPI. *Journal of Personality Assessment*, **50**, 257–264. (331)

Wang, C. (1993). *Sense and Nonsense of Statistical Inference*. New York: Dekker. (32)

Ware, J. E., and Williams, R. G. (1977). Discriminant analysis of student ratings as a means for identifying lecturers who differ in enthusiasm or information-giving. *Educational and Psychological Measurement*, **37**, 627–639. (301)

Warren, W. G. (1986). On the presentation of statistical analysis: Reason or ritual. *Canadian Journal of Forest Research*, **16**, 1185–1191. (30)

Webb, A. R. (2002). *Statistical Pattern Recognition*. London: Arnold. (11,336,385,386,387,388,389)

Wedding, D. (1983). Comparison of statistical and actuarial models for predicting lateralization of brain damage. *Clinical Neuropsychology*, **5**, 15–20. (408)

Weidemann, C. F., and Fenster, C. A. (1978). The use of chance corrected percentage agreement to interpret results of a discriminant analysis. *Educational and Psychological Measurement*, **38**, 29–35. (331)

Welch, B. L. (1939). Note on discriminant functions. *Biometrika*, **31**, 218–220. (350)

Whitcomb, K. M., and Lahiff, M. (1993). An investigation of sample influence functions for the DS estimate of the optimum error rate in discriminant analysis. *Communications in Statistics—Theory and Methods*, **22**, 41–56. (410)

White, A. P. (1994). Cross-validation of nearest-neighbour discriminant analysis—A warning to SAS users. *Journal of Statistical Computation and Simulation*, **49**, 129–140. (364)

Wilks, S. S. (1932). Certain generalizations of the analysis of variance. *Biometrika*, **39**, 471–494. (48)

Williams, B. K., Titus, K., and Hines, J. E. (1990). Stability and bias of classification rates in biological applications of discriminant analysis. *Journal of Wildlife Management*, **54**, 331–341. (312)

Williams, C. J., Lee, S. S., Fisher, R. A., and Dickerman, L. H. (1999). A comparison of statistical methods for prenatal screening for down syndrome. *Applied Stochastic Models in Business and Industry*, **15**, 89–101. (389)

Willis, W. G. (1984). Reanalysis of an actuarial approach to neuropsychological diagnosis in consideration of base rates. *Journal of Consulting and Clinical Psychology*, **52**, 567–569. (381)

Willis, W. G. (1986). Actuarial and clinical approaches to neuropsychological diagnosis: Applied considerations. In J. E. Obrzut and G. W. Hynd (Eds.). *Child Neuropsychology* (pp. 245–262). Orlando, FL: Academic. (408)

Winer, B. J. (1971). *Statistical Principles in Experimental Design*, 2rd ed. New York: McGraw-Hill. (228)

Wright, S. P. (1992). Adjusted *P*-values for simultaneous inference. *Biometrics*, **48**, 1005–1013. (30)

Wrinch, D., and Jeffreys, H. (1921). On certain fundamental principles of scientific inquiry. *Philosophical Magazine*, **42**, 6th Series, 369–390. (262,263)

Xiao, B. (1994). Multiple and unacceptable solutions in nonlinear programming discriminant analysis. *American Journal of Mathematical and Management Sciences*, **14**, 267–287. (359)

Yao, Y. (1965). An approximate degrees of freedom solution to the multivariate Behrens–Fisher problem. *Biometrika*, **52**, 139–147. (43,74)

Yarnold, P. R. (1996). Discriminating geriatric and nongeriatric patients using functional status information: An example of classification tree analysis via UniODA. *Educational and Psychological Measurement*, **56**, 656–667. (385)

Yarnold, P. R., and Soltysik, R. C. (1991). Refining 2-group multivariate classification models using univariate optimal discriminant analysis. *Decision Sciences*, **22**, 1158–1164. (357,359)

Yarnold, P. R., Soltysik, R. C., LeFevre, F., and Martin, G. J. (1998). Predicting in-hospital mortality of patients receiving cardiopulmonary resuscitation: Unit-weighted MultiODA for binary data. *Statistics in Medicine*, **17**, 2405–2514. (371)

Yarnold, P. R., Soltysik, R. C., and Martin, G. J. (1994). Heart rate variability and susceptibility for sudden cardiac death: An example of multivariate optimal discriminant analysis. *Statistics in Medicine*, **13**, 1015–1021. (359)

Yoon, Y., Swales, G., and Margavio, T. M. (1993). A comparison of discriminant analysis versus artificial neural networks. *Journal of the Operational Research Society*, **44**, 51–60. (386)

Young, F. W., DeLeeuw, J., and Tokane, Y. (1976). Regression with qualitative and quantitative variables: An alternating least squares method with optimal scaling features. *Psychometrika*, **41**, 505–529. (369)

Answers to Exercises

CHAPTER 2

1. **(a)** $n \times p$
   (b) $p \times n$
   (c) $p \times n$

2. $$\mathbf{SSCP}_2 \doteq \begin{bmatrix} 160.773 & 191.318 \\ 191.318 & 918.955 \end{bmatrix} \qquad \mathbf{S}_2 \doteq \begin{bmatrix} 7.656 & 9.110 \\ 9.110 & 43.760 \end{bmatrix}$$

3. $$\begin{aligned} |\mathbf{S}_2| &\doteq (7.656)(43.760) - (9.110)^2 \\ &\doteq 252.034 \end{aligned}$$

4. $$\begin{aligned} \mathbf{S}_2^{-1} &\doteq \frac{1}{252.034} \begin{bmatrix} 43.760 & -9.110 \\ -9.110 & 7.656 \end{bmatrix} \\ &\doteq \begin{bmatrix} .175 & -.036 \\ -.036 & .031 \end{bmatrix} \end{aligned}$$

5. $$\sum_{j=1}^{2} \mathbf{SSCP}_j = \mathbf{E} \doteq \begin{bmatrix} 484.637 & 709.591 \\ 709.591 & 2216.41 \end{bmatrix}$$
$$\begin{aligned} \mathbf{S}_e &\doteq \frac{1}{44-2} \begin{bmatrix} 484.637 & 709.591 \\ 709.591 & 2216.41 \end{bmatrix} \\ &\doteq \begin{bmatrix} 11.539 & 16.895 \\ 16.895 & 52.772 \end{bmatrix} \end{aligned}$$

Applied MANOVA and Discriminant Analysis, Second Edition, by Carl J. Huberty and Stephen Olejnik
Copyright © 2006 John Wiley & Sons, Inc.

6.

$$|\mathbf{S}_e| \doteq (11.539)(52.772) - (16.895)^2$$
$$\doteq 323.495$$

7.

$$\mathbf{S}_e^{-1} \doteq \frac{1}{323.495}\begin{bmatrix} 52.772 & -16.895 \\ -16.895 & 11.539 \end{bmatrix}$$
$$\doteq \begin{bmatrix} .163 & -.052 \\ -052 & .036 \end{bmatrix}$$

8.

$$\mathbf{S}_e\mathbf{S}_e^{-1} \doteq \begin{bmatrix} 11.539 & 16.895 \\ 16.895 & 52.772 \end{bmatrix}\begin{bmatrix} .163 & -.052 \\ -052 & .036 \end{bmatrix}$$
$$\doteq \begin{bmatrix} (11.539)(.163)+(16.895)(-.052) & (11.539)(-.052)+(16.895)(.036) \\ (16.895)(.163)+(52.772)(-.052) & (16.895)(-.052)+(52.772)(.036) \end{bmatrix}$$
$$\doteq \begin{bmatrix} 1.002 & .008 \\ .009 & 1.021 \end{bmatrix}$$

9.

$$|\mathbf{A} - \lambda\mathbf{I}| = \lambda^2 - 16\lambda + 15$$
$$\lambda = \frac{16 \pm \sqrt{16^2 - 4(15)}}{2}$$
$$= (16 \pm 14)/2$$
$$= 1 \quad \text{and} \quad 15$$

10. $$D^2 \doteq \begin{bmatrix} 7.77 - 6.68 & 43.45 - 42.05 \end{bmatrix}\begin{bmatrix} .163 & -.052 \\ -.052 & .036 \end{bmatrix}\begin{bmatrix} 7.77 - 6.68 \\ 43.45 - 42.05 \end{bmatrix}$$
$$\doteq .106$$

CHAPTER 3

1. For variable Y_1:

$$t \doteq \frac{4.40 - 5.50}{\sqrt{(1.174^2 + 1.269^2/10}} \doteq -2.011$$

For variable Y_2:

$$t \doteq \frac{13.00 - 15.20}{\sqrt{(2.582^2 + 2.616^2/10}} \doteq -1.893$$

There is insufficient evidence to indicate that the observed differences in anxiety are generalizable to the populations they represent.

2. For variable Y_1:

$$r_{\mathrm{pb}}^2 \doteq \frac{-2.011^2}{-2.011^2 + 18} \doteq .183$$

The correlation could be either plus or minus depending the order of subtraction.

$$d \doteq \frac{4.40 - 5.50}{\sqrt{\dfrac{(1.174^2 + 1.269^2)}{2}}} \doteq -.900$$

For variable Y_2:

$$r_{\mathrm{pb}}^2 \doteq \frac{-1.893^2}{-1.893^2 + 18} \doteq .166$$

$$d \doteq \frac{13.00 - 15.20}{\sqrt{\dfrac{(2.582^2 + 2.616^2)}{2}}} \doteq -.846$$

3. (a)

$$\mathbf{SSCP}_{\mathrm{IG}} \doteq \begin{bmatrix} 12.402 & 4.014 \\ 4.014 & 60.001 \end{bmatrix} \qquad \mathbf{SSCP}_{\mathrm{CG}} \doteq \begin{bmatrix} 14.490 & -4.986 \\ -4.986 & 61.587 \end{bmatrix}$$

and

$$\mathbf{E} \doteq \begin{bmatrix} 26.892 & -.972 \\ -.972 & 121.690 \end{bmatrix} \qquad \mathbf{S}_e = \frac{1}{18}\mathbf{E} \doteq \begin{bmatrix} 1.494 & -.054 \\ -.054 & 6.761 \end{bmatrix}$$

$$|\mathbf{S}_e| \doteq (1.494)(6.761) - (-.054)^2 \doteq 10.098$$

$$\mathbf{S}_e^{-1} \doteq \frac{1}{10.098}\begin{bmatrix} 6.761 & .054 \\ .054 & 1.494 \end{bmatrix} \doteq \begin{bmatrix} .669 & .005 \\ .005 & .148 \end{bmatrix}$$

(b)

$$T^2 \doteq \frac{(10)(10)}{10 + 10}\begin{bmatrix} 4.40 - 5.50 & 13.00 - 15.20 \end{bmatrix} \times \begin{bmatrix} .669 & .005 \\ .005 & .148 \end{bmatrix}\begin{bmatrix} 4.40 - 5.50 \\ 13.00 - 15.20 \end{bmatrix} \doteq 7.750$$

$$F \doteq \frac{18 - 2 + 1}{2(18}(7.75) \doteq 3.660$$

(c) $$D^2 \doteq \begin{bmatrix} 4.40 - 5.50 & 13.00 - 15.20 \end{bmatrix}\begin{bmatrix} .669 & .005 \\ .005 & .148 \end{bmatrix}\begin{bmatrix} 4.40 - 5.50 \\ 13.00 - 15.20 \end{bmatrix} \doteq 1.550$$

4. The mutlivariate test is more powerful than the univariate test.

5.

$$\mathbf{E} = \sum_{j=1}^{3} \mathbf{SSCP}_j \doteq \begin{bmatrix} 330.4 & 220.4 & 203.9 \\ 220.4 & 297.6 & 206.2 \\ 203.9 & 206.2 & 478.4 \end{bmatrix}$$

$$\mathbf{S}_e = \frac{1}{42}\mathbf{E} \doteq \begin{bmatrix} 7.867 & 5.248 & 4.855 \\ 5.248 & 7.086 & 4.910 \\ 4.855 & 4.910 & 11.390 \end{bmatrix}$$

6. Using Eq. (3.15) and with $\mathbf{y}_{..} \doteq \begin{bmatrix} 14.30 \\ 14.87 \\ 19.23 \end{bmatrix}$ the individual **SSCP** are computed as:

$$\mathbf{SSCP}_1 \doteq \begin{bmatrix} 38.400 & 13.680 & -90.720 \\ 13.680 & 5.875 & -32.319 \\ -90.720 & -32.319 & 214.326 \end{bmatrix}$$

$$\mathbf{SSCP}_2 \doteq \begin{bmatrix} 21.600 & 32.940 & -7.740 \\ 32.940 & 50.250 & -11.804 \\ -7.740 & -11.804 & 2.774 \end{bmatrix}$$

$$\mathbf{SSCP}_3 \doteq \begin{bmatrix} 2.400 & -7.620 & -19.980 \\ -7.620 & 24.194 & 63.437 \\ -19.980 & 63.437 & 166.334 \end{bmatrix}$$

$$\mathbf{H} = \sum_{j=1}^{3} \mathbf{SSCP}_j \doteq \begin{bmatrix} 62.400 & 39.000 & -118.440 \\ 39.000 & 79.319 & 19.314 \\ -118.440 & 19.314 & 381.434 \end{bmatrix}$$

7. For variables Y_1 and Y_2 the relevant matrices are

$$\mathbf{H} \doteq \begin{bmatrix} 62.400 & 39.000 \\ 39.000 & 79.319 \end{bmatrix} \qquad \mathbf{E} \doteq \begin{bmatrix} 330.4 & 220.4 \\ 220.4 & 297.6 \end{bmatrix}$$

$$\Lambda = \frac{|\mathbf{E}|}{|\mathbf{E}+\mathbf{H}|} \doteq \frac{4.98 \times 10^4}{8.05 \times 10^4} \doteq .619$$

8. Because three groups are being compared, Λ is transformed to an F statistic using Eq. (3.18):

$$F \doteq \frac{1 - .619^{1/2}}{.619^{1/2}} \frac{42 - 2 + 1}{2}$$

$$\doteq 5.556$$

9. **(a)**

$$\Lambda \doteq \frac{1}{2.694} \frac{1}{1.312}$$

$$\doteq .283$$

(b)

$$U \doteq \frac{1.694}{2.694} + \frac{.312}{1.312}$$
$$\doteq .867$$

(c)

$$\Theta \doteq \frac{1.694}{2.694}$$
$$\doteq .629$$

(d)

$$V \doteq 1.694 + .312$$
$$\doteq 2.006$$

10. (a)

$$F \doteq \frac{1 - .283^{1/2}}{.283^{1/2}} \frac{42 - 3 + 1}{3}$$
$$\doteq 11.730$$
$$\nu_1 = 2(3) = 6 \quad \text{and} \quad \nu_2 = 2(42 - 3 + 1) = 80$$

(b)

$$F \doteq \frac{.867}{2 - .867} \frac{42 - 3 + 2}{3}$$
$$\doteq 10.458$$
$$\nu_1 = 3(2) = 6 \quad \text{and} \quad \nu_2 = 2(42 - 3 + 2) = 82$$

(c)

$$F \doteq \frac{(45 - 3 - 1)1.694}{3}$$
$$\doteq 23.151$$
$$\nu_1 = 3 \quad \text{and} \quad \nu_2 = 45 - 3 - 1 = 41$$

(d)

$$F \doteq 2.004 \frac{2(42 - 3 - 1) + 2}{2^2 3}$$
$$\doteq 13.026$$
$$\nu_1 = 3(2) = 6 \quad \text{and} \quad \nu_2 = 2(42 - 3 - 1) + 2 = 78$$

11. The sample sizes are: Black n_1=182, Hispanic n_2=169, and White n_3= 194.

12. (a) $M \doteq 146.740$

(b) $\chi^2(90) \doteq 1.43.335$

(c) $\ln|\boldsymbol{S}_1| \doteq 23.85$, $\ln|\boldsymbol{S}_2| \doteq 24.48$, $\ln|\boldsymbol{S}_3| \doteq 23.37$, and $\ln|\boldsymbol{S}_e| \doteq 24.14$.

(d) No, because the sample sizes are similar and the log determinants of the covariance matrices do not differ much.

13. The two variables that have the highest correlation are Student–Faculty Effort (qefac) and Interstudent Effort (qestacq). These two variables are correlated .485.

14.

$$\Lambda \doteq .863$$
$$U \doteq .142$$
$$V \doteq .153$$

15. Yes, there is sufficient evidence to conclude that the observed differences among the group centroids do generalize to the populations they represent, $\Lambda = .863$, $F(18, 1068) = 4.533$, $P = .000$.

CHAPTER 4

1. No, because the number of eigenvalues is $\min(p, \mathrm{df}_h)$. For this context $p = 4$ and $\mathrm{df}_h = 3$, so the minimum is 3.

2. (a)

$$\Lambda \doteq \frac{1}{5.764}\frac{1}{2.237}\frac{1}{1.044} \doteq .074$$

(b)

$$U \doteq \frac{4.764}{5.764} + \frac{1.237}{2.237} + \frac{.044}{1.044} \doteq 1.422$$

(c)

$$\Theta \doteq \frac{4.764}{5.764} \doteq .827$$

(d)

$$V \doteq 4.764 + 1.237 + .044 \doteq 6.045$$

3. (a)

$$F \doteq \frac{1 - .074^{1/2.646}}{.074^{1/2.646}} \frac{(43)(2.646) - 4(3)/2 + 1}{4(3)} \doteq 15.175$$

$$\nu_1 = 4(3) = 12$$

$$\nu_2 \doteq (43)(2.646) - \frac{4(3)}{2} + 1 \doteq 108.778$$

(b)

$$F \doteq \frac{1.422}{3 - 1.422} \frac{44 - 4 + 3}{4} \doteq 9.687$$

$$\nu_1 = 4(3) = 12$$

$$\nu_2 = 3(44 - 4 + 3) = 129$$

(c)

$$F \doteq \frac{(48-4-1)4.764}{4}$$
$$\doteq 51.213$$
$$\nu_1 = 4$$
$$\nu_2 = 48-4-1 = 43$$

(d)

$$F \doteq 6.045\frac{3(44-4-1)+2}{3^2 4}$$
$$\doteq 19.982$$
$$\nu_1 = 4(3) = 12$$
$$\nu_2 = 3(44-4-1)+2 = 119$$

4. **(a)** $\eta^2_{\text{Mult}} \doteq 1 - .074 \doteq .926$
(b) $\tau^2 = 1 - .074^{1/3} \doteq .580$
(c) $\xi^2 \doteq \dfrac{1.422}{3} \doteq .474$
(d) $\zeta^2 \doteq \dfrac{6.045}{3+6.045} \doteq .668$
(e) $\omega^2_{\text{Mult}} \doteq 1 - \dfrac{48(.074)}{(48-3-1)+.074} \doteq .919$

5. **(a)**

$$\tau^2_{\text{adj}} \doteq 1 - \frac{48-1}{48-4-1}(1-.580)$$
$$\doteq .541$$

(b)

$$\zeta^2_{\text{adj}} \doteq 1 - \frac{48-1}{48-4-1}(1-.668)$$
$$\doteq .637$$

(c)

$$\xi^2_{\text{adj}} \doteq 1 - \frac{48-1}{48-4-1}(1-.474)$$
$$\doteq .425$$

6. The first contrast, 1 0 0 −1, compares the centroids of Mathematics students with Business students. The second contrast 0 1 0 −1, compares the centroids of English students with Business students, and the third contrast compares the centroids of Psychology students with Business students.

7. One eigenvalue is associated with each contrast because each contrast has $\text{df}_h = 1$.

8.

$$\mathbf{H}_{\hat{\psi}} \doteq \frac{12}{(1)^2 + (-1)^2} \begin{bmatrix} 3.6 \\ 2.2 \\ -3.8 \\ -8.9 \end{bmatrix} [3.6 \quad 2.2 \quad -3.8 \quad -8.9]$$

$$\doteq \begin{bmatrix} 77.76 & 47.52 & -82.08 & 193.74 \\ 47.52 & 29.04 & -50.16 & -117.48 \\ -82.08 & -50.16 & 86.64 & 202.92 \\ 193.74 & -117.48 & 202.92 & 475.26 \end{bmatrix}$$

9. **(a)** $\Lambda \doteq \dfrac{1}{1 + 3.248} \doteq .235$

(b) $F \doteq \dfrac{1 - .235}{.235} \dfrac{44 - 4 + 1}{4} \doteq 33.3676$

(c) $\nu_1 = 4, \quad \nu_2 = 44 - 4 + 1 = 41$

10. **(a)** $\xi^2 \doteq \dfrac{3.248}{1 + 3.248} \doteq .765$

(b) $\xi^2_{\text{adj}} \doteq 1 - \dfrac{48 - 1}{48 - 4 - 1}(1 - .765) \doteq .743$

11. **(a)** $\eta^2_{\text{Mult}} \doteq 1 - .863 = .137$

(b) $\tau^2 \doteq 1 - .863^{1/2} \doteq .071$

(c) $\xi^2 \doteq .071$

(d) $\zeta^2 \doteq .071$

12. **(a)** $\tau^2_{\text{adj}} \doteq 1 - \dfrac{545 - 1}{545 - 9 - 1}(1 - .071) \doteq .055$

(b) $\xi^2_{\text{adj}} \doteq .055$

(c) $\zeta^2_{\text{adj}} \doteq .055$

13. **(a)** $\Lambda \doteq .939$, $U \doteq .061$, and $V \doteq .065$

(b) $F \doteq 3.885$, $\nu_1 = 9$, and $\nu_2 = 534$

(c) Yes, there is evidence to indicate that the centroids of Hispanic and black students differ, $P \doteq .000$.

14. **(a)** $\Lambda \doteq .919$, $U \doteq .080$, and $V \doteq .087$

(b) $F \doteq 5.160$, $\nu_1 = 9$, and $\nu_2 = 534$

(c) Yes, there is evidence to indicate that the mean centroid of Hispanic and Black students differs from the centroid of White students, $P \doteq .000$.

15. **(a)** **i.** For the contrast comparing Hispanic with Black students $\xi^2 \doteq .061$.

ii. For the contrast comparing Minority students with White students $\xi^2 \doteq .080$.

(b) **i.** For the contrast of Hispanic and Black students,

$$\xi^2_{\text{adj}} \doteq 1 - \frac{545 - 1}{545 - 9 - 1}(1 - .061) \doteq .045.$$

ii. For the contrast of Minority students with White students,

$$\xi^2_{\text{adj}} \doteq 1 - \frac{545 - 1}{545 - 9 - 1}(1 - .080) \doteq .065.$$

CHAPTER 5

1. The squared first canonical correlation is $4.764/(1 + 4.764) \doteq .827$. The squared second canonical correlation is $1.237/(1 + 1.237) \doteq .553$. The squared third canonical correlation is $.044/(1 + .044) \doteq .042$.

2. The first construct explains $4.764/(4.764 + 1.237 + .004) \doteq 78.8$ percent of the variance in the system. The second construct explains $1.237/(4.764 + 1.237 + .004) \doteq 20.5$ percent of the variance in the system. And the third construct explains $.044/(4.764 + 1.237 + .004) \doteq .74$ percent of the variance in the system. Based on these results two constructs may be necessary to describe group separation.

3. Null hypothesis test for no separation on any dimension

$$\begin{aligned}
\Lambda &\doteq \frac{1}{1 + 4.764}\,\frac{1}{1 + 1.237}\,\frac{1}{1 + .044} \\
&\doteq .074 \\
F &\doteq \frac{1 - .074^{1/2.646}}{.074^{1/2.646}}\,\frac{(43)(2.646) - 4(3)/2 + 1}{4(3)} \\
&\doteq 15.175 \\
\nu_1 &= 4(3) = 12 \\
\nu_2 &\doteq (43)(2.646) - 4(3) = 108.778
\end{aligned}$$

Null hypothesis test for separation on at most one dimension

$$\begin{aligned}
\Lambda_1 &\doteq \frac{1}{1 + 1.237}\,\frac{1}{1 + .044} \\
&\doteq .428 \\
F &\doteq \frac{1 - .428^{1/2}}{.428^{1/2}}\,\frac{(44) - 3 + 1}{3} \\
&\doteq 7.406
\end{aligned}$$

$$\nu_1 = 3(2) = 6$$
$$\nu_2 = 2(44 - 3 + 1) = 84$$

Null hypothesis test for separation on at most two dimensions

$$\Lambda_2 \doteq \frac{1}{1 + .044}$$
$$\doteq .958$$
$$F \doteq \frac{1 - .958}{.958} \frac{(44) - 2 + 1}{2}$$
$$\doteq .946$$
$$\nu_1 = 2$$
$$\nu_2 = 44 - 2 + 1 = 43$$

Based on these results, we would conclude that at most two dimensions are necessary to describe group separation.

4. An LDF is a weighted composite of a set of response variables.

5. Linear discriminant functions are used to determine group centroids in LDF space. These centroids are then used to describe group separation.

6. Teacher and Effort appear to define the first LDF. Luck appears to define the second LDF.

7. **(a)** For Mathematics majors:

$$\text{LDF}_1 \doteq -.187(11.5) + -.044(25.3) + .027(22.7) + .227(32.3)$$
$$\doteq 4.681$$
$$\text{LDF}_2 \doteq .096(11.5) + .035(25.3) + .249(22.7) + .139(32.3)$$
$$\doteq 12.132$$

(b) For English majors:

$$\text{LDF}_1 \doteq -.187(22.2) + -.044(31.5) + .027(30.5) + .227(19.0)$$
$$\doteq -.401$$
$$\text{LDF}_2 \doteq .096(22.2) + .035(31.5) + .249(30.5) + .139(19.0)$$
$$\doteq 13.469$$

(c) For Psychology majors:

$$\begin{aligned}\text{LDF}_1 &\doteq -.187(17.9) + -.044(28.6) + .027(30.3) + .227(22.4)\\ &\doteq 1.297\\ \text{LDF}_2 &\doteq .096(17.9) + .035(28.6) + .249(30.3) + .139(22.4)\\ &\doteq 13.378\end{aligned}$$

(d) For Business majors:

$$\begin{aligned}\text{LDF}_1 &\doteq -.187(14.3) + -.044(26.4) + .027(34.1) + .227(31.3)\\ &\doteq 4.190\\ \text{LDF}_2 &\doteq .096(14.3) + .035(26.4) + .249(34.1) + .139(31.3)\\ &\doteq 15.138\end{aligned}$$

8.

	Math	Eng	Psy	Bus
LDF_1	4.68	−.40	1.30	4.19
LDF_2	12.13	13.47	13.38	15.14

English and Psychology majors appear to differ from Mathematics and Business majors on the first LDF. English and Psychology majors appear to be similar on the second LDF but differ from both Business and Mathematics majors. In addition, Mathematics majors appear to differ from Business majors on the second LDF.

9. The squared canonical correlation is $\dfrac{1.257}{1 + 1.257} \doteq .557.$

10. **(a)** $\Lambda \doteq \dfrac{1}{1 + 1.257} \doteq .443$

(b) $\tau^2 \doteq 1 - .443 \doteq .557$

11. The contrast between Mathematics and Business majors seems to be defined by the variable luck.

12. The first construct (LDF) explains 62.1 percent of the variance, and the second construct explains 37.9 percent of the variance in the system.

13. Based on the dimension reduction analysis two constructs are needed to describe the group separation. The test for at least one dimension resulted in $\Lambda \doteq .863$, $F(18, 1068) = 4.533$, $P \doteq .000$. The test of at least two dimensions resulted in $\Lambda \doteq .945$, $F(8, 535) = 3.883$, $P \doteq .000$.

14. The squared canonical correlation between race and the first and second constructs are .087 and .055, respectively.

15. The first construct is defined by counsum (.404), learnsum (.535), qelib (.522), qewrite (.408) and qesci(−.437). The second construct is defined by qefac (.453), qestacq (.389), and qeamt (−.402). The values in parentheses are the structure r's.

16. Using the DISCRIMINANT program in SPSS, the group LDF centroids are

	Function	
Ethnicity	1	2
Black	.370	.178
Hispanic	.026	−.358
White	−.370	.145

17. For the contrast between Black and Hispanic students the construct that separates the two groups is defined by learnsum (.466), qefac (−.499), qestacq (−.470), and qewrite (.439).

18. For the contrast between Black and Hispanic students and White students the construct is defined by counsum (−.448), learnsum (−.414), qelib (−.490), and qesci (.472).

CHAPTER 6

1. An analysis for determining sample-based best outcome variable subsets of all possible sizes.

2. Ten (at most) best outcome variable subsets of each size.

3. **(a)** They both determine the best subset of a given size.
   (b) The McCabe analysis determines the 10 (at most) best outcome variable subsets of each size, while the McHenry analysis determines only the best subset of each size. The McCabe analysis is based on raw data input, while the McHenry analysis is based on matrix input.

4. **(a)** Initial outcomes variable selection is based on study-related substantive theory.
   (b) Multiple correlation, canonical correlation, principal component analysis, and factor analysis.

5. All outcome variables may not be included in the analysis.

6. **(a)** $Y3$, $Y5$; $Y1$, $Y2$, $Y3$, $Y5$, $Y6$, $Y9$
   (b) Yes. Yes.
   (c) $Y1$, $Y2$, $Y3$, $Y4$, $Y5$, $Y6$, $Y9$ or $Y1$, $Y2$, $Y3$, $Y5$, $Y6$, $Y8$, $Y9$
   (d) $Y7$, $Y8$

(e) $Y5, Y9$

(f) $Y3, Y4; Y7, Y5$

CHAPTER 8

1. (a) Three, there would be one set of LDFs for each main effect and one set of LDFs for the interaction.

   (b) There would be two LDFs for the main effect having three levels, three LDFs for the main effect having four levels, and six LDFs for the interaction.

2. (a) $$\mathbf{y}_{.11} \doteq \begin{bmatrix} 6.9 \\ 9.2 \\ 12.0 \\ 15.1 \end{bmatrix}$$

   (b) $$\mathbf{y}_{..1} \doteq \begin{bmatrix} 9.07 \\ 10.87 \\ 13.10 \\ 16.10 \end{bmatrix}$$

   (c) $$\mathbf{y}_{.1.} \doteq \begin{bmatrix} 8.55 \\ 9.35 \\ 10.75 \\ 12.85 \end{bmatrix}$$

   (d) $$\mathbf{y}_{...} \doteq \begin{bmatrix} 10.967 \\ 11.283 \\ 11.683 \\ 13.683 \end{bmatrix}$$

3.

$$\mathbf{H}_{G_1} \doteq \begin{bmatrix} -1.897 \\ -.413 \\ 1.417 \\ 2.417 \end{bmatrix} \begin{bmatrix} -1.897 & -.413 & 1.417 & 2.417 \end{bmatrix}$$

$$\doteq \begin{bmatrix} 3.599 & .783 & -2.688 & -4.585 \\ .783 & .171 & -.585 & -.998 \\ -2.688 & -.585 & 2.008 & 3.425 \\ -4.585 & -.998 & 3.425 & 5.842 \end{bmatrix}$$

$$\mathbf{H}_{G_2} \doteq \begin{bmatrix} 1.903 \\ .417 \\ -1.413 \\ -2.413 \end{bmatrix} \begin{bmatrix} 1.903 & .417 & -1.413 & -2.413 \end{bmatrix}$$

$$\doteq \begin{bmatrix} 3.621 & .794 & -2.689 & -4.592 \\ .794 & .174 & -.589 & -1.006 \\ -2.689 & -.589 & 1.997 & 3.410 \\ -4.592 & -1.006 & 3.410 & 5.823 \end{bmatrix}$$

$\mathbf{H}_G = 30[\mathbf{H}_{G_1} + \mathbf{H}_{G_2}]$

$$\doteq \begin{bmatrix} 216.600 & 47.310 & -161.310 & -273.510 \\ 47.310 & 10.350 & -35.220 & -60.120 \\ -161.310 & -35.220 & 120.150 & 205.050 \\ -273.510 & -60.120 & 205.050 & 349.950 \end{bmatrix}$$

4. **(a)** $\Lambda \doteq \dfrac{1}{1+.080}\,\dfrac{1}{1+.005} \doteq .921$

(b) $F \doteq \dfrac{1-.921^{1/2}}{.921^{1/2}}\,\dfrac{84-4+1}{4} \doteq .851$

(c) $\nu_1 = 2(4) = 8, \quad \nu_2 = 2(84-4+1) = 162$

5. $.988 \doteq \dfrac{1}{1+\lambda}$, $\lambda \doteq .012$. The squared canonical correlation equals $\dfrac{.012}{1+.012} \doteq .012.$

6. Based on the results provided, the three levels of variable B appear to be best described on one dimension. The first construct explains 97.6 percent of the variation in the system while the second construct explains only 2.54 percent of the variation. Furthermore the results of the Dimension Reduction Analysis indicates that at most only one dimension is needed to separate the groups being compared, $\Lambda \doteq .971$, $F(3, 82) \doteq .827$, $P \doteq .483$.

7. For the B main effect,

$$\xi^2 \doteq \left(\frac{1.291}{1+1.291} + \frac{.030}{1+1.030}\right)\Big/ 2 \doteq .289$$

$$\xi^2_{\text{adj}} \doteq 1 - \frac{90-1}{90-4-1}(1-.289) \doteq .256$$

8. $\dfrac{.576}{1+.576} \doteq .365$

9. $\Lambda \doteq \dfrac{1}{1+.576} \doteq .635$

$U \doteq \dfrac{.576}{1+.576} \doteq .365$

$V \doteq .576$

$\Theta \doteq \dfrac{.567}{1+.576} \doteq .365$

All four test criteria will result in the same F statistic and degrees of freedom.

$$F \doteq \frac{1-.635}{.635}\frac{84-4+1}{4} \doteq 11.640,\ \nu_1 = 4,\ \nu_2 = 81.$$

10. /lmatrix "cell a_2b_1 vs. cell a_2b_2" B 1 −1 0 A*B 0 0 0 1 −1 0

11. **(a)** $M \doteq 917.412$

(b) $\chi^2 \doteq 808.726$, $\nu_1 = 630$

(c) There is a positive relationship between sample size and the log determinants, $r \doteq .496$.

(d) Because the sample sizes and log determinants are positively related, the reported P values will overestimate the actual P value. If the evidence provided in the test is judged to be statistically significant, the conclusions will be statistically valid.

12. **(a)** $\Lambda \doteq .842$, $F(72, 3182.78) \doteq 1.270$, $P \doteq .065$

(b) $\xi^2 \doteq .021$ and $\xi^2_{adj} \doteq 1 - \frac{545-1}{545-9-1}(1-.021) \doteq .005$

Based on these results, there is little evidence to indicate the interaction is statistically or meaningfully significant.

13. **(a)** $\Lambda \doteq .885$, $F(18, 1044) \doteq 3.641$, $P \doteq .000$

(b) $\xi^2 \doteq .059$ and $\xi^2_{adj} \doteq 1 - \frac{545-1}{545-9-1}(1-.059) \doteq .043$

Based on these results, there is evidence to indicate that the Race centroids differ both statistically and meaningfully.

14. **(a)** The first construct explains 69.3 percent of the variance in the system, while the second construct explains 30.7 percent of the variance.

(b) The Race variable explains 8 percent of the variation in the first construct and 3.7 percent of the variation in the second construct.

(c) Based on the Dimension Reduction Analysis two dimensions are needed to describe group differences on the Race variable, $\Lambda_1 \doteq .963$, $F(8, 532) \doteq 2.53$, $P \doteq .010$.

15. **(a)** $\Lambda \doteq .888$, $F(36, 1957.91) \doteq 1.746$, $P \doteq .004$

(b) $\xi^2 \doteq .029$ and $\xi^2_{adj} \doteq 1 - \frac{545-1}{545-9-1}(1-.029) \doteq .013$

Based on these results, there is evidence to indicate that the Race centroids differ statistically, but the meaningfulness of the relationship may be questionable.

16. **(a)** The first construct explains 46.8 percent of the variation in the system. The second construct explains 26.0 percent of the variance. The third construct explains 21.9 percent of the variance in the system. And the fourth construct explains 5.2 percent of the variance.

(b) The Grade variable explains 5.3 percent of the variation in the first construct, 3.1 percent of the variation in the second construct, 2.6 percent of the variation in the third construct, and .6 percent of the variation in the fourth construct.

(c) Only one dimension is needed to describe group differences among levels of the Grade variable $\Lambda_0 \doteq .888$, $F(36, 1957.91) \doteq 1.746$, $P \doteq .004$; $\Lambda_1 \doteq .939$, $F(24, 1517.46) \doteq 1.400$, $P \doteq .096$.

17. For the Grade variable the construct that separates the groups is defined based on the structure r's by qefac (−.616) primarily and to a lesser extent by qeamt (−.374), qewrite (−.356), and qesci (−.347).

18. **(a)** $\Lambda \doteq .966$, $F(9, 522) \doteq 2.036$, $P \doteq .034$

(b) $\xi^2 \doteq .034$ and $\xi^2_{\text{adj}} \doteq 1 - \dfrac{545 - 1}{545 - 9 - 1}(1 - .034) \doteq .0178$

If it is assumed that four contrasts would be of interest (e.g., comparing each Grade group with the A student group), then a Bonferroni adjusted P value for this contrast would be .136 (4 × .034), and the results would provide little evidence that the group centroids differ statistically or meaningfully.

19. **(a)** $\Lambda \doteq .907$, $F(36, 1957.91) \doteq 1.442$, $P \doteq .044$

(b) $\xi^2 \doteq .024$ and $\xi^2_{\text{adj}} \doteq 1 - \dfrac{545 - 1}{545 - 9 - 1}(1 - .024) \doteq .008$

(c) The first construct explains 59.9 percent of the variance in the system. The second construct explains 24.4 percent of the variance in the system. The third construct explains 11.2 percent of the variance, and the fourth construct explains 4.6 percent of the variance in the system.

(d) The Grade variable within the Black student sample explains 5.7, 2.4, 1.1, and .5 percent of the variation in constructs 1 through 4, respectively.

(e) Only one dimension is needed to the Black student sample $\Lambda \doteq .907$, $F(36, 1957.91) \doteq 1.442$, $P \doteq .044$; $\Lambda_1 \doteq .961$, $F(24, 1517.46) \doteq .874$, $P \doteq .639$.

(f) Based on the structure r's, group separation is defined by learnsum (−.361), qelib (.542), qefac (.452), qestacq (.394), and qeamt (.479).

20. **(a)** There is no evidence to indicate that for the Black student population the centroids for students earning A grades differ from those earning grades of C or C−; $\Lambda \doteq .986$, $F(9, 522) \doteq .803$, $P \doteq .613$.

(b) $\tau^2 \doteq 1 - .986 \doteq .014$, $\tau^2_{\text{adj}} \doteq 1 - \dfrac{545 - 1}{545 - 9 - 1}(1 - .012) \doteq -.003 \doteq .000$.

CHAPTER 9

1.

Group 1	Group 2
$b_{Y_1\|X_1} \doteq 261.7/4424.5 \doteq .059$	$b_{Y_1\|X_1} \doteq 409.5/7729.5 \doteq .053$
$b_{Y_2\|X_1} \doteq 140.1/4424.5 \doteq .032$	$b_{Y_2\|X_1} \doteq 209.7/7729.5 \doteq .027$
$b_{Y_3\|X_1} \doteq 424.1/4424.5 \doteq .096$	$b_{Y_3\|X_1} \doteq 553.6/7729.5 \doteq .072$

2.

$$\mathbf{E} \doteq \begin{bmatrix} 90.5 & 34.1 & 86.9 & 671.2 \\ 34.1 & 23.9 & 40.3 & 349.8 \\ 86.9 & 40.3 & 129.2 & 977.7 \\ 671.2 & 349.8 & 977.7 & 12154.0 \end{bmatrix}$$

3.

$$\begin{aligned} b_{Y_1|X_1} &\doteq 671.2/12154.0 \doteq .055 \\ b_{Y_2|X_1} &\doteq 349.8/12154.0 \doteq .029 \\ b_{Y_3|X_1} &\doteq 977.7/12154.0 \doteq .080 \end{aligned}$$

4. One.

5. Because $J = 2$, Eq. (3.12) can be used.

$$F \doteq \frac{1 - .959}{.959} \frac{36 - 3 + 1}{3} \doteq .485,$$

with $\nu_1 = 3$ and $\nu_2 = 34$.

6.

$$\mathbf{E}^* \doteq \begin{bmatrix} 90.5 & 34.1 & 86.9 \\ 34.1 & 23.9 & 40.3 \\ 86.9 & 40.3 & 129.2 \end{bmatrix} - .00008 \begin{bmatrix} 450509.44 & 234785.76 & 656232.24 \\ 234785.76 & 122360.04 & 341999.46 \\ 656232.24 & 341999.46 & 955897.29 \end{bmatrix}$$

$$\doteq \begin{bmatrix} 54.459 & 15.317 & 34.401 \\ 15.317 & 14.111 & 12.940 \\ 34.401 & 12.940 & 52.728 \end{bmatrix}$$

7. $\Lambda \doteq 1 - .905^2 \doteq .181$

8. $F = \dfrac{1 - .181}{.181} \dfrac{37 - 3 + 1}{3} \doteq 52.790$ with $\nu_1 = 3$ and $\nu_2 = 35$.

9. The construct is definded by variables Y_1 and Y_3.

10. For Group 1 LDF $\doteq -.284(15.728) - .097(8.114) + .321(7.640) \doteq -2.801$.
For Group 2 LDF $\doteq -.284(9.022) - .097(6.336) + .321(13.910) \doteq 1.288$.

11. **(a)** $\Lambda \doteq .976$
(b) $F \doteq .713$
(c) $\nu_1 = 18$ and $\nu_2 = 1062$
(d) $P \doteq .800$

12. The conclusion does not change when Time is added to the model as a covariate. From Chapter 3, Exercise 15, $\Lambda \doteq .863$, $F(18, 1068) \doteq 4.533$, and $P \doteq .000$. With Time as a covariate, $\Lambda \doteq .867$, $F(18, 1066) \doteq 4.375$, and $P \doteq .000$.

13.

Without Time

Construct 1		Construct 2	
Variable	Structure r	Variable	Structure r
counsum	.404	qefac	.453
learnsum	.535	qestacq	.389
qelib	.522	qeamt	−.402
qewrite	.408		
qesci	−.437		

With Time

Construct 1		Construct 2	
Variable	Structure r	Variable	Structure r
counsum	−.371	qefac	.427
learnsum	−.563	qestacq	.419
qelib	−.492	qeamt	−.350
qewrite	−.428		
qesci	.421		

The variables defining the constructs are the same when the covariate is included in the model and when it is excluded.

14. For Black students:

$$\begin{aligned}\text{LDF}_1 &\doteq -.103(4.7) - .024(18.3) - .088(19.5) - .084(15.6) - .008(16.6)\\ &\quad - .067(12.5) + .076(8.9) - .024(22.9) + .114(15.0)\\ &\doteq -3.08\end{aligned}$$

$$\begin{aligned}\text{LDF}_2 &\doteq -.211(4.7) - .097(18.3) + .046(19.5) - .025(15.6) + .122(16.6)\\ &\quad + .114(12.5) - .183(8.9) + .039(22.9) + .057(15.0)\\ &\doteq 1.31\end{aligned}$$

For Hispanic students:

$$\begin{aligned}\text{LDF}_1 &\doteq -.103(4.6) - .024(18.2) - .088(18.1) - .084(14.8) - .008(15.3)\\ &\quad - .067(11.2) + .076(9.5) - .024(21.4) + .114(15.4)\\ &\doteq -2.66\end{aligned}$$

$$\begin{aligned}\text{LDF}_2 &\doteq -.211(4.6) - .097(18.2) + .046(18.1) - .025(14.8) + .122(15.3)\\ &\quad + .114(11.2) - .183(9.5) + .039(21.4) + .057(15.4)\\ &\doteq .844\end{aligned}$$

For White students:

$$\begin{aligned} \text{LDF}_1 &\doteq -.103(4.1) - .024(17.2) - .088(17.5) - .084(13.8) - .008(15.9) \\ &\quad - .067(11.6) + .076(8.7) - .024(21.2) + .114(17.1) \\ &\doteq -2.22 \\ \text{LDF}_2 &\doteq -.211(4.1) - .097(17.2) + .046(17.5) - .025(13.8) + .122(15.9) \\ &\quad + .114(11.6) - .183(8.7) + .039(21.2) + .057(17.1) \\ &\doteq 1.40 \end{aligned}$$

White students appear to differ from Black and Hispanic students on the first dimension. On the second dimension Hispanic students appear to differ from White and Black students.

15. **(a)** $\Lambda \doteq .929$
(b) $F \doteq 4.550$
(c) $\nu_1 = 9$ and $\nu_2 = 533$
(d) $P \doteq .000$
(e) The variables that separate the groups are counsum (−.453), learnsum (−.403), qelib (−4.24), and qesci (.513).

CHAPTER 10

1. **(a)** 4×5

(b) $\mathbf{A} = \begin{bmatrix} 1 & -1 & 0 & 0 & 0 \\ 1 & 0 & -1 & 0 & 0 \\ 1 & 0 & 0 & -1 & 0 \\ 1 & 0 & 0 & 0 & -1 \end{bmatrix}$

(c) The contrasts are independent but they are not orthogonal.

2.

$$\mathbf{H} \doteq 20 \begin{bmatrix} 1.48 \\ 3.18 \\ 3.78 \\ -.92 \\ -7.52 \end{bmatrix} \begin{bmatrix} 1.48 & 3.18 & 3.78 & -.92 & -7.52 \end{bmatrix}$$

$$\doteq \begin{bmatrix} 43.8 & 94.1 & 111.9 & -27.2 & -222.6 \\ 94.1 & 202.2 & 240.4 & -58.5 & -478.3 \\ 111.9 & 240.4 & 285.8 & -69.6 & -568.5 \\ -27.2 & -58.5 & -69.6 & 16.9 & 138.4 \\ -222.6 & -478.3 & -568.5 & 138.4 & 1131.0 \end{bmatrix}$$

3. $\Lambda \doteq \dfrac{1 - .326}{.326} \dfrac{19 - 4 + 1}{4} \doteq 8.267$

4. **(a)** $(-2)(2) + (-1)(-1) + (0)(-2) + (1)(-1) + (2)(2) = 0$

(b) For the first contrast $w \doteq \sqrt{1/10} \doteq .316$

For the second contrast $w \doteq \sqrt{1/14} \doteq .267$

For the third contrast $w \doteq \sqrt{1/10} \doteq .316$

For the fourth contrast $w \doteq \sqrt{1/70} \doteq .120$

$$\mathcal{A} \doteq \begin{bmatrix} -.632 & -.316 & .000 & .316 & .632 \\ .535 & -.267 & -.535 & -.267 & .535 \\ -.316 & .632 & .000 & -.632 & .316 \\ .120 & -.478 & .717 & -.478 & .120 \end{bmatrix}$$

5. It appears that the sphericity assumption is violated. The variances differ by as much as a factor of 3.8 (121.2/31.4) and the covariances vary between −35.0 and 8.2.

6. **(a)** $W \doteq \dfrac{4.27 \times 10^6}{\left[\dfrac{(121.2 + 45.8 + 38.6 + 31.4)}{4}\right]^4} \doteq .347$

(b)

$$d \doteq 1 - \frac{2(5)^2 - 3(5) + 3}{6(20 - 1)(5 - 1)} \doteq .917$$

$$\chi^2 \doteq -(20 - 1).917(\ln .347) \doteq 18.441.$$

(c) $\text{df} \doteq \dfrac{(5(5 - 1)}{2} - 1 = 9.$

There is some evidence to indicate that the sphericity assumption is violated. This is consistent with the observations made in Exercise 5.

7. **(a)** $\varepsilon^* = \dfrac{1}{5 - 1} = .25$

(b) $\varepsilon' = \dfrac{(237)^2}{(5 - 1)(22930.44)} \doteq .612$

(c) $\tilde{\varepsilon} \doteq \dfrac{20(5 - 1).612 - 2}{(5 - 1)[(20 - 1) - (5 - 1).612]} \doteq .709$

8. $\text{MS}_{S \times T} \doteq \dfrac{(121.2 + 45.8 + 38.6 + 31.4)}{4} \doteq 59.25$

9. **(a)**

$$\begin{aligned} \hat{\psi}_{\text{linear}} &\doteq -.632(252.2) - .316(253.9) + .000(254.5) \\ &\quad + .316(249.8) + .632(243.2) \\ &\doteq -6.984 \end{aligned}$$

(b) $s_{\hat{\psi}_{\text{linear}}} \doteq \sqrt{121.2 \dfrac{(-.632)^2 + (-.316)^2 + .000^2 + .316^2 + .632^2}{20}} \doteq 2.46$

(c) $t \doteq \dfrac{-6.984}{2.46} \doteq -2.839$

(d) df $= 20 - 1 = 19$

(e)
$$\begin{aligned}\hat{\psi}_{\text{quadratic}} &\doteq .535(252.2) - .267(253.9) - .535(254.5)\\ &\quad - .267(249.8) + .535(243.2)\\ &\doteq -5.606\end{aligned}$$

(f) $$s_{\hat{\psi}_{\text{quadratic}}} \doteq \sqrt{45.8\frac{.535^2 + (-.267)^2 + (-.535)^2 + (-.267)^2 .535^2}{20}} \doteq 1.514$$

(g) $t \doteq \dfrac{-5.606}{1.514} \doteq -3.705$

(h) df $= 20 - 1 = 19$

10. If the sphericity assumption had been met, the univariate approach would use the average contrast variance (i.e., $\text{MS}_{S\times T}$) rather than each individual contrast variance when computing the standard error of a contrast.

11. (a) **i.** $W \doteq .140$

ii. $\chi^2 \doteq 34.265$

iii. df $= 9$

iv. $P \doteq .000$

(b) $\tilde{\varepsilon} \doteq .557$

12. Yes, there is statistical evidence to indicate that behavior changed over the 5-week period.

(a) $\Lambda \doteq .044$

(b) $F \doteq 86.998$

(c) $\nu_1 = 4$ and $\nu_2 = 16$

(d) $P \doteq .000$

13. $\xi^2 \doteq .956$ and $\xi^2_{\text{adj}} \doteq 1 - \dfrac{100-1}{100-4-1}(1 - .956) \doteq .952$

14. (a) For a fourth-degree polynomial $t(19) \doteq .873$, $P \doteq .394$.

(b) For a cubic polynomial $t(19) \doteq 1.488$, $P \doteq .153$.

(c) For a quadratic polynomial $t(19) \doteq -6.888$, $P \doteq .000$.

(d) For a linear model $t(19) \doteq 16.292$, $P \doteq .000$.

Using a Bonferroni-adjusted P' (.05/4=.0125), the results provide evidence for a quadratic relationship between the repeated-measure variable and the outcome variable.

CHAPTER 11

1. Three; there is one covariance matrix for each of the three weight groups.

2. $d \doteq 1 - \left[\dfrac{2(4)^2 - 3(4) + 3}{6(45-3)(4-1)}\right] \doteq .970$

3. $\chi^2 \doteq -(45-3).970[\ln(.728)] \doteq 12.933$

4. $\tilde{\varepsilon} \doteq \dfrac{45(4-1).840 - 2}{(4-1)[(45-3)-(4-1).840]} \doteq .941$

5. $F \doteq \dfrac{1 - .818^{1/2}}{.818^{1/2}} \dfrac{42-3+1}{3} \doteq 1.409$ with $\nu_1 = 2(3) = 6$ and $\nu_2 = 2(42-3+1) = 80$

6.

$$\mathbf{H}_{\text{Minutes}} \doteq 45 \begin{bmatrix} 1.4 \\ .1 \\ -.7 \\ -.8 \end{bmatrix} \begin{bmatrix} 1.4 & .1 & -.7 & .8 \end{bmatrix}$$

$$\doteq \begin{bmatrix} 88.2 & 6.3 & -44.1 & -50.4 \\ 6.3 & .45 & -3.15 & -3.6 \\ -44.1 & -3.15 & 22.05 & 25.2 \\ -50.4 & -3.6 & 25.2 & 28.8 \end{bmatrix}$$

7. $\xi^2 \doteq .548$ and $\xi^2_{\text{adj}} \doteq 1 - \dfrac{180-1}{180-3-1}(1 - .548) \doteq .540$

8. $s_{\hat{\psi}_{G\times\text{linear}}}$

$$\doteq \sqrt{2.70 \frac{[-.671^2 + -.224^2 + \cdots + .671^2] + [-.671^2 + -.224^2 + \cdots + .671^2]}{15}}$$

$\doteq .6$

9. **(a)** **i.** $\hat{\psi}_{\text{linear}} \doteq -.671(14.8) + (-.224)(13.5) + .224(12.7) + .671(12.6)$
$\doteq -1.655$

ii. $s_{\text{linear}} \doteq \sqrt{2.70 \dfrac{-.671^2 + (-.224)^2 + .244^2 + .671^2}{45}} \doteq .245$

iii. $t \doteq \dfrac{-1.655}{.245} \doteq -6.755$

(b) **i.** $\hat{\psi}_{\text{quadratic}} \doteq .500(14.8) + (-.500)(13.5) + (-.500)(12.7) + .500(12.6) \doteq .600$

ii. $s_{\text{quadratic}} \doteq \sqrt{1.62 \dfrac{.500^2 + (-.500)^2 + (-.500)^2 + .500^2}{45}} \doteq .190$

iii. $t \doteq \dfrac{.600}{.190} \doteq 3.158$

(c) There is a quadratic relationship between muscle soreness and warm-up time.

10. **(a)** $M \doteq 22.903$

(b) $\chi^2 \doteq 20.046$

(c) $\text{df} = 30$

(d) $P \doteq .915$

11. **(a)** **i.** $W \doteq .116$

ii. $\chi^2 \doteq 119.371$

iii. $\text{df} = 9$

iv. $P \doteq .000$

(b) $\tilde{\varepsilon} \doteq .579$

12. **(a)** $\Lambda \doteq .037$

(b) $F \doteq 56.577$

(c) $\nu_1 = 8$ and $\nu_2 = 108$

(d) $P \doteq .000$

(e) $\xi^2_{\text{adj}} \doteq 1 - \dfrac{300 - 1}{300 - 8 - 1}(1 - .784) \doteq .778$

13. For the interaction contrast comparing changes in behavior for Group 1 with Group 3.

(a) $F \doteq 55.02$

(b) $v_1 = 4$, $v_2 = 228$

(c) $P \doteq .000$

The test comparing the quadratic trends for Group 1 with Group 3 indicates a statistical difference, $t(57) \doteq -6.141$, $P \doteq .000$.

14. Averaged over the 5-week observation period, there is evidence to indicate that the groups differ:

(a) $F \doteq 245.01$

(b) $\nu_1 = 2$ and $\nu_2 = 57$

(c) $\eta^2_G \doteq \dfrac{2317.76}{2317.76 + 269.61 + 936.34} \doteq .658$

15. **(a)** There is evidence of a difference between Group 1 and Group 3.

i. $t \doteq 21.719$

ii. $\nu = 57$

iii. $P \doteq .000$

(b) There is evidence of a difference between Group 2 and Group 3.

i. $t \doteq 7.153$

ii. $\nu = 57$

iii. $P \doteq .000$

CHAPTER 12

1.

$$\begin{aligned} d^2 &= 4^2 + 6^2 \\ &= 52 \\ d &\doteq 7.21 \end{aligned}$$

2. $\Delta^2 = [-2-2 \quad 1-7] \begin{bmatrix} 2 & -\frac{3}{2} \\ -3 & \frac{5}{2} \end{bmatrix} \begin{bmatrix} -4 \\ -6 \end{bmatrix} = [10 \quad -9] \begin{bmatrix} -4 \\ -6 \end{bmatrix} = 14$

3. (a)

$$[6-10 \quad 7-12] \begin{bmatrix} 2 & -\frac{3}{2} \\ -3 & \frac{5}{2} \end{bmatrix} \begin{bmatrix} -4 \\ -5 \end{bmatrix} = [7 \quad -6.5] \begin{bmatrix} -4 \\ -5 \end{bmatrix} = 4.5$$

$$[6-2 \quad 7-3] \begin{bmatrix} 2 & -\frac{3}{2} \\ -3 & \frac{5}{2} \end{bmatrix} \begin{bmatrix} 4 \\ 4 \end{bmatrix} = [-4 \quad 4] \begin{bmatrix} 4 \\ 4 \end{bmatrix} = 0$$

(b) $[10-2 \quad 12-3] \begin{bmatrix} 2 & -\frac{3}{2} \\ -3 & \frac{5}{2} \end{bmatrix} \begin{bmatrix} 8 \\ 9 \end{bmatrix} = [-11 \quad 10.5] \begin{bmatrix} 8 \\ 9 \end{bmatrix} = 6.5$

4. (a) Distance between Leah and Joe:

$$\begin{aligned} \Delta^2 &= [30-30 \quad 20-30] \left(\frac{1}{144} \begin{bmatrix} 9 & -9 \\ -9 & 25 \end{bmatrix} \right) \begin{bmatrix} 0 \\ -10 \end{bmatrix} \\ &= \frac{1}{144} [90 \quad -250] \begin{bmatrix} 0 \\ -10 \end{bmatrix} \doteq 17.36 \end{aligned}$$

(b) Distance between Leah and John:

$$\begin{aligned} \Delta^2 &= [30-20 \quad 20-25] \left(\frac{1}{144} \begin{bmatrix} 9 & -9 \\ -9 & 25 \end{bmatrix} \right) \begin{bmatrix} 10 \\ -5 \end{bmatrix} \\ &= \frac{1}{144} [135 - 215] \begin{bmatrix} 10 \\ -5 \end{bmatrix} \doteq 16.84 \end{aligned}$$

(c) Distance between Joe and John

$$\begin{aligned} \Delta^2 &= [30-20 \quad 30-25] \left(\frac{1}{144} \begin{bmatrix} 9 & -9 \\ -9 & 25 \end{bmatrix} \right) \begin{bmatrix} 10 \\ 5 \end{bmatrix} \\ &= \frac{1}{144} [45 \quad 35] \begin{bmatrix} 10 \\ 5 \end{bmatrix} \doteq 4.34 \end{aligned}$$

5.

$$\boldsymbol{\Sigma}_Y = \begin{bmatrix} 25 & 0 \\ 0 & 9 \end{bmatrix} \qquad \boldsymbol{\Sigma}_Y^{-1} = \frac{1}{225}\begin{bmatrix} 9 & 0 \\ 0 & 25 \end{bmatrix}$$

$$\boldsymbol{\Sigma}_X = \begin{bmatrix} 1 & 0 \\ 0 & 1 \end{bmatrix} = \mathbf{I}$$

Δ_Y^2 of Sherrie and Kama:

$$[20-30 \quad 6-3]\left(\frac{1}{225}\begin{bmatrix} 9 & 0 \\ 0 & 25 \end{bmatrix}\right)\begin{bmatrix} -10 \\ 3 \end{bmatrix}$$

$$= \frac{1}{225}[-90 \quad 75]\begin{bmatrix} -10 \\ 3 \end{bmatrix} = 5$$

Δ_Y^2 of Sherrie and Sandy:

$$[20-25 \quad 6-9]\left(\frac{1}{225}\begin{bmatrix} 9 & 0 \\ 0 & 25 \end{bmatrix}\right)\begin{bmatrix} -5 \\ -3 \end{bmatrix}$$

$$= \frac{1}{225}[-45 \quad -75]\begin{bmatrix} -5 \\ -3 \end{bmatrix} = 2$$

Δ_Y^2 of Kama and Sandy:

$$[30-25 \quad 3-9]\left(\frac{1}{225}\begin{bmatrix} 9 & 0 \\ 0 & 25 \end{bmatrix}\right)\begin{bmatrix} 5 \\ -6 \end{bmatrix}$$

$$= \frac{1}{225}[45 \quad -150]\begin{bmatrix} 5 \\ -6 \end{bmatrix} = 5$$

d_X^2 of Sherrie and Kama:

$$\begin{bmatrix} \frac{20}{5}-\frac{30}{5} & \frac{6}{3}-\frac{3}{3} \end{bmatrix}\begin{bmatrix} -2 \\ 1 \end{bmatrix} = 5$$

d_X^2 of Sherrie and Sandy:

$$\begin{bmatrix} \frac{20}{5}-\frac{25}{5} & \frac{6}{3}-\frac{9}{3} \end{bmatrix}\begin{bmatrix} -1 \\ -1 \end{bmatrix} = 2$$

d_X^2 of Kama and Sandy:

$$\begin{bmatrix} \frac{30}{5}-\frac{25}{5} & \frac{3}{3}-\frac{9}{3} \end{bmatrix}\begin{bmatrix} 1 \\ -2 \end{bmatrix} = 5$$

6.

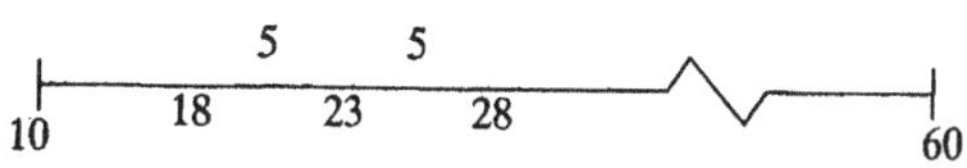

7.

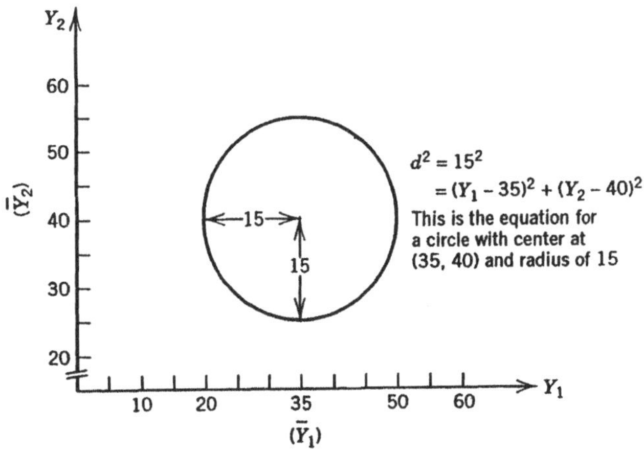

8. **(a)** Unknown. But, assume that $P(\mathbf{x}|2) = .55$.

(b) Group 2, because $.55 > .30$.

(c)

$$P(1|\mathbf{x}) = \frac{.30}{.30 + .55} \doteq .35$$

$$P(2|\mathbf{x}) = \frac{.55}{.85} \doteq .65$$

(d) 1.0; yes.

(e) $P(2) = 1 - .70 = .30$

(f)

$$P(1|\mathbf{x}) = \frac{.70(.30)}{.70(.30) + .30(.55)} \doteq .56$$

$$P(2|\mathbf{x}) \doteq 1 - .56 = .44$$

(g) Group 1.

CHAPTER 13

1. $f(1.20) \doteq .1942$

2. $f(2) \doteq .0997$
$f(6) \doteq .0605$

3. To determine predicted group membership.

CHAPTER 14

1. **(a)** Group 2

(b) Posterior probabilities

(c) Typicality probabilities

(d) $1 - (.30 + .65) = .05$

(e) About .039, using Eq. (12.6) (with equal priors); unknown for unequal priors

2. To seek in-doubt and outlying units.

3. **(a)** Briefly, there is a reuse of data; data idiosyncrasies are capitalized on.

(b) Same

4. **(a)** $\hat{P}(3|\mathbf{x}_{24}) \doteq .286$
$\hat{P}(\mathbf{x}_{24}|3) \doteq .477$

(b) obs/unit 1,137 (from SAS or SPSS)

(c) obs/unit 4,59 (from SPSS)

(d) For group 2: 17.19 percent
For group 3: 67.21 percent
For all five groups: $(11 + 22 + 82 + 10 + 8)/545 \doteq .244$

CHAPTER 15

1. Apparent hit rate

2. Linear internal

3. Probability associated with distance that analysis unit is from centroid of assigned group.

4. Large N/p ratio

5. **(a)** Use new priors.

(b) Yes; simply use new priors in rule, which otherwise is built on 2004 data.

6. Internal (external): Build rule on a data set; then use same (different) data set to assess rule.

7. **(a)** $D^2 \doteq 1.2398, \qquad \tilde{D}^2 \doteq 0.9844$

(b)

	Hit Rate		
	G_1	G_2	Overall
Internal	.26	.93	.67

(c)

	Hit Rate		
	G_1	G_2	Overall
L-O-O	.21	.89	.63

(d)

	Hit Rate		
	G_1	G_2	Overall
M-P-P/I	.26	.90	.69
M-P-P/L-O-O	.26	.90	.69

8. (a) $\hat{P}(1|\mathbf{x}_{21}) \doteq .580$
$\hat{P}(2|\mathbf{x}_{21}) \doteq .261$
$\hat{P}(3|\mathbf{x}_{21}) \doteq 1 - (.580 + .261) = .159$

(b) $\hat{P}(\mathbf{x}_{21}|2) \doteq .692$

(c) See Table 13.1.

9.

Group	**Linear L-O-O Hit Rate**
1	.145
2	.172
3	.672
4	.065
5	.121

Total group hit rate $\doteq$ .244

10. Answers depend on located fence riders.

11. (a) Total group hit rate = 23/91 $\doteq$ .253

Actual Group	Predicted Group 1	2	3	4	5	Predicted Cases	In-doubt Students	Grand Total
G_1	2 (.026)	4	8	0	0	14	62	76
G_2	3	2 (.016)	7	0	0	12	116	128
G_3	1	3	19 (.156)	0	1	24	98	122
G_4	0	4	25	0 (.000)	1	30	123	153
G_5	1	1	9	0	0 (.000)	11	55	66
Total	7	14	68	0	2	91	454	545

(b) $5/24 \doteq .208$

12. (a) $q_2 = .20$, $q_3 = .30$

(b) $\hat{P}(1|\mathbf{x}_{88}) \doteq .060, \quad \hat{P}(3|\mathbf{x}_{88}) \doteq .156, \quad \hat{P}(25|\mathbf{x}_{88}) \doteq .255$
(c) Group 3

13. A significant F or χ^2 value with $\alpha < .01$, or "wide" differences among the six covariance matrix log determinants, or among the six matrix traces.

CHAPTER 16

1. (a) $z \doteq 5.219$, $P \doteq .000$; therefore, yes.
(b) $I \doteq .594$
(c) $e = 150$, $z \doteq .816$, $P \doteq .205$; therefore, no.

2. (a) Is an obtained hit rate (statistically) better than the corresponding chance hit rate?
(b) Are the results of one classification rule (statistically) better than those of a second rule?

3. Linear composite/combination.

4. (a) $e_1 = .15(76) = 11.4$; $e_2 = 25.6$; $e_3 = 36.6$; $e_4 = 30.6$; $e_5 = 9.9$
(b) $e = 114.1$; $o = 118$
(c) $z \doteq 0.411$, $P \doteq .342$
LL $\doteq 102.4$
(d) The observed total-group hit-rate (118/545) is not statistically different from chance. Similarly for the separate-Group hit rates except for Group 3.
(e) $I \doteq .009$
(f) Proportional chance

5. (a) Linear L-O-O Group 1 hit rate: .145
(b) Quadratic L-O-O Group 1 hit rate: .158
(c)

		Quad. Rule Hit	Quad. Rule Miss	
Lin. Rule	Hit	67	66	133
	Miss	51	361	412
		118	427	545

McNemar:
$\chi^2(1) \doteq 1.923$
$P \doteq .165$
Therefore, no.

6. Note that in terms of hit rates, Eq. (16.4) may be expressed as:

$$z = \frac{H_o - H_e}{\sqrt{H_e \cdot (1 - H_e)/N}}$$

Similarly for Eq. (16.6).
(a) Group 1, $z \doteq 0.947$; Group 2, $z \doteq 2.167$; Group 3, $z \doteq 9.149$; Group 4, NA; Group 5, $z \doteq 0.844$; total-group, $z \doteq 7.493$.

(b) Group 1, $I \doteq .046$; Group 2, $I \doteq .096$; Group 3; $I \doteq .542$; Group 4, NA; Group 5, $I \doteq .044$; total-group, $I \doteq .165$.

CHAPTER 17

1. Deleting some predictor(s) may *increase* the hit rate of interest (separate-group or total-group).

2. (a) X_5, X_7, X_8, X_9

(b) X_4, X_7, X_8, X_9

(c) X_2, X_7, X_8, X_9

3. X_5, X_7, X_3 & X_8, X_9, X_6

4. (a) Three best subsets of size

8	X_1,	X_3,	X_4,	X_5,	X_6,	X,	X_8,	X_9	(.136)
	X_1,	X_2,	X_3,	X_4,	X_5,	X_6,	X,	X_9	(.132)
	X_1,	X_2,	X_3,	X_4,	X_5,	X_6,	X_7,	X_8	(.130)
7	X_1,	X_3,	X_4,	X_5,	X_7,	X_8,	X_9	(.143)	
	X,	X_3,	X_5,	X_6,	X_7,	X_8,	X_9	(.143)	
	X_1,	X_2,	X_4,	X_5,	X_7,	X_8,	X_9	(.141)	
6	X_1,	X_2,	X_4,	X_5,	X_6,	X_9	(.145)		
	$X1$,	X_2,	X_3,	X_4,	X_5,	X_7	(.141)		
	X_1,	X_2,	X_5,	X_6,	X_7,	X_8	(.141)		
5	X_1,	X_3,	X_4,	X_5,	X_7	(.147)			
	X_1,	X_2,	X_3,	X_5,	X_8	(.145)			
	X_1,	X_3,	X_5,	X_7,	X_8	(.145)			
4	X_5,	X_6,	X_8,	X_9	(.149)				
	X_2,	X_4,	X_5,	X_8	(.147)				
	X_2,	X_5,	X_7,	X_9	(.147)				
3	X_1,	X_3,	X_5	(.149)					
	X_1,	X_5,	X_9	(.147)					
	X_2,	X_5,	X_9	(.147)					
2	X_5,	X_6	(.149)						
	X_3,	X_5	(.143)						
	X_5,	X_9	(.143)						
1	X_5	(.147)							
	X_8	(.106)							
	X_3	(.106)							

(b) X_5, X_1, X_7, X_3, X_4

CHAPTER 18

1.

	(a) Regression Weight	(b) LCF Difference	(c) Ratio (a/b)
X_1	.037	−.1697	−.22
X_2	.004	−.0204	−.20
X_3	.010	−.0472	−.21
X_4	.004	−.0197	−.20
X_5	−.028	.1276	−.22
X_6	.004	−.0171	−.23
X_7	.003	−.0157	−.19
X_8	.007	−.0303	−.23
X_9	−.012	.0562	−.21

(e) $R^2 \doteq .093$
$D^2 \doteq 2.368$

2. $D \doteq 1.54$; $n_1 = 76$ is OK (with $p = 9$) for hit rate estimate to be within .05.

3.

(Internal)

		Pred. Group 1	Pred. Group 2	
Actual Group	1	66	10	76
	2	95	27	122
		161	37	198

(L-O-O)

		Pred. Group 1	Pred. Group 2	
Actual Group	1	61	15	76
	2	97	25	122
		158	40	198

CHAPTER 19

1. **(a)** Interger scaling: Poor (1), Fair (2), Good (3), Excellent (4)
   (b) 0–1 scoring: Initial (0), Advanced (1)
   (c) Fisher–Lancaster scaling

2. For the lone categorical predictor, define *two* variables with scores:
   v1: 1 if in the first category
   v2: 0 if in second category
   So, the original (categorical) predictor is replaced with *two* predictors.

Index

Note: Pages in bold indicate a figure or a table; an *f* indicates a footnote.

Applied MANOVA and Discriminant Analysis, Second Edition, by Carl J. Huberty and Stephen Olejnik
Copyright © 2006 John Wiley & Sons, Inc.

WILEY SERIES IN PROBABILITY AND STATISTICS

ESTABLISHED BY WALTER A. SHEWHART AND SAMUEL S. WILKS

Editors: *David J. Balding, Noel A. C. Cressie, Nicholas I. Fisher, Iain M. Johnstone, J. B. Kadane, Geert Molenberghs. Louise M. Ryan, David W. Scott, Adrian F. M. Smith, Jozef L. Teugels*

Editors Emeriti: *Vic Barnett, J. Stuart Hunter, David G. Kendall*

The ***Wiley Series in Probability and Statistics*** is well established and authoritative. It covers many topics of current research interest in both pure and applied statistics and probability theory. Written by leading statisticians and institutions, the titles span both state-of-the-art developments in the field and classical methods.

Reflecting the wide range of current research in statistics, the series encompasses applied, methodological and theoretical statistics, ranging from applications and new techniques made possible by advances in computerized practice to rigorous treatment of theoretical approaches.

This series provides essential and invaluable reading for all statisticians, whether in academia, industry, government, or research.

† ABRAHAM and LEDOLTER · Statistical Methods for Forecasting
AGRESTI · Analysis of Ordinal Categorical Data
AGRESTI · An Introduction to Categorical Data Analysis
AGRESTI · Categorical Data Analysis, *Second Edition*
ALTMAN, GILL, and McDONALD · Numerical Issues in Statistical Computing for the Social Scientist
AMARATUNGA and CABRERA · Exploration and Analysis of DNA Microarray and Protein Array Data
ANDĚL · Mathematics of Chance
ANDERSON · An Introduction to Multivariate Statistical Analysis, *Third Edition*
* ANDERSON · The Statistical Analysis of Time Series
ANDERSON, AUQUIER, HAUCK, OAKES, VANDAELE, and WEISBERG · Statistical Methods for Comparative Studies
ANDERSON and LOYNES · The Teaching of Practical Statistics
ARMITAGE and DAVID (editors) · Advances in Biometry
ARNOLD, BALAKRISHNAN, and NAGARAJA · Records
* ARTHANARI and DODGE · Mathematical Programming in Statistics
* BAILEY · The Elements of Stochastic Processes with Applications to the Natural Sciences
BALAKRISHNAN and KOUTRAS · Runs and Scans with Applications
BALAKRISHNAN and NG · Precedence-Type Tests and Applications
BARNETT · Comparative Statistical Inference, *Third Edition*
BARNETT · Environmental Statistics
BARNETT and LEWIS · Outliers in Statistical Data, *Third Edition*
BARTOSZYNSKI and NIEWIADOMSKA-BUGAJ · Probability and Statistical Inference
BASILEVSKY · Statistical Factor Analysis and Related Methods: Theory and Applications
BASU and RIGDON · Statistical Methods for the Reliability of Repairable Systems
BATES and WATTS · Nonlinear Regression Analysis and Its Applications
BECHHOFER, SANTNER, and GOLDSMAN · Design and Analysis of Experiments for Statistical Selection, Screening, and Multiple Comparisons

*Now available in a lower priced paperback edition in the Wiley Classics Library.
†Now available in a lower priced paperback edition in the Wiley–Interscience Paperback Series.

BELSLEY · Conditioning Diagnostics: Collinearity and Weak Data in Regression
† BELSLEY, KUH, and WELSCH · Regression Diagnostics: Identifying Influential Data and Sources of Collinearity
BENDAT and PIERSOL · Random Data: Analysis and Measurement Procedures, *Third Edition*
BERRY, CHALONER, and GEWEKE · Bayesian Analysis in Statistics and Econometrics: Essays in Honor of Arnold Zellner
BERNARDO and SMITH · Bayesian Theory
BHAT and MILLER · Elements of Applied Stochastic Processes, *Third Edition*
BHATTACHARYA and WAYMIRE · Stochastic Processes with Applications
† BIEMER, GROVES, LYBERG, MATHIOWETZ, and SUDMAN · Measurement Errors in Surveys
BILLINGSLEY · Convergence of Probability Measures, *Second Edition*
BILLINGSLEY · Probability and Measure, *Third Edition*
BIRKES and DODGE · Alternative Methods of Regression
BLISCHKE AND MURTHY (editors) · Case Studies in Reliability and Maintenance
BLISCHKE AND MURTHY · Reliability: Modeling, Prediction, and Optimization
BLOOMFIELD · Fourier Analysis of Time Series: An Introduction, *Second Edition*
BOLLEN · Structural Equations with Latent Variables
BOLLEN and CURRAN · Latent Curve Models: A Structural Equation Perspective
BOROVKOV · Ergodicity and Stability of Stochastic Processes
BOULEAU · Numerical Methods for Stochastic Processes
BOX · Bayesian Inference in Statistical Analysis
BOX · R. A. Fisher, the Life of a Scientist
BOX and DRAPER · Empirical Model-Building and Response Surfaces
* BOX and DRAPER · Evolutionary Operation: A Statistical Method for Process Improvement
BOX and FRIENDS · Improving Almost Anything, *Revised Edition*
BOX, HUNTER, and HUNTER · Statistics for Experimenters: Design, Innovation, and Discovery, *Second Editon*
BOX and LUCEÑO · Statistical Control by Monitoring and Feedback Adjustment
BRANDIMARTE · Numerical Methods in Finance: A MATLAB-Based Introduction
BROWN and HOLLANDER · Statistics: A Biomedical Introduction
BRUNNER, DOMHOF, and LANGER · Nonparametric Analysis of Longitudinal Data in Factorial Experiments
BUCKLEW · Large Deviation Techniques in Decision, Simulation, and Estimation
CAIROLI and DALANG · Sequential Stochastic Optimization
CASTILLO, HADI, BALAKRISHNAN, and SARABIA · Extreme Value and Related Models with Applications in Engineering and Science
CHAN · Time Series: Applications to Finance
CHARALAMBIDES · Combinatorial Methods in Discrete Distributions
CHATTERJEE and HADI · Sensitivity Analysis in Linear Regression
CHATTERJEE and PRICE · Regression Analysis by Example, *Third Edition*
CHERNICK · Bootstrap Methods: A Practitioner's Guide
CHERNICK and FRIIS · Introductory Biostatistics for the Health Sciences
CHILÈS and DELFINER · Geostatistics: Modeling Spatial Uncertainty
CHOW and LIU · Design and Analysis of Clinical Trials: Concepts and Methodologies, *Second Edition*
CLARKE and DISNEY · Probability and Random Processes: A First Course with Applications, *Second Edition*
* COCHRAN and COX · Experimental Designs, *Second Edition*
CONGDON · Applied Bayesian Modelling
CONGDON · Bayesian Models for Categorical Data
CONGDON · Bayesian Statistical Modelling

*Now available in a lower priced paperback edition in the Wiley Classics Library.
†Now available in a lower priced paperback edition in the Wiley–Interscience Paperback Series.

CONOVER · Practical Nonparametric Statistics, *Third Edition*
COOK · Regression Graphics
COOK and WEISBERG · Applied Regression Including Computing and Graphics
COOK and WEISBERG · An Introduction to Regression Graphics
CORNELL · Experiments with Mixtures, Designs, Models, and the Analysis of Mixture Data, *Third Edition*
COVER and THOMAS · Elements of Information Theory
COX · A Handbook of Introductory Statistical Methods
* COX · Planning of Experiments
CRESSIE · Statistics for Spatial Data, *Revised Edition*
CSÖRGŐ and HORVÁTH · Limit Theorems in Change Point Analysis
DANIEL · Applications of Statistics to Industrial Experimentation
DANIEL · Biostatistics: A Foundation for Analysis in the Health Sciences, *Eighth Edition*
* DANIEL · Fitting Equations to Data: Computer Analysis of Multifactor Data, *Second Edition*
DASU and JOHNSON · Exploratory Data Mining and Data Cleaning
DAVID and NAGARAJA · Order Statistics, *Third Edition*
* DEGROOT, FIENBERG, and KADANE · Statistics and the Law
DEL CASTILLO · Statistical Process Adjustment for Quality Control
DEMARIS · Regression with Social Data: Modeling Continuous and Limited Response Variables
DEMIDENKO · Mixed Models: Theory and Applications
DENISON, HOLMES, MALLICK and SMITH · Bayesian Methods for Nonlinear Classification and Regression
DETTE and STUDDEN · The Theory of Canonical Moments with Applications in Statistics, Probability, and Analysis
DEY and MUKERJEE · Fractional Factorial Plans
DILLON and GOLDSTEIN · Multivariate Analysis: Methods and Applications
DODGE · Alternative Methods of Regression
* DODGE and ROMIG · Sampling Inspection Tables, *Second Edition*
* DOOB · Stochastic Processes
DOWDY, WEARDEN, and CHILKO · Statistics for Research, *Third Edition*
DRAPER and SMITH · Applied Regression Analysis, *Third Edition*
DRYDEN and MARDIA · Statistical Shape Analysis
DUDEWICZ and MISHRA · Modern Mathematical Statistics
DUNN and CLARK · Basic Statistics: A Primer for the Biomedical Sciences, *Third Edition*
DUPUIS and ELLIS · A Weak Convergence Approach to the Theory of Large Deviations
EDLER and KITSOS · Recent Advances in Quantitative Methods in Cancer and Human Health Risk Assessment
* ELANDT-JOHNSON and JOHNSON · Survival Models and Data Analysis
ENDERS · Applied Econometric Time Series
† ETHIER and KURTZ · Markov Processes: Characterization and Convergence
EVANS, HASTINGS, and PEACOCK · Statistical Distributions, *Third Edition*
FELLER · An Introduction to Probability Theory and Its Applications, Volume I, *Third Edition*, Revised; Volume II, *Second Edition*
FISHER and VAN BELLE · Biostatistics: A Methodology for the Health Sciences
FITZMAURICE, LAIRD, and WARE · Applied Longitudinal Analysis
* FLEISS · The Design and Analysis of Clinical Experiments
FLEISS · Statistical Methods for Rates and Proportions, *Third Edition*
† FLEMING and HARRINGTON · Counting Processes and Survival Analysis
FULLER · Introduction to Statistical Time Series, *Second Edition*
FULLER · Measurement Error Models

*Now available in a lower priced paperback edition in the Wiley Classics Library.
†Now available in a lower priced paperback edition in the Wiley–Interscience Paperback Series.

GALLANT · Nonlinear Statistical Models
GEISSER · Modes of Parametric Statistical Inference
GELMAN and MENG · Applied Bayesian Modeling and Causal Inference from Incomplete-Data Perspectives
GEWEKE · Contemporary Bayesian Econometrics and Statistics
GHOSH, MUKHOPADHYAY, and SEN · Sequential Estimation
GIESBRECHT and GUMPERTZ · Planning, Construction, and Statistical Analysis of Comparative Experiments
GIFI · Nonlinear Multivariate Analysis
GIVENS and HOETING · Computational Statistics
GLASSERMAN and YAO · Monotone Structure in Discrete-Event Systems
GNANADESIKAN · Methods for Statistical Data Analysis of Multivariate Observations, *Second Edition*
GOLDSTEIN and LEWIS · Assessment: Problems, Development, and Statistical Issues
GREENWOOD and NIKULIN · A Guide to Chi-Squared Testing
GROSS and HARRIS · Fundamentals of Queueing Theory, *Third Edition*
* HAHN and SHAPIRO · Statistical Models in Engineering
HAHN and MEEKER · Statistical Intervals: A Guide for Practitioners
HALD · A History of Probability and Statistics and their Applications Before 1750
HALD · A History of Mathematical Statistics from 1750 to 1930
† HAMPEL · Robust Statistics: The Approach Based on Influence Functions
HANNAN and DEISTLER · The Statistical Theory of Linear Systems
HEIBERGER · Computation for the Analysis of Designed Experiments
HEDAYAT and SINHA · Design and Inference in Finite Population Sampling
HEDEKER and GIBBONS · Longitudinal Data Analysis
HELLER · MACSYMA for Statisticians
HINKELMANN and KEMPTHORNE · Design and Analysis of Experiments, Volume 1: Introduction to Experimental Design
HINKELMANN and KEMPTHORNE · Design and Analysis of Experiments, Volume 2: Advanced Experimental Design
HOAGLIN, MOSTELLER, and TUKEY · Exploratory Approach to Analysis of Variance
* HOAGLIN, MOSTELLER, and TUKEY · Exploring Data Tables, Trends and Shapes
* HOAGLIN, MOSTELLER, and TUKEY · Understanding Robust and Exploratory Data Analysis
HOCHBERG and TAMHANE · Multiple Comparison Procedures
HOCKING · Methods and Applications of Linear Models: Regression and the Analysis of Variance, *Second Edition*
HOEL · Introduction to Mathematical Statistics, *Fifth Edition*
HOGG and KLUGMAN · Loss Distributions
HOLLANDER and WOLFE · Nonparametric Statistical Methods, *Second Edition*
HOSMER and LEMESHOW · Applied Logistic Regression, *Second Edition*
HOSMER and LEMESHOW · Applied Survival Analysis: Regression Modeling of Time to Event Data
† HUBER · Robust Statistics
HUBERTY · Applied Discriminant Analysis
HUBERTY and OLEJNIK · Applied MANOVA and Discriminant Analysis, *Second Edition*
HUNT and KENNEDY · Financial Derivatives in Theory and Practice, *Revised Edition*
HUSKOVA, BERAN, and DUPAC · Collected Works of Jaroslav Hajek—with Commentary
HUZURBAZAR · Flowgraph Models for Multistate Time-to-Event Data
IMAN and CONOVER · A Modern Approach to Statistics

*Now available in a lower priced paperback edition in the Wiley Classics Library.
†Now available in a lower priced paperback edition in the Wiley–Interscience Paperback Series.

† JACKSON · A User's Guide to Principle Components
JOHN · Statistical Methods in Engineering and Quality Assurance
JOHNSON · Multivariate Statistical Simulation
JOHNSON and BALAKRISHNAN · Advances in the Theory and Practice of Statistics: A Volume in Honor of Samuel Kotz
JOHNSON and BHATTACHARYYA · Statistics: Principles and Methods, *Fifth Edition*
JOHNSON and KOTZ · Distributions in Statistics
JOHNSON and KOTZ (editors) · Leading Personalities in Statistical Sciences: From the Seventeenth Century to the Present
JOHNSON, KOTZ, and BALAKRISHNAN · Continuous Univariate Distributions, Volume 1, *Second Edition*
JOHNSON, KOTZ, and BALAKRISHNAN · Continuous Univariate Distributions, Volume 2, *Second Edition*
JOHNSON, KOTZ, and BALAKRISHNAN · Discrete Multivariate Distributions
JOHNSON, KEMP, and KOTZ · Univariate Discrete Distributions, *Third Edition*
JUDGE, GRIFFITHS, HILL, LÜTKEPOHL, and LEE · The Theory and Practice of Econometrics, *Second Edition*
JUREČKOVÁ and SEN · Robust Statistical Procedures: Aymptotics and Interrelations
JUREK and MASON · Operator-Limit Distributions in Probability Theory
KADANE · Bayesian Methods and Ethics in a Clinical Trial Design
KADANE AND SCHUM · A Probabilistic Analysis of the Sacco and Vanzetti Evidence
KALBFLEISCH and PRENTICE · The Statistical Analysis of Failure Time Data, *Second Edition*
KARIYA and KURATA · Generalized Least Squares
KASS and VOS · Geometrical Foundations of Asymptotic Inference
† KAUFMAN and ROUSSEEUW · Finding Groups in Data: An Introduction to Cluster Analysis
KEDEM and FOKIANOS · Regression Models for Time Series Analysis
KENDALL, BARDEN, CARNE, and LE · Shape and Shape Theory
KHURI · Advanced Calculus with Applications in Statistics, *Second Edition*
KHURI, MATHEW, and SINHA · Statistical Tests for Mixed Linear Models
* KISH · Statistical Design for Research
KLEIBER and KOTZ · Statistical Size Distributions in Economics and Actuarial Sciences
KLUGMAN, PANJER, and WILLMOT · Loss Models: From Data to Decisions, *Second Edition*
KLUGMAN, PANJER, and WILLMOT · Solutions Manual to Accompany Loss Models: From Data to Decisions, *Second Edition*
KOTZ, BALAKRISHNAN, and JOHNSON · Continuous Multivariate Distributions, Volume 1, *Second Edition*
KOTZ and JOHNSON (editors) · Encyclopedia of Statistical Sciences: Volumes 1 to 9 with Index
KOTZ and JOHNSON (editors) · Encyclopedia of Statistical Sciences: Supplement Volume
KOTZ, READ, and BANKS (editors) · Encyclopedia of Statistical Sciences: Update Volume 1
KOTZ, READ, and BANKS (editors) · Encyclopedia of Statistical Sciences: Update Volume 2
KOVALENKO, KUZNETZOV, and PEGG · Mathematical Theory of Reliability of Time-Dependent Systems with Practical Applications
LACHIN · Biostatistical Methods: The Assessment of Relative Risks
LAD · Operational Subjective Statistical Methods: A Mathematical, Philosophical, and Historical Introduction
LAMPERTI · Probability: A Survey of the Mathematical Theory, *Second Edition*

*Now available in a lower priced paperback edition in the Wiley Classics Library.
†Now available in a lower priced paperback edition in the Wiley–Interscience Paperback Series.

LANGE, RYAN, BILLARD, BRILLINGER, CONQUEST, and GREENHOUSE · Case Studies in Biometry
LARSON · Introduction to Probability Theory and Statistical Inference, *Third Edition*
LAWLESS · Statistical Models and Methods for Lifetime Data, *Second Edition*
LAWSON · Statistical Methods in Spatial Epidemiology
LE · Applied Categorical Data Analysis
LE · Applied Survival Analysis
LEE and WANG · Statistical Methods for Survival Data Analysis, *Third Edition*
LePAGE and BILLARD · Exploring the Limits of Bootstrap
LEYLAND and GOLDSTEIN (editors) · Multilevel Modelling of Health Statistics
LIAO · Statistical Group Comparison
LINDVALL · Lectures on the Coupling Method
LIN · Introductory Stochastic Analysis for Finance and Insurance
LINHART and ZUCCHINI · Model Selection
LITTLE and RUBIN · Statistical Analysis with Missing Data, *Second Edition*
LLOYD · The Statistical Analysis of Categorical Data
LOWEN and TEICH · Fractal-Based Point Processes
MAGNUS and NEUDECKER · Matrix Differential Calculus with Applications in Statistics and Econometrics, *Revised Edition*
MALLER and ZHOU · Survival Analysis with Long Term Survivors
MALLOWS · Design, Data, and Analysis by Some Friends of Cuthbert Daniel
MANN, SCHAFER, and SINGPURWALLA · Methods for Statistical Analysis of Reliability and Life Data
MANTON, WOODBURY, and TOLLEY · Statistical Applications Using Fuzzy Sets
MARCHETTE · Random Graphs for Statistical Pattern Recognition
MARDIA and JUPP · Directional Statistics
MASON, GUNST, and HESS · Statistical Design and Analysis of Experiments with Applications to Engineering and Science, *Second Edition*
McCULLOCH and SEARLE · Generalized, Linear, and Mixed Models
McFADDEN · Management of Data in Clinical Trials
* McLACHLAN · Discriminant Analysis and Statistical Pattern Recognition
McLACHLAN, DO, and AMBROISE · Analyzing Microarray Gene Expression Data
McLACHLAN and KRISHNAN · The EM Algorithm and Extensions
McLACHLAN and PEEL · Finite Mixture Models
McNEIL · Epidemiological Research Methods
MEEKER and ESCOBAR · Statistical Methods for Reliability Data
MEERSCHAERT and SCHEFFLER · Limit Distributions for Sums of Independent Random Vectors: Heavy Tails in Theory and Practice
MICKEY, DUNN, and CLARK · Applied Statistics: Analysis of Variance and Regression, *Third Edition*
* MILLER · Survival Analysis, *Second Edition*
MONTGOMERY, PECK, and VINING · Introduction to Linear Regression Analysis, *Third Edition*
MORGENTHALER and TUKEY · Configural Polysampling: A Route to Practical Robustness
MUIRHEAD · Aspects of Multivariate Statistical Theory
MULLER and STOYAN · Comparison Methods for Stochastic Models and Risks
MURRAY · X-STAT 2.0 Statistical Experimentation, Design Data Analysis, and Nonlinear Optimization
MURTHY, XIE, and JIANG · Weibull Models
MYERS and MONTGOMERY · Response Surface Methodology: Process and Product Optimization Using Designed Experiments, *Second Edition*
MYERS, MONTGOMERY, and VINING · Generalized Linear Models. With Applications in Engineering and the Sciences

*Now available in a lower priced paperback edition in the Wiley Classics Library.
†Now available in a lower priced paperback edition in the Wiley–Interscience Paperback Series.

† NELSON · Accelerated Testing, Statistical Models, Test Plans, and Data Analyses
† NELSON · Applied Life Data Analysis
NEWMAN · Biostatistical Methods in Epidemiology
OCHI · Applied Probability and Stochastic Processes in Engineering and Physical Sciences
OKABE, BOOTS, SUGIHARA, and CHIU · Spatial Tesselations: Concepts and Applications of Voronoi Diagrams, *Second Edition*
OLIVER and SMITH · Influence Diagrams, Belief Nets and Decision Analysis
PALTA · Quantitative Methods in Population Health: Extensions of Ordinary Regressions
PANKRATZ · Forecasting with Dynamic Regression Models
PANKRATZ · Forecasting with Univariate Box-Jenkins Models: Concepts and Cases
* PARZEN · Modern Probability Theory and Its Applications
PEÑA, TIAO, and TSAY · A Course in Time Series Analysis
PIANTADOSI · Clinical Trials: A Methodologic Perspective
PORT · Theoretical Probability for Applications
POURAHMADI · Foundations of Time Series Analysis and Prediction Theory
PRESS · Bayesian Statistics: Principles, Models, and Applications
PRESS · Subjective and Objective Bayesian Statistics, *Second Edition*
PRESS and TANUR · The Subjectivity of Scientists and the Bayesian Approach
PUKELSHEIM · Optimal Experimental Design
PURI, VILAPLANA, and WERTZ · New Perspectives in Theoretical and Applied Statistics
† PUTERMAN · Markov Decision Processes: Discrete Stochastic Dynamic Programming
QIU · Image Processing and Jump Regression Analysis
* RAO · Linear Statistical Inference and Its Applications, *Second Edition*
RAUSAND and HØYLAND · System Reliability Theory: Models, Statistical Methods, and Applications, *Second Edition*
RENCHER · Linear Models in Statistics
RENCHER · Methods of Multivariate Analysis, *Second Edition*
RENCHER · Multivariate Statistical Inference with Applications
* RIPLEY · Spatial Statistics
* RIPLEY · Stochastic Simulation
ROBINSON · Practical Strategies for Experimenting
ROHATGI and SALEH · An Introduction to Probability and Statistics, *Second Edition*
ROLSKI, SCHMIDLI, SCHMIDT, and TEUGELS · Stochastic Processes for Insurance and Finance
ROSENBERGER and LACHIN · Randomization in Clinical Trials: Theory and Practice
ROSS · Introduction to Probability and Statistics for Engineers and Scientists
ROSSI, ALLENBY, and McCULLOCH · Bayesian Statistics and Marketing
† ROUSSEEUW and LEROY · Robust Regression and Outlier Detection
* RUBIN · Multiple Imputation for Nonresponse in Surveys
RUBINSTEIN · Simulation and the Monte Carlo Method
RUBINSTEIN and MELAMED · Modern Simulation and Modeling
RYAN · Modern Regression Methods
RYAN · Statistical Methods for Quality Improvement, *Second Edition*
SALEH · Theory of Preliminary Test and Stein-Type Estimation with Applications
* SCHEFFE · The Analysis of Variance
SCHIMEK · Smoothing and Regression: Approaches, Computation, and Application
SCHOTT · Matrix Analysis for Statistics, *Second Edition*
SCHOUTENS · Levy Processes in Finance: Pricing Financial Derivatives
SCHUSS · Theory and Applications of Stochastic Differential Equations
SCOTT · Multivariate Density Estimation: Theory, Practice, and Visualization
† SEARLE · Linear Models for Unbalanced Data
† SEARLE · Matrix Algebra Useful for Statistics

*Now available in a lower priced paperback edition in the Wiley Classics Library.
†Now available in a lower priced paperback edition in the Wiley–Interscience Paperback Series.

† SEARLE, CASELLA, and McCULLOCH · Variance Components
SEARLE and WILLETT · Matrix Algebra for Applied Economics
SEBER and LEE · Linear Regression Analysis, *Second Edition*
† SEBER · Multivariate Observations
† SEBER and WILD · Nonlinear Regression
SENNOTT · Stochastic Dynamic Programming and the Control of Queueing Systems
* SERFLING · Approximation Theorems of Mathematical Statistics
SHAFER and VOVK · Probability and Finance: It's Only a Game!
SILVAPULLE and SEN · Constrained Statistical Inference: Inequality, Order, and Shape Restrictions
SMALL and McLEISH · Hilbert Space Methods in Probability and Statistical Inference
SRIVASTAVA · Methods of Multivariate Statistics
STAPLETON · Linear Statistical Models
STAUDTE and SHEATHER · Robust Estimation and Testing
STOYAN, KENDALL, and MECKE · Stochastic Geometry and Its Applications, *Second Edition*
STOYAN and STOYAN · Fractals, Random Shapes and Point Fields: Methods of Geometrical Statistics
STYAN · The Collected Papers of T. W. Anderson: 1943–1985
SUTTON, ABRAMS, JONES, SHELDON, and SONG · Methods for Meta-Analysis in Medical Research
TAKEZAWA · Introduction to Nonparametric Regression
TANAKA · Time Series Analysis: Nonstationary and Noninvertible Distribution Theory
THOMPSON · Empirical Model Building
THOMPSON · Sampling, *Second Edition*
THOMPSON · Simulation: A Modeler's Approach
THOMPSON and SEBER · Adaptive Sampling
THOMPSON, WILLIAMS, and FINDLAY · Models for Investors in Real World Markets
TIAO, BISGAARD, HILL, PEÑA, and STIGLER (editors) · Box on Quality and Discovery: with Design, Control, and Robustness
TIERNEY · LISP-STAT: An Object-Oriented Environment for Statistical Computing and Dynamic Graphics
TSAY · Analysis of Financial Time Series, *Second Edition*
UPTON and FINGLETON · Spatial Data Analysis by Example, Volume II: Categorical and Directional Data
VAN BELLE · Statistical Rules of Thumb
VAN BELLE, FISHER, HEAGERTY, and LUMLEY · Biostatistics: A Methodology for the Health Sciences, *Second Edition*
VESTRUP · The Theory of Measures and Integration
VIDAKOVIC · Statistical Modeling by Wavelets
VINOD and REAGLE · Preparing for the Worst: Incorporating Downside Risk in Stock Market Investments
WALLER and GOTWAY · Applied Spatial Statistics for Public Health Data
WEERAHANDI · Generalized Inference in Repeated Measures: Exact Methods in MANOVA and Mixed Models
WEISBERG · Applied Linear Regression, *Third Edition*
WELSH · Aspects of Statistical Inference
WESTFALL and YOUNG · Resampling-Based Multiple Testing: Examples and Methods for *p*-Value Adjustment
WHITTAKER · Graphical Models in Applied Multivariate Statistics
WINKER · Optimization Heuristics in Economics: Applications of Threshold Accepting
WONNACOTT and WONNACOTT · Econometrics, *Second Edition*
WOODING · Planning Pharmaceutical Clinical Trials: Basic Statistical Principles

*Now available in a lower priced paperback edition in the Wiley Classics Library.
†Now available in a lower priced paperback edition in the Wiley–Interscience Paperback Series.

WOODWORTH · Biostatistics: A Bayesian Introduction
WOOLSON and CLARKE · Statistical Methods for the Analysis of Biomedical Data, *Second Edition*
WU and HAMADA · Experiments: Planning, Analysis, and Parameter Design Optimization
WU and ZHANG · Nonparametric Regression Methods for Longitudinal Data Analysis
YANG · The Construction Theory of Denumerable Markov Processes
ZELTERMAN · Discrete Distributions—Applications in the Health Sciences
* ZELLNER · An Introduction to Bayesian Inference in Econometrics
ZHOU, OBUCHOWSKI, and McCLISH · Statistical Methods in Diagnostic Medicine

Printed and bound by CPI Group (UK) Ltd, Croydon, CR0 4YY

Printed and bound by CPI Group (UK) Ltd, Croydon, CR0 4YY
18/09/2026
14965015-0001